CAMBRIDGE LIBRARY COLLECTION

Books of enduring scholarly value

Technology

The focus of this series is engineering, broadly construed. It covers technological innovation from a range of periods and cultures, but centres on the technological achievements of the industrial era in the West, particularly in the nineteenth century, as understood by their contemporaries. Infrastructure is one major focus, covering the building of railways and canals, bridges and tunnels, land drainage, the laying of submarine cables, and the construction of docks and lighthouses. Other key topics include developments in industrial and manufacturing fields such as mining technology, the production of iron and steel, the use of steam power, and chemical processes such as photography and textile dyes.

Shipbuilding in Iron and Steel

Sir Edward James Reed (1830–1906) was an innovative naval architect who, after attending the Central School of Mathematics and Naval Architecture in Portsmouth, was appointed chief constructor of the Navy in 1863. This was a crucial time for the navy, as traditional wooden sailing ships began to be armoured or rebuilt in iron in response to more powerful weaponry. Reed pioneered the methodical use of scientific calculations to determine a ship's weight, strength and stability, and was responsible for a number of revolutionary designs. He later founded his own consultancy and designed ships for other countries including Germany, Chile and Brazil. This textbook, first published in 1869, contains a comprehensive overview of the design and construction of the various classes of iron ships of the time, and the new features they incorporated. It also includes descriptions of procedures in Royal Dockyards and various civilian shipyards.

Shipbuilding in Iron and Steel

A Practical Treatise

Edward James Reed

Cambridge
University Press

CAMBRIDGE UNIVERSITY PRESS

Cambridge, New York, Melbourne, Madrid, Cape Town, Singapore,
São Paolo, Delhi, Dubai, Tokyo, Mexico City

Published in the United States of America by Cambridge University Press, New York

www.cambridge.org
Information on this title: www.cambridge.org/9781108026451

This edition first published 1869
This digitally printed version 2010

ISBN 978-1-108-02645-1 Paperback

SHIPBUILDING

IN

IRON AND STEEL.

A PRACTICAL TREATISE,
GIVING FULL DETAILS OF CONSTRUCTION, PROCESSES OF MANUFACTURE, AND BUILDING ARRANGEMENTS;
WITH RESULTS OF EXPERIMENTS ON IRON AND STEEL, AND ON THE STRENGTH AND WATERTIGHTNESS OF RIVETED WORK.

By E. J. REED, C.B.,
CHIEF CONSTRUCTOR OF THE NAVY,
VICE-PRESIDENT OF THE INSTITUTION OF NAVAL ARCHITECTS, AND HONORARY MEMBER OF THE LIVERPOOL LITERARY AND PHILOSOPHICAL SOCIETY.

By order of the Lords Commissioners of the Admiralty, the Examinations in Practical Iron Shipbuilding of Candidates for Promotion in H.M. Dockyards will be mainly based upon this Work.

LONDON:
JOHN MURRAY, ALBEMARLE STREET.
1869.

LONDON: PRINTED BY WILLIAM CLOWES AND SONS, DUKE STREET, STAMFORD STREET,
AND CHARING CROSS.

TO

VICE-ADMIRAL ROBERT SPENCER ROBINSON,

CONTROLLER OF THE NAVY,

IN ACKNOWLEDGMENT OF

THE MANY AND IMPORTANT IMPROVEMENTS

IN THE

CONSTRUCTION OF IRON SHIPS

WHICH HAVE BEEN MADE

BY HIS DESIRE AND UNDER HIS AUTHORITY,

This Work

IS RESPECTFULLY DEDICATED

BY HIS OBLIGED AND OBEDIENT SERVANT,

THE AUTHOR.

PREFACE.

My principal object in writing this book has been to furnish to shipbuilders, and to shipbuilding officers of all grades, fuller information respecting the details of ship construction in iron and steel than any previous work records. In the body of the book I have borne repeated testimony to the merits of the existing writings of Mr. Grantham, Mr. Scott Russell, Dr. Fairbairn, Professor Rankine, and others; but none of these either gives, or professes to give, that copious detailed information of which those who have to superintend the practical operations of shipbuilding, and even cultivated workmen, have often felt the want.

The book is without pretension as a record of the history of iron shipbuilding; but I have nevertheless given, under the several heads, a brief account of the methods of construction adopted in early iron ships, and of the successive improvements which have led up to existing methods. In all cases I have endeavoured to give to the treatise an essentially practical character,—the descriptions and criticisms comprised in it being based on what are universally recognised as the first principles of ship construction. The reader who seeks for theoretical instruction is referred to the works of Mr. Scott Russell, Dr. Fairbairn, and Professor Rankine, named above.

It must not be supposed, because my daily duties connect me most intimately with the building of Government ships, that I have restricted this volume to that branch of ship construction; still less that the following pages are devoted exclusively to iron-clad vessels. On the contrary, I have thought it well to treat largely of the practice of mercantile shipbuilders, and have given copious descriptions of the systems of work adopted on the Mersey, the Clyde, the Tyne, and the Thames, taking pains, in all cases, to represent as accurately as possible the general practice of the shipbuilding firms upon those rivers. At the same time, I have considered it desirable to bring together information, useful to shipbuilders, from numerous sources which must be more or less difficult of access to many whom I hope to have among my readers, —such as the *Proceedings* of the Institution of Civil Engineers; the *Transactions* of the Royal Society, of the Institution of Naval Architects, and of the Scottish Shipbuilders' Association; and the Specifications of Patents,—nor have I hesitated in a few cases, to make mention of improvements which have yet to find a place in actual practice.

The construction of war ships is now so important a part of the shipbuilder's art, even in private establishments, that I have described at length the novel features introduced into recent iron-clad ships, in so far as the arrangement and combination of their frames, plating, decks, and bulkheads are concerned; bringing the information down to the date of the 'Invincible' class. The methods of testing the materials used in the construction of the ships of the Navy; the building system adopted in the Royal Dockyards at Chatham and Pembroke; and the operations connected with preparing, fitting, and fastening armour plates, are also fully explained.

There are four chapters of this work upon which great labour has been bestowed, in order not only to bring together much scattered information, but also to set forth facts and considerations which have hitherto been to some extent disregarded or misapprehended. These are the chapters on Outside Plating, Steel Plates for Shipbuilding, Rivets and Rivet Work, and Systems of Work. The chapter on Outside Plating, taken in connection with that on Rivets and Rivet-work, will be found valuable, I trust, in aiding the shipbuilder to avoid the employment of an unnecessary weight of material, and to obtain the utmost structural strength out of the material and labour which are employed. I have no hesitation in saying that riveted work, both in its general relations and in its special adaptations to shipbuilding purposes, is more fully considered in this volume than in any other work of like character. The information gathered from numerous works on Bridge construction and from many published accounts of experiments, is supplemented by further accounts of experiments recently made in the Royal Dockyards which have not been previously published. The investigations and calculations of the strengths of butt fastenings are also original, and are based upon examples taken from actual practice and upon the most trustworthy experiments. The chapter on Steel Plates brings together the principal results of the experience we have had of the use of this material in shipbuilding. It necessarily partakes somewhat of the tentative and partial character by which the employment of steel for this purpose has hitherto been marked; but the fulness of the information afforded will, I hope, tend nevertheless to facilitate the extended application of the material. The chapter on Systems of Work, to the substance of which I have already referred, will well repay all the pains taken in its compilation, if it promotes greater simplicity and uniformity in the various shipbuilding establishments throughout the country.

In preparing this volume I have, at times, drawn largely from the writings and experiments of others; but as I have in all cases acknowledged the sources of my information, it is unnecessary to further acknowledge them here. I must, however, observe that I have received assistance from Messrs. Barnaby, Barnes, and Crossland, Assistant Constructors of the Navy, especially in working out the details of the novel systems of construction which have been embodied in the recent ships of H.M. Navy, and described in the following pages. Mr. W. H. White, late of the Royal School of Naval Architecture, has given me valuable assistance during the preparation of several of the chapters of this book, and in the correction of the press.

I have only to add that, as one of my principal objects in writing this book has been to extend a knowledge of the details of their business among the officers and men of H.M. Dockyards, it is a great satisfaction and pleasure to me to have received the order of the Lords of the Admiralty to employ it as the text-book upon which the examinations in Practical Iron Shipbuilding of candidates for promotion in those Dockyards will hereafter be mainly based.

E. J. REED.

LONDON, *Oct.* 1868.

CONTENTS.

CHAPTER I.

PRACTICAL CONSIDERATIONS ON THE STRENGTH OF IRON SHIPS.

CHAPTER II.

KEELS, KEELSONS, AND GARBOARD-STRAKES.

CHAPTER III.

STEMS.

CHAPTER IV.

STERN POSTS.

CHAPTER V.

TRANSVERSE AND LONGITUDINAL SYSTEMS OF FRAMING.

CHAPTER VI.

COMBINED TRANSVERSE AND LONGITUDINAL SYSTEM OF FRAMING—FRAMING OF 'WARRIOR,' 'NORTHUMBERLAND,' &c.

CHAPTER VII.

BRACKET-PLATE SYSTEM OF FRAMING—FRAMING OF 'BELLEROPHON,' 'HERCULES,' &c.

CHAPTER VIII.

DECK FRAMING AND PILLARING.

CHAPTER IX.

DECK STRINGERS AND PLATING.

CHAPTER X.

OUTSIDE PLATING.

CHAPTER XI.

BULKHEADS.

CHAPTER XII.

TOPSIDES.

CHAPTER XIII.

RUDDERS.

CHAPTER XIV.

IRON MASTS.

CHAPTER XV.

MISCELLANEOUS DETAILS.

CHAPTER XVI.

STEEL PLATES FOR SHIPBUILDING.

CHAPTER XVII.

RIVETS AND RIVET-WORK.

CHAPTER XVIII.

ON TESTING IRON AND STEEL.

CHAPTER XIX.

LLOYD'S AND THE LIVERPOOL RULES FOR IRON SHIPBUILDING.

CHAPTER XX.

SYSTEMS OF WORK.

CHAPTER XXI.

ARMOUR PLATING.

APPENDIX.

LIST OF ILLUSTRATIONS.

PLATES.

WOOD ENGRAVINGS.

SHIPBUILDING IN IRON AND STEEL.

CHAPTER I.

PRACTICAL CONSIDERATIONS ON THE STRENGTH OF IRON SHIPS.

THE Art of Shipbuilding owes both its difficulty and its dignity to the fact that the comprehension of great scientific principles and the power of wisely arranging a multitude of minute details are alike essential to its complete mastery.

We all know that Land Architecture, or the art of constructing edifices upon the fixed and solid earth, has absorbed a large measure of the thought and invention of the races of men, and has been the means by which many splendid reputations have been worthily erected. But it may be permitted to us, who are engaged in the Architecture of the Sea, or the construction of edifices which shall stand upon and traverse the unsteady and yielding deep, to claim for our art both the greater difficulty and the greater honour; or, if this be denied, we may at least assure ourselves that we are devoted to an extremely arduous and honourable employment.

The construction of a fixed suspended bridge, like that across the Menai Straits for example, was a work requiring an immense amount of investigation and invention, as those know who have read the writings of Dr. Fairbairn, Mr. Edwin Clark, and Professor Hodgkinson on the subject; and that neither the theory nor the practice of bridge construction is yet complete must be obvious to those who have, on the one hand, made themselves acquainted with the recent experiments of Sir Charles Fox, or those of Mr. Crampton, or, on the other hand, have studièd that extremely able paper, 'On the Strains in the Interior of Beams,' which Mr. Airy, the Astronomer Royal, contributed to the Royal Society a short time ago. But the scientific construction of a ship involves all that is comprised in the construction of a fixed beam or girder, together with a mass of other facts and circumstances peculiar to its own purposes and uses.*

It is to Dr. Fairbairn that we chiefly owe the repeated enforcement of the fact that a ship is in many respects to be regarded in this way as a huge beam or girder; and it must be acknowledged that ships have repeatedly been placed in positions which, although exceptional, fully justify even the extreme examples with which he has illustrated his argument, by showing ships sometimes supported wholly at the middle and sometimes wholly suspended by the extremities. A remarkable instance of the former occurred recently, when an iron ship, laden with a cargo of iron, got across a stone causeway; and, as an extraordinary case of suspension by the ends, we may mention that of the 'Prince of Wales,' described by Mr. Clark in his work above referred to. The incident took place in the launching of the vessel at Blackwall, many years ago, at the works of Messrs. Miller and Ravenhill, and was considered at the time so demonstrative of the extraordinary strength of iron ships that Mr. Miller published an account of it. She was an iron boat 180 feet long, and, by the giving way of the bolts of the launching cleat, she was let down till the bilge bore on the wharf. She was ultimately forced off, cutting her way deeper into the concrete and planking of the wharf as she went, until she attained a steeply inclined position in which she was supported by the water at the stern and by the wharf at the bow, when the distance from the face of the wharf to the point of contact of the vessel with the surface of the water was 110 feet. Although the whole of the deck in the centre of the vessel was left unfastened for the reception of the machinery, it was found, when she was completely afloat, that her sheer was not broken, and that she had received no injury beyond that of the twisting of the bow by the set of the tide against the side. Three of the angle-iron frames were broken, and one of the plates cracked, the repair of these defects being effected in four days. Several cases of a similar kind occurred on the Mersey in the early days of iron shipbuilding; among others the 'Nun,' a

* The next few pages contain the substance of a Paper read by the Author at the Institution of Naval Architects, April 11, 1867, 'On Certain Cases of Weakness in Iron Ships,' and published *in extenso*, with illustrative diagrams, in the *Transactions* of that year.—E. J. R.

vessel of 65 H.P. built by Messrs. Laird, which got aground on the end of a stone pier, and was there left by the ebb tide resting by the stern on the pier and by the bow on a hard stone bottom. This vessel was 105 feet long, having in the centre an engine weighing 65 tons; and although she remained for many hours in this position, with a distance of 81 feet between the points of support, no visible deflection could be observed in the keel. It is unnecessary to multiply such examples, of which hundreds might probably be cited.

But a ship would obviously be most imperfectly constructed if designed as a beam or girder only, for she has to endure forces, and to undergo deteriorating influences, to which a fixed beam, or a bridge, is not at all subjected. A ship has, of course, to be propelled through the sea, either by the wind acting from without, or by steam generated within her; she has to be largely immersed in corrosive and vegetating salt water; she has to be lifted from the hollows of waves to their summits, and pitched thence into their hollows again; she has to endure being rolled violently through extreme angles; she has to undergo all this under the burden of weights often far greater than her own weight; and she has, in the mercantile marine, to withstand the deteriorating effects of cargoes that often tend greatly to injure and destroy her; while in the Royal Navy she now is often expected to withstand not only the simultaneous discharge of several heavy guns, but also the shock of missiles weighing many hundred weights, propelled by enormous quantities of powder, and containing explosive charges of considerable force. Besides all this, she is to be made capable of withstanding, as far as possible, all the trials of collisions, storms, groundings, and a multitude of other evils and mischances.

As it is our intention to give to these remarks as practical a character as possible, we shall not enter into those general investigations, respecting the nature and amount of the forces to which ships are liable to be subjected, which have been very ably discussed by other persons. It will be sufficient for our purpose to bear in mind that a ship is substantially of the nature of a hollow beam or girder, narrowed away to nothing in breadth at the ends; that this beam is in a sea-way supported, more or less, by the ends and by the middle alternately, the supports shifting incessantly, and completely changing positions many times in a minute, thus throwing the top and bottom into states of tension and compression successively; that the strains of the masts and sails, and of the sea, tend to produce sudden and frequent changes of angle between the decks and sides; and in cases of steam ships, the thrust of the propelling shafts and the resistance of the water, being exerted usually at different heights, tend also to rack and strain the structure longitudinally. This outline of the subject will suffice to keep before our minds, for occasional reference, the nature and circumstances of the structure we have to consider, viz., an iron ship's hull. Instead of seeking to fill up this outline by theoretical enquiries and expositions, we shall endeavour to do so by pointing out some of the weaknesses and defects which iron ships have practically been found to possess; because, however sound one's theoretical principles may be, the true requirements of a structure subject to so many exigencies as the hull of a ship, cannot be thoroughly understood without a large resort to practical experience.

As the primary object to be kept in view in the construction of a fixed tubular girder is to adapt the top and bottom to receive the principal strains, of tension or compression as the case may be, the longitudinal strength of the upper and lower parts of a ship must obviously require to be very considerable. But as the construction of fixed girders was but very imperfectly understood prior to the building of the Britannia Bridge, and as the resemblance of a ship to such a structure is even now but partially understood, or, to say the least, admitted, among shipbuilders, it is not surprising that both the upper parts and the bottoms of ships have been in many cases too weakly constructed. We propose, before describing the various details of construction which we shall have to explain, to mention a few instances of this kind which have at various times come under our notice; and in order to familiarise the reader with details of work, and to complete our information, we shall in these, as in all future illustrations, explain in general terms, so far as we may be able, the arrangement and devices which have been adopted for making good the deficiencies of strength that we shall have occasion to point out.

We will, in the first place, instance the case of a large Atlantic mail paddle-steamer. This ship was not by any means deficient in the quantity of material put into her: on the contrary, the weight of iron in her hull was unusually great. But she was a very long ship, and after making a few passages across the Atlantic it was found that she had not sufficient longitudinal strength to properly withstand the strains to which she was subjected. On examination it was found that one of the plates, and the strap over the adjacent butt, in the topside above the spar-deck in wake of the paddle-box, were broken; several of the rivets in the plating had worked loose in the neighbourhood of this fracture; and other slack rivets were found in the bottom plating under the engines and boilers, and in the hollow of the bows,—a certain amount of leakage resulting, of course.

Additional strength was supplied in this case in the following manner. An inch plate 2 feet 3 inches broad was worked on the frames as a doubler to the ½-inch strake immediately above the spar-deck; and above this, one ⅝-inch and one ¾-inch plate, these plates completing the side up to the rail, which was formed as a continuous ¾-inch plate, connected to the uppermost strake by an angle-iron. A second rail-plate 15 inches broad was worked a few inches below the upper rail, being let partly in between the frames to meet the outer plating; and below this and connected to it by an angle-iron, a ¾-inch plate 18 inches broad was worked on the inside of the frames, stiffened by an angle-iron on its lower edge. This inner plate extended for a length of 103 feet; the outer plating for 112 feet; and the plate-rails for 180 feet 6 inches, the ship being 374 feet long. This completed the strengthenings of what we may call the top of the girder. Below the spar-deck a ¾-inch clamp-plate 2 feet broad, stiffened by an angle-iron on its upper edge, was worked for a length of 240 feet. The bottom was strengthened by doubling the whole of the inner plates up to the turn of the bilge for 50 feet in wake of the engines, thus making the bottom plating a flush surface at that part.*

This mode of strengthening the bilge is one easy of application, and has been adopted in many cases by Messrs. Laird of Birkenhead, who have had great experience in the actual indications of weakness in ocean-going steam ships, owing in some degree to the circumstance of their works being easily accessible to vessels which have suffered from stress of weather in the Atlantic. In some vessels it has formed a part of their original design; in others it has been added where signs of weakness have become

* For illustrations of the strengthenings applied in this and the following cases see the Paper referred to in the footnote, p. 2.—E. J. R.

apparent after the vessels have been at work for some time, and have gone to them for repair; and in others it has been adopted when vessels originally designed for general trade have been fitted out for carrying special cargoes of dead weight, such as machinery, or telegraph cables, or have been lengthened in the midship body.

An example of the first may be found in the Holyhead mail-boats 'Ulster,' 'Munster,' and 'Connaught,' and in many of the long screw-steamers now employed in the merchant service, trading from the port of Liverpool. The system is provided for in one of the Rules of the Liverpool Underwriters' Registry for Iron Vessels, when the length of the vessels exceeds certain proportions of breadth and depth. A specimen of it may be found in the S.S. 'Queen' of 3250 tons, recently built by Messrs. Laird for the National Steam Navigation Company, of which vessel a view of the amidship framing is given in Plate 1. In the cases of vessels which have been strengthened in this way, where weakness has become apparent after some amount of service, it has usually had the effect of preventing any extension of injury. Among vessels altered for special service may be mentioned the 'Imperador,' 'Imperatriz,' and 'Bahiana,' originally intended for the South American trade, but afterwards fitted for carrying the electric telegraph cable to the Red Sea, when very heavy weights were placed in small sections of the vessel's length, and at a considerable distance before and abaft the centre of displacement.

The plates used are of the same thickness as the other parts of the outside plating, and care is taken to shift the butts of the doubling-strake from the butts of the adjoining strakes. They extend from a half to two-thirds the length of the vessel. At each butt of the doubling-strake there is a butt-strap placed inside of the skin-plate, and the rivets pass through both. All the butts and seams are made close and caulked tight, so that no moisture can get between the plates. Sometimes a layer of canvas saturated with red-lead is placed between the surfaces, but often there is nothing but red-lead paint, or a mixture of red and white lead, which does perfectly well if the plates are carefully fitted.

Another case which may be mentioned instructively is that of a large mail paddle-steamer, built for ocean-service, which encountered a gale in the English Channel on her first voyage out of port. On being taken into dock it was reported that the whole of the butts of the flat keel and bottom plating for about 180 feet amidships were very much strained; and several of the butts in the upper strakes, in the wake of the paddle-boxes and sponsons, were likewise much strained, indicating such a deficiency of strength in the vessel, especially in the longitudinal direction, as to render her unseaworthy, and to fully account for the leakage which had taken place, and which was said to have required the whole of the engine-pumps and bilge-injections to keep it under. An experienced surveyor, who inspected this vessel on behalf of the owners, reported that in his opinion the structural weakness of this ship was so great that she could not proceed to sea without the risk of foundering. The strengthenings which were applied to her before she was again sent out on service were as follows:—An external iron keel made up of several thicknesses of plates, was applied, and connected to the original flat keel by garboard-plates, which of course served as doublers to the main flat keel. The plating of the bottom was made flush from this point to above the turn of the bilge over a length of about 240 feet, by plates worked between the lapped edges of the outer over-lapping plates of the bottom. An external bilge-keel was worked on the turn of the bilge, consisting of two plates on edge, connected to the bottom by two large angle-bars. A large central box-keelson completed these lower strengthenings, which involved an addition of more than 150 tons to the weight of the hull. Besides this, an increase of 15 tons was made to the orlop-deck stringer; a stringer-plate and clamp on the maindeck was converted, by an addition of 64 tons of plate and angle-iron, into a box waterway or girder. A double sheer-strake and stringer-plate weighing over 100 tons, were applied to what were before rounded beam ends, thus giving a new top, so to speak, to the ship herself viewed as a girder. Other additions were made to the ship, in the form of bulkheads, bulwarks, a forecastle, &c.; and much of the bottom was riveted anew; but the above facts will suffice to indicate where this ship proved weak, and how she was strengthened by a very experienced firm.

Another case of a very similar kind, but presenting sufficient points of difference to justify a reference to it, occurred with a ship built for the same service. In the former case no middle line keelson above the floors was fitted in the ship originally. In the present case there was a keelson with intercostal plates below between the floors, but these intercostal plates were not riveted to the flat keel-plates. There were also other differences in the original construction, the nature of which may be gathered from what follows. It should be stated here, however, that some of the longitudinal ties of this ship were broken at the bulkheads; some of the butt-straps were considered of insufficient width to ensure good work, and some of the work itself was not of the highest class. The ship made a voyage or two across the Atlantic in safety, but she was found to strain considerably, and she was consequently taken into dock and received the following repairs and additions: The shell was re-riveted throughout, the projecting strakes of plating being nearly all taken off, and also about half the sunken strakes, the countersinks of the rivets being enlarged and the holes made fair. An external keel and doubling garboard-strakes were worked, as in the former case, throughout nearly the whole length of the ship. The box-keelson was removed, and intercostal plates were fitted and secured below to the inner garboards, and at top to the new bottom plate of the box-keelson, which last was then put together again and re-riveted. The four sunken strakes of bottom and bilge on each side of the keel were doubled with plates of equal thickness, for about 250 feet of length, the strake nearest the keel extending the whole length of the ship. The sunken strake near the main-deck beams was also doubled for about 200 feet amidships, and a short doubling-plate was worked under each paddle-beam. A bilge-keel, formed as in the previous case, was worked on the turn of the bilge. A box waterway or stringer on the main deck was re-fitted, extended to the ends of the vessel, and secured on top to the outside plating by a plate and short angle-irons between the frames, and a box-stringer was fitted on the lower deck for about 240 feet of length. The sheer-strake was doubled with a steel plate $\frac{1}{2}$ inch thick and a stringer-plate 3 feet wide by $\frac{5}{8}$ inch riveted to the beams, and doubled by an 18-inch plate, secured to the sheer-strake by angle-steels. The deck finished on this stringer against angle-steels forming a sunk waterway. A girder 2 feet 6 inches wide and $\frac{3}{4}$ inch thick, intended to aid in distributing the thrust of the paddle-shaft, was wrought above the sheer-strake, and over the paddle-beams. Other additions were made, but need not be mentioned here.

From the foregoing illustrations, it will be seen that the practical experience gained with ships at sea shows the extreme necessity of giving great longitudinal strength to them, especially at the top and at the bottom. We do not consider that precisely the best means of securing that strength was taken in every case before described, but the object aimed at in each instance is apparent enough.

In endeavouring to secure the necessary longitudinal strength, either in building a new ship or in repairing a weak ship, one thing is obviously essential to an effective use of the iron applied, viz., as near an approach to continuity of uniform strength as is possible. We have often been astonished at the extent to which this has been neglected in many iron ships that have come under our notice. It is, or certainly has been, a very common practice on the part of some builders to break their longitudinal strengthenings in a most remarkable manner. One common form of this defect is the practice of stopping short the longitudinal ties at a watertight bulkhead. In building a certain ship the central angle-iron keelson was stopped short against a bulkhead. The plating forming the seat for the engines, and extending nearly to the turn of the bilge on the tops of the floors, was likewise stopped short at a frame just before a bulkhead, no longitudinal tie existing originally between this plating and the keelson. To remedy this defect, which was discovered by a surveying officer, the builders afterwards applied a short scarphing keelson-piece, formed of a plate between two angle-irons, and carried watertight through the bulkhead; and also a $\frac{3}{4}$-inch plate, 4 feet wide, extending from the bulkhead to the plating forming the seat for the engines. The connection between the plating and the keelson was thus effectually completed. A similar arrangement was carried out at the other engine-room bulkhead, where also the central keelson was broken off. In the same ship the side keelsons also were originally stopped at the engine-room bulkheads, and had to be strengthened in a similar manner, except that the plating was not necessary, and the scarphing angle-irons had no plate between them, but were fitted directly back to back. In order to show that this ship was not by any means an exceptional case, we will make brief reference to a few other examples of similar weakness, taken from reports of surveys with which we have had to do, or which we have had to consider. In one ship of somewhat recent construction, and built for ocean mail service, we found that the butts of the angle-irons on the top of the centre keelson-plate had no butt-straps to connect them together, thereby considerably reducing, and, in fact, almost destroying the value of the longitudinal strength of these angle-irons. The gutter-plates on the top of the floors, forming the flat central keelson, were found to be badly fitted, and several rivets in the short angle-irons immediately under them were defective. The butt-straps of these gutter-plates were not made to completely cover the ends of the plates, thereby introducing a serious source of weakness. The butts of the angle-irons and bulb-irons forming the side and bilge keelsons were not sufficiently connected together, and it was recommended to cover the butts of these angle-irons with straps 24 inches long, and to introduce separate straps for the bulb-irons. The bilge keelsons and lower-deck stringers were found to be severed at some of the transverse bulkheads, and means had to be taken afterwards to preserve the continuity of their longitudinals. The main-deck sheer-strake, which was formed of two thicknesses, had butt-straps to the butts in the inner thickness only, the butts in the outer being riveted to the inner thickness of plating. This, it will be observed, was very objectionable, as it to a large extent destroyed the usefulness of the outer thickness, and, besides, with covers to one thickness only, the butts of the outer strakes should have had them in preference to those of the inner, as it was the thicker of the two. To get the greatest strength, however, both the inner and outer thickness should of course have been supported with straps, as the increased thickness of plating forming the sheer-strake was for the purpose of increasing the longitudinal strength of the topsides, and without straps the extra plating used was merely adding weight to the vessel without obtaining the advantage aimed at.

In another vessel, built for the same service as the last named, the butts of the angle-irons forming the fore and aft bilge-stringers, were not sufficiently connected, requiring additional strapping in many places. Another example of the same defect was observed in a new ocean-going vessel. It was found that her main and side keelsons were stopped short at the fore and after part of the engines, completely destroying their longitudinal strength at these parts. They were afterwards made of continuous strength there by the introduction of two deep angle-irons riveted to each, carried watertight through the bulkheads, and extended from 5 to 6 feet on each side.

It is unnecessary to further multiply instances of this kind, which are exceedingly numerous. It may be well, however, to state here that one very frequent source of longitudinal weakness in our ships—a more frequent source than would be supposed probable—is the single-riveting of the butt-straps, especially where there are but one or two passing strakes between. Out of twelve sea-going ships, whose construction we were examining not long ago, no less than five were single-riveted at the butts of the bottom plating.

Having now practically illustrated, as fully as appears to be necessary, the great importance of securing longitudinal strength in iron ships, and especially at the top and bottom, we proceed to state, that when this has been sufficiently secured, there should be provided such intermediate strength of frames, clamps, beams, stringers, waterways, &c., as will insure rigidity throughout the skin-plating of the ship under all circumstances. Flexibility in the skin is a great source of weakness and rapid deterioration, as we will show by again referring to practical experience.

The first case we shall mention—and we intend to refer but to three—is that of a remarkably well-formed ship of about 1200 tons. She had made her first voyage to India; and upon her arrival in the Thames, the captain of her engaged a dry dock for the purpose of cleaning and painting her bottom. When docked, the owner, with the captain and her builder, marked forty-two rivets which leaked, and which were consequently cut and punched out. Fourteen of these rivets were at the junction of the fore bulkhead with the ship's sides. The rivets were carefully replaced, but when the water was let into the dock to float the ship, it was found that several of them at the bulkhead leaked. The ship was kept in dock, and every suspected rivet was taken out and replaced by new. A stream of water was then thrown upon the re-riveted parts with the force of a fire-engine—a device by means of which innumerable leaks in ships are discovered—every rivet being tested, and not one of them was found to "weep." The water was again let into the dock, and just as the ship floated several rivets started again, and with the new rivets two old rivets which had not started before. The ship was retained in the dock, and the owner then took the opportunity of securing the bulkhead to the skin more firmly by means of brackets. When the spring tides returned she was floated out of dock, took in a heavy cargo at the Victoria Dock, and proceeded to sea. When near the Western Islands she sprang a leak, and had to put back to Liverpool, where a survey was held upon her. The new rivets with several others had started, and the external plating was considerably torn in the way of the rivet-holes. The owners then brought an action against the firm who performed the new riveting work, and the matter was thoroughly sifted by a legal referee. The evidence adduced on both sides was of the most important character, and the fact was undoubtedly established that a want of rigidity in the skin of the vessel was the cause of the mischief. Instances were adduced of ships "panting" in their fore compartments; and it was proved beyond doubt that iron ships have in many cases expanded when dry on the blocks, and collapsed when sustained by the water. In this instance, so strikingly was this the case that the fore bulkhead, which was perfectly tight when she was upon the blocks, buckled into an irregular curve when she was afloat. This ship now makes her voyages satisfactorily, having had more rigidity imparted to her by an entire range of orlop-beams being put into her, with a stringer on each side at their ends; by some additional reverse irons; and by a double angle-iron with plates being added at the sides of her bulkheads.

The second case is that of a ship of 1000 tons built for the Bombay trade, and admirably constructed in every respect but one, viz., insufficient beams and pillars. A ship rolling about with a heavy cargo will alter her form, as regards its transverse section, very much, if she is built of iron, and is not sufficiently strutted and tied with beams,—for beams act, of course, both as struts and ties, according to circumstances. The captain of this ship on his passage out was much alarmed by a noise on board the ship resembling the explosion of some combustible substance, and he concluded that a portion of the cargo was of this nature, and that the ship would be very soon in flames. No such result took place however; but on arriving at India and discharging his cargo, he found that some of the beams had been fractured and their pillars bent. The weight betwee decksn was moderate, and therefore the accident could not be attributable to that. But the beam-end bracket-plates appeared to have strained considerably, and the beams developed an error in their construction of an important description. Instead of being made of "bulb" iron, they were made with two pieces of half-round iron at their lower edge, so that when a violent strain was brought upon them they "buckled," the half-round iron broke, and tore off the countersinks of its rivets, and the hole for the rivet became the commencement of a fracture across the beam. The butts of this ship showed no indications of weakness whatever, and all that was done to her was to refasten the gusset bracket-plates at her beam ends, clamp the beams with a plate on each side, and introduce new pillars with a different mode of securing them, so that the lower-deck and orlop beams should be compelled to yield together, if at all. One collateral circumstance may be referred to: the rivets in the garboard-strake were in such a condition, from the alteration of the ballast and the action of the bilge-water upon their heads, that some were ordered out. It was found, however, that new ones could not be put in without disturbing the remainder of the old, and the result was that all the butt-straps were removed, refitted, and entirely riveted anew. This wearing away of the rivet heads is a source of great injury to iron ships, and has led to the very general use of cement for their protection.

The third case is that of a ship which went into dock to be painted, but as the water was leaving her, it was observed that her plates appeared to be cracked. At about twenty feet from the stem two projecting plates, and the under or sunk plate situated between them, were evidently broken across vertically. Knowing that with flexible ships the edge of the bulkhead was a sort of node to the flexure, and that the rivets were very liable to become loose, the gentleman who told us of this case asked the owner to accompany him to the larboard side, and they there found that precisely the same thing had taken place. There could be no doubt then that the bulkhead edge had to do with the mischief, and he asked the owner if he had done anything to secure the bulkhead to the ship's side. It turned out that some brackets had been put, which no doubt relieved the rivets of the bulkhead frame, but they checked the bending of the plate and it broke at the point where its flexure was stopped. The broken plates were cut out and replaced by new ones; stringers were added to the fore body to give it rigidity; additional reverse irons were fitted to the frames, and a few additional beams introduced. Here again are weakness and flexibility in the fore body, where iron-ship-builders and Lloyd's are supposed by some persons to put too much iron.

"To me it appears clear," says a practical man of great experience in writing to us on the subject, "that rigidity of the skin "of an iron ship is the most important element of strength. It is "impossible to see the broken frames, and the *odd* bows and sides "which are occasionally brought under the notice of practical men, "without coming to this conclusion. Just imagine a ship's fore "compartment full of a heavy cargo, and contemplate the force "with which it is lifted out of, and immediately afterwards dashed "into the water. What section near the neutral point in mid- "ships is required to sustain a strain like that to which this "compartment is subjected? It appears to me that a transverse "section forward is forced into a more acute or flattened form when "the ship pitches, unless adequate beams and stringers are placed "in its vicinity to prevent it. On the other hand, when the end "is thrown up, the skin pressed upon by a heavy cargo has a "tendency to expand, and mischief results in the opposite di- "rection." There is no doubt much force in these remarks.

We will conclude these general illustrations of the practical weaknesses of iron ships, by referring to two cases of ships grounding, and suffering injury from the strains thus brought upon them.

The first case is that of an ocean steamer which went on shore upon a sandbank at the entrance to a river, and broke across between the foremast and the funnel. The fracture was of such a character as to suggest at once a source of longitudinal weakness, the mention of which has been reserved till now, for obvious reasons, *viz.*, the placing of the butts of the plating and other longitudinal ties in too close proximity to each other. A butt is usually (not always and necessarily) a weak place, and it is of course essential to a uniformity of strength longitudinally that these weak places should not fall in or near the same vertical line, or the same transverse section.

This important consideration was not attended to as thoroughly as was desirable in some of the earlier portions of the Conway and Britannia Tubular Bridges, but we doubt if it was ever so completely disregarded as in the ship which is now under our notice. On the port side there was a butt of the second strake of plating and a butt of the clamp behind it, both falling in the same frame space as the port, and close above it; and immediately over these butts of the outside plate and inside clamp was a butt of the deck stringer-plate. In the strake next below the port was a butt immediately under the port, and a butt of the main-deck stringer fell exactly in the same place. In the second strake below these fell another butt of the outer plating, in accordance with the common brick-fashion arrangements of plating adopted in this and so many other ships; and close to the line of weakness formed by this astonishing succession of butts falling vertically above and below the port, the beam-stringer of the hold was broken by the bulkhead at the fore part of the boiler-room. It seems doubtful if the most evil ingenuity could have devised a worse or weaker disposition of material than was thus presented, and the strain of the ship on grounding naturally enough found out a weak place. On the starboard side also the fracturing force found for itself a somewhat similar place of weakness, and broke the side down through a butt of the deck stringer-plate, a scuttle, a butt of outer plating, a butt of inner clamp, and through other butts below. The lower part of this fracture was diverted (by some cause not observable) away from the butts of the outer plating and the hold beam-stringer, and broke through two frames.

It will add to the interest of this case if it is stated that we happen to be aware that the clamp between the upper and main decks, and the main-deck stringer, were not in the vessel originally. The clamp was added afterwards, and at the same time the original stringer to the main deck, which had been cut considerably, was removed, and a new one substituted for it on that account. As this addition and this alteration were made expressly with the view of giving the vessel increased longitudinal strength, of which she was supposed to be deficient, it seems astonishing that in arranging the butts of the new work care should not have been taken to succour the weak parts, instead of placing them in exactly the same places as the existing butts, and consequently rendering the additional plates of the least possible service.

We often hear of iron ships being practically of one piece, while wooden ships are not less often spoken of as "bundles of sticks;" but any one who will study this very instructive example will see that unless a careful disposition of the butts of a ship's plating and stringers is made, it is quite easy to fall into arrangements which will justify the comparison of her hull to a series of short trunks or tubes, very imperfectly joined at their extremities, or to a hollow beam or girder, half broken through in several places before the strain it is to bear is put upon it.

The other case of a ship being injured by grounding to which we have to refer, is that of a ship of 1500 tons, which got on shore across a stone causeway, in a river where there is a very great rise and fall of tide. Her draft of water when taking the ground was 21 feet, and the breadth of the slip or causeway was 28 feet. She had on board at the time of the accident a cargo largely composed of iron, weighing about 2100 tons. The weight of cargo and ship was estimated at 3180 tons. When on the slip-way her fore foot was a few inches in the sand, but certainly not in any way supported by it, all the strain being in the middle of the vessel. We have had an opportunity of seeing a report upon this case, written by an experienced surveyor, who says:—

"The fracture to various parts will be hereafter enumerated, "but it is worthy of remark, that during the time of her being on "the ground, the pumps on the aft side of the mainmast were "forced up through the deck about 15 inches. The mainmast "appeared unchanged until she floated off, when it settled down "and the rigging became slack, showing that the keelson had "forced itself up into the heel of the mast, taking the iron step "with it, and when the vessel floated the bottom dropped 5 to 7 "inches and the mast followed. All the damage done to the out-"side plating was confined to the flat of bottom and to the height "of the upper part of bilges, and showing no strains in the upper "works. I believe that in all cases of vessels being supported at "the middle with their ends free, as in this case, unless the base of "support is of sufficient length, or the vessel of very extreme "dimensions, the bottom must crush up, and thereby prevent the "great amount of strain to the top that would otherwise take place, "and I am disposed to think that sand is the worst description of "ground for a ship to set on as it forms a curved base, caused by "the sharp ends of vessels settling in it, and by not yielding in the "middle, communicates the strain to the top, and the breaking "must of necessity take place at the sheer-strake, deck-stringer, "&c., unless the bottom is of a very weak construction. It is "evident from this damage, that had the ship double the amount "of stringer, and any quantity of strength given to the upper works, "over and above that at present in her, the damage would have "been the same, so that there can be no doubt about the upper "part of such vessels being sufficiently strong for all practical pur-"poses. Again the base of support being only 28 feet, it is really "surprising that the floors in way of the same did not crack or "bend, as well as break down, and that the vessel ever floated off, "and that the bottom did not break into a hole."

This case of damage is certainly very interesting and instructive. The writer of the above remarks considered that it was a favourable illustration of the merits of Lloyd's Rules, as no defect resulted, in his opinion, from imperfect construction. In this case also the butts proved, as was to be expected, the weakest points. We will only add, on this case, that we concur in the tribute of praise which the writer claims for Lloyd's Rules, in conformity with which this ship was fortunately built, and consequently exhibited no such defects as the former vessel; but at the same time it appears pretty evident, as we shall see further on, that the effects of the local upward pressure upon the keel in this case were aggravated by the absence of an intercostal middle-line keelson-plate. This, however, is no reflection whatever upon Lloyd's, because their Rules are very favourable to the use of such keelsons with bar-keels.

In concluding this chapter it is only just to mention that the cases of weakness herein described were in some instances foreseen by Mr. Luke, the Admiralty Surveyor of mail and contract-built ships, to whose professional skill, and fearless fulfilment of a most difficult public duty, the art of iron ship-building owes much of its successful development in the merchant service.

CHAPTER II.

KEELS, KEELSONS, AND GARBOARD-STRAKES.

COMING now to notice the various details of iron-ship construction in succession, we will commence, as is usual, with keels and keelsons, taking it for granted that they are desirable features of a hull, and that their primary object is to give longitudinal strength and rigidity to it; to distribute along the floors pressures and strains which would otherwise be too localized in their action, such as those of pillars, deck blocks, &c.; and when external, to act as checks to lee-way and to rolling motions.

The keels of iron ships were originally external, and not unfrequently of wood. Sometimes they were formed hollow, of iron, and filled with wood. The keelsons also were of wood. It was soon found, however, that the bolting of a wooden keel to an iron plate by through bolts was a dangerous practice, because in the case of the 'Iron Duke,' and also in some other vessels that took the ground, when the keel was stripped off, the ship escaping with no other damage, the water entered through the bolt-holes and nearly caused her to founder. M. Dupuy de Lôme called attention to this circumstance as long ago as 1842, in his very able Report on the Iron Shipbuilding of this country, observing that if in some particular cases it becomes desirable to put a keel or false keel of wood on an iron vessel, it is necessary first to fix upon the inside of the plates of the bottom an opposite keel, bedded in cement or felt, and secured with bolts having their heads countersunk in the outside of the plate, then to apply the outside wood keel, and bolt it through both the plate and inner wood keel. At the same time he judiciously recommends the avoidance of such a combination wherever that is practicable. Mr. Grantham also, in his useful book on Shipbuilding,* repeats the opinion that wooden keels are imperfect and dangerous appendages to iron ships. The subject is

* 'Iron Shipbuilding, with Practical Illustrations,' by John Grantham, C.E. London, John Weale.

well worth mention here, because there appears to be a strong tendency to fall again into a similar system of construction and fastenings for keels in what are now known as composite ships. In getting rid of the wooden keel, builders very naturally fell into the use of a solid iron keel, formed with a rabbet on each side to receive the garboard-strakes of plates; the rivets of these strakes (through the keel in the vertical flange, and through the angle-irons of the frames or floor-plates in the horizontal flange) forming the only connections between the keel and the ship. The planing of the rabbet was obviously an expensive process, and the connection just mentioned was clearly not of the most satisfactory character. But this form of keel has not wholly disappeared: it is to be found, for example, in the Peninsular and Oriental Company's steam-ship 'Malta,' which was built as a paddle-steamer in 1847, and subsequently turned into a screw-ship by Mr. Laird. The Atlantic Mail steam-ship 'Persia,' built in 1855 by Messrs. Napier of Glasgow, has also a solid rabbeted keel of the form and dimensions shown in Fig. 1. It must be admitted, we think, that with very accurate workmanship, where a keel is formed with a rabbet in this manner, the keel rivets are subjected to less strain when the ship is docked, or goes on shore, than if the rabbet did not exist; but the expense of the plan, and the obvious necessity for careful work in fitting, have led to its general abandonment, and the resort to the plain bar-keel, Fig. 2.

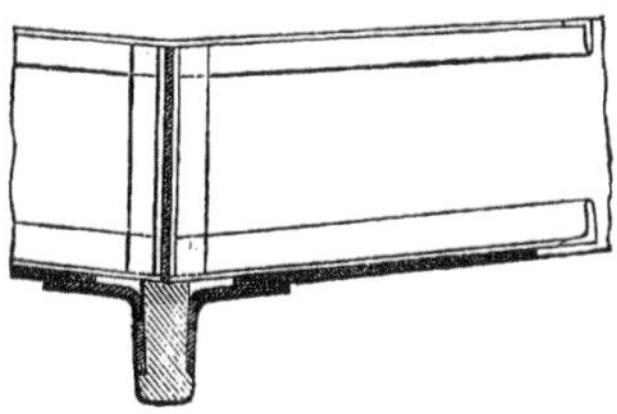

Fig. 1.

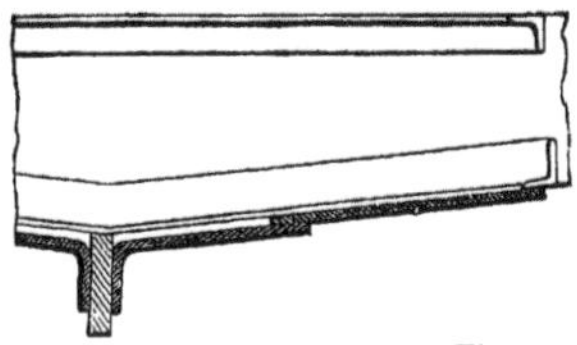

Fig. 2.

Before offering any remarks upon this last form of keel, or upon the modifications it sometimes undergoes, it will be well to refer to the hollow iron keels which were formerly much in vogue, but which are not often to be met with now. The keel of the mail-packet 'Dover,' built for the Admiralty in 1839, by Mr. Laird, was formed in this way, of $\frac{5}{8}$-inch iron plate bent to shape as shown in Figs. 3 and 4, these being a section and elevation at midships, showing also the attachment of a wooden keelson, and Fig. 5 a section near the bow. Such a keel can only be formed, of course, out of plate of good quality, especially if the angles are to be made at all sharp, and the iron is no doubt greatly distressed and injured in the bending process. Hence it would obviously be better to roll the gutter-flanged plate to the required form, if that could be done, and Mr. Grantham mentions that the Oakfarm Iron Company patented the plan of rolling it in the form shown in section in Fig. 6.

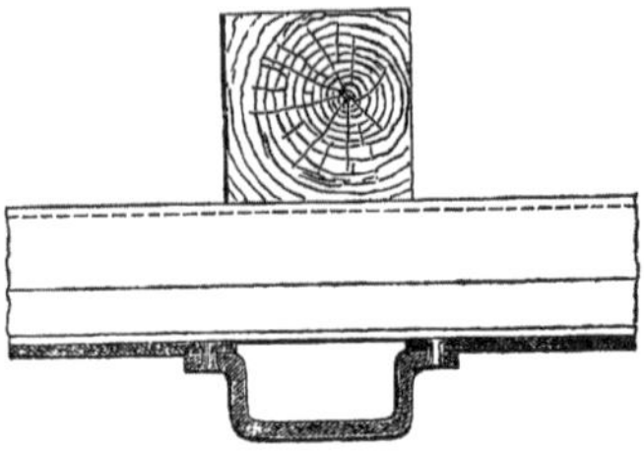

Fig. 3.

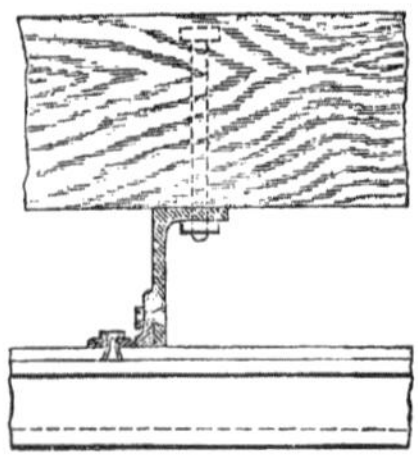

Fig. 4.

Fig. 5.

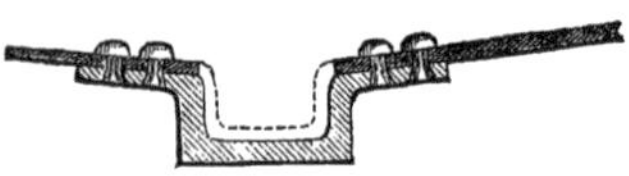

Fig. 6.

"It was rolled," he says, "with "much success, and was an interest-"ing specimen of what may be done "by machinery. Its principal dis-"advantage arose from the short "lengths in which it was rolled, and "the difficulty of welding such a "peculiar form into long lengths. It "was made of three slabs previously "prepared in separate rolls, and "then finally welded by passing the "whole together through other rolls, "and thus took the form shown in "the drawing. I have not heard of "its being used for several years."

The improvements in rolling machinery, and in the manipulation of large masses of iron at the rolls, have been so great of late years, and especially since the introduction of armour-plates, and of armour shelf-plates like those of the 'Warrior' (which are immense angle-irons, in fact), that such a section of iron could no doubt now be rolled of much greater lengths than formerly, and thus part of what must have been a great weight of butt-straps in the Oakfarm plan might be saved. The weight of these butt-straps, and the obstruction which they offer to the flow of bilge-water, along the gutter which the keel forms, was avoided by a very peculiar arrangement in the vessel already mentioned, the 'Dover.' It consisted in sinking a tapering butt-

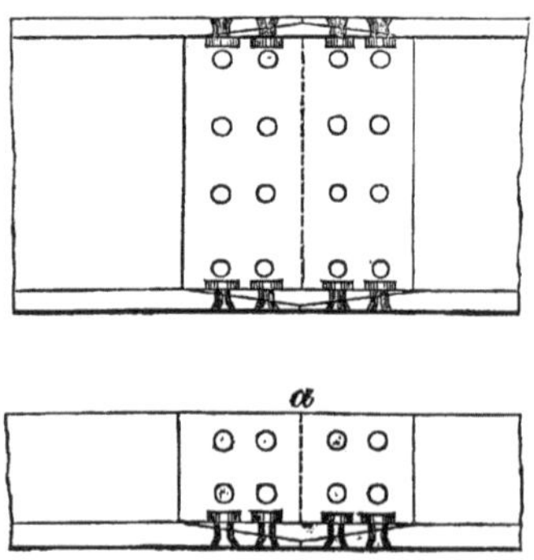

Fig. 7.

strap into the abutting ends of the keel-plates as shown in plan and elevation in Fig. 7, an arrangement which is plainly inconsistent with either due strength or economy of workmanship. Were such keels employed now, we should no doubt either weld the successive lengths together, as far as practicable, or strap the butts with treble-riveted straps, and cement the channel up to the level of the rivet-heads, or rather so as to bury them beneath a flush surface of cement.

The difficulty of manufacturing the hollow iron keel led in some cases to the adoption of curious devices. One of the most singular of these, perhaps, was that which Mr. Laird employed in the 'Birkenhead,' a ship built for the Admiralty in 1843 as a steam-frigate, but used by them as a troop transport, and wrecked in a memorable manner. This keel was formed, as shown in Fig. 8, of two angle-irons and a shallow gutter-plate, connected by single-riveted joint-straps.

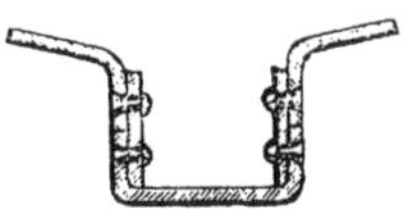

Fig. 8.

Perhaps the simplest of all the forms of hollow keels was that of a simple plate, with an easy curved depression in the middle of it—a form adopted, we remember, in the Empress of Russia's yacht 'Nevka,' and in the U.S. monitor 'Dictator,' designed by Captain Ericsson, as shown in Fig. 9.

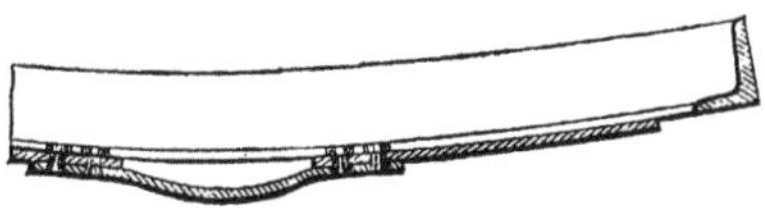

Fig. 9.

In some cases hollow keels were filled with wood, which, of course, decayed long before the iron, and could not be renewed without removing the floors or keel. In other cases the builders ran the garboard-plates across the hollow keel, as shown in Fig. 10, and thus made the latter merely an external channel, as security to the ship in the event of the keel being injured. But this, of course, led to the rapid corrosion of the keel and garboards, and at the same time rendered the repairs, which consequently became necessary, exceedingly difficult and expensive.

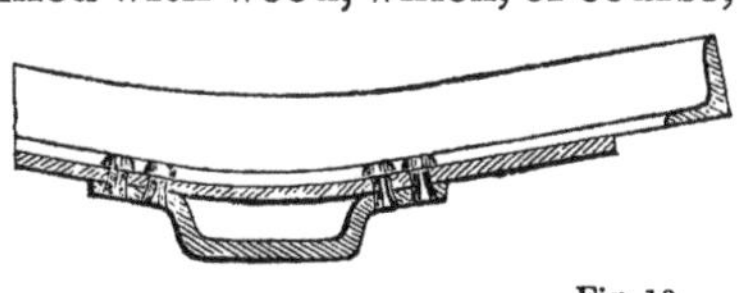

Fig. 10.

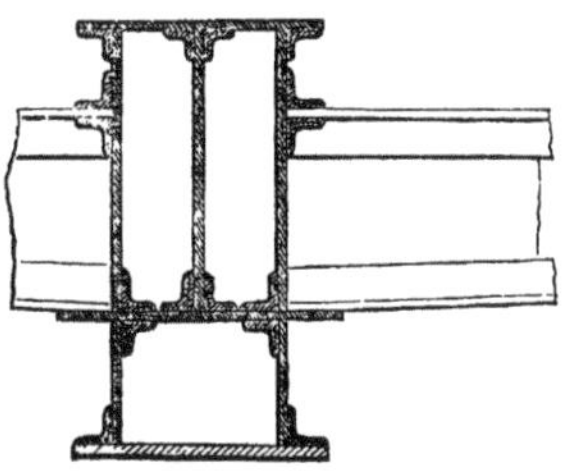

Fig. 11.

Another form of a hollow external closed keel is given in Fig. 11, which is taken from an able article on Shipbuilding in the *Encyclopædia Britannica** by Mr. Andrew Murray, who says of it:—

"In the keel, according to the above sketch, it will be observed "that the plating is carried right across it, so that it might be very "much injured, and probably even torn away in parts, without "causing any leak into the ship, its edge being made purposely "weaker than the bottom plates to which it is attached. Cross-"plates with flanges, or with angle-irons, may be riveted across it, "at any distance that may be desired, so as to stiffen it; and the "plates can be made thicker, or additional strengthening plates "may be added inside or outside the side-plates, at the stern-post "or at the fore-foot."

Mr. Murray also gives the section of a vessel designed by Mr. Bowman, in which the keel is formed box-fashion as shown in Fig. 12, the floors abutting against the sides of the box-keel; and

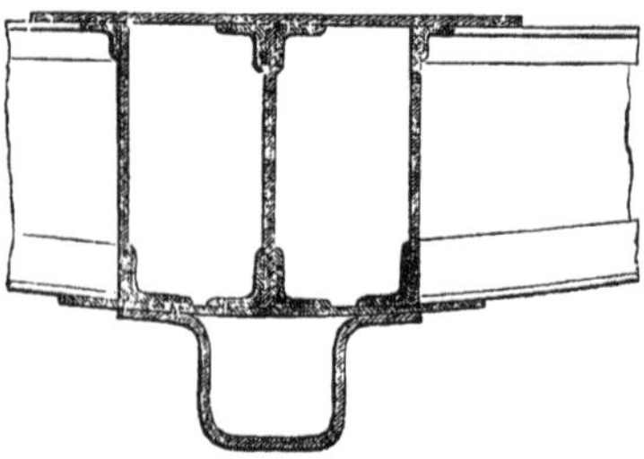

Fig. 12.

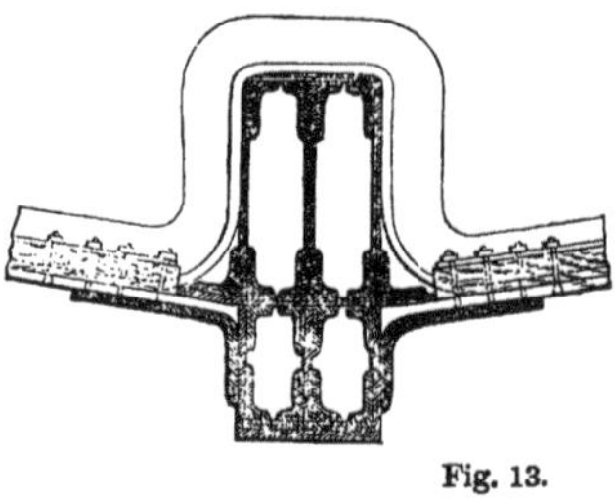

Fig. 13.

the section of a form of keel adopted by Messrs. Taylerson and Co. of Port Glasgow, as shown in Fig. 13.

The rabbeted bar-keel has been succeeded, as already stated, by the plain bar-keel, a simple bar of iron rectangular in section,

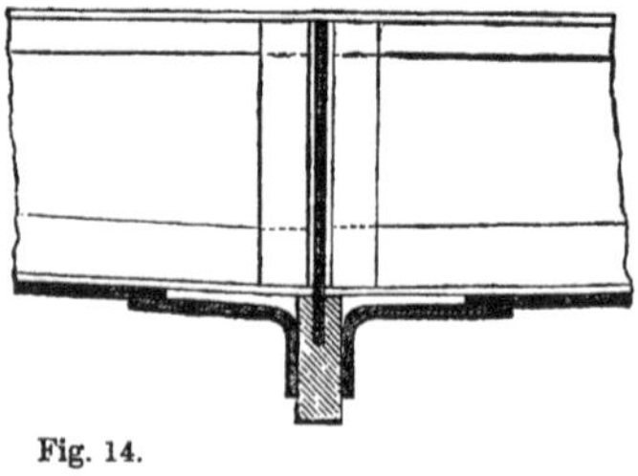

Fig. 14.

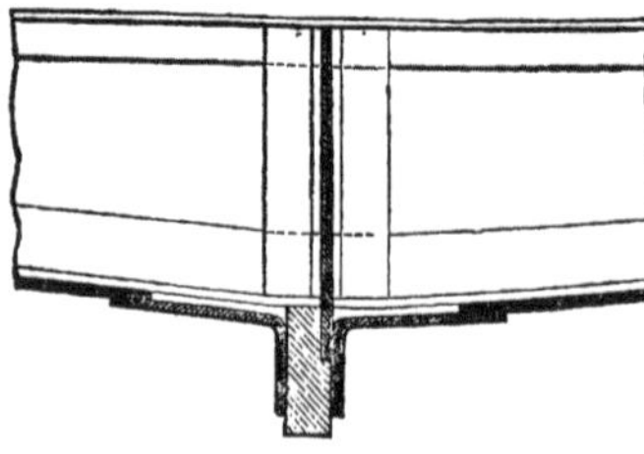

Fig. 15.

which is now very common in the merchant service. In its usual form this keel, like the rabbeted solid keel, has no other

* Reprinted in a separate treatise, and published by Black of Edinburgh.

connection with the ship than that which it derives from the garboard rivets upon which it hangs; there have been cases, however, in which a direct attachment of the bar-keel to a keelson-plate has been effected by means of a groove in the keel, as in Fig. 14, or by means of a rabbet on the side, as in Fig. 15. In these cases, the keel ceases, of course, to be a plain bar-keel, the machining of the groove or the rabbet being indispensable and expensive. There have been instances, however, of the direct attachment being effected without a rabbet or groove, the keelson-plate being simply lapped upon the keel, as will be seen from Fig. 16, which represents a middle-line section of the Peninsular and Oriental steam-ship 'Nemesis,' built at Glasgow in 1857. It will be observed that the only connection between the two is a single row of rivets, and the keel being 12 inches by 3 inches, the arrangement obviously throws the keelson $1\frac{1}{2}$ inch out of the middle line—a matter of little or no moment probably. As far as the single row of rivets can accomplish the object, this plan, like all other plans which connect the keel with a central keelson, has the important advantage of opposing the racking of the floor-plates longitudinally. Comparing this arrangement with that of either of the plans shown in Figs. 17 and 18, which

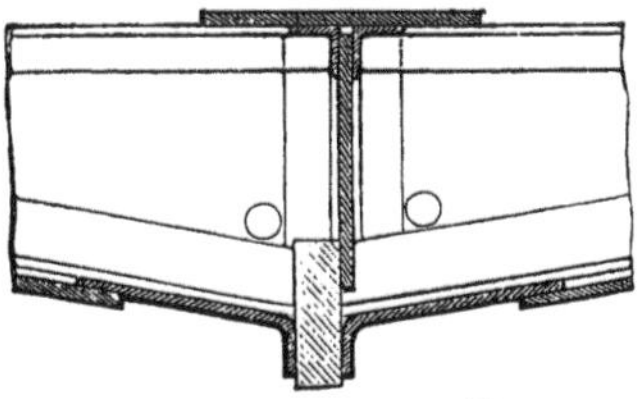

Fig. 16.

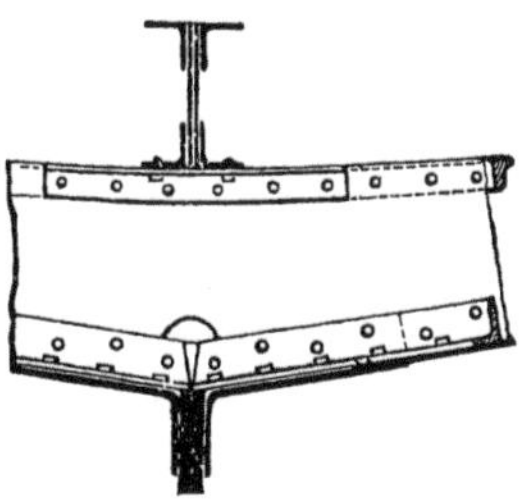

Fig. 17.

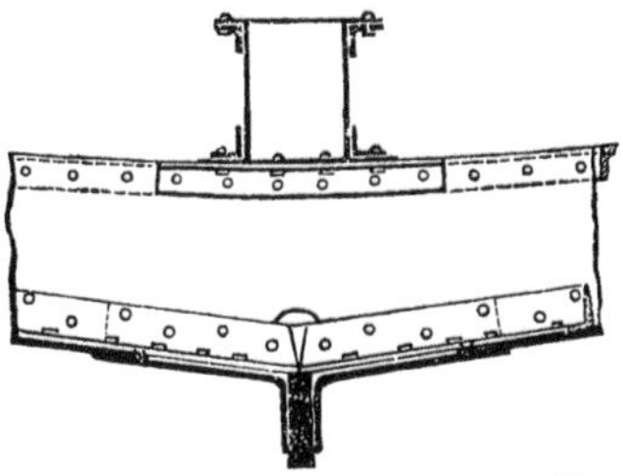

Fig. 18.

show combinations of common bar-keels and through-floors with a plate-keelson and a box-keelson respectively (as given in Lloyd's plate of illustrations for their keels for building iron ships), it will be seen that while in both the latter cases the floors are comparatively free to trip, by the keelson riding along the keel, this cannot happen in the 'Nemesis' arrangement (the floors being riveted to the keelson-plate) until the rivets connecting the keel and keelson have first been stripped. That the resistance which a single row of rivets can present to any great and violent strain—such as would ensue from the ship striking the ground at speed, for example—is not great, is very evident; but it is obviously a step in the right direction, and it gives to the plain bar-keel an advantage which it is impossible for it to possess while hung upon the garboard rivets alone. It is only necessary to add, in reference to the bar-keel, that it is not uncommon to weld the several lengths into one continuous bar; and where that is not done, it has to be carefully scarphed together by vertical scarphs, the length of which must, according to Lloyd's Rules, be in length at least 8 times the thickness of the solid keel. The Liverpool Rules require the length of the scarph to vary with the size of the keel, and on the average the length they prescribe is about 9 times the thickness.

The solid-bar keel has in some instances had an assemblage of plates laid side by side substituted for it, the butts of the several layers being carefully shifted. An early instance of this is to be found in the case of an iron brig called the 'Recruit,' built many years ago for the Admiralty. Fig. 19 is a section of her keel and garboards. It will be observed that, in this case, the built keel merely takes the place of the solid keel, and is not connected to the floors, except by the garboards. This arrangement of keel has recently been adopted in the construction of the War Department store-ship 'Earl de Grey and Ripon,' and has frequently been applied in other cases.

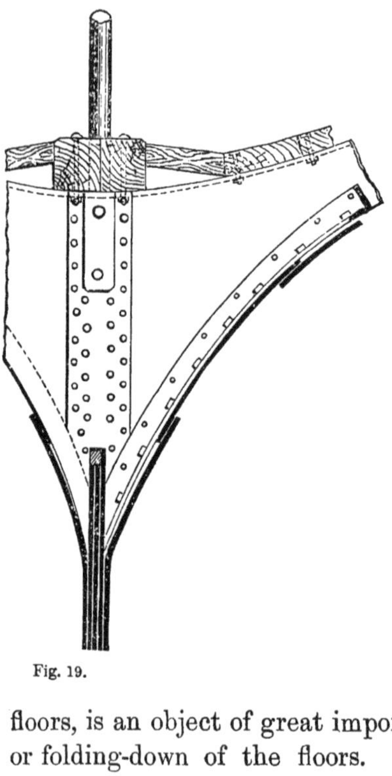

Fig. 19.

We have already seen that the connection of the keel with the middle-line keelson-plate, which is connected by angle-irons to the floors, is an object of great importance, as it obstructs the racking or folding-down of the floors. We also see how difficult it is to effect this connection satisfactorily with a solid plate. But if we make the keel of two edge-bars instead of one only, we are obviously enabled to pass the keelson-plate down between them, and by riveting the garboards through all, we shall get the keel and keelson directly combined by several rows of rivets, in addition to the indirect connection resulting from the floors being riveted to both the garboards and the keelson. This is the arrangement known as the "side-bar keel," and a very excellent arrangement it is for external iron keels. It is illustrated in combination with a central through-plate keelson, and a flat keelson-plate in Fig. 20,—a section of the Union Mail Steam-ship, 'Roman,' built by Mr. Lungley, of Deptford Green; and is favourably provided for in Lloyd's Rules, which only require the same total thickness for the keel-bars and keelson-plates as is required for a solid keel, and provide that the butts of the several plates of which the keel is formed shall be carefully shifted from each other, and from the butts of the garboard-strakes, which also must be carefully shifted so as not to be opposite or nearer each other than two spaces of frames. The Liverpool Rules also provide favourably for this keel, and state that it is preferred to the bar-keel. It was first employed, we believe, in a little vessel named the 'Vulcan,' built in 1818 for the Forth and Clyde Canal from the designs of Sir John Robison.

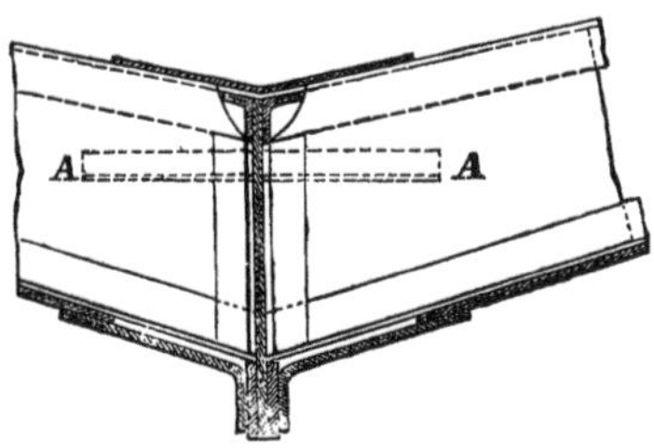

Fig. 20.

In some instances a horizontal plate has been worked in places beneath the side-bar keel, and connected thereto by a couple of angle-irons, as shown in Fig. 21. Mr. Murray gives a specimen of this arrangement in his book before referred to, the example being taken from the specification of a steam-ship built for the Peninsular and Oriental Company. The vertical flanges of the angle-irons were bolted through all. This additional plate and pair of angle-irons extended 50 feet from the after sternpost. A similar arrangement is adopted at the after end of the keel of H.M. Troop-ship 'Orontes.' We will only add at this point, that side bars, like keels themselves, are sometimes

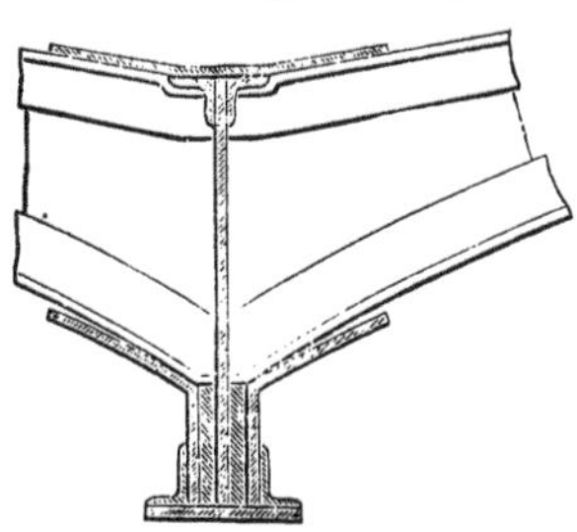

Fig. 21.

formed of more than one thickness, as we shall show hereafter by giving a section of the 'Ulster's' keel and keelson.

Before proceeding to remark upon horizontal or flat-plate iron keels, it becomes necessary to observe the forms of middle-line keelsons which builders have associated with the bar-keel and the side-bar keel. The only keelsons employed by some builders with the bar-keel have been situated upon the tops of the floor-plates, which have crossed unbroken beneath them and above the keel, as already shown in Figs. 17 and 18. Sometimes this surmounting keelson has been formed with a single plate and a pair of angle-irons at top and bottom, or with two plates placed together and a like arrangement of angle-irons as in Fig. 17, and at other times it has been formed as a box-keelson, as in Fig. 18, the lower pair of angle-irons being riveted to the angle-irons at the top of the floors. But as there is an entire absence of intercostal plates between the floors, it is obvious, for the reasons already stated, that neither of these arrangements of keel and keelsons, taken alone, is fully satisfactory.

Reverting to the case of the ship that got across the stone causeway (page 15), it may be observed that she was an instance of this kind, having a solid-bar keel and a box-keelson on the floors, with nothing between the floors, and it is easy to see that an intercostal keelson-plate would have been the very thing to oppose that bowing up of the keel which took place, and which led to the fracture both of the transverse floors and of the longitudinal keelson. An actual and considerable racking of the floor-plates obviously took place, and the strain was allowed to concentrate itself upon the *top* of the keelson and of the floors, and tear them downwards; whereas had the form of the rectangular spaces between the floors, the keel, and the keelson, been preserved, this tearing action could not have taken place, and all the sectional strength both of the keelson and of the floors would have been brought into play. For reasons which will be apparent, the solid bar is now generally associated with a middle-line keelson, consisting of short intercostal plates between the floors, connected to them by short pieces of angle-iron. Sometimes these intercostal plates are worked flush with the throats of the floors, and at other times they stand up above them. In either case, a pair of angle-irons is usually worked fore and aft upon the throats of the floors, to give a longitudinal tie above the floors as well as below them, and frequently a bulb-iron is introduced between the angle-irons, to add still further to the longitudinal tie. When the keelson-pieces are shallow, and do not rise above the floors, this bulb-iron is scored down between the floors sufficiently deep to lay hold of the keelson-pieces with a double row of rivets, as shown in Fig. 22. When the keelson-plates stand up above the floors, this scoring down of the bulb-iron is unnecessary and the arrangement then assumes the form shown in Fig. 23.

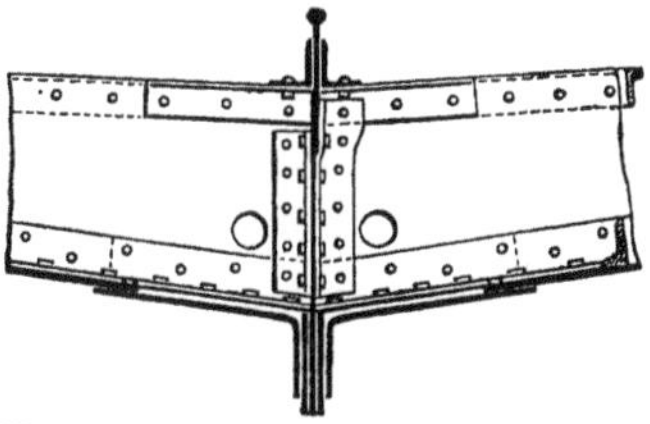

Fig. 22.

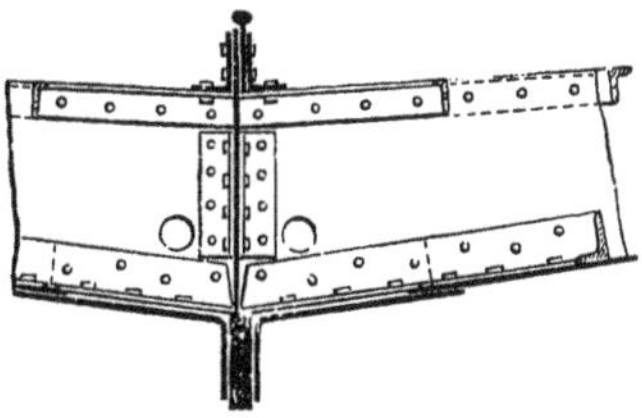

Fig. 23.

Builders are often content, however, to do without this bulb-iron, especially when the keelson-plates stand above the floors, and then the arrangement assumes the simple form shown in Plate 2, which shows, in section, the keel and keelson of the fine Atlantic Mail Steamship 'China,' built by Messrs. Napier and Sons of Glasgow.

It will be observed in this last figure that a flat keelson-plate, lying along the tops of the floors, is worked on each side of the upright keelson-plate, beneath the angle-irons, adding, of course, both to the longitudinal and the transverse strength. As the floors all cross the keel and keelson, this addition scarcely seems necessary as far as regards the transverse strength; but, on the whole, the arrangement no doubt presents a very satisfactory combination, the only weak feature of it being the absence of any direct connection between the bar-keel and the keelson.

Having thus completed all that it seems necessary to say, respecting the combination of solid single-bar keels with suitable keelsons, we have to turn to the combination of side-bar keels with their keelsons, of which we have already had one example before us in Fig. 20. In that example the upright keelson-plate is a continuous plate dividing all the floors; and a little consideration will show that the division of the floors at the middle line, and the use of a continuous plate, although not absolutely essential to the direct attachment of side keel-bars to a central keelson-plate, are by far the most convenient arrangement. As the side bars and the central keelson-plate must be effectually caulked, the plate cannot be scored down over the floors, not only because it would be exceedingly difficult to work a plate scored so deeply, but also because, if that were done, a hole or space would be left below every floor-plate. It would be practicable to score the keelson-plate and let it up between the floors, or which would be the same thing, to let the floors down into it, leaving the plate continuous below the floors, to receive the side bars, and to caulk against them. This plan is, however, not adopted, because the deep scoring of the keelson-plate is most objectionable in practice, and it is deemed simpler and better work, and more favourable to longitudinal strength, to make the central upright keelson-plate continuous through the floors, and to abut the floors against it, making good the transverse strength by other means.

Presuming the upright keelson to be a continuous or through-plate, then we may do one of two things,—either keep its upper edge level with the throating of the floors, or bring it up above them. In the first case—that in which it is stopped at the height of the floor-throat—it is usually associated with a horizontal keelson-plate worked on the floors, and connected to it by means of two angle-irons lying against both, as shown in Fig. 20 and as shown also in the accompanying section of the 'Ulster's' keel, Fig. 24 (before referred to), which likewise illustrates the use of side bars in two thicknesses, and of certain angle-iron connections of the floors and keelsons A and B to which we may hereafter revert. In this case the angle-irons connecting the flat and upright keelson-plates are continuous, but in other cases the reverse irons of the floors are made continuous, being scored down into the keelson, and then the longitudinal angle-irons of the keelsons are worked in short lengths between the floors. Some builders occasionally have cross-strapped the floors, in addition to working the horizontal keelson-plate. This was done, we remember, in the 'Tasmania,' a vessel built at Port Glasgow in 1858, for the conveyance of the European and Australian Mails. Fig. 25 is a section of her keel, keelson, and cross-straps,

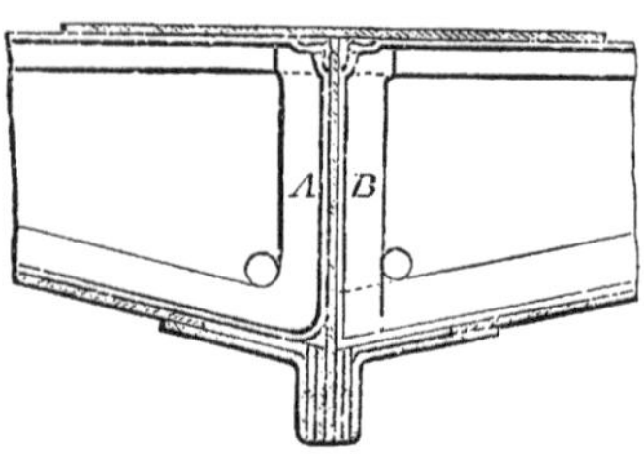

Fig. 24.

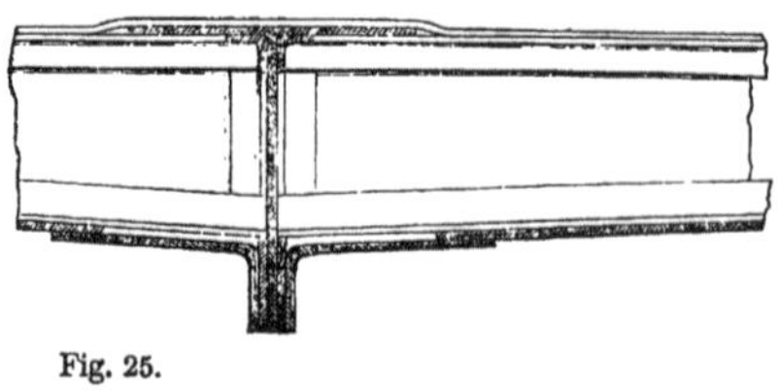
Fig. 25.

or tie-plates, these last being 9 feet long, 22 inches broad, ¾ inch thick, and worked on every second pair of floors. In other cases, again, the angle-irons, instead of being below the flat keelson-plate, are on top of it, close together and back to back, as shown in Fig. 26.

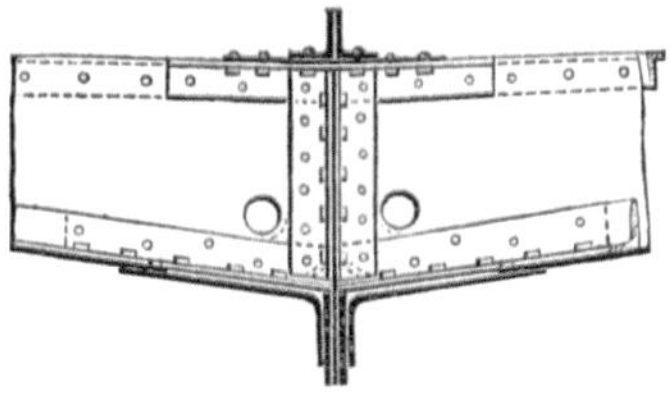

Fig. 26.

The second plan—that of carrying the upright keelson-plate above the floors—is, however, usually preferred. When this is done, a flat keelson-plate and an angle-iron are usually worked on each side of it, as shown in Plate 1, which is a view of the amidship framing of the 'Queen,' a fine screw-ship, built recently by Messrs. Laird for a Trans-Atlantic Steam Company; an elevation of this keel, with its fastenings, being given in Fig. 27, and the disposition of its butts being shown in elevation and plan in Fig. 28. The *stations* are drawn in dotted lines. The butts in dotted lines refer to the port side in the elevation, and to the garboard-strakes in the plan. These figures illustrate the shift of butts and the fastenings adopted by one of our leading private firms

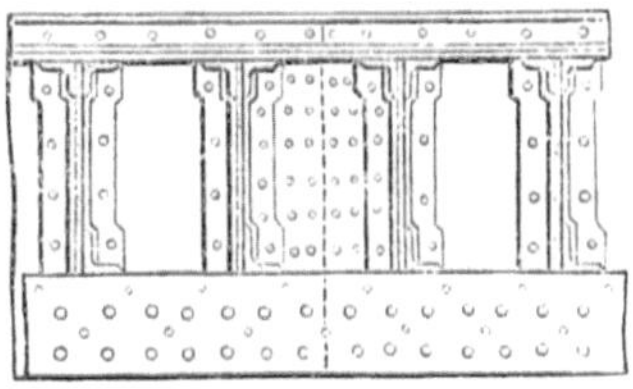

Fig. 27.

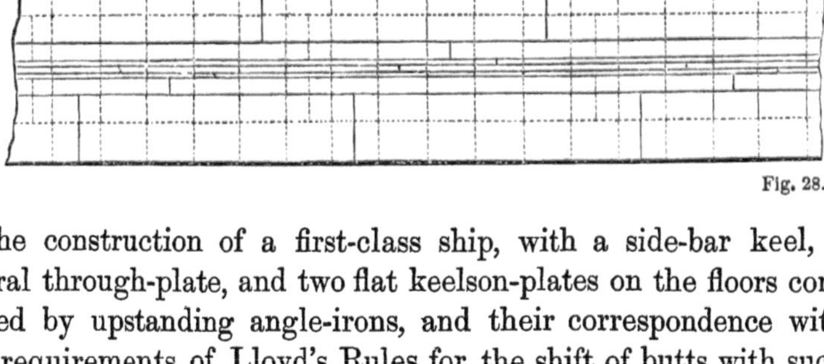

Fig. 28.

in the construction of a first-class ship, with a side-bar keel, a central through-plate, and two flat keelson-plates on the floors connected by upstanding angle-irons, and their correspondence with the requirements of Lloyd's Rules for the shift of butts with such keels, already quoted.

In Lloyd's sheet of illustrations provision is made for the central continuous plate rising up to a sufficient height to receive two pairs of angle-irons, as shown in Fig. 29; thus forming an upright-plate keelson above the floors, an arrangement which may be useful in some ships, but which, as a rule, would provide an excess of longitudinal and too little transverse strength. It is, of course, in very long and fine ships, with comparatively small beam, that so much longitudinal and so little transverse strength would be most appropriate.

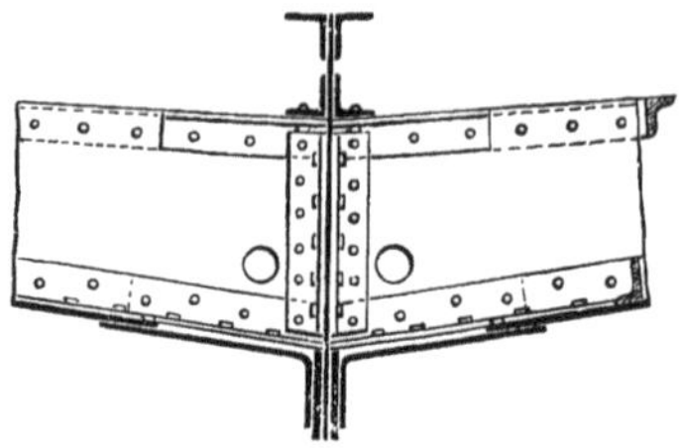
Fig. 29.

We have now considered, as completely as appears to be desirable, the cases of external central keels, and have next to consider flush keels, or internal keels, which are now frequently used in mercantile ships, and almost universally in H.M. ships. Comparing these with the keels and central keelsons already considered, we may almost say that ships fitted with them have no keels at all, for if we take away the external bar or side bars, and make the two garboard-strakes into one central plate, we have pretty nearly all that constitutes what is known as the "flat-plate keel" of the merchant service, a couple of continuous longitudinal angle-irons, to connect this single garboard with the upright plate (whether the latter be "intercostal" or "continuous") being all that are necessary to complete the combination. With this form and arrangement of keel, most of the arrangements of keelsons previously described may be combined. Fig. 30 illustrates the case in which the flat-plate keel is combined with a centre continuous plate, and with a flat-plate keelson with the angle-irons below it; and Fig. 31

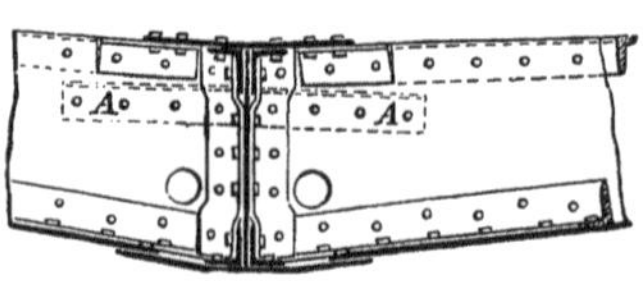
Fig. 30.

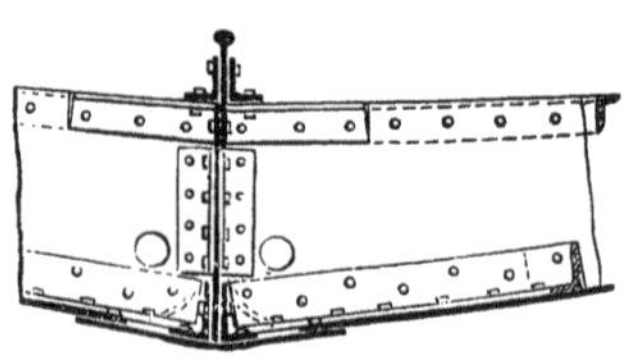
Fig. 31.

illustrates the case in which it is combined with an intercostal middle-line keelson, and a bulb-iron and pair of angle-irons running along on top of the floors.

Before passing to the ships of the Royal Navy, in which the

arrangements are somewhat different from those usually adopted in merchant ships, it will be proper to add that Lloyd's Rules provide that, where flat-plate keels are used, the intercostal keelson-plates are to be fitted close down on, and connected to, the keel by double angle-irons riveted all fore and aft to the keel and keelson ; and that in all cases, where centre continuous plates are applied, they are to be extended to the stem and stern post, and connected thereto, where practicable. As these Rules regulate the greater part of the practice of the mercantile shipbuilders, we shall not enter here or indeed, at all, upon the details of the dimensions and proportions to be given to the various parts of merchant vessels, but shall give the Rules themselves, and the well-known Table G, in the form of an Appendix. We shall also add the Rules of the Liverpool Underwriters' Registry.

The flat-keels, which have been given to H.M. iron-clad frigates differ from those already described chiefly in the fact that the flat-keel plates are doubled in them, the garboard-strakes meeting the inner-keel plate, flush with it, and the outer-keel plate being worked broad enough to lap upon the garboards, as shown in Fig. 32, which is a section of the 'Warrior's' keel and keelson-plates. Here the continuous centre-plate is 40 inches deep and $\frac{3}{4}$ inch thick, and is connected below to the two keel-plates by 6-inch angle-irons 1 inch thick, and above by two smaller irons ($3\frac{1}{2}$ inches by $3\frac{1}{2}$ inches by $\frac{5}{8}$ inch) to a flat-keelson plate $\frac{3}{4}$ inch thick and 3 feet broad. These latter angle-irons are in short lengths, between the floors, as the upper pair of the floor angle-irons and the shallow plate between them run continuous across the centre plate, into which they are scored. The outer angle-irons of the floors turn up against the centre continuous plate. The outer and inner keel-plates have each a short butt-strap worked on each side of the keel angle-irons, and the butts of the angle-irons are also covered in the usual way. The keel-plates are in lengths equal to the other plates of the bottom, as are also the vertical

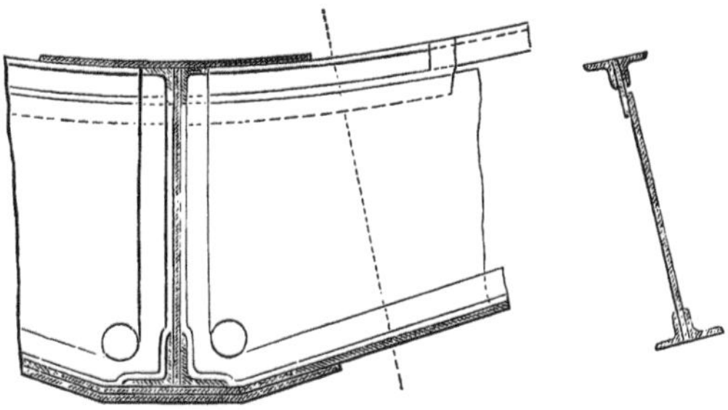

Fig. 32.

middle-line plates. The angle-irons are in length from two to four lengths of the outside plating.

The keel arrangements of the 'Northumberland' are of substantially the same character, but are given in greater details in section, plan and elevation in Figs. 33 and 34, in which are also shown the butt-straps of the outer and inner keel-plates, the garboard-strakes, the ver-

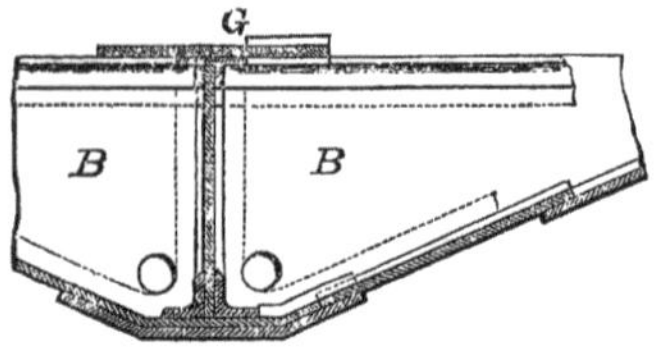

Fig. 33.

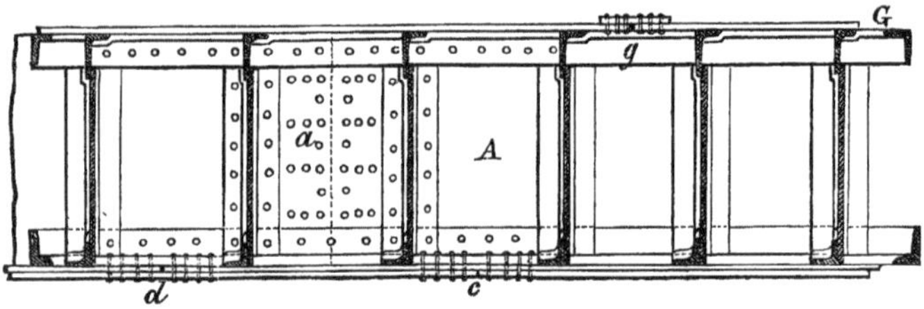

Fig. 34.

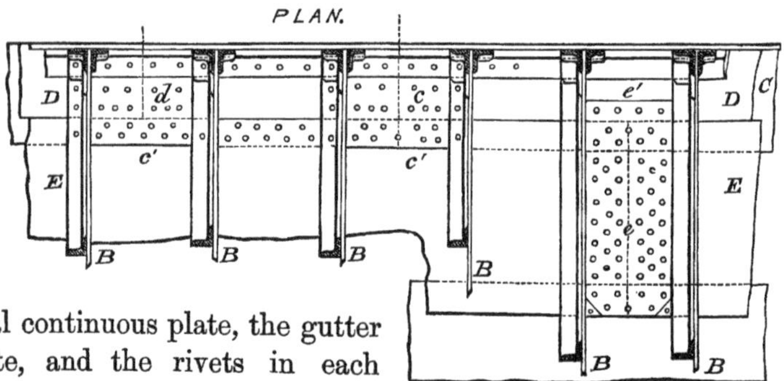

tical continuous plate, the gutter plate, and the rivets in each case. The thicknesses of the plates are as follows:—

Vertical keel-plate	$\frac{7}{8}$ inch thick.
Butt-straps to ditto, double, each	$\frac{1}{2}$,,
Outer flat-keel plate	$1\frac{1}{4}$,,
Butt-strap to ditto, single	$1\frac{1}{4}$,,
Inner flat-keel plate	$1\frac{1}{4}$,,
Butt-straps to ditto, single	$1\frac{1}{4}$,,
Garboards, and adjacent strakes	$1\frac{1}{8}$,,
Butt-straps to ditto, single	$1\frac{1}{16}$,,
Gutter or flat keelson-plate	$\frac{3}{4}$,,
Butt-straps to ditto, double, each	$\frac{7}{16}$,,
Vertical keelson-plate	$\frac{3}{4}$,,
Butt-straps to ditto, double, each	$\frac{7}{16}$,,
Floors	$\frac{9}{16}$,,
Angle-irons connecting vertical and flat-keel plates	6 inches by 6 inches by 1 inch
Angle-irons connecting vertical keel and gutter plate	$3\frac{1}{2}$,, $3\frac{1}{2}$,, $\frac{5}{8}$,,

Angle-irons connecting vertical keel and floor-plates	4	inches by	$3\frac{1}{2}$	inches by	$\frac{5}{8}$	inch
Angle-irons connecting vertical keelson-plate and gutter-plate	$3\frac{1}{2}$	,,	$3\frac{1}{2}$	,,	$\frac{9}{16}$	,,
Angle-irons on top of vertical keelson-plates ..	$3\frac{1}{2}$	,,	$3\frac{1}{2}$	,,	$\frac{1}{2}$	,,
Angle-irons forming continuous transverse frame ..	7	,,	$3\frac{1}{2}$	,,	$\frac{5}{8}$	,,

It will be seen from the Figs. 33 and 34 that the vertical keel-plate A passes through, with the floors B,B, butting against it, and the transverse angle-irons on the top of the floors scoring through it. These transverse angle-irons form continuations of the frames of the ship, and running across the keel, terminate alternately on opposite sides of it, at a distance of about 6 feet from it, thus giving shift to each other.

The vertical keel-plate A is strapped with a butt-strap, *a*, on each side, treble-chain riveted, as shown in the elevation in Fig. 34. This butt-strap, *a*, is 18 inches long and fills the whole space between the angle-irons of the adjacent floors.

Both the outer and inner keel-plates, C and D, are strapped at the butts with treble-riveted straps, *c*, *d*, respectively; both sets of straps lying on the inside of the inner keel-plate, as there are many objections, of course, to placing butt-straps on the outside of the keel and bottom plating. The straps of the outer flat keel-plate butts, *c*, are of the same thickness as the plating ($1\frac{1}{4}$ inch), and extend in breadth from the keel angle-iron across the inner keel-plate overlapping upon the garboard-strake E, at *c*′, to the same extent as the outer keel-plate itself underlaps the garboard. They receive four rows of rivets ($1\frac{3}{8}$ inch), two rows through the double keel-plates, C, D, and two rows through the outer keel-plate C, and garboard E. The straps, *d*, of the inner keel-plate are of the same length, breadth, and thickness as those, *c*, of the outer keel-plate, and riveted in the same way; these flat keel-plate straps, like those of the vertical keel-plate, usually filling up the space between adjacent floor angle-irons. The keel angle-irons, all of which are properly butt-strapped themselves, may at first sight seem to complete the strapping of both the vertical and the horizontal keel-plates, with which they are made to give a shift of butts; but this is not strictly the case, as we shall see presently.

The butts of the garboards, E, are strapped similarly with plates, *e*, $1\frac{3}{16}$ inch thick, riveted by $1\frac{1}{4}$-inch rivets. They extend from one floor angle-iron to another, and overlap the inner keel-plate sufficiently to receive a row of rivets passing through it, as shown at *e'*. This overlap of the garboard butt-strap upon the inner keel-plate, while it no doubt makes very strong and good work, appears to involve an excess of weight for the use of which there is no sufficient reason; as the treble-riveted butt-strap, if of the width of the plate to be strapped and no more, would make the butt as strong as the plate itself is in the line of the rivet-holes that connect it to the floor angle-iron. The same remark will not apply to the overrunning of the inner keel-plate butt-straps upon the garboard-strakes, because that tends to make up for the manifest deficiency of strength at the flat keel-plate butts. It is true, as already remarked, that the keel angle-irons seem to complete the strapping of the keel-plates; but it is obvious that they have only the same strength at the keel-plate butts as elsewhere, and that the strength would require to be increased in order to give the same total sectional strength at the keel-butts as away from them. At the butts of the outer keel-plate, therefore, where there is no extension of the breadth of the butt-straps beyond the edge of the plate itself, there must be places of permanent comparative weakness. There are many practicable methods, however, of succouring these points, and restoring to them the uniform strength of section which is elsewhere maintained, or very nearly that amount. One way would be to joggle the butt-strap over the keel angle-iron; another, to work additional strap-pieces of angle-iron there upon the keel angle-irons, these additional pieces being short pieces on one side of the vertical keel, and prolongations of the actual keel angle-iron butt-strap on the other side; for, as we shall soon see, there is an outer flat keel-plate butt in a floor space adjacent to every butt of the keel angle-irons. But the readiest and probably the best mode of doing it is that which was adopted in the 'Bellerophon,' 'Hercules,' and other recent ships, and which consists in increasing the thickness of the butt-strap, keeping it of the same breadth as in the 'Northumberland.'* If one or other of these methods is not adopted, it is difficult to see why the strength of the keel angle-irons themselves need be made continuous throughout by means of butt-straps, because while a loss

* A slight but insufficient increase in thickness was contemplated in the design of this ship, for the specification provided for keel-plate butt-straps $1\frac{3}{8}$ inch in thickness.

of 16 square inches of section occurs at the butts of the outer flat keel-plate, a loss of 11 square inches only would occur by the omission of the angle-iron butt-strap; and it is useless to make the section so much stronger in one place than in another. It is also obvious that the extra weight incurred in the butt-straps of the inner keel-plate is unavailing (except locally) while the outer keel-plate butts are imperfectly connected. Somewhat similar observations may also be applied in a less degree to the butt-strapping of the vertical keel-plate, which can only be done effectually by carrying the butt-straps down over the vertical flange of the keel angle-irons, or by increasing their thickness. To double those angle-irons themselves at these points would in this case be to incur an excess of strength and weight, because the only loss in the present arrangement is that of $4\frac{1}{2}$ square inches of section. Unless the strength of the lower keel-plate were made continuous, however, as already suggested, there clearly would be no advantage in thus strengthening the butts of the vertical keel-plate, as the section there is already, though weak, stronger than the section at the lower keel-plate butts, being double-strapped with $\frac{1}{2}$-inch plates, and treble-riveted, as shown at *a*.

The gutter-plate, G, is also strapped by double butt-straps, each $\frac{7}{16}$-inch thick, and treble zig-zag riveted.

The disposition of the butts of the keel-plates, garboard-strakes, &c., in the 'Northumberland,' and their relation to the spacings of the frames, are shown in Fig. 35.

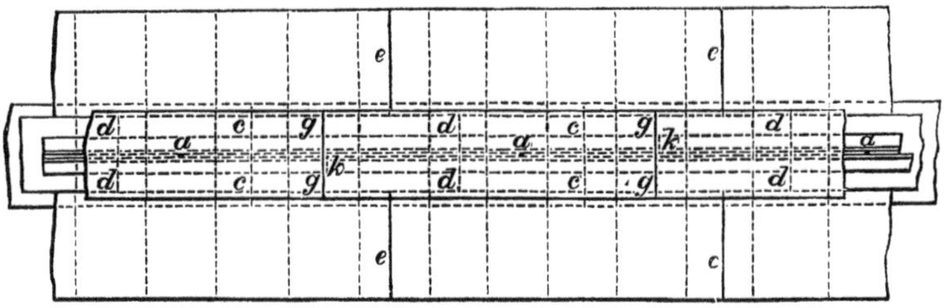

Fig. 35.

The usual frame space is 2 feet 4 inches; but at regulated intervals—*viz.* at the ports—pairs of spaces, each of 1 foot 11 inches, are thrown in to ensure a good arrangement of port-frames. The lengths of the keel, gutter, and garboard-plates are about 11 feet, and those of the keel angle-irons about 22 feet. The vertical keel-butts are marked *a a;* the outer flat keel-plate butts, *c c;* the inner, *d d;* the garboard-butts, *e e;* and those of the gutter-plate, *g g*. The butts, *c c*, of the outer flat keel-plate, and the butts, *d d*, of the inner flat keel-plate, each give shift by one frame space to the vertical keel-butts, *a a*, on opposite sides of them. The garboard-butts, in like manner, give shift by one frame space to the inner keel-butts, *d d*, a similar shift to the gutter-plate butts, and a shift of two frame spaces to the outer keel-butts, *c c*. These garboard-butts are exactly opposite to each other on opposite sides of the keel, having the three keel-plates, the keel angle-irons, and the gutter-plate unbroken between them. There are no adjacent plate-butts occurring in the same frame space throughout the arrangement. In fact, the only butts that come in the same space are the butts of the gutter-plate on top of the vertical keel, and those of the keel angle-irons, *k k*, at the bottom of it. This coincidence is obviously unavoidable, unless a worse arrangement is resorted to; for every other frame space is already occupied by a butt-strap, either of one of the three keel-plates, or of the garboard-strakes. It was, of course, out of the question to make the butt of the keel angle-iron coincident with the butt of either of the keel-plates themselves, and consequently the only choice was to make it coincide either with the garboard-butts, or with the gutter-plate butt; and of these, we think the latter was properly preferred. We shall have occasion to consider the general question of the disposition of plate-butts hereafter; but this illustration of the 'Northumberland's' keel and garboard-butts will serve to indicate the general considerations which have to be borne in mind, observing that we defer our remarks upon the necessity of making certain keel work watertight, and the means by which this is accomplished, until we come to consider, as we are now about to do, other cases in which water-tightness has to be more largely regarded.

The iron-clad frigate 'Bellerophon' was the first ship in which was carried out the bracket-frame system of construction, which has since been applied to the 'Penelope,' 'Hercules,' 'Monarch,' 'King William,' five large steam troop-ships built for the Indian Service, and other vessels. In this system the general keel arrangements of the 'Warrior,' 'Northumberland,' and other such ships were preserved; but were associated with an internal bottom, and with short brackets instead of floor-plates, as shown in the engraving given in Plate 4.

The 'Bellerophon' being a much smaller ship than the 'Northumberland,' her general scantlings (apart from the armoured side) were less. Her outer keel-plate was $1\frac{1}{8}$, and her inner $1\frac{1}{16}$ inch thick; and, in order to compensate for the absence of strapping in the centre, the side straps were made $1\frac{3}{8}$ inch thick. We do not purpose, however, to dwell upon the 'Bellerophon's' arrangements, but to pass at once to those of the 'Hercules,' which are of substantially the same kind, and fully represented in the accompanying engravings.

Fig. 36 is a side elevation of the central keel, showing a butt

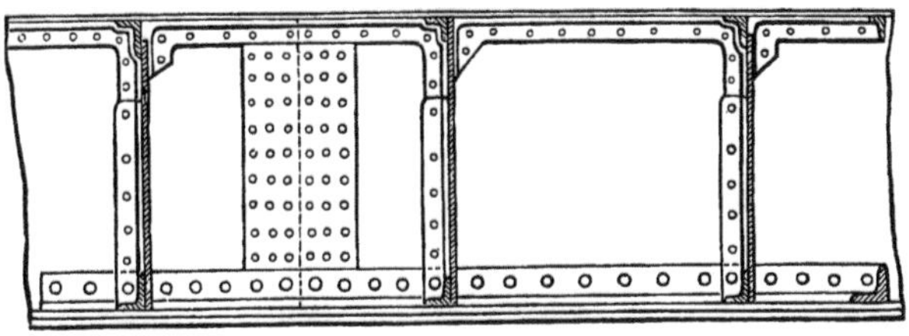

Fig. 36.

and the floor angle-irons; Fig. 37 is a section through one of the

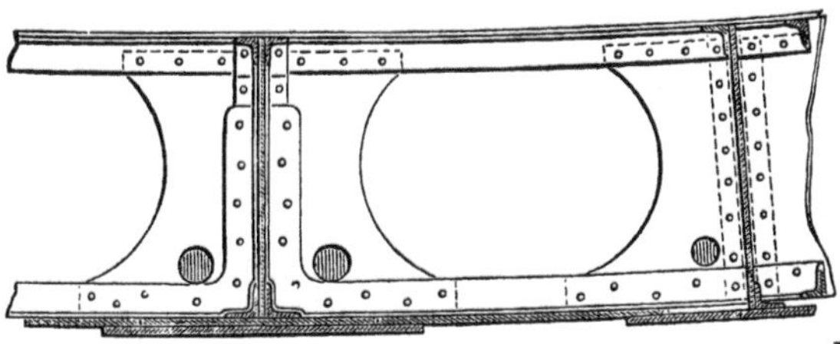

Fig. 37.

bracket frames; Fig. 38, a section through a watertight frame

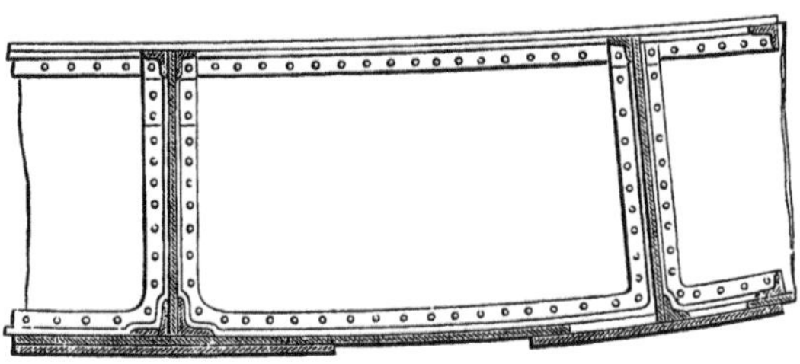

Fig. 38.

(observing that such a frame occurs every 20 feet, dividing the double bottom space into watertight compartments); and Fig. 39 is a plan view, showing the arrangements of the butt-straps of the keel-plates and garboards.

The vertical keel-plate is a $\frac{3}{4}$-inch plate; the outer flat-keel $1\frac{1}{8}$ inch thick; the inner flat keel and garboards 1 inch; the

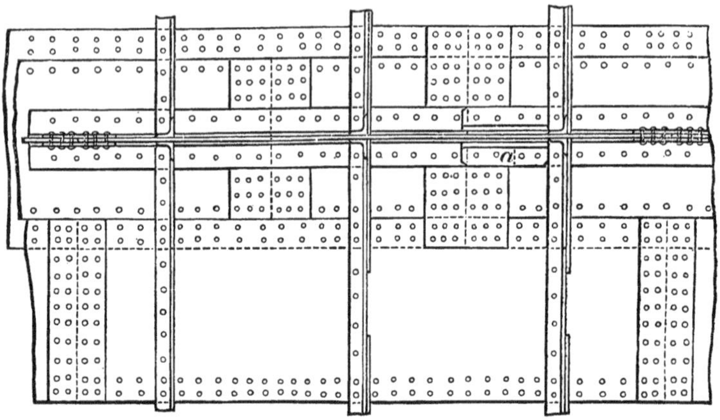

Fig. 39.

keel angle-irons, 6 inches each flange, and 1 inch thick. The plates are about 12 feet long, and as the frames are spaced 4 feet apart, it is of course impossible to separate the butts of all these plates and angle-irons by a frame space. The arrangement adopted is that of bringing the butts of the vertical keel-plate and garboards in the same frame space, the former being strapped with double straps, each $\frac{7}{16}$ inch thick, and treble-chain riveted with $1\frac{1}{8}$-inch rivets; and the latter united with single straps $\frac{15}{16}$ inch thick, double-chain riveted with $1\frac{1}{8}$-inch rivets. We shall hereafter discuss the general question of butt-strapping plates, and will therefore only observe here that there is good reason for this employment of treble-riveted butt-straps thicker than the plates to be strapped in the one case (the vertical keel), and double-riveted straps thinner than the plates in the other (the garboards); for while in the former case the deficient breadth of the butt-strap as compared with that of the plate has to be compensated for, in the latter the section of the butt-strap and keel-plates, through a row of rivet-holes in the garboard-butt, is obviously equal in strength to a section of the garboard-strakes and keel-plates taken through the rivet-holes of the frame angle-irons.

In the spaces adjacent to that containing the butts of the vertical keel and garboards come on the one side the butt of the outer keel-plate, and on the other the butt of the inner keel-plate. In both cases the butt-straps extend in width from the keel angle-irons to the edge of the plate to be strapped; and in both cases these straps are $1\frac{3}{8}$ inch in thickness, and treble-chain riveted with $1\frac{3}{8}$-inch rivets. The same thickness of butt-straps is preserved in both cases for the sake of simplicity, although the plates differ by $\frac{1}{8}$ inch in thickness, and the excess of thickness is adopted as a compensation for the deficiency of breadth, which results from the butt-straps stopping at the sides of the keel angle-irons. Taking the case of the inner keel-plate, for example, we have a plate 3 feet 1 inch broad, and 1 inch thick, its butt therefore having a sectional area of 37 square inches. The butt-straps (taking the two together) are $24\frac{1}{4}$ inches broad, and $1\frac{3}{8}$ inch thick, and therefore have a sectional area of $33\frac{3}{8}$ inches. But the inner flat keel-plate is pierced at the frames by 2 rivets, each $1\frac{3}{8}$ inch in diameter, and 4 of $1\frac{1}{8}$ inch in diameter, and therefore has its section of 37 inches reduced by $7\frac{1}{4}$ inches, while there are 6 rivets, each $1\frac{3}{8}$ inch, in the weakest line of the butt-strap, which reduce its sectional area by $11\frac{3}{8}$ inches; so that the true relative strengths of the strap and plate (considering these alone) are in the proportion of 22 : $29\frac{3}{4}$.

In the same frame space as the butts of the outer keel-plate come the butts of the keel angle-irons, which are strapped by double-riveted butt-straps, as shown at *a* in Fig. 39.

As the two flat keel-plates are bound together by all the rivets passing through the horizontal flanges of the keel angle-irons, as well as by an additional rivet in each floor-plate, a single row of rivets is obviously sufficient to connect the outer edge of the inner keel-plate to the outer plate, while the laps of the outer keel-plate and garboards, of course, require the usual double row.

In the system of construction now under consideration, the vertical keel is made water-tight. At the bottom of the plate this is easily accomplished, because the keel angle-bars are continuous, and when riveted to the vertical keel and flat plates, have only to be caulked in the usual way; the transverse floor angle-irons, which lie along the bottom, and turn up the vertical keel-plate, being worked afterwards, and joggled over them, as shown in Figs. 37 and 38. The upper transverse angle-irons or frames, however, which receive the inner bottom plating and the heads of the brackets, run unbroken across the keel, scoring down into it. To make the keel water-tight at the upper part, therefore, it becomes necessary to turn down both ends of the short pieces of longitudinal angle-iron that connect the vertical keel-plate to the inner bottom plating, and to joggle one end over the transverse through-angle frame, abutting it also carefully against the upper end of the large angle-iron which has been turned up from below, and working the ends of the adjacent short pieces of longitudinal angle-iron close against each other, as well as against the transverse frame. All this angle-iron work is well caulked, and caulked also against the vertical keel-plate wherever the two come together, and the whole is thus made water-tight before the bracket plates are brought on.

It is only at the water-tight divisions of the double bottom, as shown at Fig. 38, that the transverse frames require to be made water-tight away from the keel. In this case the brackets are replaced by solid-plate frames which are carried home to the vertical keel-plate between the ends of the short upper longitudinal angle-irons, and these latter are caulked against them and against the keel.

The system of framing adopted in the turret-ship 'Captain,' now being built by Messrs. Laird, is identical with that just described for the 'Hercules,' but the disposition of the butts of the middle-line work and the arrangements of the riveting are different. The vertical keel-plate is 42 inches deep and $\frac{5}{8}$ inch thick; the outer flat keel is 1 inch thick, and the inner flat keel and garboards each $\frac{7}{8}$ inch thick; the keel angle-irons have each flange 5 inches wide and are $\frac{3}{4}$ inch thick, and the gutter-plate is $\frac{1}{2}$ inch thick. The plates forming the vertical keel are worked in 24 feet lengths; all the other plates in the keels, garboards, and gutter-plate in 12 feet lengths; the keel angle-irons are made up of 36 feet lengths. The spacing of the frames is identical with that of the 'Hercules,' and here also it is absolutely impossible to prevent more than one butt of a plate or angle-iron coming in the same frame space.

The butts of the inner keel-plate and the butts of the garboard-strake on each side are brought into the same frame space, and have a distance between them of about 19 inches. The butts of the outer flat keel come in the space on one side of that in which the butts of the inner flat keel and garboards are placed, and the butts of the vertical keel-plate come in the frame space on the other side. The butts of the gutter-plate give a shift of 2 feet to those of the vertical keel, at their shortest distance; but on account of the lengths of the plates in the vertical keel being double those in the gutter-plate, there are two butts of the gutter-plate between every two butts of the vertical keel. A similar remark applies to the butts of the flat keels and garboards in relation to the butts of the vertical keel. The butts of the keel angle-irons are placed midway between the butts of the outer and inner flat keel, falling in the same frame space as the butt of the outer flat keel; the shift between the butts of the keel angle-irons on opposite sides of the vertical keel is 12 feet.

The butt-straps to the vertical keel-plates are double, each being $\frac{3}{8}$ inch thick, and are treble-chain riveted with $\frac{13}{16}$-inch rivets. These butt-straps only extend from the upper edge of the keel angle-irons to the lower edge of the short longitudinal angle-irons on the upper edge of the keel plate, and are $14\frac{1}{2}$ inches wide. The butt-straps to the inner and outer flat keels are single, their thickness being $1\frac{1}{4}$ inch and their breadth 20 inches. They are treble-chain riveted with $1\frac{1}{8}$-inch rivets, and extend from the keel angle-irons to the edge of the plates they connect. The inner flat keel is 2 feet wide, and the outer 3 feet $2\frac{1}{2}$ inches, and, as in the 'Hercules,' the edges of the inner flat keel are single-riveted, and the garboards double-riveted to the outer flat keel, the rivets in this ship being $1\frac{1}{8}$ inch and their pitch 4 inches. The butts of the garboard are connected by $\frac{7}{8}$-inch butt-straps, which extend the whole width of the strake, and are $17\frac{1}{2}$ inches wide; they are treble-chain riveted with 1-inch rivets. The butt-straps of the gutter-plate are 15 inches wide and $\frac{1}{2}$ inch thick, and are treble-chain riveted with $\frac{13}{16}$-inch rivets. The butts of the keel angle-irons are connected by double-riveted covering straps 2 feet long. These angle-irons are secured to the flat keels by $1\frac{1}{4}$-inch rivets, and to the vertical keel-plate by $1\frac{1}{8}$-inch rivets, the pitch in both flanges being 6 inches.

It will be remembered that a comparison between the strengths of the inner flat keel-plate and its butt-strap has been made for the 'Hercules,' and a similar comparison for this ship, the 'Captain,' may be of interest. In this ship the total width of the inner flat keel is 2 feet, and at a frame the plate is weakened by two $1\frac{1}{4}$-inch rivets in the keel angle-irons, and two $1\frac{1}{8}$-inch rivets in the edges, thus reducing its effective width to $19\frac{1}{4}$ inches, and its effective sectional area to $19\frac{1}{4}$ by $\frac{7}{8} = 16\frac{7}{8}$ sq. inches. The width covered by the keel angle-irons and vertical keel is $10\frac{5}{8}$

inches, and thus the total width of strap is $13\frac{3}{8}$ inches, and the effective width is less than the total width by four $1\frac{1}{8}$-inch rivet-holes; and therefore the effective width equals $8\frac{7}{8}$ inches, and the effective sectional area equals $8\frac{7}{8}$ by $1\frac{1}{4} = 11\frac{1}{8}$ sq. inches. We thus see that the relative strength of the plate and its butt-strap along their weakest lines are in the proportions of 3 : 2 nearly.

The vertical keel of the ship is made water-tight in a somewhat different manner from that of the 'Hercules.' The angle-irons on the upper edge are forged staple fashion, and on both sides of the water-tight frames, or on one side of the bracket frames, they butt on the vertical parts of the frame angle-irons (where they are turned up against the keel) and are connected to them by short covering angle-irons.

There have been instances in which internal keels, resembling those which we have lately been describing, have been combined with an external bar-keel and turned down garboards. This arrangement was carried out in the large iron-clad frigate 'Victoria,' built by the Thames Shipbuilding Company for the Spanish Government. Fig. 40 is a section of it.

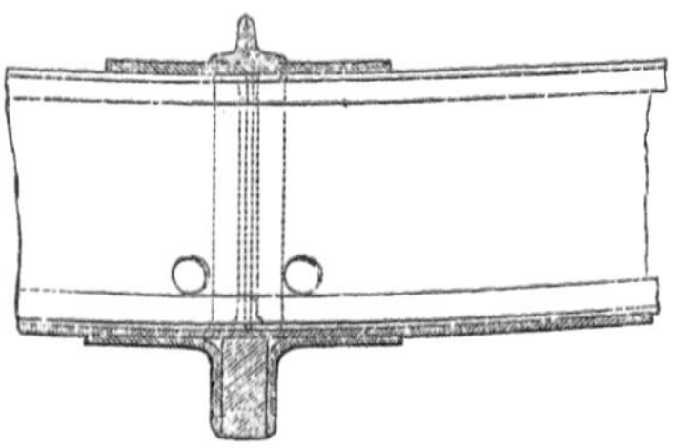
Fig. 40.

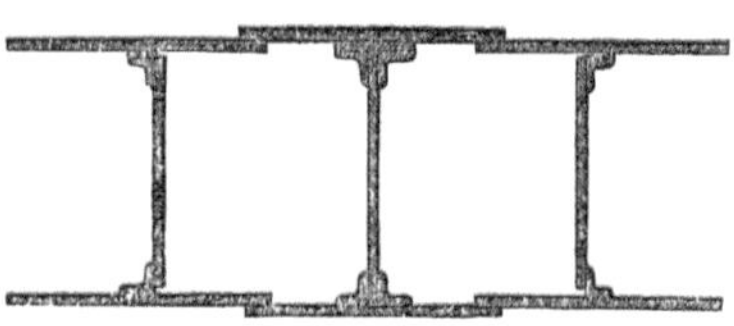
Fig. 41.

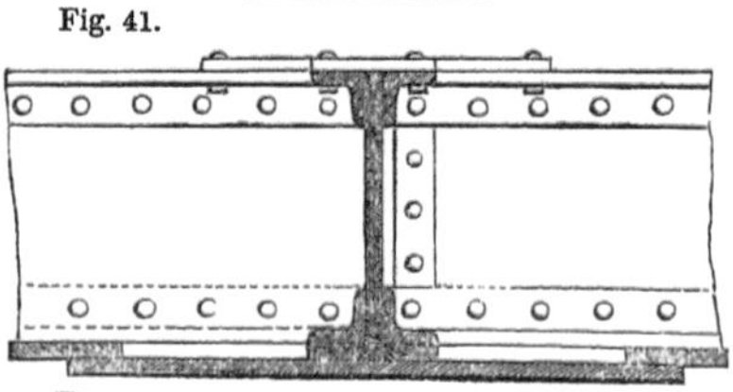
Fig. 42.

In Mr. Scott Russell's longitudinal system, which preceded the Admiralty system of combined longitudinal and transverse frames, a vertical keel is employed with a single flat keel-plate, as shown in Fig. 41, which is a section of the 'Great Eastern's' keel, and in Fig. 42, which is a section of a vessel named the 'Annette,' built upon the same system, but without an inner bottom. These constructions will be further illustrated when we come to speak of methods of *framing* iron ships.*

* See Chapter V.

It is not at all an uncommon thing to employ side-keels, which are variously called by that designation, or known as "drift-keels," "auxiliary keels," "bilge-keels," and so forth. They no doubt check the tendency of a ship to make lee-way when under canvas; they tend to check rolling; and, if suitably fitted, add greatly to the chances of safety for a ship in the event of her going ashore.

The 'Great Britain' is an early and remarkable instance of an iron ship having no external middle-line keel, but a keel on each side. In his Report on Iron Shipbuilding in England, M. Dupuy de Lôme, speaking of this ship, makes the following remarks :—"At 6 feet from the middle line, measured athwartship "upon the plates, are fixed two side-keels, composed of plates 1 inch "thick, and angle-irons of 5 inches. These keels extend for one-"half the length of the vessel. They are only 8 inches deep in "midships, but as their under sides are horizontal, at some distance "from the middle they are 14 inches deep from the bottom." And in a note he says, "The builders of the 'Great Britain' con-"sider this application of the keels as altogether a new idea, but it "has been employed many years in France. Builders have there "used this means to increase the lateral resistance of vessels "without giving them a large draught of water. In 1811 it was "adopted by M. Bonard, our present Inspector-General of Marine, "in projecting a floating battery of 28 guns for the protection of "the river Bordeaux." He also instances several vessels, built in some cases for the Nile, in others for the coasting trade of Algiers, in which three keels were used for reducing lee-way.

In the Royal Navy side-keels have become very common since the introduction of iron-clad ships. They were used in the 'Warrior,' formed as shown in section in Plate 3, the angle-irons being riveted through the bottom plating. Such keels only extend over a portion of the length of the ship, at the midship part, stopping where the rise of the body forward and aft renders them no longer necessary. Similar keels were applied to our other iron-clad ships when built of iron, until a change was introduced in the 'Bellerophon.' As that ship has a flatter floor than her predecessors, and as we wished to give her all the security possible, we attached no less than four side-keels to her bottom, forming them as shown in section in Plate 4. They are each composed of two angle-irons, tap-riveted or screwed (and not through-riveted) to the bottom plating, immediately beneath a longitudinal frame, the rivet-holes not being formed completely through the plate, in order that if these angle-irons should get torn off by rocks or otherwise leakage should not result. Between the falling flanges of these angle-irons are bolted timber keels, to which again false keels are lightly attached, after the manner of the false keels of wooden ships, these keels and false keels being sheathed with thin iron plating. Similar side-keels have been applied to the 'Penelope,' 'Hercules,' 'Monarch,' the five new troop-ships for India, and other ships. In the 'Penelope,' a bilge-keel of which ship is shown in section in Fig. 43, the angle-irons for connecting it to the bottom (4 inches by 4 inches by $\frac{5}{8}$ inch) are secured to the plating by 1-inch screws tapped through it, with a nut on the inside, it being presumed that in the event of the angle-irons getting stripped off, part at least of the screw would be left in the plating, and prevent leakage. The screws used for the purpose are placed 6 inches apart, and have countersunk heads, with a square projection formed on each for heaving it up, which projection is afterwards cut off. Lugs are formed on the outer angle-iron, to enable the bolts through the wood keel to take a better hold of it, and to pass more squarely through it. A false keel is worked outside of all, as in the 'Bellerophon.'

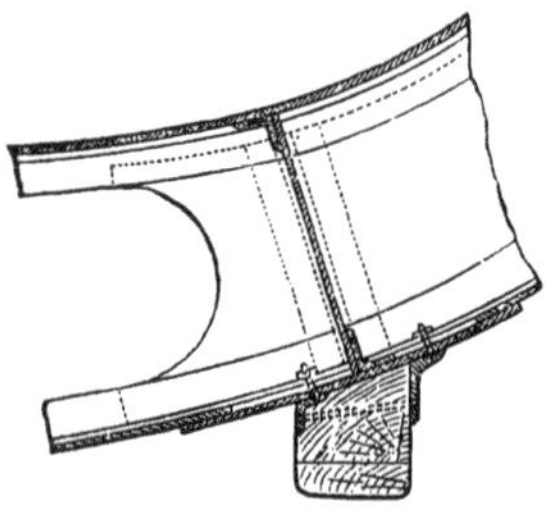

Fig. 43.

Fig. 44 illustrates in section and plan the manner in which the bilge-keels are fitted to the 'Malabar,' one of the five large Indian troop-ships just referred to. In fitting these keels, their positions

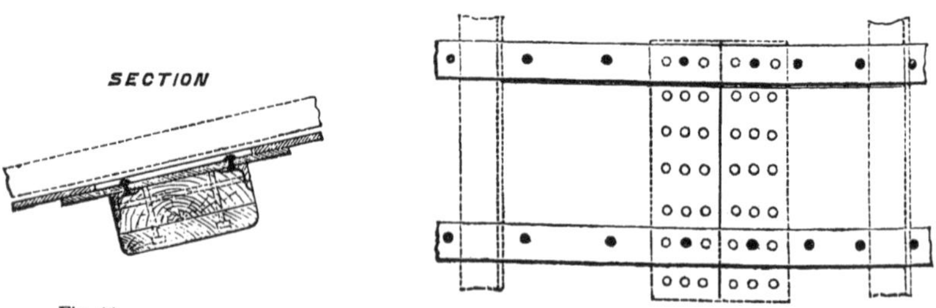

Fig. 44.

were lined off upon the frames of the ship; and when an account was being taken of the plates upon which they were to come. these lines were transferred to the template (Messrs. Napier, who built this ship, employing templates for this purpose, as will be explained hereafter) in order that the rivets in the butts of these plates might be so arranged as to admit of tap-rivets being placed as shown by the black rivets of the figures. By these means the butts of the plates can be efficiently caulked before the angle-iron is secured to the bottom. This angle-iron is also carefully caulked on both sides, the tap-rivets which secure it being made sufficiently long to admit of the points being clenched over and caulked.

In fixing the positions and directions of external side or bilge keels, it is desirable, of course, to so place them that they shall move truly endwise through the water when the ship is steaming or sailing ahead without encountering pressure upon either side, as that would check the progress of the ship. With this view, it is usual to place them in planes parallel to the direction of the motion, or at right angles to the midship section. It is also customary to reduce the width and depth of these keels at the fore and after ends, in order to avoid the direct resistance of the fluid.

The forms of keelsons and sister-keelsons used in iron ship-building have been almost as various as the forms of keels. In early days it was not at all an uncommon thing to make them of wood. This was the case in the mail-packet 'Dover' and in the brig 'Recruit.' The keelsons of the former vessel, with the mode of fastening them, are shown in Figs. 3 and 4, and in Fig. 70, page 74, and those of the 'Recruit' in Fig. 19.

The iron ship 'Birkenhead' (which like the two preceding vessels has before been adverted to) had keelsons formed of iron, those under the engines being box-keelsons, formed with the upper angle-irons inside, and those under the boilers single-plate keelsons, having double angle-irons on the upper and lower edges, with doubling plates introduced between the vertical flanges of the angle-irons,—the construction of beams being less understood twenty-seven years ago than it now is.

In considering keels we have already become so familiar with the commoner forms of central keelsons that little remains to be said respecting them. All that Lloyd's Rules provide for is, that the middle-line keelson, if of single plate, shall be of the same thickness as the garboard-strakes, and in depth equal to two-thirds the depth of the floors above which it is to stand, and to which it is to be well fitted and riveted. Double angle-irons at both top and bottom, and extending fore and aft, are insisted upon, and the horizontal flanges of the lower pair of angle-irons have to be riveted to double reversed angle-irons on the floors. If a box-keelson is employed, it has to be formed with a foundation plate, the depth the same as the single-plate keelson, and the breadth of the box two-thirds its depth. The Liverpool Rules require that the centre keelson standing upon the floors shall be made either box-shaped with top, bottom, and two sides, or of a double centre-plate with top and bottom plates. They also require double reversed angle-irons in wake of the keelsons.

Messrs. Napier have sometimes used a box or hollow keelson with a curved top, as shown in Fig. 45, which is a section of the keelson of the 'Colombo,' a ship built for the Peninsular and Oriental Company. The main keelson of the 'Columbian,' a ship built at Whiteinch, near Glasgow, is formed, as shown in Fig. 46. This form is not uncommon, and only differs from one of those shown in the illustrations of Lloyd's Rules by having the keelson-plates carried up to lay hold of the bulb-iron plate, instead of having the latter carried down between the floors to lay hold of the keelson-plates. In fact, precisely the same arrangement is shown in these illustrations in association with a solid bar-keel.

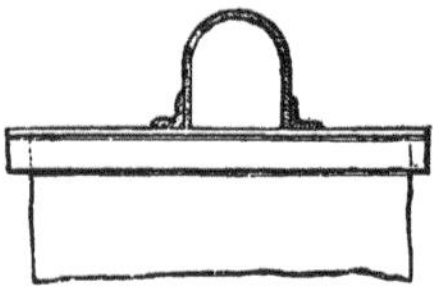

Fig 45.

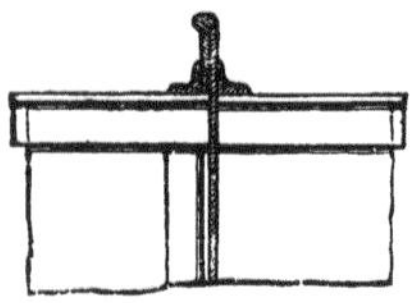

Fig. 46.

Side keelsons usually consist, when situated at the deeper portions of the floors, of intercostal plates carried up and laid hold of by a pair of angle-irons running along the floors, and when situated near the outer ends of the floors, of a pair of angle-irons only. But different builders vary the number and description of the side keelsons. The sketch given in Plate 2, shows the arrangement adopted by Messrs. Napier in the 'China' Atlantic Mail-steamer; this is unquestionably a good and strong arrangement, double angle-irons being used to connect the lower edge of the intercostal plate to the bottom-plating, where a single iron only is ordinarily employed. Plate 1 shows a different arrangement adopted by Messrs. Laird in a large Atlantic passenger-steamship, from which it will be seen that pairs of angle-irons are chiefly used as keelsons, and that external keels, similarly formed, are also applied under the bilge. The intercostal side keelson is associated with a standing box-keelson in wake of engines. Lloyd's Rules require that at least one intercostal keelson shall be fitted on each side of all ships of 1000 tons and upwards, and be carried as far forward and aft as practicable. It is to be placed about midway between the middle-line keelson and the bilge-keelson, with a double angle-iron riveted on the top of the floor-plates. All vessels of 500 tons and upwards must have fitted between the bilge-keelsons and the hold beams, at the upper part of the turn of the bilge, strong angle-irons, as stringers, extending all fore and aft, riveted back to back and to the reversed iron frames, the size of them not to be less than that of the angle-irons used for the middle-line keelson. And Lloyd's likewise very properly provide for that continuity of strength of which we have previously shown the necessity, by requiring that: "In all cases of "middle-line, side, and bilge keelsons, and, where practicable, the "stringers, are to be carried fore and aft, without being cut off at "the bulkheads, the latter being made watertight around them;" and where such parts of the ship are necessarily separated, the longitudinal strength is to be efficiently maintained to the satisfaction of the surveying officer.

The Liverpool Rules require an intercostal keelson to be put at half-floor, in vessels exceeding 32 feet beam, for two-thirds of the vessel's length where practicable, to be fastened through the skin and the floors, and to project above the floors to form a keelson. All vessels must have two stringers, formed of double angle-iron, one at the lower and one at the upper turn of the bilge. Special regulations are made for vessels of different dimensions, as will be seen in the Appendix. Similar care is enjoined in preserving the continuity of the strength of these keelsons and stringers as is required by Lloyd's Rules.

The iron-clad frigates of our Navy, built of iron, have numerous continuous side and bilge keelsons, which practically form a series of longitudinal frames, and will be considered more particularly hereafter as part of the frames of the ship. The same remark applies to the 'Great Eastern,' and to other vessels built on the longitudinal system of Mr. Scott Russell.

CHAPTER III.

ON STEMS.

THE stem of an iron ship, like that of a wooden ship, is usually a prolongation of the keel; but in iron ships of war, which are now most frequently formed to act as rams, a very different construction to that of the keel has become necessary. The construction of such ram-stems, and their connection with the keel, will afford us very interesting and instructive matter for consideration; but before attending to it, we propose to notice, briefly, some of the more ordinary forms of stems.

The mail-steamer 'Dover,' whose hollow-iron keel we have seen in Fig. 3, furnished a good example of the old hollow form of iron stem, and of its combination with the bottom plating, and with the knee of the head above; but it is not of sufficient interest, now, to justify the expenditure of so much space and trouble as would be required for its description. The 'Birkenhead' also had a hollow iron stem, which was formed of two plates of iron each $\frac{3}{4}$ inch thick, bent to shape, and riveted together as shown in Fig. 47. Considering the difficulty and expense of forming such stems, the plates of which could only be bent in short lengths; the great nicety of workmanship required in putting them together; and the punishment which the iron of necessity underwent in being bent to the necessary sharpness for the stem of a ship of moderate size, and which rendered it liable to split under a blow, it is not at all surprising that they were speedily replaced by solid stems in the same manner as the hollow keels became replaced by solid. Like the solid keel, the stem also was rabbeted, in the first instance, to receive the bottom plating; but the rabbeting has now been almost universally dispensed with, for the sake of simplicity and economy, and the iron stem has become simply a curved solid bar of uniform section, or nearly so, generally forming the contour of the bow, even where a projecting knee forms an ornamental head. Lloyd's Rules simply provide, that the keel and stem shall be scarphed or

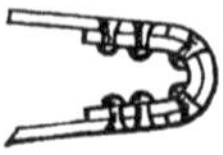

Fig. 47.

welded together; if scarphed, the length of the scarphs must be eight times the thickness. The Liverpool Rules require that the feet of stems shall be extended so as to form part of the keel, not less than four and a half feet long. The various devices which we have previously seen resorted to for connecting external keels with the vertical keelson-plates—grooves, rabbets, simple laps, and side-bars—have all been repeated in the case of stems. Even with side-bar keels, however, the stem is frequently solid, and when this is the case, either the central through-plate is scored for some distance into the stem, or the stem is formed to run along the side of the through-plate on one side, and thus form a scarph with the side-bar of the keel on the opposite side. In other cases of side-bar keels, the side-bars scarph with the stem on each side of the centre through-plate, as shown in the sketches in Fig. 48, which

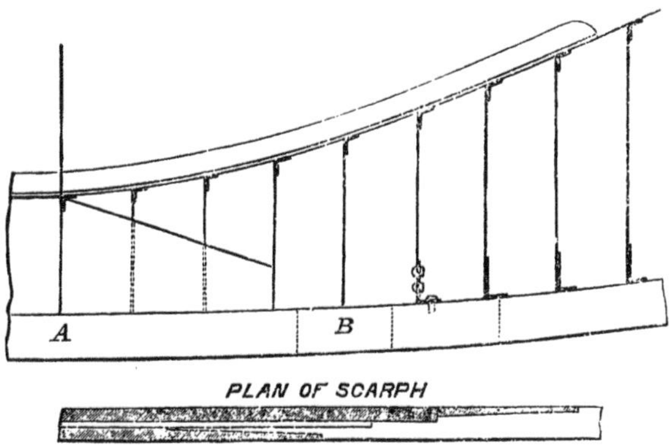

Fig. 48.

represent the scarphing of the stem and keel of H.M. troop-ship 'Orontes.' It will be seen, also, that in this case the continuous plate is tapered down in depth from station A, and at the station on the after side of B is cut down to the same depth as the side-bars. At A there is an athwartship bulkhead, at which the gutter-plate ends, but the angle-irons at the middle line are run through and secured to the deep-throated floor-plates on the fore side. The length of the plane scarph made by the side-bar and the stem is 20 inches, and the butt of the centre plate is 8 inches from each termination of the scarphs, the whole length of the after part of the stem directly connected with the keel-plates being 4 feet 8 inches.

Where internal or flat-plate keels are used, the solid stem runs down inside and is simply riveted to the keel-plates, garboards, and bottom plating, as shown in Fig. 49, which represents the junction of the keel and stem of an 1100-ton steamer, built for blockade running during the late American War.

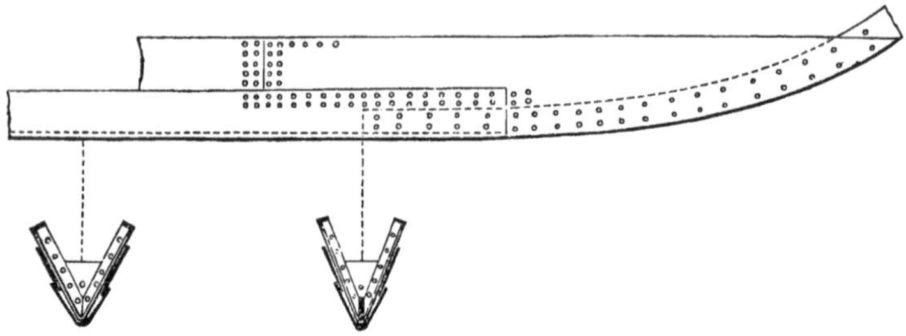

Fig. 49.

The manner in which the stems of the five Indian troop-ships are formed and connected to the keels is shown in Fig. 50. It will be remembered that these vessels are built with double flat-plate keels, a central water-tight through-keel or keelson, and an inner bottom, like the 'Bellerophon.' The stem is a common rabbeted stem, as shown in the sections B,C, and is connected to the keelson-plate by means of a pair of angle-irons screwed on to it, and receiving the keelson-plate between, as shown at D and at E.

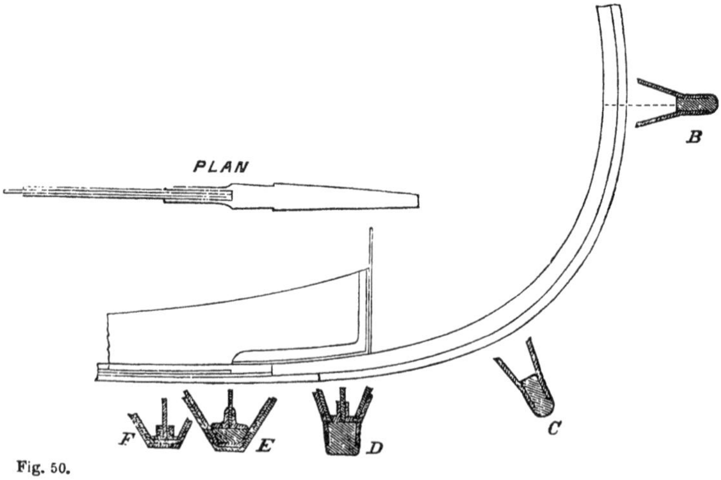

Fig. 50.

Abaft this it is forked, and embraces the vertical flanges of the main keel angle-irons as shown in the section at F, and in the plan. At F, and at E also, the flat keel-plates are both worked under the heel of the stem, but before the point where the section E is taken they are stopped in succession, the one stopping altogether, and the other rising up into the rabbet, as shown at D, and as will be shown with greater particularity presently, in reference to the stems of other ships similarly connected.

The manner in which the stem of a large iron-clad frigate is formed and connected to a flat-plate keel is shown in Figs. 51 and 52,

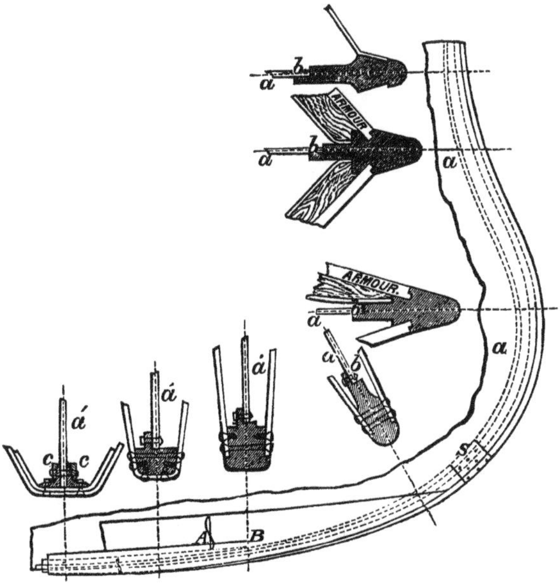

Fig. 51.

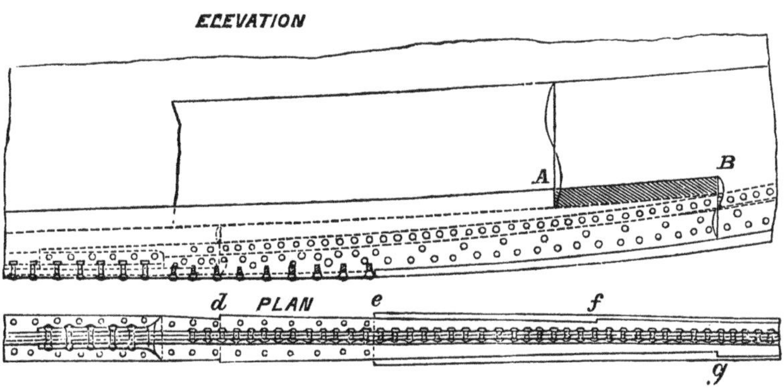

Fig. 52.

which represent the stem (and its connections) of the 'Northumberland,' with the details of the keel of which ship we are already familiar. In this figure we have given cross sections on a large scale at so many points, that a very few words will suffice by way of further explanation. As this stem is formed and fitted with special regard to its adaptation for forcing or ramming in the sides of other ships, the consideration of expense, which so largely and so properly controls the designs of mercantile vessels, is here subordinated to other considerations, and the forging and planing of the stem into any required form is held to be justifiable. The first thing to be accomplished is to give to such a stem the support of all the bow bottom plating and armour plating in delivering a horizontal blow. For this purpose all such plating is let into the substance of the stem, abutting squarely and closely against the fore side of the rabbet; the stem being made deep enough in front of the plate-ends to form a sufficiently stout abutment for them, and deep enough behind the rabbet-line—or, in other words, affording sufficient surface for the skin-plating—to receive a double row of bolts through that plating. In the wake of armour, the stem has to be formed sufficiently deep to receive not the armour only but the skin-plating behind it; and as it is not desirable to provide for this by deepening the stem suddenly at that part, the increased depth (measured in a fore and aft direction) is carried upward and downward, narrowing gradually as shown. In order still further to support the stem against a blow, a middle-line web-plate, *a*, is worked (extending back many feet, and supported by frames, decks, and deck-hooks, as will be explained hereafter), and to receive this plate, a rabbet is formed on the inside of the stem at *b*, from top to bottom, as shown in the sections. The lower part or heel of the stem is formed with a fork, the arms of which, *c c*, receive the vertical flanges of the keel angle-bars and the vertical keel or keelson, *a′* (see sectional view), as already explained with reference to the stems of the Indian transports, the fork being long enough to receive six $1\frac{1}{4}$-inch rivets or bolts, as will be seen clearly from the enlarged plan in Fig. 52. The keel angle-irons butt, of course, against the inner fore end of the fork, and the vertical keel-plate there steps up upon the inside of the stem, and stands against the tongue formed to connect with it. From this point forward the two are connected by a row of rivets, the keel-plate merging itself above into a stem-plate; the two practically forming the termination of the middle-line bulkhead, and the row of rivets ($1\frac{1}{8}$ inch) extending completely up and down at *b, b*. At about 18 inches before the butts of the keel angle-irons and the step of the vertical keel-plate, the inner flat keel-plate terminates, an abutment, *d*, Fig. 52, being formed in the stem to receive the end of it, three pairs of tapped rivets connecting the two from below as shown. At about 3 feet before the butt of the inner flat keel-plate, the outer plate is joggled up at *e*, Fig. 52, into the under side of the stem (which now comes to the front and continues there to the top) and falls into the stem-rabbet, abutting at its fore end into the garboard-strake, which is thinned away to receive it at the shaded part, marked A B; observing that while the outer keel-plate is $1\frac{3}{16}$ inch thick, the garboard-strake is 1 inch thick only. It will be remarked also in the plan that the stops for the fore ends of the garboards, marked *f* and *g*, are arranged so as to give shift to each other. All these arrangements will be made plain to the practised eye by the figures, and especially by the enlarged sketches of the connection given in Fig. 52. It is proper, however, to state that the stem was formed in two pieces, connected by a carefully-fitted hook-scarph, shown in the engraving at S, Fig. 51, eight 1-inch rivets or bolts passing through the scarph.

In the 'Bellerophon' we did away with the middle-line bulk-head, and made the bottom-plating double for about 45 feet from the stem, rabbeting each thickness separately into the stem below the armour as shown in Fig. 53, and dealing with the double-plating in wake of armour as shown in Fig. 54. The inside or

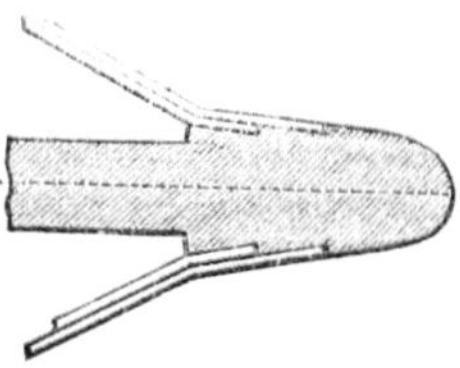

Fig. 53.

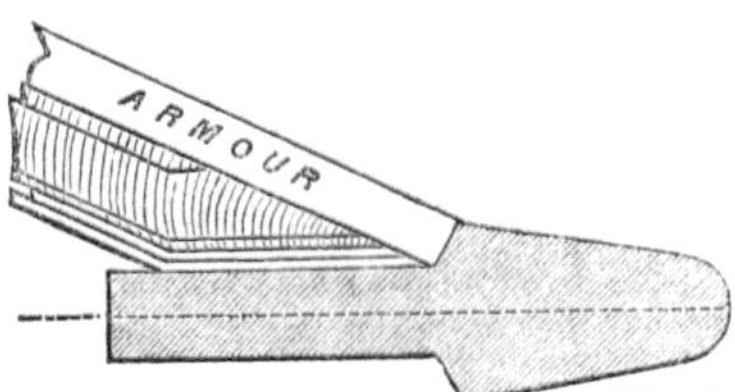

Fig. 54.

back of the stem was formed square, a series of breast-hooks ending against it. The connection of the stem with the keel was formed in substantially the same manner as in the 'Northumber-land,' with the exception that, in order to diminish the athwart-ship thickness of the fork of the stem, we carried the keel angle-irons only (back to back) into the fork; the vertical keel-plate being withdrawn from between them, and stepped upon them before they entered the fork.

The stem of the 'King William,' the large Prussian iron-clad, is formed in a manner similar to that of the 'Bellerophon.' The same observation applies to the stem of the 'Penelope.'

The stems of all H. M. iron-clad frigates and of the 'King William,' are formed of the best scrap-iron under the steam-hammer, commencing with a number of flat bars made from that description of iron (say about 12 feet long, 8 inches wide, and $1\frac{1}{4}$ inch thick, or larger), which bars are welded together to form a staff, or foundation piece. The end of this staff is hammered down as at A to receive the pile of slabs B, as shown in Fig. 55, and is built upon by piles of slabs in this way until it becomes of the size required for the stem, the end being left sufficiently large to receive another pile of slabs, which, like the former, are brought on at a carefully regulated welding heat. In this way the forging is gradually built along, being controlled and handled while under the hammer, and moved about by means of the other end of the staff. Great care is taken with the heats, and to bring the grain of the iron in the right direction; also to pile the slabs opposite to each other, so that the scarphs or welds may not be all on one side of the forging. The slabs used are forged from the best hammered scrap-iron, as follows:—a rough wooden frame is first made to about a foot square, and the scrap, after being cut up in lengths of not more than a foot, but of all sizes below that, is piled upon it in layers; care being taken to break all joints and butts, and every (or nearly every) alternate layer being piled across the other until it becomes about a foot deep. The pile is then put into the furnace and brought to a welding heat, and then hammered out to the size required, the weight being about 2 cwt.; two of these piles are brought from the furnace at the same time and welded together, the two forming one of the slabs previously mentioned.

Fig. 55.

The stem having been forged as above described, is sometimes bent to form and planed afterwards, and at other times planed first and then bent. The best method is to bend it first and plane it afterwards; but the planing in that case occupies a long time, and is very costly, owing to the planing-tool having to be made to travel round the varying curvature of the stem. Where the curvature is very great, as in the middle-piece of the 'Penelope's' stem, for example, it is almost a necessity to plane it after the bending; and this was done, both in the case of the 'Penelope' and in that of the 'Agincourt' (which were made at the Mersey Iron-works), while the upper and lower pieces were first planed and then bent. For the sake of economy, it is now the practice to plane the stem-piece before it is bent, wherever that can be done consistently with its form and character.

The mode of bending varies in different works. At the Millwall Works and the Thames Iron-works, the stems of the 'Northumberland' and 'Minotaur' respectively were bent to shape on the cast-iron slabs used for bending ship-frames, a coke fire being made round a length of about 8 feet at a time; and when the heat was sufficient, the fire was removed, and the bending effected by means of wedge-setts, a tackle and crab, and other like appliances. This operation was repeated until the whole length was brought to the required shape.

At the Mersey Works, where such appliances as the above do not exist, the bending of the 'Agincourt's' stem and of the 'Penelope's' also, was effected as follows: the stem was slung on edge by a crane, as shown in Fig. 56, and a portion of it, from A to A, was brought to the requisite heat; a patent lift-jack was then placed in the centre of the heat, as shown, its upper part pressing against the chain sling, and was worked until the stem was bent to the required curvature; this operation being repeated as often as was necessary.

Fig. 56.

CHAPTER IV.

STERN POSTS.

THE stern posts of iron ships admit of the same variety as the keels and stems. Hollow stern posts were at first used, in conjunction with hollow keels and stems, and were, like them, open to the objections of weakening the iron by excessive bending, and of being made up of short lengths, generally averaging about 6 feet. In this arrangement the groove or gulleting on the afterside of the rudder post to receive the rudder was obtained by riveting on a solid piece of iron with a hollow in it, or by hollowing the plates themselves. An illustration of this is found in the 'Dover,' the keel of which ship has been before described.

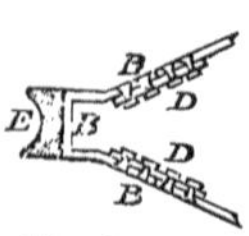
Fig. 57.

The construction of the stern post is illustrated in section by Fig. 57. In the sketch, B shows the stern post, and E what was then known as the rudder post. This rudder post was fitted to the rudder at the forge; when in place, it was secured by nut and screw bolts. The strap-plates D were fitted in order to secure the after ends of the bottom plating to the stern post.

In Fig. 58 a section of the 'Birkenhead's' stern post is given.

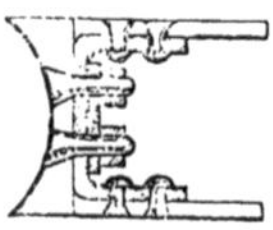
Fig. 58.

It was formed of plates $\frac{3}{4}$ inch thick, and had a wrought-iron rudder post $2\frac{1}{2}$ inches thick, similar to that of the 'Dover.' The screw bolts securing the rudder post were $1\frac{1}{2}$ inch diameter, tapped through.

These hollow-plate stern posts gave place to solid bars, of which the feet are connected with the keel in a manner suited to its character. If the keel is formed by a solid bar, the stern post is scarphed or welded to it in the same manner as the stem is secured. In either case, Lloyd's Rules now require that in a vessel so constructed, the stern post, and the after end of the keel, shall be double the thickness of, or double the sectional area of the adjoining length of keel, and be tapered fair into that length, the siding in no case being less than the thickness of keel amidships. If the stern post be scarphed to the keel the length of the scarph must be eight times the thickness of the keel. The Liverpool Rules require the feet of stern posts to be extended so as to form part of the keel, not less than $4\frac{1}{2}$ feet long. When a vessel is built with a side-bar keel, the connection of the stern post to it is similar to that of the stem. The details of this connection, as carried out in the steam-ship 'Queen,' are given in Fig. 59. In this case the fore end of the sole piece on the post

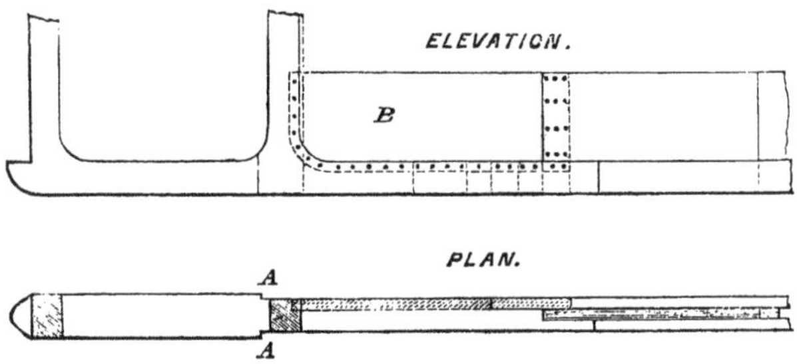

Fig. 59.

was formed so as to successively butt first one side-bar, then the centre keel-plate, and then the other side-bar. The plate marked B was lapped on the after end of the centre keel-plate, and was double riveted to it. Its lower edge and after end were rabbeted into the post, and riveted to it, thus completing the connection of the keel and post. The stops for the garboard-strakes are

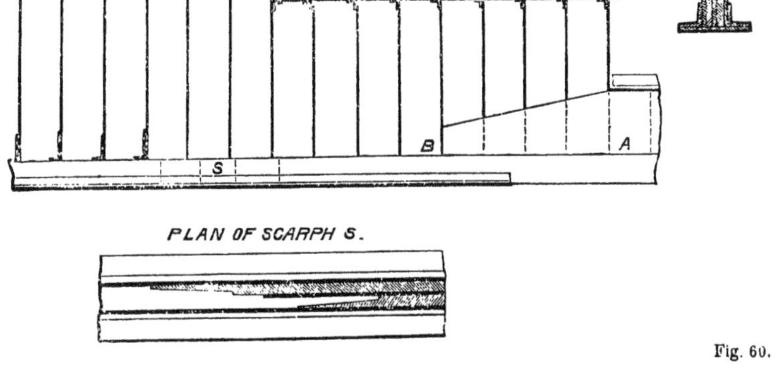

Fig. 60.

marked A in the sketch. The connection of the stern post and side-bar keel of H.M. ship 'Orontes' is given in Fig. 60. The side-bars terminated in plane scarphs, which fitted against corresponding scarphs in the fore end of the sole or keel-piece on the post, and were through riveted. The butts of the bars gave shift to each other, and to the butt of the centre keel-plate, as shown in the plan. In this vessel the depth of the centre plate was the same as that of the side-bars, for about 11 feet of its after part from the station marked B. The transverse frames abaft the stuffing-box bulkhead were formed of deep-throated solid plates and angle-irons. In order to keep up the longitudinal strength and make a good connection, a horizontal stiffening-plate, 1 inch thick, was worked underneath the sole-piece and keel, and extended about 27 feet before the post. It was connected to the sole-piece, keel, and garboards, by angle-irons, 6 by 4½ by 1 inches, riveted through them, as shown in the section at B.

The great expense involved in making the large forgings for the stern posts of screw ships, led to the proposal to make them of several thicknesses of thin plates riveted together. It was found on trial that, in consequence of the amount of vibration in the ship when under full steam, these stern posts, combined with the stiffeners in wake of the screw then in use, were not sufficient to bear the heavy strains brought upon them.

The stern posts now in universal use are solid forgings. The body post in screw ships is fashioned in wake of the shaft, to receive the engineer's shaft-tube, &c., and the rudder post has the lugs for carrying the rudder, either forged upon it, or secured to it by forked arms embracing it and riveted to it. In sailing ships, and in paddle-wheel steamers, the stern post, besides being secured to the outside plating of the ship's counter by an angle-iron collar, runs up to the deck above, and is connected to one or more of the beams.

In screw ships, when the weight of the two posts, with their connecting pieces, is not too great, the whole mass is forged in one piece. The first stern frame forged in one, at the Thames Iron Works, was for the Peninsular and Oriental Company's steam ship 'Pera,' and its weight was about 20 tons, while the posts, &c., of the Turkish frigate 'Sultan Mahmoud' weighed 27 tons, and are supposed to have been the largest single forging ever put into the hull of a ship—although not the largest existing in a ship, as will be seen presently when the stern posts of the 'Northumberland' are described.

In Fig. 61 the arrangements and connections of the stern posts

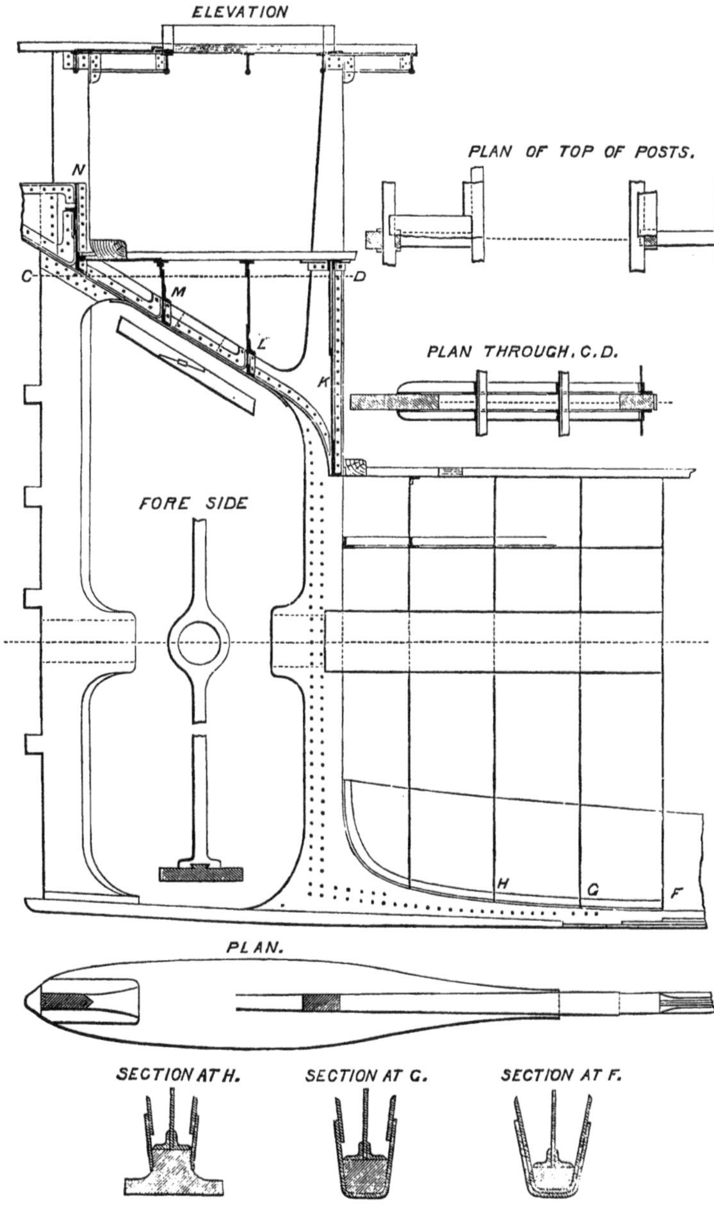

Fig. 61.

of the new Indian troop-ships are fully shown. The fore post and the sole or keel-piece were forged in one; the rudder post, with its lugs for the pintles, was forged separately, and had a large foot with a dovetail mortice cut in it, into which fitted a dovetail tenon that projected above the after end of the sole-piece. The connection of the two was secured by very large rivets driven down from the upper side, and riveted up underneath the sole-piece. In some cases the rivets used for this purpose are formed of bar-iron, cut to the length required, and turned in a lathe so as to fit the holes accurately. The rivets or bolts thus formed are made hot at one end and put into the hole, the head being formed by beating it down so as to fill the countersink; when it has cooled, the rivet is driven out, and, after its point has been heated, is replaced and knocked down so as fill the countersink in the sole-piece. The connecting piece, at the ship's counter, in this case, was forged in two parts, one being formed on each post; and when the posts were in place, the two parts were connected by a keyed scarph, as indicated in the sketch.

In nearly all the screw ships of the Royal Navy, and in many other vessels, the sole-piece is very broad and shallow in wake of the aperture. This form is adopted for the double purpose of giving stiffness to resist bending sideways, and keeping the screw as low as possible.* It will be seen, from the plan of the sole-piece, that this is done in the vessels now under notice. In them the sole-piece extends forward 18 feet into the vessel, and is connected with the keelson-plate, flat plate-keels, and keel angle-irons, in a similar manner to that previously described for the stems of those ships. The fore end of the sole-piece is forked and embraces the vertical flanges of the keel angle-irons, and the keelson-plate, the connection being completed by through riveting. The under side of the sole-piece is cut back in order to allow the horizontal flanges of the keel angle-irons to work in flush above the inner keel-plate. The fork is about 3 feet 6 inches long, and the keel angle-irons are stopped at its after end. At that

* In 'Shipbuilding Theoretical and Practical,' edited by Prof. Rankine, an account is given of a method, employed by Mr. J. R. Napier, for giving deeper immersion to the screw in shallow-draught vessels. In the case of the 'Lancefield' this object was attained by means of a curved depression of the after end of the keel, so that while the vessel floated on an even keel, at a draught of 8 feet, the depth of immersion of the screw was 10½ feet.

point the keelson-plate steps up above the sole-piece, and is taken up by angle-irons on each side, tap-riveted to the sole-piece, and through-riveted to the plate, as shown in the sections at F, G, and H. The-keel plates fold up around the sole-piece, as shown in the sections, and are tap-riveted to it. The inner keel-plate ends about 21 inches abaft the fork, and at 30 inches from its ending the outer keel-plate ceases to fold up around the sole-piece, and steps up into a rabbet in the side. The ends of the bottom plating are through-riveted to the foremost post, except in wake of the shaft, where the rivets are tapped. The posts are run up and have their heads secured to the beams, as shown in full detail in the elevation and plan of top of the posts in Fig. 61. The framing of the rudder-hole is, in these ships, made a means of strongly connecting the after post and the stern of the ship, and the transverse plate-frames K, L, and M, are taken up by angle-irons on the posts and connecting piece, thus adding greatly to the vessels' rigidity, and resisting the local vibration so generally experienced in ships where special provisions have not been made to prevent it.

The connections between the stern posts, centre keelson, flat keel-plates, &c., of the iron armour-plated frigate 'Northumberland,' are shown in Fig. 62. As in the case of the stem and its connections, the elevation, plan, and sections illustrate fully the details of the arrangements, which are, in general, similar to those above described for the Indian troop-ships. At the section, marked A in the sketch, the fork of the sole-piece ends; at B the inner flat keel-plate stops, and at C the outer flat keel-plate steps up into the rabbet in the post, the garboard-strake in the part from E to F being snaped away in order to make flush work. One difference between this arrangement and that of the Indian troop-ships is that a vertical flange is forged on the upper side of the sole-piece, and the centre keelson is through-riveted to it, instead of being taken up by angle-irons on each side. The after end of the centre keelson-plate is secured to the stern post by a pair of angle-irons tap-riveted to the post. The keel-plates are tap-riveted to the under side of the sole-piece, the arrangement of the rivets being shown in the sections, and their diameter being $1\frac{3}{8}$ inch. The garboard-strakes and bottom plating are secured to the post by $1\frac{1}{8}$-inch tapped rivets, the after row of rivets in the ends of the bottom plating, between the upper edge of the garboard and the swell of the post in wake of the screw-shaft, being $1\frac{1}{8}$-inch through-rivets, so as to make a strong connection. The rudder post with the lugs, &c., was forged separately, and connected by the usual arrangement of

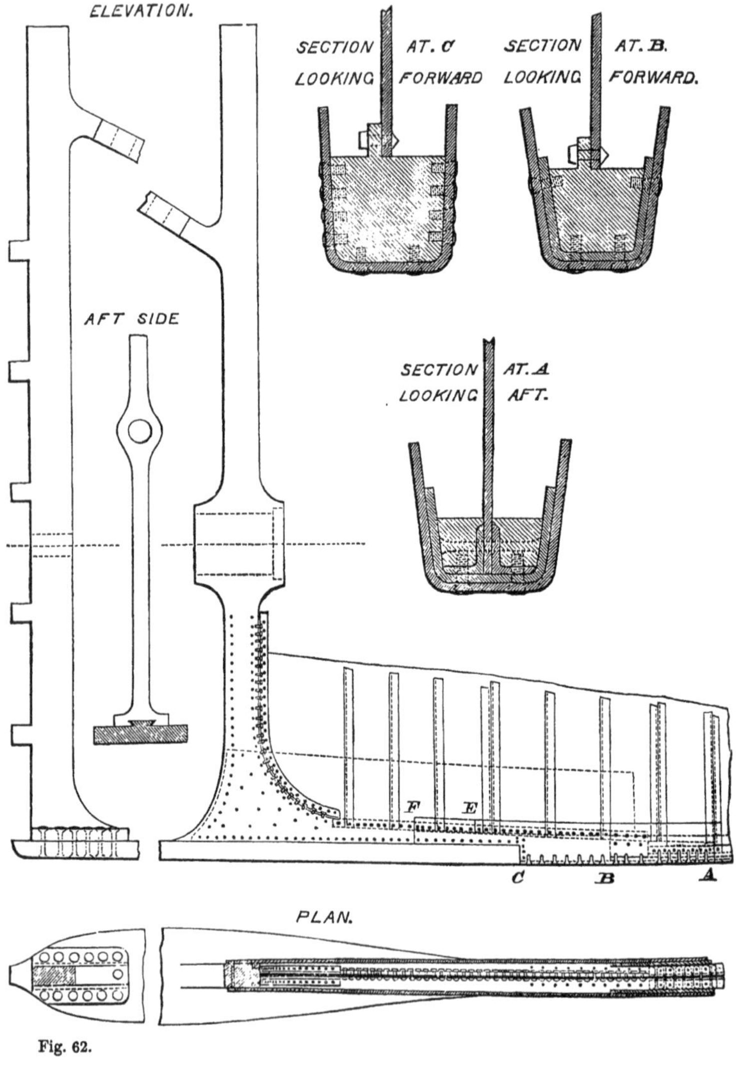

Fig. 62.

dovetail mortice and tenon, to the after end of the sole-piece; the rivets completing the connection being $2\frac{1}{2}$ inches in diameter, and arranged as shown in the plan.

In the case of the stern post there is not the same necessity for protecting the edge of the plating by burying it in a rabbet, as exists at the stem of an iron-clad which has a ram-bow. The sole-piece forms a rabbet for the lower edge of plating, and protects it from the chances of injury if the ship grounds; and on the sides of the post itself, where there is little likelihood of the plating being ripped off, it is worked plain, and a strong connection made by the arrangement of rivets described above.

The process of forging the stern posts of a large iron frigate is conducted as follows:—In the first place a staff is formed similar to that used in forging the stem, and piles of slabs are added so as to form a forging about 4 feet 6 inches broad, 2 feet 6 inches thick, and 5 feet long. This forging is for the boss of the body post. Large slabs are then piled on each side of the forging, each slab being about 5 feet long, 1 foot 6 inches broad, and 5 inches thick, and a forging about 5 feet 6 inches long and 4 feet 6 inches square is formed, and, when welded into a solid mass, is drawn out at each end, to form the boss and a portion of the upper and lower part of the post. The large slabs are piled on each side of the mass, in the direction of the upper and lower part of the post, for the purpose of giving the greatest amount of strength to the sides of the boss when the shaft-hole is bored out. The boss is then taken to the fitting shop, and its fore and after sides having been planed, it is fixed on the bed of a lathe and has the shaft-hole bored in it, and the fore and after sides turned off. This turning having been completed, the boss is placed on the table of the slotting machine, and its sides and a portion of the upper and lower parts of the post are slotted. It is then taken to the forge, and the upper part of the post is finished. The lower part of the fore post and the foot are then forged, the operation being conducted as before described. The first pile of slabs forming the lower part of the post, and the fore and after sides of the foot, is arranged as shown in Fig. 63. Slabs marked B are then laid on each side of the bottom of the post so as to form the foot or sole-piece, the arrangement of the slabs for the fore part being similar to that shown in Fig. 64. A staff is then welded to the fore part of

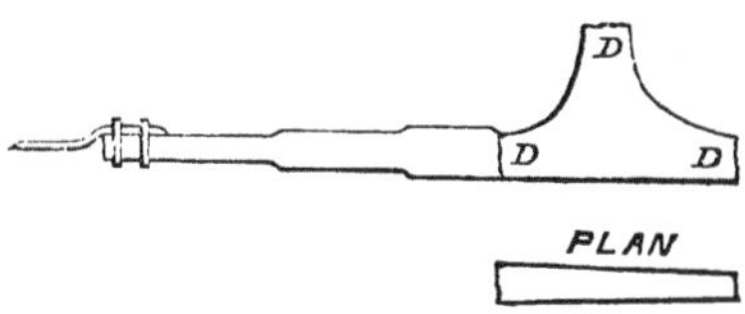

Fig. 63.

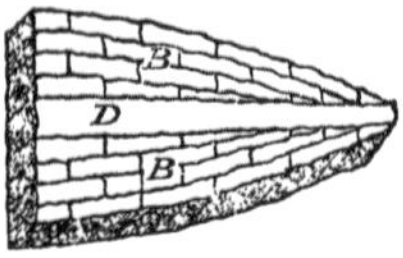

Fig. 64.

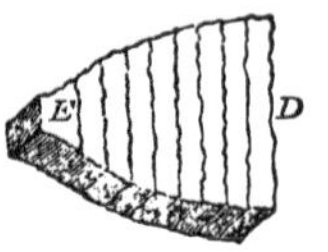

Fig. 65.

the foot and the forging completed from the point marked D to that marked E in Fig. 65, by laying slabs across the foot, commencing at D. A staff is next fixed on the after part of the foot, and the forging of the fore end, which is connected to the keel, is completed. The foot is thus formed in one solid forging, and being taken to the planing machine, all the planing required before welding it to the other part of the post, is performed. In the case of the 'Northumberland' this welding was performed in the manner illustrated by Fig. 66. The foot or sole-piece with the stump of the body post forged on it and V'd, was first placed in its proper position on the blocks on which the ship was being built. The body post was then hoisted

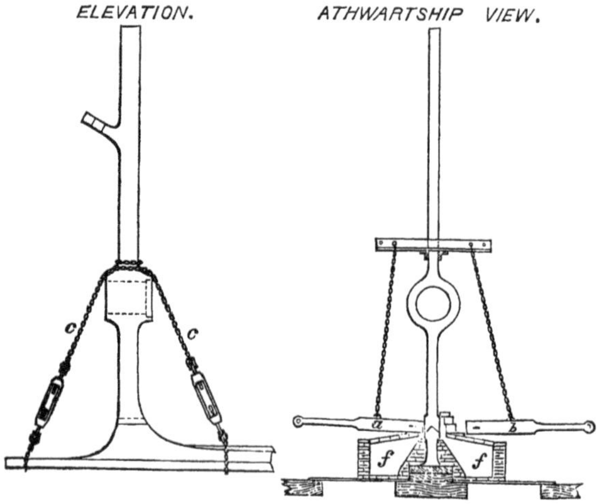

Fig. 66.

into place, and the parts V'd were brought together as shown by the sketches. Two small coke furnaces, *f f*, were then built, one on each side of the stern-post, with a blast leading to them, as shown in the athwartship view. A coke fire was next made, and when a good welding heat was obtained, the upper piece of the body post was pressed heavily down by two screw-rods and chains *c*, in addition to its own weight, while in the fire, until it had contracted about 3 inches in length. The crowns of the furnaces were then suddenly removed, and two monkeys or iron battering-rams *a* and *b*, one on each side, were brought to bear upon the heated parts simultaneously, until they were welded. The crowns of the furnaces were then rebuilt and another welding heat obtained, when the operation of striking the welded parts with the battering-rams was repeated, care being taken this time to get the stern post to its proper length, it having been ascertained by the first welding heat how much it would contract in cooling between the centre of the boss and the keel-piece. This having been satisfactorily performed, the furnaces were removed and the surplus iron chipped off.

The rudder post of an iron screw steam-ship is forged in the same way as the stem; and when the forging has been completed it is taken to the machine, and the heel, sides, lugs, &c., are planed. This course was followed in the 'Northumberland;' and when the welding of the body post had been completed, in the manner just described, the rudder post was hoisted into place and secured. The weight of the two posts and their connecting pieces, exceeded 40 tons. It will be seen that in this ship the connecting piece, at the counter, was a separate forging, and was connected to the projecting pieces forged on the posts by two hooked scarphs, whose positions are shown in the elevation in Fig. 62. The slabs used in forging the stern posts of this vessel were made of the best selected scrap-iron, such as the shearings of rolled plates, and other new, clean scrap kept for the purpose of forging important work.

The arrangement proposed for the 'King William,' Prussian iron-clad frigate, previous to her being supplied with a balanced rudder, is illustrated by the elevation and view of the after side of

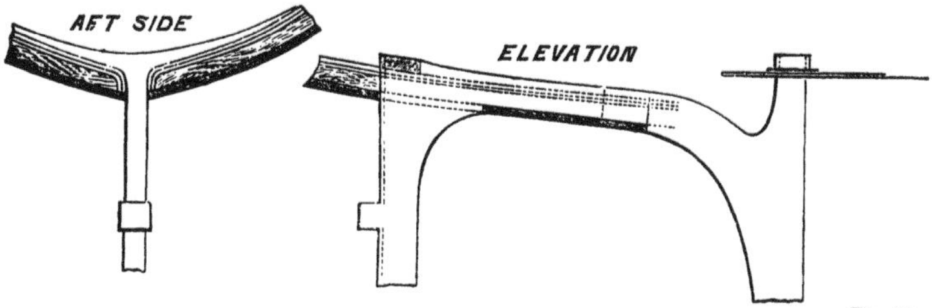

Fig. 67.

the rudder post in Fig. 67. The connections of the keels, &c., with the sole-piece are of an identical character with those of the 'Northumberland,' and it is, consequently, unnecessary to show the lower part of the stern frame, the peculiarity of the mode of securing a firm connection between the head of the post and the stern being that which constitutes its interest. In the common arrangement an attempt is made to increase the connection of the heavy stern posts with the body of the ship by bringing the heads of the posts up to a deck and securing them there. In the case before us we intended to combine the upper part of the rudder post with the frame of the vessel by forging horns on it. These horns were formed as shown in the view of the aft side, and extended 5 feet up the side, being tapered, in thickness, from 6 inches at the middle line to $1\frac{1}{2}$ inch at their extremities. They were bolted to the outside plating, and so an increased extent of direct connection was effected between the post and the hull of the ship; and this seems preferable to the usually indirect connection between the two, made by the beams and deck-plating, while it also does away with the very local character of the usual combination by spreading the fastenings over a larger area.

An arrangement, having the same object in view, but differing in the mode of accomplishing it, was made, we have since found, in the screw steam-ship 'Barwon,' of 485 tons, built in 1854 by Mr. J. Bourne,* for the Australian coasting trade. The stern frame is thus described in the specification :—"To be forged in one "piece of the best scrap-iron, with an aperture of a size adequate to "admit the most approved screw, and the forging to be without "scarphs or joinings in it; but the frame is to be scarphed to the "keel with a long scarph with planed joint, which is to be riveted "with turned rivets, the holes being accurately rimelled out and "the rivets driven in so as to fit accurately throughout their "length. The scantling of the stern frame is to be 8 inches "broad by 3 inches thick, with a projecting spur for the reception "of the rudder heel, and a tapered piece is to be welded to the "keel so as to bring up the thickness of the keel gradually to that "of the stern frame, so as to enable the plates to lie fair over the "joint. On the foremost upper corner of the frame a palm is to "be forged on, which is to be riveted to a strong breast-hook plate, "so as to enable the stern frame to obtain a firm hold of the ship "near the water-line where the breadth is sufficient to resist the "lateral strains which the stern frame has to withstand." The palm here spoken of was shaped like the palm of a vice, and was run in underneath the iron flat of the lower saloon, and riveted

* The well-known author of works on the steam-engine, &c.

to it. The rudder post was run up to the upper deck and secured in the usual manner.

The arrangements of the stern posts of H.M. twin-screw armour-plated ship 'Penelope,' which is constructed with two after dead-woods, are shown in Fig. 68. She is fitted with two screw-wells so

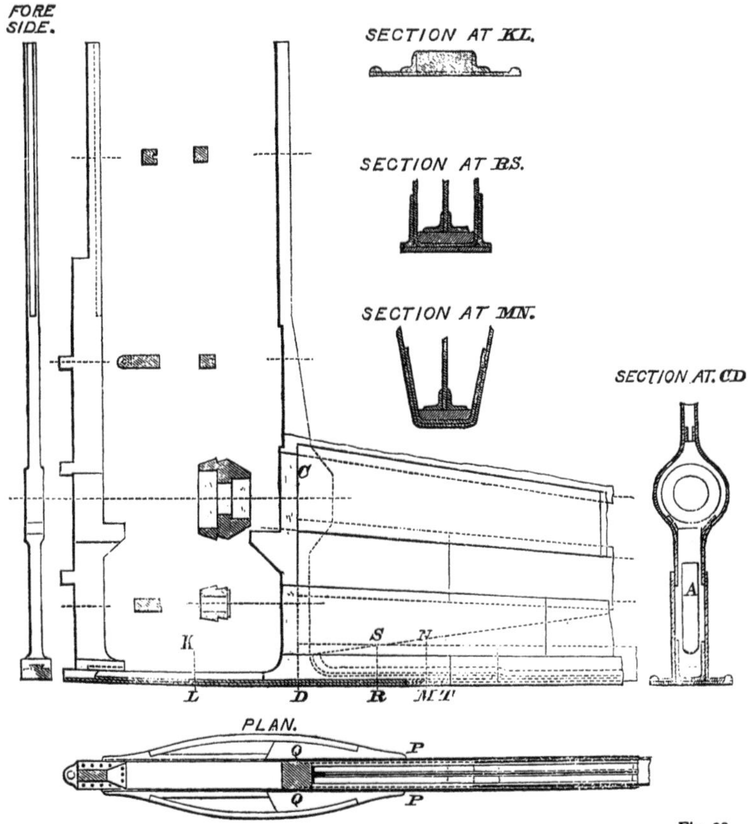

Fig. 68.

as to lift the screws inboard when necessary, and consequently the posts are run up to the upper deck. Their scantlings are reduced on account of their having less strain to bear than those of single screw vessels, and, as shown in the various sketches, every means has been taken to lighten them. The principal modes of doing this are, the reduction of both the heads of the posts to a siding and moulding of 6 inches; the cutting of a groove 11 feet long, 4 inches wide, and 2 inches deep in the fore side of the head of the rudder post, as shown in the view of its fore side; the cutting out the hole A in the lower part of the body post, as shown in the section at C D; and the unusual arrangements of the boss and bearings shown in the sections. It may appear to the reader that the siding of the lower part of the body post is excessively great; but this was purposely designed in order to give more room for working in the extreme after part of the ship. For with thinner posts it has been found very difficult, and in some cases almost impossible, to complete the riveting of the after end of the vertical keelson-plate and its angle-irons, and of the plating of this part of the ship, in a proper manner. This difficulty was got rid of in the 'Bellerophon' by increasing the siding of the body post to 22 inches, as is shown in the sketch of her stern post given further on. In the 'Penelope' the siding is 14 inches, and as this is far above that required by the strains experienced by the stern frame, the hole marked A in the section at C D is cut out and a piece of plate riveted on the inside to keep out the water.

The fore end of the sole-piece is not forked like that of the Indian troop-ships, but is run off to a thin edge at about 11 feet before the post, the keel angle-irons and centre keelson-plate running along on its upper side, and the keel-plates folding up around it as shown in the section at M N in Fig. 68. The keel angle-irons are secured to the sole-piece by $1\frac{1}{8}$-inch tapped rivets, and to the keelson-plate by $\frac{7}{8}$-inch rivets. The inner keel-plate is $\frac{5}{8}$ inch thick, and the outer $\frac{3}{4}$ inch, their connection being made by 1-inch rivets. In this vessel the outer keel-plate is not rabbeted up into the sole-piece, but is run out underneath and tap-riveted to it, the shape of the after part being that shown in the plan. It will be seen on reference to this plan that from the points P aft, a flanged plate Q is worked, the horizontal part fitting upon the outer flat keel and being riveted to it, and the vertical flange being tap-riveted to the heel of the post; this arrangement is illustrated by the section at R S. On the after side of the plates marked Q the flat keel is connected to the sole-piece by angle-irons on each side, as shown in the plan and the section at K L, and the outer edges are stiffened by bars of half-round iron riveted through the plate. By means of this arrangement the cost of forging the sole-piece, and its weight, are both considerably reduced while the strength of the after part is amply sufficient to resist the strains tending to bend it side-ways. In the elevation the butt marked T is the point at which the outer keel-plate ceases to fold up around the sole-piece, and on the aft side of T the vertical flange of the plate Q completes the double thickness of plating, in the manner shown by the section at M N. The after part of the outside plating is worked flush, internal longitudinal strips being fitted to take the edge riveting. The after ends of the strakes of plating are double chain-riveted to the body post.

In ships fitted with the ordinary rudder the after post is essential, and, in addition to serving as a means of hanging the rudder, it is of use in taking the thrust of the propeller when the vessel is going astern, in those ships which are fitted with gear for lifting the screw inboard. In ships which have balanced rudders (as for instance the 'Bellerophon,' whose stern arrangements are given in great detail in Fig. 87) the after post is sometimes dispensed with, the rudder being kept in position by the counter of the ship aloft, and by a massive pintle in the heel, which fits into a socket in the sole-piece.

The stern post of the 'Bellerophon' is of a very novel form, and its connections with the keel-plates and with the hull of the vessel are of an entirely different character from those previously illustrated, as will be seen on reference to the profile view in Fig. 87, page 120, there is no sole-piece forged on this body post, only a short toe being forged on the after side at the heel. The sole-piece is formed by a prolongation of the keel-plates arranged and stiffened as shown in the plan and in the section at E F, and the comparative slightness of the arrangement is justified by the considerations that there is no after post to connect with it, and that the weight of the rudder is taken inboard, as will be described hereafter. At the station marked S in the plan, the outer flat keel ceases to fold up around the frames, and is worked out so as to form the lower plate of the sole-piece; but in order to keep up the double thickness of plating at the heels of the frames, and make a good connection between the sole-piece and the hull, a flanged-plate marked D $\frac{7}{8}$-inch thick, is worked on each side, upon the upper thickness of plating in the sole-piece, and against the vertical flange of the inner flat keel as shown in the section at E F. The sole-piece on the after side of the plates D, is formed by a double thickness of plating, amounting to $2\frac{3}{16}$ inches, and in order to give strength to resist transverse bending, the extreme width of the sole-piece is increased to 6 feet, and along the edges on its under side a forged plate 8 inches wide is worked, which serves as a stiffener to the plating. The hole C is made in order to allow the free passage of water when the ship is pitching or 'scending, and so relieve the sole-piece from the great pressure thus brought on it. At the after end is a circular hole which takes the pintle in the heel of the rudder, its position being shown in the elevation and plan.

The siding of the stern post is 22 inches, and its moulding 21 inches, the former being adopted in order to give room for working in the extreme after part of the ship, as was before stated, and it succeeded perfectly in this respect, as there was not a single rivet in the stern of the 'Bellerophon,' the knocking down or the tapping of which presented any difficulty or caused any delay. In addition to this, the large dimensions of the stern post allowed the introduction of a great mass of iron in the short post, and so tended to resist the vibration consequent on the immense engine-power, and with the stern framing (which will be fully described further on), effectually accomplished this object.

The boss on the post was forged in the usual manner, but the head was formed by welding on a plate $4\frac{1}{2}$ inches thick, marked M in the profile, of which the shape is shown in plan in Fig. 87. It will be seen on reference to the profile view, that the plate M ran in underneath the flat A, which was formed of $\frac{3}{4}$-inch iron plate, and was riveted to it, thus strongly connecting the head of the post with the body of the ship.* This combination was further strengthened by the connection with the post of the iron flat B, and the double wrought-iron shaft tubes. The whole of this longitudinal framing was run forward and connected with the stuffing-box bulkhead. The connection of the stern post with the outside plating was made by treble chain-riveting, the plating being worked plain on the sides of the post, and the rivets being tapped into it. Above the stern post, two thicknesses of plates were bent to the shape of the stern, and had their lower edges worked into the rabbets shown in the head of the post, so that the outer surface of the plating was flush with the after side of the post. These plates were secured to the head of the post by double tap-riveting and the outer plate was narrower than the inner, so that the after ends of the bottom plating lapped on the inner plate and were

* It will be seen that this arrangement substantially resembles that adopted by Mr. Bourne in the 'Barwon.' When the 'Bellerophon' was designed, however, I had no knowledge of the latter vessel, or of a like plan ever having been adopted.—E. J. R.

double riveted to it, and fitted flush against the edge of the outer plate. The connection of the foot of the post with the sole-piece was made by the flanged plates D, and the short pieces of angle-iron worked around the sides of the toe, as shown in the plan, together with large rivets through the plate sole-piece; the aftermost rivets were driven through the thin part of the toe, and knocked down in a countersink as usual, but those in the foremost part were tapped up from beneath into the post. The after end of the centre keelson-plate was taken up by double angle-irons, tap-riveted to the post; thus the connections of the latter were completed, and, though novel, they have stood the test of actual service most satisfactorily.

The stern post of the Prussian iron-clad frigate 'King William' as fitted in connection with a balanced rudder, combines the mode of connection of the sole-piece with the flat keels, &c., adopted in the 'Northumberland,' with a slightly different mode of connecting the head of the post to the hull, from that described above for the 'Bellerophon.' The head of the post is 7 feet above the centre of shaft, or about 3 feet higher than the 'Bellerophon's' extends, and it is secured to an iron flat by a frame of angle-iron enclosing it, the flat being strengthened to receive the fastenings by a doubling plate worked on it, and extended to the aftermost transverse plate frame. The wrought-iron tube which takes the engineer's shaft tube, is in this vessel made of a single thickness of plate, and the fore side of the boss is forged differently to that of the 'Bellerophon,' so as to rivet the tube to it. The flat corresponding to that marked B in the sketch of the 'Bellerophon,' Fig. 87, is omitted in this vessel, the transverse plate frames being continuous from the keel to the flat at the head of the post. The longitudinal strength is kept up, however, by the fore end of the sole-piece being run 4 feet before the stuffing-box bulkhead.

In the 'Hercules,' the connections of the sole-piece with the flat keels, plates, &c., and the arrangements of the after end of the outer flat keel, are very similar to those described in detail for the 'Penelope.' There is no fork at the fore end, but the sole-piece is run off to a thin edge, and the vertical keel and keel angle-irons run along upon it. This stern post differs from all the preceding in being scarphed, instead of welded, to the sole-piece. The connection of the head of the stern post with the plate-iron flat is identical in character with the corresponding connection in the 'Bellerophon.' It should be added that the body post, while retaining a very large siding, has a comparatively small moulding, except in the neighbourhood of the boss, differing in this respect from the 'Bellerophon's' post.

A very novel arrangement of the stern frame, designed by Mr. Mackrow, for the 'Pervenetz,' a Russian iron-clad battery built at the Thames Iron Works, is shown in Fig. 69. The stern

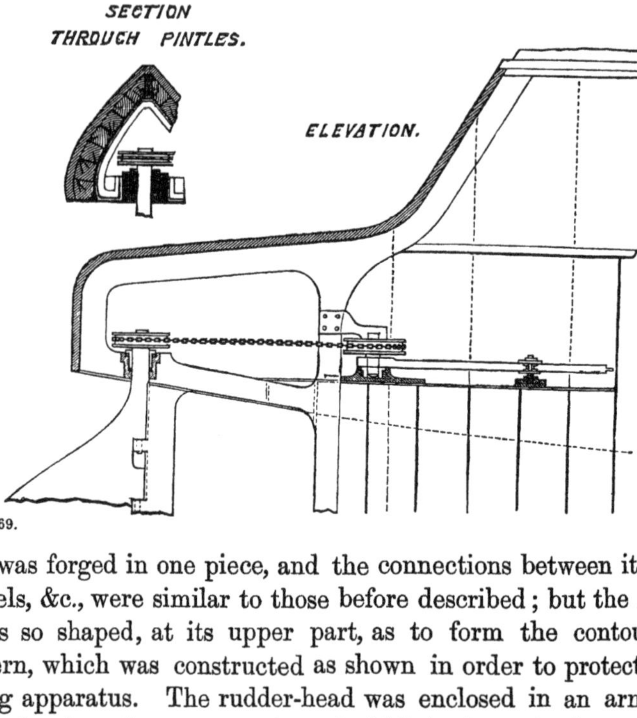

Fig. 69.

frame was forged in one piece, and the connections between it and the keels, &c., were similar to those before described; but the forging was so shaped, at its upper part, as to form the contour of the stern, which was constructed as shown in order to protect the steering apparatus. The rudder-head was enclosed in an armour-plated chamber, the construction of which is shown in the section. The whole of the weight of this box or chamber was attached to the forging directly, and through it and the skin-plating behind armour to the hull of the ship. The shape of the section was such as must cause the impact of projectiles to be usually oblique, and the weight of protecting armour was reduced to a minimum by the small height of the chamber. The connection of the head of the post to the hull was made by the skin-plating behind armour, and by the armour plates being bent around it, and lying in direct contact with it.

CHAPTER V.

TRANSVERSE AND LONGITUDINAL SYSTEMS OF FRAMING.

THE systems of framing adopted in iron ships may be classed under three heads—1. The transverse system; 2. The longitudinal system introduced by Mr. Scott Russell; and 3. The systems followed in the construction of H. M. iron ships, which may be regarded as combinations of the other two systems.* This classification represents the order of introduction of the respective systems, and is that which will be followed in their illustration. The transverse and longitudinal systems only will be described in this chapter; the systems adopted in H. M. ships will form the subject of further chapters. The transverse frames of the early iron vessels were formed either by a single angle-iron, or by a frame angle-iron, and a reversed frame riveted back to back, the rivets being spaced from 5 to 6 inches apart. The addition of the reverse iron added greatly to the strength of the frame to resist bending, and at the same time it served to receive the fastenings of the internal planking. In the case of a single angle-iron forming the frame, the internal planking was secured to short pieces of angle-iron, or to wooden scantling fastened to the sides of the frame angle-irons. The frames extended from gunwale to gunwale, and were formed of several lengths of angle-iron either scarphed or welded to each other. When a single angle-iron frame was used, it was either made up of two lengths scarphed to each other,

* This classification is intended only to include the systems of framing which have been employed to some considerable extent in the construction of iron ships, and, consequently, does not embrace the special modes of framing which have been adopted in particular cases, such as, for instance, the arrangement in which the frames are placed diagonally, &c. It may be added that iron ships have been built which have no frames, and, as an illustration of such a vessel, we may refer to a collier described by Mr. Henderson, in a discussion on 'Steam and Sailing Colliers,' which is recorded in the *Proceedings* of the Institution of Civil Engineers for 1855. This ship was designed by Mr. Hodgson of Liverpool, and was built of stout iron plates only. Along each bilge a compartment was formed for water ballast, and the coal was carried in the central space between and above the compartments.

the scarphs of adjacent frames being on opposite sides of the keel, or of three pieces, one of which crossed the keel and extended up the bilge on each side, where the other two lengths were scarphed to it. An illustration of the latter arrangement is given in Fig. 70

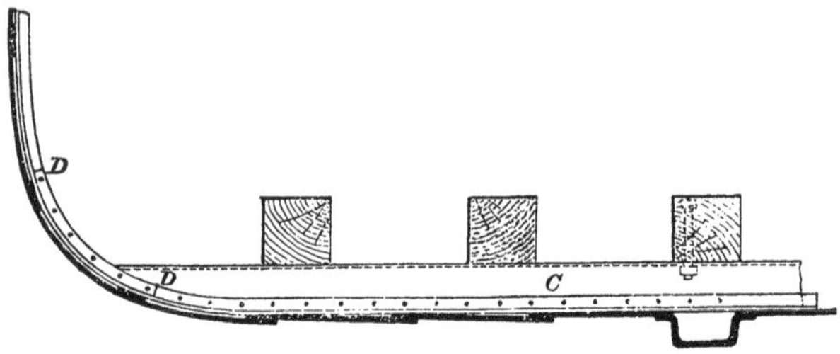

Fig. 70.

which shows a part section of the iron packet 'Dover,' before alluded to. In this vessel the frame angle-iron was 3 by 3 by $\frac{5}{8}$ inches, and was scarphed as shown at D D, the length of the scarph being 2 feet 6 inches.

The number of lengths of angle-iron usually employed in forming a double frame—that is, a frame having a reversed angle-iron—was five, and the scarphs of adjacent frames were carefully shifted. These frames possessed such great advantages, in strength and convenience, over the single angle-iron frames just described, as to lead to their almost universal employment. M. Dupuy de Lôme states, that in 1842 "The usual scantlings of angle-iron, in "the frames of sea-going vessels, were from 3 to 6 inches, the "flanges to which the bottom plating was attached never exceed- "ing $3\frac{1}{2}$ inches. The space between the frames was variable, "according to the character of the vessels, but in sea-going ships, "in the midship part, they were never nearer than 15 inches, nor "further apart than 20 inches. These intervals were augmented "gradually toward the extremities, where they varied from 2 to "3 feet." The outer flanges of the frame angle-irons were then, as now, turned aft in the fore body, and forward in the after body, so as to always have an obtuse angle between the two flanges, and thus reduce the strength of the angle-iron less than would be the case if the angle were acute, and in addition give greater facilities for riveting the bottom plating.

Another illustration of the old transverse single frames is given in Fig. 71, which represents the midship section of the 'Birkenhead.' It will be seen that in wake of the beam-ends short pieces of reversed angle-iron were worked, for the purpose of receiving the fastenings of the waterway, plate shelf, &c. The connection of the beams to the ship's side was peculiar, and will be described hereafter. The scantlings of the frames and their spacing

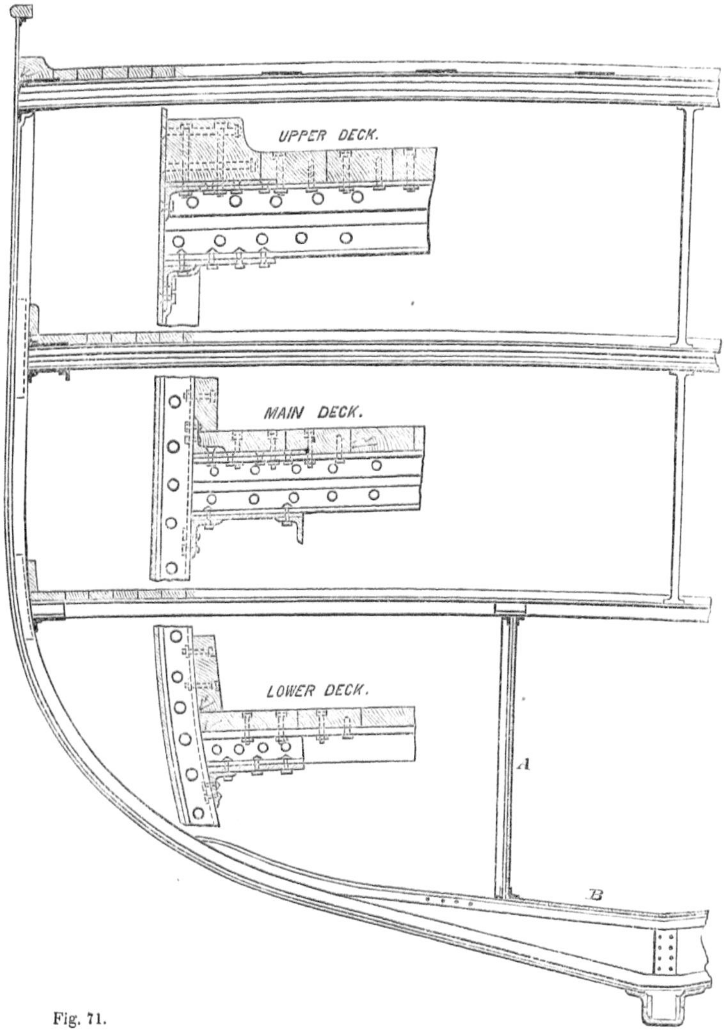

Fig. 71.

were as follows:—For 120 feet amidships, 5 by $4\frac{1}{2}$ by $\frac{1}{2}$ inches, spaced 15 inches apart; forward, 5 by $4\frac{1}{2}$ by $\frac{7}{16}$ inches, spaced 18 inches apart; and aft 5 by $4\frac{1}{2}$ by $\frac{7}{16}$ inches, spaced 20 inches apart from centre to centre. The total length of the ship was 210 feet, and it will be seen from the above that the framing of the extremities was considerably lighter than that of the amidship portion. The longitudinal bulkheads marked *A* in the sketch were fitted in the fore and after holds, and were so arranged as to form the sides of the magazines, the water-tight plating B being worked on the floors between them as shown in the sketch. In wake of the engines and boilers the lower deck was discontinued, and the longitudinal bulkheads forming the sides of the coal bunkers were continuous from the floors to the upper deck, being stiffened by vertical angle-irons, worked upon the plating of the bulkheads, and supported by stiffeners formed of bar and angle-iron, which extended from the bulkheads to the frames. A water-tight flat was fitted at the bottom of the coal space, and secured to the frames, to the bottom plating, and to the bulkhead, thus making the bunker a water-tight compartment.

H. M. iron brig 'Recruit,' the details of the framing of which are given in section in Fig. 72, had a reversed angle-iron riveted on every other frame, and wooden quartering bolted to the sides of the single frames to receive the fastenings of the internal planking. The port timbers were of wood, and were secured to the outside plating, and planked over inside the vessel, as shown in the sketch. The iron frames were, in general, extended up to the plank sheer, those only in wake of the ports or of the port timbers being stopped short.

The use of floor-plates dates from the commencement of iron shipbuilding, and is illustrated in each of the preceding sections. The arrangements of the 'Dover's' floors are shown in Fig. 70. The plates marked C were flanged in the midship part of the ship, to receive the bolts of the engine and boiler bearers, being, under those timbers, 9 inches deep, $4\frac{1}{2}$ inches flange, and $\frac{3}{8}$ inch thick. Forward and aft the floor-plates were 9 by $\frac{5}{16}$ inches at the middle line, and there were no reversed angle-irons worked on them. The floors of the 'Birkenhead' are shown in Fig. 71, and are thus described in the specification:—To be 2 feet deep amidships, and to extend across until they die away with rise of floor, being formed of plates $\frac{1}{2}$ inch thick riveted to the frames, and having a $3\frac{1}{2}$ by $3\frac{1}{2}$ by $\frac{7}{16}$ inches angle-iron riveted to the upper edges, and under the engines there are to be two angle-irons on the upper edge for better securing the keelson. In the 'Dover' the floor-plates and frames ran across the middle line; but in the 'Birkenhead,' the floor-plates abutted at the middle line, and were connected by a single riveted strap as shown in the sketch.

In the section of the 'Recruit' given in Fig. 72, it will be seen that the floor-plates butted at the middle line, and were notched down over the keel; the upper part of the score above the keel served as a watercourse, and as the frame angle-irons stopped at the keel, it was necessary to secure the two pieces of the floor-plate by a double riveted butt-strap. The details of this arrangement are shown clearly in Fig. 19, p. 24. The rivets which are there shown on the middle line served to secure the heel of the iron pillar to the floor-plate. In this vessel the reversed angle-irons on the upper edge of the floor-plates were run across the middle line, and took the bolts which secured the wood keelson.

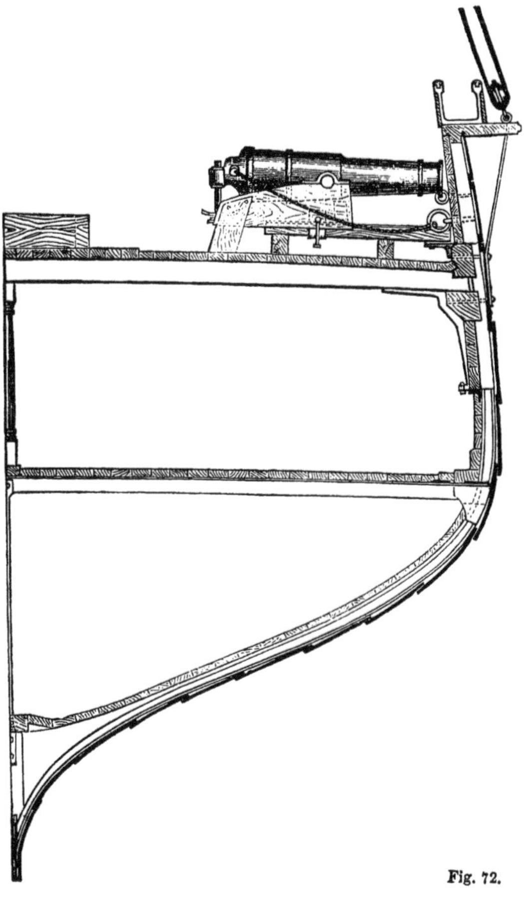

Fig. 72.

In very small and light vessels, the floor-plates were frequently omitted, and a short piece of reversed angle-iron was riveted to the frame angle-iron to receive the fastenings of the keelsons.

In more modern vessels built on the transverse system, the framing always consists of three distinct parts, the frame angle-iron, the reversed frame, and the floor-plate. The first of these does not differ from that used in earlier vessels, and is formed in lengths which, if necessary, are either scarphed or welded to each other, care being taken to give shift to the scarphs or welds of adjacent frames. Lloyd's rule with respect to frames, is, that they "are to be in as "great lengths as possible, fitted close to the upper edge of keel, "and in all cases to extend to the gunwale; and when butted on "the keel (except when double frames or centre continuous keels "are adopted), and wherever else butted, to have not less than "four feet lengths of corresponding angle-iron, fitted back to back, "to cover and support the butts, and receive the plating. If "welded together, the welds to be perfect with not less than 4 feet "shifts. The spacing from centre to centre, with single frames, is "not to exceed 21 inches, but provided an additional frame be "fitted for half the length amidships at opposite sides of the floor-"plates across the keel, and extended to the upper part of the "bilges, and riveted to the floor-plates and main frames, and to "the bottom plating similarly to the riveting required for the main "frames, the space may be increased to 23 inches in ships under "1000 tons, and to 24 inches in ships of 1000 tons and upwards." The Liverpool Rules require the frames "To be spaced so as not to "exceed 21 inches from centre to centre throughout in vessels "under 1000 tons. In vessels of 1000 tons and above, the frames "may be spaced 24 inches from centre to centre, for one-fifth the "vessel's length from each end; or may be spaced throughout so "as not to exceed 24 inches from centre to centre, provided a "double frame of the same size as the frames of the vessel be "carried from the centre line to the upper turn of the bilge, and "be properly secured to the floors and shell of the vessel for three-"fifths the length amidships. All frames should be in one length, "but when butted they must have scarph pieces same size as "frames, 4 feet long in vessels up to 900 tons, and 6 feet long in "vessels above 900 tons, with a good shift of butts. Lapping "pieces to connect heels of frames across the centre line, where "bar-keels are used, to be not less than 4 feet long, and of the "same size as the frames." It will be seen that in all the principal provisions the two sets of rules are identical

The various modes in which floor-plates are fitted, are regulated by the arrangement of keel which is adopted, and the sketches of keel arrangements already given fully illustrate the practice of different shipbuilders. In accordance with Lloyd's Rules, the floor-plates of a vessel must be fitted and riveted to every frame, and extend up the bilges to a perpendicular height of twice the depth of floors in midships, and must not be less moulded at the heads than the moulding of the frames. The depth of the floor-plates at the middle line is determined by the following rule :—" To the vessel's depth, measured from the top of the keel " to the top of the upper or spar-deck beams amidships, add the " extreme breadth of the vessel; two-fifths of that sum *in inches* " will be the depth required." The dimensions fixed by Lloyd's for the floor-plates of vessels of various sizes will be found in the Appendix. The Liverpool Rules state that floor-plates " are to be " riveted on every frame, to be half the centre depth at lower " turn of bilge, and to be carried well up into the bilge, and " finished at the depth of the moulding edge of the frames."

In vessels which have external solid-bar keels, the floor-plates usually cross the middle line, while the frame angle-iron ends at the middle line, in many cases. A strap of angle-iron about 4 feet long of the same size as the frame angle-iron, is riveted on the opposite side of the floor-plates in most vessels where this arrangement is adopted, and so keeps up the transverse strength of the frame and secures the bottom-plating. Sometimes, however, the frame angle-iron is continuous across the middle line, as shown in Fig. 2, page 19. The limber-holes in the floor-plates are, as a general rule, cut above the frame angle-iron, and, to prevent water lodging in the spaces below, they are usually filled up with cement.

When hollow-plate keels are adopted the arrangement of the floor-plates is exactly similar to that just described for a bar-keel; the frame angle-irons, however, in these cases generally run across the keel, and the hollow keel itself forms the watercourse.

In vessels which have side-bar keels and continuous centre plates the arrangements are often similar to those shown in Fig. 20, page 25, which is a part section of the screw steam-ship 'Roman,' previously referred to. The floor-plates on each side heel against the continuous centre plate, and are connected to it by two vertical angle-irons $3\frac{1}{2}$ by $3\frac{1}{2}$ by $\frac{1}{2}$ inches. The centre plate is $\frac{11}{16}$ inch thick, and has two continuous angle-iron bars $3\frac{1}{2}$ by $3\frac{1}{2}$ by $\frac{1}{2}$ inches riveted to its upper edge, the floor plates being cut back, as shown, to allow them to pass. In order to keep up the transverse strength the gutter plate is worked above the floors and riveted to the reversed angle-irons in wake of it, which are $3\frac{1}{2}$ by $3\frac{1}{2}$ by $\frac{7}{16}$; and in addition the angle-iron bar A A is reeved through a score cut in the centre plate, and riveted to the floor-plates. This bar is 3 by 3 by $\frac{7}{16}$ inches, and extends about 18 inches on each side of the middle line. As the score in the centre plate is cut very near the neutral axis of the girder formed by the plate and its angle-irons, &c., the reduction made in the longitudinal strength is comparatively small, while at the same time the bar A is in a most favourable position to act as a transverse tie to the floor-plates.

In some vessels the continuous centre plate is run up above the floors and taken up by angle-irons on two gutter plates as is shown in Plate 1. In such cases the transverse tie-bar A is dispensed with, but the short piece of angle-iron worked underneath the gutter plates to receive their fastenings is reeved through a score, and this leaves sufficient longitudinal strength, as the depth of the centre plate has been increased. The sketches in Fig. 73 are taken from a screw vessel of 2500 tons, and represent a floor section and an elevation of the keel, the latter being drawn so as to represent the condition of the keel when put upon the blocks. It will be seen that in this vessel there are double frames, and one of the angle-irons at each transverse frame is reeved through a score cut just above the upper edge of the side-bar, in addition to the reeving through of the angle-iron marked B B, so that a very strong transverse connection is made.

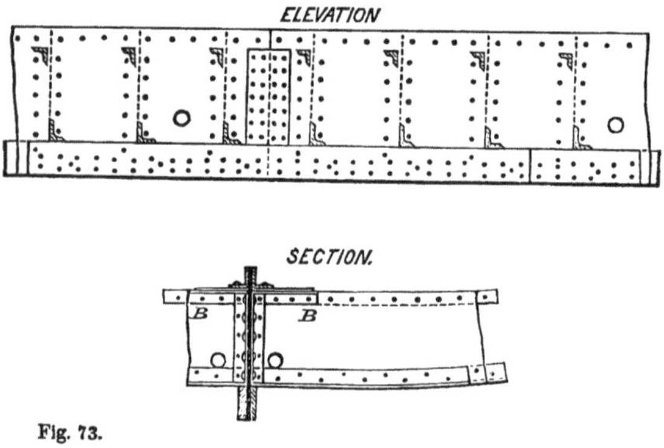

Fig. 73.

When the continuous centre plate of a side-bar arrangement is run up above the floors and its upper part riveted to two pairs of angle-irons, so as to form an upright I-shaped plate keelson similar to that shown in Fig. 29, page 30, the transverse connection of the floor-plates is made in a similar manner to that described above.

When flat-plate keels are adopted, and associated with intercostal middle-line keelsons, as shown in Fig. 31, page 30, the floor-plates are continuous across the middle line and the intercostal plates are connected to them by double angle-irons, as there shown. If, however, as is usual, a continuous centre plate is employed in connection with the flat-plate keel, the floor-plates are heeled against the centre plate, and connected to it in a similar manner to that described for a side-bar keel and continuous centre plate. An illustration of the arrangement usually followed in the construction of vessels belonging to the mercantile marine has already been given in Fig. 30, page 30. It will be seen on reference to the sketch that the pairs of angle-irons at the upper and lower edges of the continuous centre plate are run fore and aft in continuous bars. The frame angle-irons and the reversed frames are stopped on each side, and a bar of angle-iron, A A, is reeved through a score near the middle of the continuous centre plate, and, being riveted to the floor-plates, completes the transverse connection previously made by riveting the reversed angle-irons and the angle-irons at the upper edge of the centre plate to the flat keelson or gutter plate, and the flat keel plate to the frame angle-irons on each side.

An unusual arrangement proposed by Mr. Mackrow is shown in section and elevation in Fig. 74. It is designed for a very flat bottomed vessel with a flat-plate keel and a continuous centre plate run up so as to form an upright-plate keelson. The floor-plates butt against the centre plate, and are cut away as shown at their lower corners in order to form limber holes. The tie-bar A passes through a score in the centre plate, and the double frame B B is curved at the middle line, and reeved through another score. The two angle-iron bars, together with the vertical angle-irons C, complete the transverse connection, while the scores in the centre plate are some distance from the edges, and thus leave ample longitudinal strength. The object in view in making this arrangement was the avoidance of the great weight of cement which would have been required if the limber holes had been cut above the frame angle-irons, as is very often the case. It need hardly be said that the curving of the tie-bar B B is a disadvantage as regards strength, but is probably justified in special cases by the considerations above mentioned.

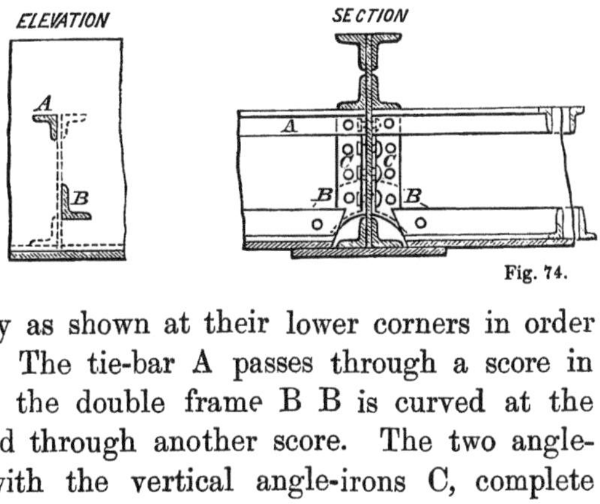

Fig. 74.

The early introduction of the reversed angle-iron into the frame will appear from the remarks previously made on the framing of the first iron vessels, as will also the great advantages resulting from its employment, which have led to its general adoption. It may not, however, be out of place to regard more closely the part played by the reversed frames in resisting alteration in the shape of the transverse sections of a vessel. We may suppose the ship to be cut completely through by two transverse planes of which the distances from a frame equal half the room and space on either side. The section of the plating and deck so cut off may be regarded as forming (with the frame and beam) a hoop, the shape of which tends to alter under the strains resulting from the pressure of the fluid, the weight of the hull and cargo, and other causes. The strength of the frame to resist those strains is measured by the moment of inertia of the section of the frame, reversed frame, and plating made by a plane perpendicular to the ship's surface at any point, such as, for instance, the section at *a b* in Fig. 75. As the centre of gravity, G, of such a section lies very close to the outside plating, it follows that the reversed frame, being the most distant from the centre of gravity of the section and neutral axis of the hoop at that point, is that which is most effective in resisting alteration of form. We may also conclude from this fact, that having provided a sufficiently strong frame angle-iron to receive the fastenings of the outside plating, and of the reversed angle-iron, the most judicious employment of surplus weight in the frame of a ship would be the increase of the size of the reversed frames. In making this statement it is supposed that proper regard has been had to the longitudinal strength of the ship, which is independent of the transverse strength derived from the framing under consideration.

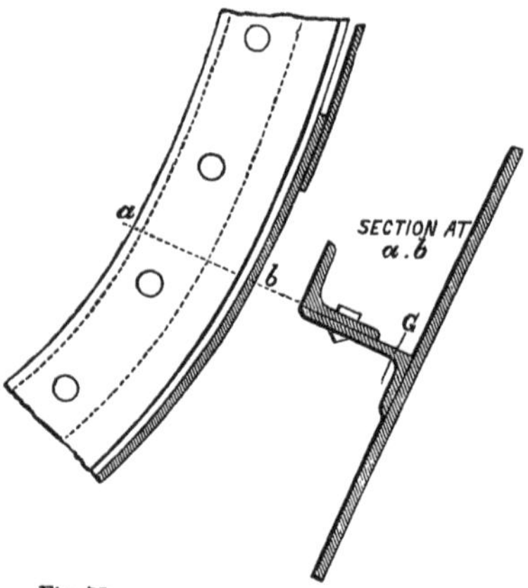

Fig. 75.

Lloyd's Rules provide that all vessels are to have reversed angle-iron riveted to every frame and floor-plate across the middle-line to the height of the upper part of the bilges, and to have double reversed angle-iron in way of all keelsons and stringers in hold; and in addition, all vessels of 300 tons and upwards are to have reversed angle-iron extended from the bilges to the upper deck beam stringer on alternate frames; and vessels of 800 tons and upwards to have reversed angle-iron extended on every frame from the bilges to above the lower deck or hold beam stringer, if the vessel has two decks or tiers of beams, and to above the height of the middle deck beam stringer if the vessel has three decks or tiers of beams. The rivets securing the reversed angle-iron to the frames and floor-plates are not to be spaced more than 8 diameters apart, and the butts of the angle-iron are to be secured with butt-straps.

The Liverpool Rules are similar to Lloyd's in requiring double reversed angle-irons in wake of keelsons and hold stringers, and in addition it is stipulated that reversed frames are to be riveted to every frame, and to be carried to the upper parts of the bilge and gunwale alternately, in vessels under 12 feet in depth of hold; to the upper bilge stringer and the gunwale alternately in vessels with two tiers of beams; and in vessels with three decks to the main and upper deck alternately. Where much closing bevil is required in the short pieces of angle-iron used under keelsons and stringers at the ends of vessels, the pieces requiring closing are to be left out, and the keelsons or stringers are to be fastened to the reversed frames only.

In most of the ships built on the transverse system the framing of the bow and stern is of a similar character to that of the midship part, the only difference requiring notice being that the floor-plates are considerably increased in depth at the bow and stern frames. A very common practice amongst shipbuilders is to fit transverse plate-ties from side to side, about midway between the lower deck beams and the throating of the floors in the fine parts of the vessel. These ties are riveted to the transverse flanges of the frame angle-irons, and strongly connect the two sides of the ship.

Some shipbuilders, however, cant a few of the foremost and aftermost frames so as to make them stand square to the side, or nearly so, as is the case in the corresponding frames of a wooden ship; their object being, not to save material, as is the case there, but to render the angle-iron frames less fatigued by reducing the bevilling of their flanges, and thus preserving their strength unim-

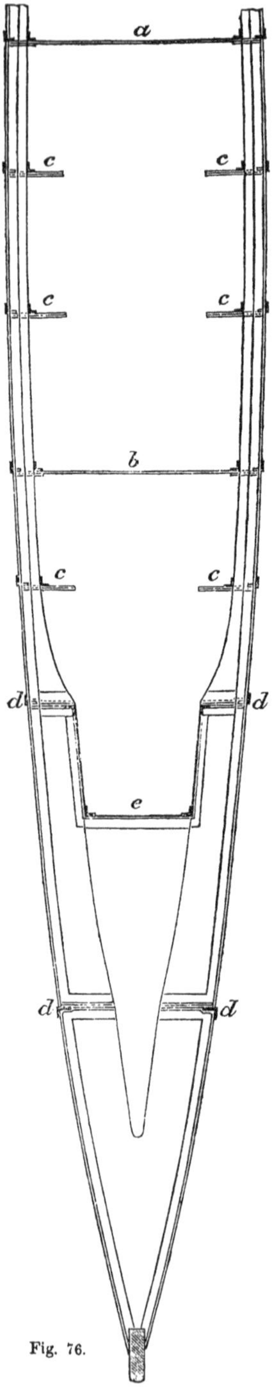

Fig. 76.

paired. The difficulties experienced in framing the deck, and attaching the beams to those frames have, however, rendered the practice of making all the frames transverse almost universal. For it will easily be seen that in places where the frames are considerably canted the beam knees must be flanged in order to fit against them, and thus the transverse tie is somewhat reduced. In H. M. S. 'Bellerophon,' where a considerable number of the bow and stern frames were canted, the full-length beams were dispensed with, and a different mode of framing the deck adopted, the details of which will be described hereafter.

In some ships all the frames are transverse except those at the bow, which are arranged so as to form a series of diagonal breast-hooks. An instance of this is found in the British and North-American mail steamer 'Persia,' built by Messrs. Napier, in which ship the frames are inclined almost square to the stem, and their after ends are secured to the collision bulkhead. The support thus given to the stem and bow, in case of collision, is evidently much greater than that possessed by a ship where the bow is framed transversely, unless a large additional weight of iron is put into her in the form of breasthooks. A proof of the satisfactory nature of the 'Persia's' arrangement is given in the fact, which Mr. Grantham states, that she encountered a small iceberg when at full speed, and split it in two without receiving any injury except that caused by the fragments floating to the paddle-wheels and breaking several floats.

The practice of strengthening the bows of vessels which are transversely framed by working breasthooks formed of continuous or intercostal plates and angle-irons has been already alluded to. These breasthooks are very commonly formed by joining the fore ends of the stringer-plates on the various tiers of beams, and stiffening them for some distance aft by angle-iron stringers on the edges. In most of the vessels in the mercantile marine this is the only means adopted for strengthening the bow longitudinally, but in some of the larger iron ships additional breasthooks have been fitted between the various decks. In ships which have a very fine entrance, the breasthook plates are not run right forward to the stem, but are stopped at transverse plate frames, and their fore ends are connected by transverse horizontal plates. The outer edges of the plates are stiffened and secured to the frames by continuous angle-irons. On the fore side of the frames on which the breasthooks end, the transverse frames are completed across the bow. The primary object of this arrangement is obviously to thoroughly connect the two sides and not to support the stem, the decks and side plating furnishing all the strength that is considered necessary for that purpose. The bows of the new Indian troopships are strengthened by a series of breast-hooks, the arrangement of which is shown in section in Fig. 76. The main deck is marked *a*, the lower deck *b*, and the breast-hooks *c*; while the longitudinal frames which extend forward to the bow are marked *d*, and the platform below the lower deck is marked *e*. The details of a portion of one of the breasthooks *c* are given in plan in Fig. 77, from which it will be seen that the plates are scored in between the frames, and connected with the frames and outside plating by staple angle-irons. An illustration of the arrangement of the breasthooks of the vessels of the 'Northumberland' class is given in Fig. 82, page 108, and will be described further on. It may be remarked here, that one great difference between it and the arrangement first described is that the breasthook

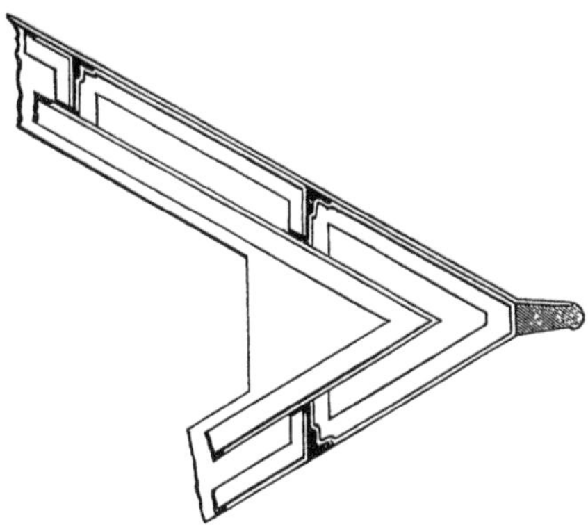

Fig. 77.

plates are fitted in between the frames, and so a direct connection is made between them and the outside plating similar to that obtained in the troop-ships just described, while at the same time the folding-back of the frames in case of the vessel being used as a ram, is effectually prevented.

The employment of diagonal ties on the frames of iron vessels has been proposed by many individuals, and patents have been taken out for various modes of fitting them. In a wooden ship the necessity for their adoption is evident, as they serve to prevent longitudinal bending and vertical racking in the structure, and especially to resist the change in the relative positions of the planks and frames, which the former kind of straining tends to produce. But in an iron ship which has the plates of the skin riveted to each other, so as to form an almost perfectly united mass, there is no possibility of the sliding of edge on edge which would take place in the skin of a wooden ship that had no diagonal strengtheners. Another objection to the use of such ties in an iron ship is the additional work which their use entails. For these reasons they have never come into general use.

The provisions made to ensure longitudinal strength in a vessel framed on the transverse system, might, with propriety, be included here; but as they have already been described, in part, while treating of keels and keelsons, and will be further spoken of while describing the modes of framing decks and the details of plating, they are only alluded to here in order to note the extreme importance of keeping up longitudinal connections, failures in which respect, combined with ill-designed strengtheners, have caused many of the weaknesses which were mentioned in the commencement of this work.

In Plates 1 and 2 are given perspective views of a portion of the amidship framing of the steam-ship 'Queen,' built by Messrs. Laird, and of the Atlantic mail steamer 'China,' built by Messrs. Napier, which ships may be taken as instances of well-built vessels framed on the transverse system.

The 'Queen' is 400 feet long and 42 feet in extreme breadth, and in consequence of the large proportion which the length bears to the breadth, special arrangements are made to give longitudinal strength, as will be seen on reference to Plate 1. The dimensions of the various parts of the framing are as follows :—

The "Queen"

Plate. 1.

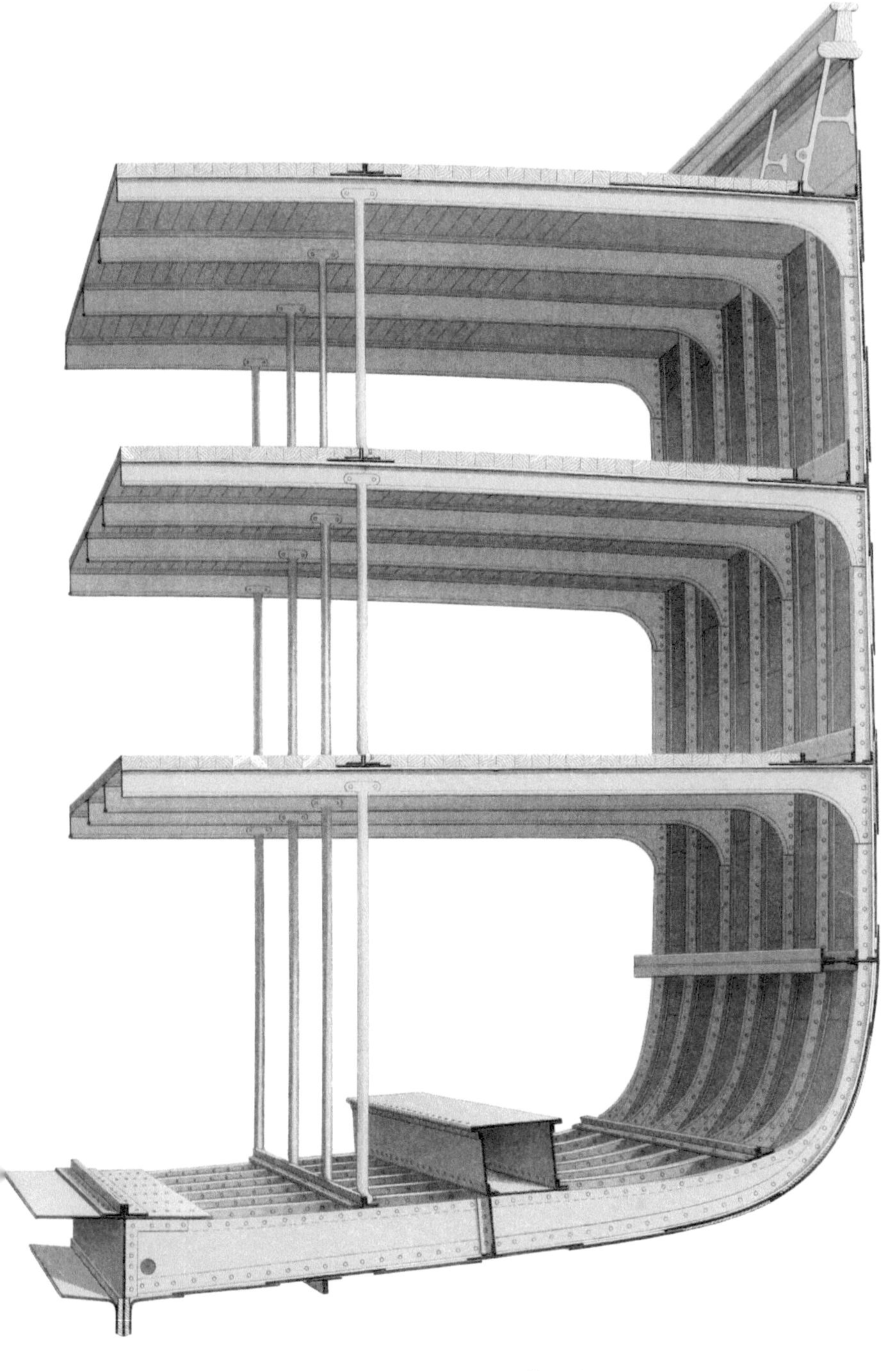

Drawn by J. Maxton. John Murray, Albemarle Street, Novr 1868. Engraved by J. W. Lowry

Angle-irons forming frames	$5\frac{1}{2}$ by $3\frac{1}{2}$ by $\frac{5}{8}$ inches.
Angle-irons forming reversed frames	4 by $3\frac{1}{2}$ by $\frac{9}{16}$,,
Angle-irons connecting the floors to continuous centre plate	$3\frac{1}{2}$ by $3\frac{1}{2}$ by $\frac{5}{8}$,,
Angle-irons connecting the floors to intercostal plates	$3\frac{1}{2}$ by $3\frac{1}{2}$ by $\frac{1}{2}$,,
Angle-irons forming stringers on tie-plates to decks Angle-irons connecting gutter-plates to centre plate Angle-irons forming stringer on turn of bilge .. Angle-irons at heels of pillars .. .. Angle-irons forming bilge-keel for 210 feet ..	6 by 4 by $\frac{5}{8}$,,
Angle-irons worked on inner edges of deck-stringers	5 by 3 by $\frac{9}{16}$,,
Angle-irons worked on outer edges of deck-stringers	4 by 4 by $\frac{5}{8}$,,
Angle-irons worked on inner edge of I-shaped hold-stringer	4 by 4 by $\frac{9}{16}$,,
Angle-irons worked on frames at I-shaped hold-stringer	3 by 3 by $\frac{1}{2}$,,
Angle-irons worked on intercostals at I-shaped hold-stringer	$3\frac{1}{2}$ by $3\frac{1}{2}$ by $\frac{9}{16}$,,
Continuous centre plate	44 by $\frac{7}{8}$ inches.
Side-bars	12 by 1 ,,
Floor-plates at middle line	27 by $\frac{9}{16}$,,
Gutter-plates	30 by $\frac{5}{8}$,,
Plating in sides of box-girder on the floors	24 by $\frac{1}{2}$,,
Plating in top of box-girder on the floors	31 by $\frac{1}{2}$,,
Plating in bottom of box-girder on the floors	24 by $\frac{1}{2}$,,
Intercostal plates at box-girder on the floors	19 by $\frac{1}{2}$,,
Plate in I-shaped hold-stringer	12 by $\frac{1}{2}$,,
Upper-deck plating at sides	84 by $\frac{5}{8}$,,
Upper-deck stringer-plate	24 by $\frac{5}{8}$,,
Upper-deck tie-plate	36 by $\frac{5}{8}$,,
Middle-deck stringer-plate	36 by $\frac{5}{8}$,,
Middle-deck vertical stringer, or clamp	15 by $\frac{5}{8}$,,
Middle-deck tie-plate	18 by $\frac{5}{8}$,,
Lower-deck stringer-plate	30 by $\frac{5}{8}$,,
Lower-deck vertical stringer, or clamp	15 by $\frac{5}{8}$,,
Lower-deck tie-plate	18 by $\frac{5}{8}$,,

All deck beams, except those at principal hatchways, 10 by $6\frac{1}{2}$ by $\frac{9}{16}$ inches bulb T-iron.

The garboard-strakes are $\frac{3}{4}$ inch thick forward and $\frac{7}{8}$ inch aft; the general thickness of bottom plating is $\frac{5}{8}$, and the sheer-strake is $\frac{9}{16}$ inch thick, and is doubled by a $\frac{9}{16}$-inch plate worked inside it throughout the entire length. In addition to this doubling plate others are worked throughout the length on the second strake below the sheer strake, and on the strake in wake of the beam-ends on the lower deck, their thickness being $\frac{9}{16}$ inch. On the turn of the bilge two sunken strakes and one outer strake are doubled for two-thirds of the length with $\frac{5}{8}$-inch plating on the plan previously described. All this, as has been stated, is done to give longitudinal strength, and this object is further attained by the working of box and I-shaped hold stringers, combined with intercostal plates; and by the bilge keel outside, which extends from 100 feet abaft the midship section to 110 feet before it. The deck stringer and tie-plates are also specially stiffened by the angle-irons worked on them, and a great addition of longitudinal stiffness is given to the ship by the partial iron upper deck, or very broad stringer, by the vertical stringers on the two lower tiers of beam-ends, and by a box stringer at the ship's side on the amidship part of the lower deck. These stringer arrangements will be described hereafter. The transverse connection of the floors of this ship, and the whole arrangement of keel-bars have been previously illustrated. It will be seen that a very efficient arrangement of pillars is adopted. Those heeling on the floors and lowest tie-plate are $3\frac{1}{2}$ inches in diameter and the rest 3 inches. Their heads give direct support to the beams, and are also secured to the beam-plates by two $\frac{7}{8}$-inch rivets passing through lugs forged on them. Their heels have palms formed on them which fit against the vertical flanges of the angle-irons on the tie-plates and floors, and are riveted to them. The details of the topside fittings will be described further on.

The 'China' is 323 feet long and 40 feet 4 inches broad, her tonnage B. M. being 2575. The principal dimensions of her framing are as follows:—

Angle-irons forming frames	6 by 4 by $\frac{5}{8}$ inches.
Angle-irons forming reversed frames	4 by 3 by $\frac{1}{2}$,,
Angle-irons connecting floors to intercostal keelsons	4 by 4 by $\frac{1}{2}$,,
Angle-irons connecting gutter-plate to intercostal keelson	8 by 4 by $\frac{5}{8}$,,
Angle-irons on upper edge of floors at the lower side-keelson	6 by 4 by $\frac{5}{8}$,,
Angle-irons on floors at upper side-keelson .. .. Angle-irons on lower-deck stringer-plate	5 by 4 by $\frac{1}{2}$,,
Angle-irons on middle-deck stringer-plate .. .. Angle-irons on upper-deck stringer-plate	5 by 4 by $\frac{5}{8}$,,
Angle-irons on vertical stringer or clamp	5 by 3 by $\frac{1}{2}$,,
Angle-irons on outer edge of upper-deck stringer ..	4 by 4 by $\frac{5}{8}$,,
Angle-irons on upper edge of upper and lower deck beams	$3\frac{1}{2}$ by 3 by $\frac{7}{16}$,,
Angle-irons on upper edge of middle-deck beams ..	4 by 3 by $\frac{1}{2}$,,
Floor plates at middle line—amidships	29 by $\frac{5}{8}$ inches.
Floor-plates at middle line—forward and aft	29 by $\frac{1}{2}$,,
Gutter-plates	14 by $\frac{5}{8}$,,
Lower-deck stringer-plate	24 by $\frac{9}{16}$,,
Middle-deck stringer-plate	36 by $\frac{5}{8}$,,
Middle-deck tie-plate	24 by $\frac{1}{2}$,,
Vertical stringer or clamp between upper and middle decks	24 by $\frac{5}{8}$,,
Upper-deck stringer-plate	42 by $\frac{5}{8}$,,
Upper-deck tie-plate	24 by $\frac{9}{16}$,,
Upper and lower deck beam-plates (bulb-iron)	8 by $\frac{5}{8}$,,
Middle-deck beam-plates (bulb-iron)	10 by $\frac{5}{8}$,,

The "China"

Plate 2.

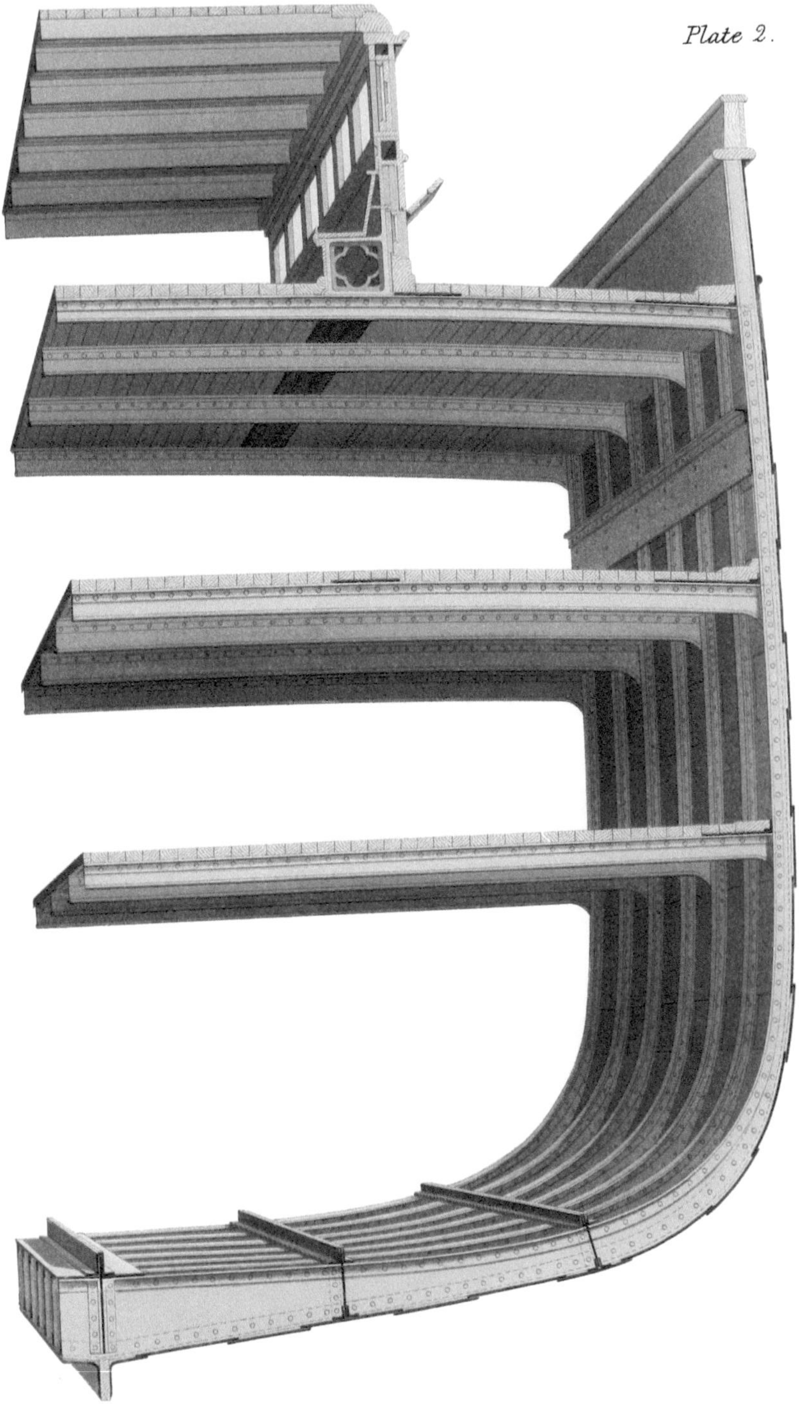

Drawn by J. Maxton. John Murray, Albemarle Street, Novr 1868. Engraved by J.W. Lowry

Forward and aft the frame angle-irons are reduced to 6 by 4 by $\frac{1}{2}$ inches, and in the engine and boiler space the reversed angle-irons are worked double and are $4\frac{1}{2}$ by $3\frac{1}{2}$ by $\frac{1}{2}$. The keel arrangements have been previously described, and it will be remarked that the side intercostal plates, which are $\frac{5}{8}$ inch thick, are secured to the bottom plating by double angle-irons 4 by 4 by $\frac{1}{2}$ inches. The importance of this mode of connection has been already alluded to, though many shipbuilders neglect it. The stringer plates are not connected to the outside plating on the lower and middle decks, but on the upper deck the stringer is run out between the frames and connected to the sheer-strake. The vertical stringer or clamp between the upper and middle decks is worked in order to give longitudinal strength. The garboard-strakes of the 'China' are 1 inch thick; the next strake is $\frac{14}{16}$ inch; the next eight strakes $\frac{13}{16}$ inch; the next five strakes $\frac{12}{16}$ inch; and the remainder, up to the sheer-strake, $\frac{10}{16}$ inch. The sheer-strake is $\frac{14}{16}$ inch thick, and has a doubling strake $\frac{10}{16}$ inch thick worked on it. The topside plating is $\frac{3}{8}$ inch thick, and it is worked directly on the frames which are run up about 4 feet above the upper deck. The deck house extends the whole length of the ship, and is framed in the manner shown in the section, the wood beams of the spar deck being 7 by 7 inches, and the planking $2\frac{3}{4}$ inches thick.

The longitudinal system of framing practised by Mr. Scott Russell, was described by him in a paper read at the meeting of the Institution of Naval Architects, in May, 1862, from which paper the following particulars are taken.* The principal arrangements of this system are thus stated by the author:—

1. " To divide the ship by as many transverse watertight iron " bulkheads as the practical use of the ship will admit. I like to " have at least one bulkhead for every breadth of the ship in her " length. In a ship eight breadths to her length, I wish to have " at least eight transverse bulkheads.

2. " I have between these bulkheads, what I call partial bulk- " heads, or the outer rim of a complete bulkhead, with the centre

* The author has since republished this description in his large work on 'Naval Architecture,' where detailed illustrative drawings of longitudinally framed ships are given.

"part omitted, so as to form a kind of continuous girder running transversely all round the ship, and not interfering with stowage.

3. "I run from bulkhead to bulkhead, longitudinal iron beams or stringers, one along the centre of every plate of the skin, so giving each strake of plates the continuous strength of an iron beam, one portion placed at right angles to another. This longitudinal forms one continuous scarph across all the butt joints of the plates, hitherto their weakest part; and adds also to the strength of the rivets of the joint, the help of a line of rivets and angle-irons along the centre of the plate. These longitudinals and the skin are therefore one.

4. "What remains over after this is done, of the superfluous iron formerly used in ribs, I make into a continuous iron deck, mainly carried by the bulkheads and by longitudinals under it; and I believe this iron is infinitely better applied in a deck than in ribs fastened to the skin."

The following table gives an idea of the difference between the new and the old systems, and is taken from ships of 600 to 700 tons scale, built on both systems.

WEIGHTS OF IRON USED IN THE GENERAL STRUCTURE.

	Old System.	New System.
In the skin	110 tons.	110 tons.
In transverse internal strengthening	130 ,,	40 ,,
In longitudinal strengthening	40 ,,	130 ,,

Mr. Russell goes on to say, that from the proportions of transverse and longitudinal strengthening in the old system, it might be supposed that the strains tending to break a ship athwartship, were greater than the longitudinal strains tending to break her through the middle; but it is notorious that the contrary is the fact, and that a ship framed on the old system is most strained and weakened lengthways. When the longitudinal system is adopted, the relative longitudinal breaking strengths obtained, as compared with that given by the transverse system (equal weights of iron being used), are in the ratio of 5 : 4.

The following description of the 'Annette,' iron auxiliary screw clipper, of the 600-ton scale, 845 ton B. O. M. classed A 1 12 years at Lloyd's, is given by him as a further illustration of the system under consideration.

COMPARISON OF 'ANNETTE,' AS BUILT ON THE LONGITUDINAL SYSTEM, WITH A SIMILAR VESSEL BUILT ON THE TRANSVERSE SYSTEM.

LONGITUDINAL SYSTEM.

Weight of the iron hull of a ship built on the 600-ton grade longitudinal principle, to class A 1 for 12 years at Lloyd's:—

Dimensions of Ship.

	Feet.	Inches.
Length	190	0
Breadth	30	0
Depth at side	18	0
Depth in hold	18	3

Tonnage, N.M., under 700 tons.

Specification.

Outer plating.
Keel-strake, $\frac{11}{16}$ inch.
From keel-strake to upper part of bilge, $\frac{10}{16}$ inch.
From upper part of bilge to sheer-strake, $\frac{9}{16}$ inch.
Sheer-strake, $\frac{10}{16}$ inch.
All longitudinal seams single riveted.
Butts of sheer-strake double ditto.
Five bottom webs, 18 inches deep by $\frac{1}{2}$-inch bars = 18 lbs. per square foot.
Twelve side webs, 13 inches deep by $\frac{1}{2}$ inch.
Angle-iron, $3\frac{1}{2}$ by $3\frac{1}{2}$ by $\frac{1}{2}$ inches, and 3 by 3 by $\frac{1}{2}$ inches.
Iron deck, $\frac{1}{4}$ inch thick; shelf, 2 feet 6 inches wide and $\frac{1}{2}$ inch thick.
Bulkheads, five watertight, $\frac{3}{4}$ inch, stiffened with 3 by 3 by $\frac{1}{2}$ inches angle-irons.
Ten partial bulkheads, 13 feet apart.

TRANSVERSE SYSTEM.

Weight of the iron hull of a ship built under Lloyd's Rules for the 600-ton grade, to class A 1 for 12 years:—

Dimensions of Ship.

	Feet.	Inches.
Length	190	0
Breadth	30	0
Depth at side	18	0
Depth in hold	17	0

Tonnage, N.M., under 700 tons.

Specification.

Keel, 7 by $2\frac{3}{4}$ inches.
Garboard-strake, $\frac{11}{16}$ inch, longitudinal seams double riveted.
From garboard to upper part of bilge, $\frac{10}{16}$ inch, ditto ditto.
From upper part of bilge to sheer-strake, $\frac{9}{16}$ inch; longitudinal seams single riveted.
Sheer-strake, $\frac{10}{16}$ inch, upper edge double riveted.
All butts double ditto.
Frames, 4 by 3 by $\frac{7}{16}$ inches, spaced 18 inches from moulding-edge to moulding-edge.
Reversed angle-iron, 3 by $2\frac{3}{4}$ by $\frac{6}{16}$ inches.
Floors, 18 by $\frac{9}{16}$ inches.
Deck beams bulb-iron, 7 by $\frac{7}{16}$ inches, with double angle-irons on top edge.
Centre keelson intercostal, 18 by $\frac{7}{16}$ inches, with double angle-irons.
Flat plate on top of centre keelson, 21 by $\frac{1}{2}$ inches.
Side keelsons, 12 by $\frac{7}{16}$ inches, with double angle-iron.
Bilge-stringers, double angle-irons, 5 by 3 by $\frac{1}{2}$ inches, back to back.
Hold ditto ditto ditto ditto.
Upper-deck shelf, 24 by $\frac{9}{16}$ inches.
Angle-iron, to connect ditto to gunwale, 5 by 3 by $\frac{1}{2}$ inches.
Two deck-stringers, 12 by $\frac{1}{2}$ inches.
Three watertight bulkheads, $\frac{3}{8}$ inch, stiffened with angle-iron 3 by 3 by $\frac{1}{2}$ inches.

Comparison of 'Annette,' as built on the Longitudinal System, &c.—*continued.*

LONGITUDINAL SYSTEM.
Minimum Section of Metal Midship Section.

	Total.	Rivet-holes.	Effective.
	sq. in.	sq. in.	sq. in.
Keel-plate	28·29	2·4	25·89
From keel plate to upper part of bilge and sheer-strake	251·6	17·0	234·6
From upper part of bilge to sheer-strake	182·4	12·0	170·4
Seventeen webs	221·87	19·2	202·67
Iron deck	82·8	3·5	79·3
T-irons under deck	28·0	5·0	23·0
Deck-webs	24·0	5·0	19·0
Gunwale angle-iron	8·0	2·25	5·75
Extra shelf	15·0	2·5	12·5
	841·96	68·85	773·11
Deduct for 7 butts single riveted, 2 ditto double ditto, butts of iron deck single ditto			124·0
Total square inches			649·11

TRANSVERSE SYSTEM.
Minimum Section of Metal Midship Section.

	Total.	Rivet-holes.	Effective.
	sq. in.	sq. in.	sq. in.
Garboard-strakes	31·6	5·6	26·0
From garboard to upper part of bilge ..	208·0	15·0	193·0
From upper part of bilge to sheer-strake	180·0	6·8	173·2
Sheer-strake	52·0	2·8	49·2
Centre keelson	10·13	0·75	9·38
Angle-iron for ditto	13·5	1·5	12·0
Side keelsons and angle-irons	40·5	4·5	36·0
Bilge-stringers	14·0	2·25	11·75
Hold ditto	14·0	2·25	11·75
Upper-deck shelf	27·0	0·75	26·25
Angle-iron for ditto	7·0	1·25	5·75
Deck stringers	12·0	1·25	10·75
Flat-plate on centre keelson	10·5	1·5	9·0
	620·23	46·2	574·03
Deduct for 9 butts double riveted, calculated at 75			54·0
Total square inches			520·03

If this comparison be correct, the ultimate measures of the strengths of the ships to resist a strain tending to hog or sag, or break them across, is in round numbers as 650 to 520, or as 5 : 4. The gain in longitudinal strength by the adoption of the new system, is thus about 25 per cent.

In a paper by B. Jensen, Esq., of Dantzic, on the "Comparative Merits of the Longitudinal and Vertical Systems of Iron Shipbuilding," it is stated that in a vessel of 600 tons O. M., the gain in weight on a similar ship built according to Lloyd's Rules, was 7 per cent., and the gain in strength 10 per cent.; and that if the plating had been made as thick as is required by Lloyd's for a vessel of the same tonnage, the gain in weight would have been 3·6 per cent., and the gain in strength 18 per cent. on a ship intended to occupy the highest class A at Lloyd's.

The 'Great Eastern' is another instance of the adoption of the longitudinal system of framing, combined with a cellular double bottom and upper deck. The longitudinal frames in this ship are 34 inches deep and ½ inch thick, with single angle-irons 4½ by 4½ by ½ inches on both edges. Up to the height of 36 feet, to which the double bottom extends, the longitudinals are placed about 5 feet apart, and the plating is so arranged, that a longitudinal comes in the centre of every alternate strake. On the bottom where the weight of the ship is taken when she grounds, the longitudinals are about 2 feet 6 inches apart, and above the double bottom the distance between them is about 8 feet, and they are arranged so as to form stringers on the various decks. A similar system of transverse and partial bulkheads to that just described is adopted in this vessel, and in order to strengthen the structure above the double bottom, the partial bulkheads are increased in number. In addition to this transverse and longitudinal framing, there are longitudinal, vertical bulkheads, which rise from the inner bottom to the upper deck, and form the sides of the engine and boiler spaces.

Among the advantages claimed for the longitudinal system by Mr. Russell and other shipbuilders, are the increased strength and simplicity of the bow and stern framing; the distribution of strength from the transverse bulkheads by means of the longitudinal frames, and a consequent increase in local strength generally; the support and connection of the plating of the bottom due to the longitudinal stringers on each strake crossing the butts, and the rivets connecting the frames to the plating acting as joint-rivets at the weak points; the convenience for stowing cargo, and for cleaning and painting the inside of the ship; the prevention of the wash of bilge-water with the coals, dirt, and *débris* of every kind, which it carries with it when the ship rolls, and the consequent wearing down of the rivet heads, and laps of plating; and the facilities for making the frames in place, and bending the angle-irons to the easier curves required by the longitudinal system, thus effecting a saving in time, materials, and workmanship.

It is evident that there is a great addition of strength in the bow framing of a ship which is built on the longitudinal system, over that of one which is transversely framed. For in the latter, in case of collision, the transverse frames do not resist the compression or collapse of the bow, especially in very fine ships; while in the former the longitudinal frames become close and numerous at the bow, and form a series of breasthooks, which lie in the line of greatest strain, and are placed in the best way for strengthening the bow. This capability of the bow to withstand collision is valuable in the mercantile marine, and much more so in war vessels built for running-down purposes, as will be shown hereafter when illustrating the arrangements of some ram-bows. In addition to these advantages, which are gained by framing the bow longitudinally, there is the further gain in the simplicity of the workmanship, and the time required to perform it. For in framing the bow of a long fine ship on the vertical system, it is found that the work is extremely difficult to execute, and that it is so circumstanced as to allow only a few men to be employed upon it, and those men work at the greatest disadvantage, and consequently cannot build quickly. This is a very important point, because if it were required to build an iron navy rapidly, no amount of pecuniary compensation could enable the builders to frame the bows of such ships quickly. It may be mentioned, in illustration of this statement, that while in the ships of the 'Northumberland' class, the bows of which were framed on the vertical system, a considerable time was taken in completing the framing forward, in the 'Bellerophon,' where the principal framing of the bow is longitudinal, it was completed and put in place in a fortnight.

The stern framing of ships built on the longitudinal system, is in some longitudinal, in some transverse, and in others a combination of the two. Whenever the longitudinal frames can be continued aft, they are well adapted to resist the vibration caused by the screw propeller, and in this respect, also, they are superior to the transverse frames. An instance of transverse framing in the after part, is, however, given in the vessel described by Mr. Jensen, and before alluded to. This vessel's longitudinal frames ended at the after engine-room bulkhead, and a space of 31 feet 6 inches on the after side was framed with 4 by 3 by $\frac{7}{16}$ inches ribs, pitched 21 inches apart. The details of this vessel will be found in the *Transactions* of the Institution of Naval Architects for 1865.

In the 'Great Eastern' the after part of the ship abaft the stuffing-box bulkhead is strengthened by a large number of horizontal flats extending between the bulkhead and a cast-iron cellular stern-post. Above the screw aperture, and the upper iron flat, the framing of the stern is completed by vertical frames, the aftermost ones being canted. The longitudinal frames are ended at various stations, three being terminated at the aftermost bulkhead, five at the bulkhead next before it, three in the compartment between those bulkheads, and the remainder at the third transverse bulkhead from aft. Particulars of the arrangements will be found in Mr. Russell's great work on 'The Modern System of Naval Architecture.'

It has been objected to the longitudinal system of framing, that a greater space of unsupported bottom plating is left between the frames than is the case in the vertical system. But it has been stated, in reply, that in case a vessel with transverse frames strikes on a rock, those transverse frames become immediately the most certain agents of destruction to the bottom of the ship; while in the longitudinal system, especially when a double bottom and inner skin are adopted, the weakness existing is precisely what is wanted, for it allows the plates between the longitudinals to be indented, or torn through, without the general structure of the ship becoming injured. A case in point is that of the 'Great Eastern,' which ran ashore on the rocks on her voyage to America, and though she had nine holes torn in her bottom, one of which was 85 feet long, by 4 or 5 feet wide, yet she continued her voyage to New York in safety, as the inner skin was not penetrated.

It has been suggested by some shipbuilders that the framing of that part of a ship which is near the middle of her depth, might, with advantage, be made vertical, even when her bottom and upper part are framed longitudinally. This is especially the case in wall-sided vessels, where the vertical strains are very considerable; but it cannot be doubted that in all vessels it is necessary to provide against these vertical strains. When the longitudinal system in its entirety is carried out, the only provisions made to give the requisite vertical strength, are the use of partial bulkheads intermediate between the complete transverse bulkheads, and the strength gained by attaching the beam-ends to the sides, when, as is sometimes the case, the deck is framed transversely. Mr. Russell, however, while denying the necessity of additional vertical strength to that thus obtained, in vessels built for ordinary purposes, admits that in exceptional cases, as, for instance, underneath a heavy gun or other concentrated weight, the number of the partial bulkheads should be increased, in order to take the vertical strains, and keep the longitudinals from buckling.

Other objections made to the longitudinal system, are the inconveniences experienced in securing the bulkheads and completing the internal fittings, which are felt by those who have been accustomed to build vessels on the transverse system; but to such objections the closing remarks of Mr. Russell in the paper before quoted, may be fairly applied:—"It may be considered a disad-"vantage of the longitudinal system that somewhat greater skill is "required in its design, greater intelligence in its construction, and "greater accuracy and excellence in its workmanship, blunders "made are less easily remedied, want of forethought in the beginning "is less easily compensated by afterthought, and blundering execu-"tion will make a mess of it; but I trust that the growing intelli-"gence among shipbuilders, the growing science among naval "architects, better information among shipowners, greater know-"ledge in ship captains, and better training among workmen, will "bring us to a point of design and execution in this country, such "that we shall never be prevented from preferring the better to "the worse, for want of science, forethought, and skill."

In the construction of the 'Sentinel,' a vessel designed by Mr. Spencer, and built by Messrs. Palmer, of Jarrow-on-Tyne, the framing out to the lower part of the bilge was made longitudinal, and above that height, transverse. The double bottom was 21 inches deep at the middle line, and 15 inches at the upper longitudinal frame. The spaces enclosed by the two bottoms and the longitudinal frames, were used for water ballast, provision for which was the principal object of the construction. The partial floor-plate was secured to the outer longitudinal frame by the double reversed angle-iron being bent down so as to fit against it. The arrangement of the frames and reversed frames in the upper part of the ship was similar to that employed in the common transverse framing. In a vessel named the 'Rouen,' designed by Mr. McIntyre previous to the building of the 'Sentinel,' the transverse framing was completed in the usual manner, and fore and aft keelsons were worked on top of the floors. Every other keelson was run down between the floors and riveted to the skin plating, and an inner bottom being worked above the keelsons, completed the space for water ballast. The arrangement of the 'Sentinel's' frame gave more room for stowage of cargo and less for water ballast than that of the 'Rouen,' while, as was said before, all the transverse framing of the bottom was dispensed with. The foregoing description of the 'Sentinel' has been given here on account of the fact that the longitudinal and transverse systems are both represented in her construction.

CHAPTER VI.

COMBINED TRANSVERSE AND LONGITUDINAL SYSTEM OF FRAMING.

FRAMING OF 'WARRIOR,' 'NORTHUMBERLAND,' &c.

IN all the iron-built armour-clad ships of the Royal Navy, and in the new Indian-troop ships, the system of framing adopted combines both of those previously described. In this chapter we shall illustrate the arrangements of the framing in the 'Warrior' and 'Northumberland,' which ships may be taken as examples of the earlier iron-clads. The 'Warrior's' midship section is given in Plate 3, which also shows a perspective view of a portion of her framing. The details of the keel arrangements have been described already, and illustrated by Fig. 32, page 31. It will be seen from the midship section, that from the armour-shelf to the keel the framing is made up of six longitudinal frames, and of short transverse plate-frames fitted between them, the athwartship connection being kept up by continuous transverse frames, made up of plates and angle-irons, in the manner shown in the section in Fig. 32. These frames pass through scores cut in the upper edges of Nos. 3 and 6 longitudinal frames, and all the other longitudinals are of such a depth as to allow the continuous transverse frames to pass above their inner edges; and consequently the angle-irons on their inner edges are continuous, while on Nos. 3 and 6 longitudinals they are worked in short lengths between the transverse frames. The following are the scantlings of the longitudinal frames:—

	Plate.	Angle-irons.	
		Inner Edge.	Outer Edge.
	inches.	inches.	inches.
No. 1 Longitudinal, or Shelf-plate ..	23 by $\frac{7}{8}$	5 by 5 by $\frac{5}{8}$	6 by 6 by $\frac{3}{4}$
No. 2 ,, ...	21 by $\frac{1}{2}$	$3\frac{1}{2}$ by 3 by $\frac{9}{16}$	4 by $3\frac{1}{2}$ by $\frac{5}{8}$
No. 3 ,, ...	27 by $\frac{5}{8}$	ditto	ditto
No. 4 ,, ...	19 by $\frac{1}{2}$	ditto	ditto
No. 5 ,, ...	19 by $\frac{1}{2}$	ditto	ditto
No. 6 ,, ...	27 by $\frac{5}{8}$	4 by $3\frac{1}{2}$ by $\frac{9}{16}$	ditto

The "Warriors."

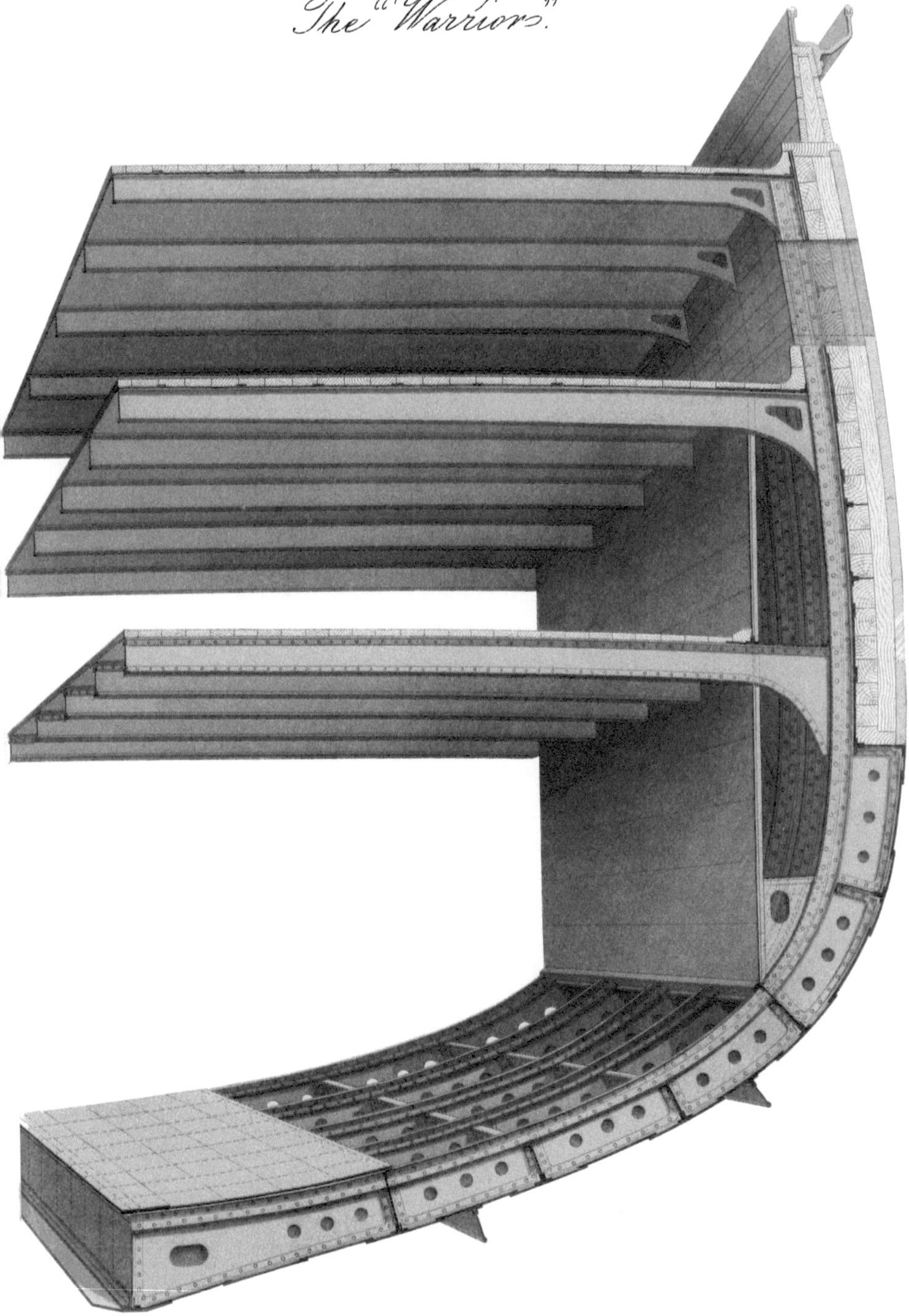

Drawn by J. Maxton. John Murray, Albemarle Street, Novr 1868. Engraved by J. W. Lowry

The angle-irons on the outer edges of all the longitudinals, except that forming the armour shelf, are double; those on the inner edges are single. The longitudinal which forms the armour-shelf is a flanged plate, or very large angle-iron, the horizontal flange being 23 inches in width, and the vertical flange 17 inches. The details of the manufacture and mode of bending this plate will be described hereafter. The longitudinal frames extend for a length of from 280 to 320 feet, and are terminated at transverse bulkheads. The fore and after parts of the ship are framed transversely, as will hereafter be explained with reference to the 'Northumberland.' The specification of this ship states that "the plates and angle-irons "composing the longitudinal frames are to be wrought in the "greatest lengths procurable, and great care is to be taken that the "butts are well fitted; the angle-irons, if required, as well as the "plates, to pass through the bulkheads, the surrounding joints "being carefully caulked and made watertight; the butts of the "plates to be secured by double butt-straps of $\frac{3}{8}$-inch plate, double "riveted; the butts of the angle-irons are also to be supported by "straps or covering angle-irons. The longitudinal frames must be "placed in such a direction throughout their length as to clear all "the longitudinal joints of the outside plates, and fillings or liners "between the angle-irons and the bottom plates are to be dispensed "with, the angle-irons being bent so as to pass over the butt-"straps."

It will be seen from Plate 3, that two continuous girders, forming a kind of external framing, are worked longitudinally on the skin-plating behind armour, between the heights of the main and lower decks, and another similar girder is worked in short lengths between the ports. The dimensions of these girders are 10 by $\frac{5}{8}$ inches plate, with double angle-irons $4\frac{1}{2}$ by $3\frac{1}{2}$ by $\frac{5}{8}$ inches on the inner edge. Intercostal plates and angle-irons are also worked just above and below the ports, and riveted to the frames and skin-plating.

As far as the longitudinal frames extend the spacing of the continuous transverse frames is usually 3 feet 8 inches, but before and abaft their termination it is 22 inches. In wake of the armour-plating the continuous frames are made up of 10 by $\frac{1}{2}$ inches plates, with an angle-iron $3\frac{1}{2}$ by $3\frac{1}{2}$ by $\frac{5}{8}$ inches on the inner edge, and double angle-irons $3\frac{1}{2}$ by $4\frac{1}{2}$ by $\frac{5}{8}$ inches on the outer edge. The plate tapers gradually to a moulding of 7 inches, and runs, with the angle-iron on the inner edge, from gunwale to gunwale; while the angle-irons on the outer edge only run down close to No. 3 longitudinal. Below this point another angle-iron is worked on the inner edge of the continuous transverse frames, as shown in the section in Fig. 32. The butts of the plates and angle-irons in these frames are not nearer than 6 feet to the middle line, and they are properly shifted and butt-strapped.

An intermediate frame is fitted between every pair of regular frames behind the armour-plating, and is run down from the upper deck to the third longitudinal. The scantlings of these intermediate frames are identical with those of the frames behind armour, except that the plates are $\frac{7}{16}$ inch thick. Plates, similar to those worked at the frames, are fitted between the second and third longitudinals, in order to secure the heels of the intermediate frames.

The short transverse plates, which, with the angle-irons on their edges, form the transverse framing in wake of the longitudinals, are $\frac{1}{2}$ inch thick, except those at the watertight bulkheads, and the floor-plates, which are both $\frac{5}{8}$ inch thick. These plate-frames are fitted between the longitudinals, and secured to them by the double frame angle-irons, 4 by $3\frac{1}{2}$ by $\frac{9}{16}$ inches, worked *staple* fashion. All the plates, with the exception of those at the watertight bulkheads, are lightened, as shown in the section in Plate 3, and their inner edges overlap the continuous transverse frames about 3 inches and are riveted to them. Intermediate floor-plates are fitted in this ship, and extend from the vertical keelson-plate to No. 6 longitudinal, which is about 10 feet 6 inches from the middle line. These floor-plates have a reversed angle-iron, 7 by $3\frac{1}{2}$ by $\frac{5}{8}$ inches, on one side of their upper edge, extending in one length right across the keel, and a single angle-iron, 4 by $3\frac{1}{2}$ by $\frac{5}{8}$ inches, on their lower edge.

The armour-plating of this ship only extends over a portion of her length (213 feet), and the unprotected parts, forward and aft, are framed differently from the protected midship part. For 30 feet before and abaft the armour-plates, the continuous frames are formed of 7 by 4 by $\frac{5}{8}$ inches angle-irons, and the reversed frames are $3\frac{1}{2}$ by 3 by $\frac{9}{16}$ inches; for the next 30 feet, before and abaft this, the frames are 6 by 4 by $\frac{9}{16}$ inches, and the reversed frames $3\frac{1}{2}$ by 3 by $\frac{9}{16}$ inches; and beyond this, the frames are 5 by 4 by $\frac{9}{16}$ inches, and the reversed frames $3\frac{1}{2}$ by 3 by $\frac{1}{2}$ inches. A floor-plate is attached to each transverse frame, and has a reversed angle-iron worked on its upper edge.

It will be remarked that this ship has a longitudinal vertical bulkhead at about 3 feet from the inside of the frames, which extends from the main deck down to the third longitudinal. The plating of this wing passage bulkhead is $\frac{5}{8}$ inch thick at the lower part, $\frac{1}{2}$ inch thick at the lower deck, and $\frac{7}{16}$ inch thick between decks, and has vertical stiffeners $3\frac{1}{2}$ by 3 by $\frac{1}{2}$ inches, worked at intervals of 22 inches. The upper edge of the bulkhead is effectually secured to the underside of the plating on the main deck, and the lower edge to the inner angle-iron on the third longitudinal. This longitudinal frame is carefully made watertight where the continuous frames pass through it, and the plating of the bulkhead is also carefully worked and caulked, so that the space thus enclosed shall form a longitudinal watertight compartment. The compartment thus formed is further subdivided, by transverse bulkheads and partial bulkheads, and by the lower-deck stringer plate; and admission is obtained to each division by watertight doors and man-holes. Sluice valves and cocks are also fitted to each compartment. The details of the watertight work thus rendered necessary will be fully described hereafter. By means of this arrangement of bulkheads an immense addition is made to the strength of the structure; for, being stiffened and connected as they are, they form longitudinal, vertical, rigid webs throughout their length, and help to prevent the change of form which the unequal distribution of weights, and the motions of the ship in a seaway, tend to produce; while at the same time they afford an additional provision for the ship's safety, in case the side should be penetrated near the water-line. As these bulkheads enclose a space on each side of the lower deck, corresponding in position with the wing passages of wooden ships of war, they have been very generally termed "wing passage bulkheads."

The internal plating shown in the sketch extending from the keelson to the nearest longitudinal on each side, only reaches from a little before the boiler space to a little abaft the engines. This plating is $\frac{5}{8}$ inch thick and is made watertight, the longitudinal compartment thus formed on each side being subdivided by continuations of the transverse watertight bulkheads.

The details of the general framing of the 'Northumberland' are in most particulars identical with those just described for the 'Warrior.' The number of the longitudinal frames is the same, and their dimensions are as follow:—

	Plate.	Angle-irons.	
		Inner Edge.	Outer Edge.
	inches.	inches.	inches.
No. 1 Longitudinal, or Shelf-plate ..	15 by $\frac{7}{8}$	5 by 5 by $\frac{5}{8}$	6 by 6 by $\frac{3}{4}$
No. 2 ,, ...	17 by $\frac{7}{16}$	$3\frac{1}{2}$ by 3 by $\frac{1}{2}$	4 by $3\frac{1}{2}$ by $\frac{5}{8}$
No. 3 ,, ...	25 by $\frac{9}{16}$	ditto	ditto
No. 4 ,, ...	18 by $\frac{7}{16}$	ditto	ditto
No. 5 ,, ...	18 by $\frac{7}{16}$	ditto	ditto
No. 6 ,, ...	27 by $\frac{9}{16}$	ditto	ditto

One difference between the longitudinal frames of the two ships is that the angle-irons on their outer edges are worked single in the 'Northumberland' and double in the 'Warrior,' and it will be seen, on a comparison of the two tables of dimensions, that the plates and angle-irons in the 'Northumberland' are lighter than those in the 'Warrior.' The decrease in breadth of the shelf-plate of this ship, from that of the 'Warrior,' is due to the fact that only one 9-inch thickness of wood backing is worked, instead of two layers, making a total thickness of 18 inches in the 'Warrior.' The same care is taken to keep up the continuity of longitudinal strength in this ship as in the 'Warrior,' the specification stating that the longitudinal frames are to be reduced in breadth at the extremities of the ship, and as many of them continued to the stem and body post as may be directed. The butts of the longitudinals are connected by double straps, treble riveted. At the points where the longitudinal frames pass through the transverse watertight bulkheads, great pains have to be taken in order to make watertight work, and the manner in which this is done in the 'Northumberland' is illustrated by Figs. 78 and 79, the former of which shows a section of No. 5 longitudinal, and the latter a section of No. 6, which is nearest to the keel, the difference in the arrangements being due to the fact that the continuous transverse frame is scored down into No. 6 longitudinal, but runs above the inner edge of No. 5, as was noticed in the 'Warrior.' In both the sketches the liners to the outer strake of plating, on which the longitudinal comes, are drawn in solid black, and the fish pieces or covering plates worked inside the liners are stroked across. The liners extend under one frame on each side of the bulkhead, so as to form a butt-strap over the weak place caused by the line of rivets in the bulkhead angle-irons, which are worked double. The fish pieces extend between the frames on each side of the bulkhead, and make the liners watertight. Canvas is used, in the immediate vicinity of the

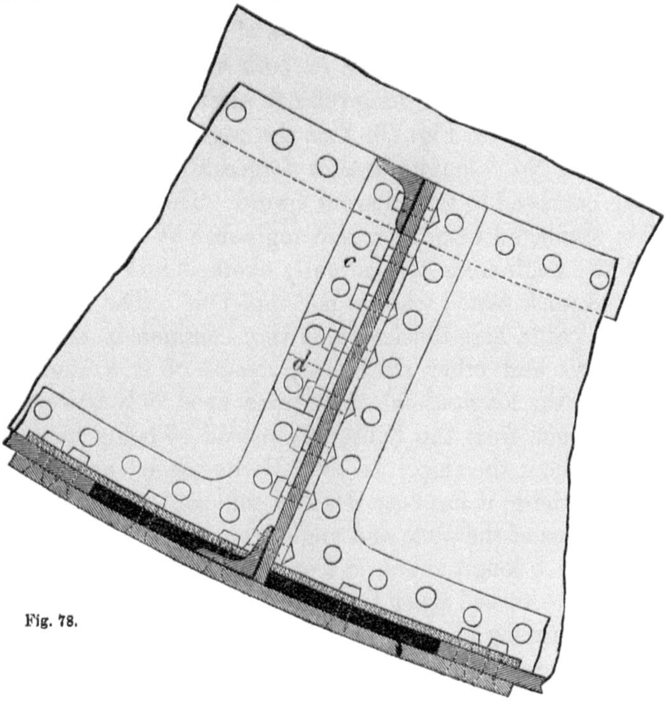

Fig. 78.

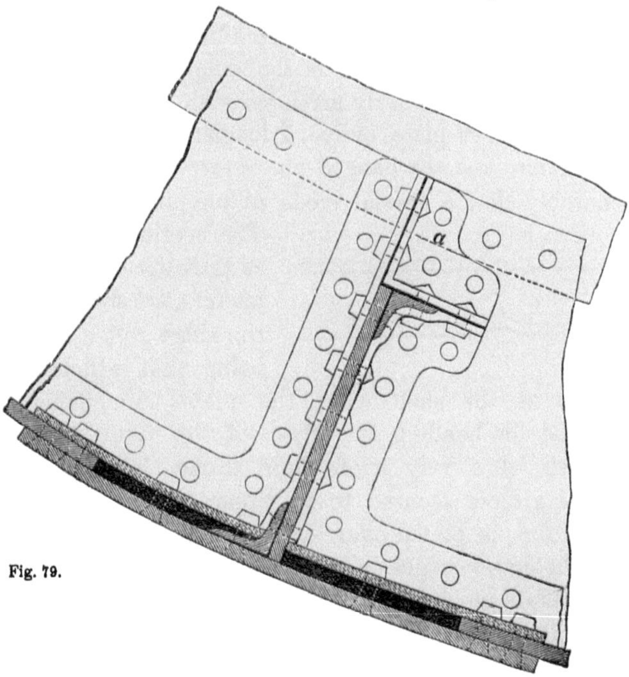

Fig. 79.

bulkheads, in order to make the joints of the plates and angle-irons watertight. Strong black lines in both sketches show where the canvas is used behind the longitudinal angle-irons, &c. It will be seen, on reference to Fig. 79, that the angle-iron marked *c* on the inner edge of No. 6 longitudinal is forged staple fashion, and run from the bulkhead to the adjacent frame. The butt of the angle-iron *c* is connected with the frame angle-iron by an angle-iron strap *d*. All the angle-irons are carefully caulked and the watertightness of the work tested when it is completed. The use of canvas steeped in paint, here illustrated, is very common in the bulkhead work of this and other ships. By means of it a stop-water is formed, and the longitudinal angle-irons need only be caulked for a short distance from the bulkhead, instead of being made watertight throughout the ship. In the 'Bellerophon' and later ships built at Chatham, it has been deemed sufficient to thickly coat the faying surfaces of the plate and angle-iron, with red lead. In the sketch of No. 5 longitudinal, in Fig. 78, it is shown that the frame angle-irons are turned up on both sides of the longitudinal, and one of them is run up the full depth of the transverse plate-frames, while the other is stopped at the upper edge of the longitudinal. A short clump angle-iron, *a*, connects these frame angle-irons, and has canvas at its back, as shown in black lines. All the other longitudinals, except that at the foot of the wing-passage bulkhead and the armour-shelf, are similarly arranged with No. 5.

The armour-shelf plate, or No. 1 longitudinal, is similar to that of the 'Warrior,' and the flanged plates forming it have to be bent so considerably that a special mode of manufacture is necessary. The sections of these plates, as they are served in by the makers, are shown in Fig. 80, the ridge rolled in the centre being that which forms the outer angle of the shelf when the plates are flanged. The machine used for bending the plates of the 'Northumberland's' shelf consists of a very strong iron frame, having a series of longitudinal girders secured to its upper part. The under side of these girders is plated over, and to this plating are attached brackets, the lower edges of which are formed to the angle to which the plate is required to be flanged, which is generally a right angle. Hydraulic presses placed below the frame sup-

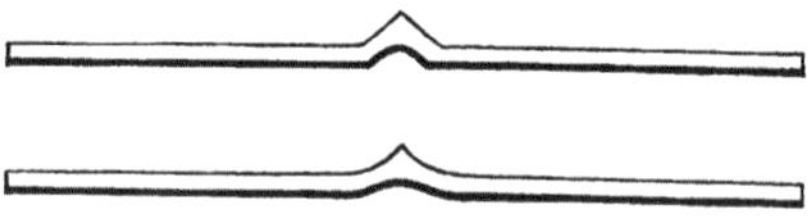

Fig. 80.

port a longitudinal ridged bed, on which the plate to be flanged is laid, after having been heated in the furnace. When all is ready, the pumps of the presses are set to work, and the bed, with the plate upon it, is forced up under the brackets, the flanges being thus turned down. Blows are struck on the flanges, by the workmen, at intervals, so as to facilitate the flanging; and when the operation is completed, the plate is allowed to cool in place, and when cold is removed to have the edges and butts planed, holes punched and countersunk, &c. The iron in these plates requires to be of the very best description, and in some cases it has been found that, notwithstanding the greatest care, the severe bending they have to undergo has caused plates to break under the press, while in other instances after a plate has been worked and riveted in place, it has been found to crack, and it has been necessary to remove it and work another. The expense consequent on the manufacture of, and the injury done by bending to, these plates, have led to the introduction of other arrangements of armour shelf, which will be fully described hereafter.

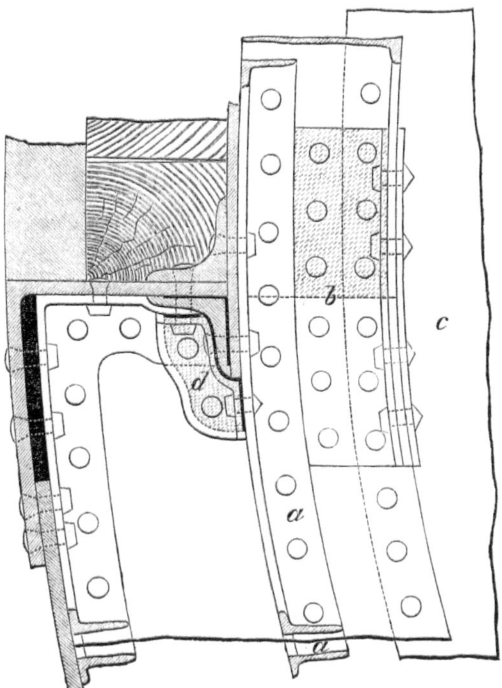

Fig. 81.

The inner edge of the shelf-plate of this ship is secured to the skin-plating behind armour by double angle-irons, as shown in Fig. 81. These angle-irons are continuous, and the lower one has consequently to pass through the transverse watertight bulkheads, and a special arrangement which was adopted for making the joints watertight is shown fully in the sketch. The angle-irons marked *a*, *a*, are those which receive the fastenings of the skin-plating behind armour. At *b* the 10-inch reversed frame is butted, and strapped and riveted as shown. The bulkhead plate *c* overlaps the transverse frame and plates, and the rivets securing it have a pitch of 3¾ inches so as to make the joint watertight. The reversed frame is only ended in this way at the watertight bulkheads. In order to make the angle-iron on the underside of the shelf-plate watertight, a short piece of angle-iron *d* is carefully fitted against the plate and angle-iron, and canvas steeped in paint is employed to form a stop-water, its position being indicated as before by strong black lines in the joints. Here also it will be seen that unless canvas, or some substitute is employed, the edges of the longitudinal angle-iron would have to be caulked throughout the ship's length, as the water might enter at some distance from the bulkhead and run along underneath the flange of the angle-iron into the adjoining compartments. The black plate-liner, shown under the vertical flange of the shelf-plate, extends to the adjacent frame on each side of the bulkhead.

The spacing of the continuous transverse frames of this ship has already been given when describing her keel arrangements, but may be repeated here for convenience. The usual frame space is 2 feet 4 inches, observing that at the ports pairs of spaces 1 foot 11 inches are thrown in, so as to ensure a good arrangement of port-frames. In wake of armour these continuous frames are formed of 10 by 3½ by ½ inches angle-irons, tapering to 7 by 3½ by ½ inches at the turn of the bilge. They extend from gunwale to gunwale, their butts being properly shifted and strapped, and no butt is nearer to the middle line than 6 feet. The heads of all frames to which beams are not attached are secured to the underside of the upper-deck stringer-plate by short angle-irons. The double angle-irons on the outer edge of the 10-inch frames are 3½ by 4½ by ⅝ inches, and they extend from the gunwale to the longitudinal next below the armour-shelf.

The transverse plate-frames of this ship are $\frac{7}{16}$ inch thick, except at the watertight bulkheads, where they are ⅝ inch. The mode of lightening is similar to that shown in the 'Warrior's' midship section in Plate 3. The frame angle-irons are worked staple fashion between the longitudinals, and are single on all except the bulkhead frames. Their dimensions are as follow:— On the floor-plates 4 by 3½ by ⅝ inches, and on all other plates 4 by 3½ by $\frac{9}{16}$ inches.

The principal differences in the transverse framing of the 'Warrior' and 'Northumberland' are, that in the latter the continuous transverse frame is formed by a reversed angle-iron, instead of a plate and angle-iron, as in the former; that the intermediate frames and floor-plates, and double-reversed frames on the plate-frames of the 'Warrior,' are dispensed with in the 'Northumberland'; that the spacing of the frames is considerably reduced in the latter vessel; that the double angle-irons on transverse frames behind armour are only run down to No. 2 longitudinal in the latter, while they extend to No. 3 in the former; and that the frame angle-irons on the common transverse frames are single in the 'Northumberland,' and double in the 'Warrior.'

The framing of the bow in vessels of this class is of a different character from that amidships. A middle-line bulkhead is fitted in the bow, and extends back into the ship about 54 feet. At the fore end it reaches from the keel to the main deck, and its height is gradually reduced by successive steps, until it is brought down to 18 feet from the keel at the after end. The plating in this bulkhead is worked flush at the edges and butts, the edge-strips being single, and the butt-straps double riveted. Vertical angle-iron stiffeners are fitted on alternate sides of the bulkhead, at intervals of from 2 feet to 2 feet 6 inches; and the bulkhead is further stiffened by the angle-irons connecting the floor-plates and transverse bulkheads to it. All these arrangements tend to make the bulkhead rigid, and to prevent it from buckling, in case the ship should be used as a ram; and in order to give additional strength to the bow for this purpose, breast-hooks are fitted at and intermediate between the decks. The details of the construction of one of these breast-hooks are given in Fig. 82, which represents a plan of the foremost end. The plates are fitted between the frames, and are secured to the outside plating by short pieces of angle-iron worked underneath them. A strong angle-iron runs along inside the reversed frames and is riveted to them, and to the transverse and middle-line bulkheads, as shown. The inner edge of the breast-hook plates is stiffened by an angle-iron worked underneath, as indicated in dotted lines. By means of this combination of bulkheads and breasthooks the transversely framed bow is made sufficiently strong to act as a ram in case of need; and, in addition, the forward part of the vessel is divided into a large number of watertight compartments, which add to the ship's safety when so employed. None of the longitudinal frames are continued forward to the stem, the upper longitudinal being worked into a flat at its fore end, and the endings of the remaining three longitudinals being shifted, in order to avoid a sudden reduction of longitudinal strength in any one transverse plane.

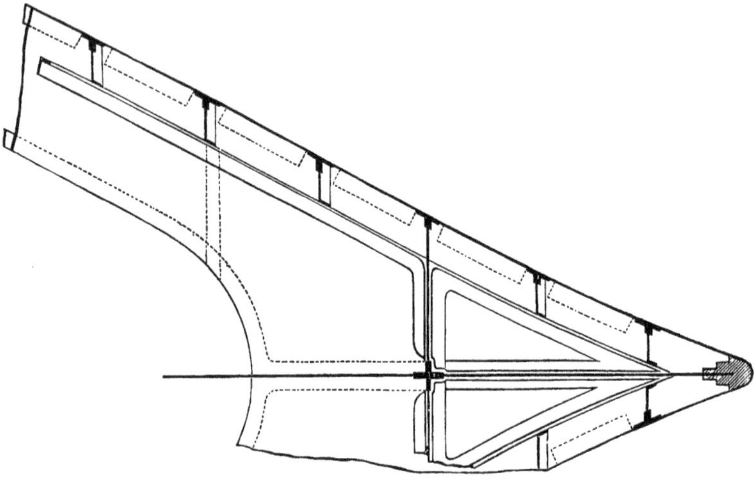

Fig. 82.

The stern framing of the vessels of this class requires a brief description. One of the longitudinals is run aft to the stern post, and the others are ended at the stuffing-box bulkhead. The loss of longitudinal strength thus involved is, however, compensated for in a measure, by the plating on the lower-deck beams, the flat below the lower deck, the shaft tube, and the middle-line bulkhead, all being connected with the post and the bulkhead. An iron flat is laid below the lower-deck beams, on the fore side of the stuffing-box bulkhead, and serves both as a platform for storerooms and a crown to the shaft-passage, in addition to adding considerably to the strength of the ship. The shaft-passage bulkheads are formed of ½-inch plates, and are stiffened by vertical bars 3 by 2½ by ⅜ inches. The whole of this work is made watertight, and so a longitudinal compartment is formed, which extends from the bulkhead at the after end of the engine-room to the stern. The communication with the engine-room is made by an opening in the transverse bulkhead, to which a watertight door is fitted, and there is a stuffing-box arrangement on the shaft where it passes through the bulkhead. The frames abaft the stuffing-box bulkhead are formed of ½-inch plates, and are secured to the shaft-tube by encircling angle-irons. On the aft side of the stuffing-box bulkhead the vertical keelson is increased in height, so as to form a middle-line bulkhead below the shaft-tube, and this bulkhead is completed above the shaft-tube by a vertical plate extending up to the flat below the lower deck.

The wing passage and its connections are similar to those of the 'Warrior,' and the flats, stiffeners, &c., are identical in arrangement, the only difference being that the plating in this ship is $\frac{1}{16}$ inch thinner than the 'Warrior's.' One important difference between the two ships, however, both as respects strength and safety, consists in the fact, that throughout the engine and boiler spaces of the 'Northumberland,' an inner bottom is formed by working watertight plating upon the bearers, and continuing it up the frames to the foot of the wing-passage bulkhead. This inside plating extends a short distance before and abaft the bulkheads, which bound the boiler and engine-rooms.

CHAPTER VII.

BRACKET-PLATE SYSTEM OF FRAMING.

FRAMING OF 'BELLEROPHON,' 'HERCULES,' &c.

THE bracket-plate system of construction, which was introduced for the first time in the design of H. M. armour-clad frigate 'Bel-'lerophon,' is fully illustrated in the perspective view of a portion of the midship framing of that vessel, which is given in Plate 4. The objects of the invention and introduction of this system were to save weight, to simplify workmanship, and to add both to the strength and safety of the ship. The characteristic features of the system are the adoption of an inner bottom, and of short angle-irons connected by bracket-plates, in place of staple and other forged angle-iron work. Many minor differences of detail in construction will be remarked hereafter. A great increase of longitudinal strength is gained by the use of much deeper longitudinal frames than those of the 'Warrior' and other of the earlier iron-clads. Another important feature resulting from the employment of deep longitudinals is that the space between the two bottoms is roomy and easy of access for cleaning and painting, operations which are essential to the preservation of an iron structure. Facilities are also offered by these arrangements for letting in water between the bottoms to serve as ballast, the space being so divided into watertight compartments as to enable the officer to regulate the trim of the vessel by filling the fore or the aft spaces. Provision is, of course, made to pump out any compartment, when required. The dimensions of the longitudinal frames of the 'Bel-'lerophon' are as follow :—

	Plate.	Angle-irons.	
		Inner Edge.	Outer Edge.
	inches.	inches.	inches.
No. 1 Longitudinal, or armour-shelf ..	$16\frac{1}{2}$ by $\frac{7}{8}$	5 by 5 by $\frac{5}{8}$	..
No. 2 ,, ...	17 by $\frac{7}{16}$	3 by 3 by $\frac{1}{2}$	$3\frac{1}{2}$ by 4 by $\frac{9}{16}$
No. 3 ,, ...	31 by $\frac{1}{2}$	ditto	ditto
No. 4 ,, ...	37 by $\frac{7}{16}$	ditto	ditto
No. 5 ,, ...	43 by $\frac{7}{16}$	ditto	ditto
No. 6 ,, ...	49 by $\frac{1}{2}$	ditto	ditto

The "Bellerophon"

Drawn by J. Maxton. John Murray, Albemarle Street, Novr 1868. J.W. Lowry sc.

The longitudinals are made of one depth of plate; the butts are carefully fitted and secured by double butt-straps each $\frac{1}{16}$ inch thicker than half the thickness of the plates they connect. All the butts are double riveted, except those of No. 3 longitudinal, which are treble riveted, as this longitudinal is a watertight division, and is therefore left solid. The angle-irons on the outer edges of the longitudinal frames are single, and continuous throughout the ship. As the longitudinals are worked directly upon the outside plating, these angle-irons have to be bent and the plates notched, in order to fit over the butt-straps to the plating. The butts of the continuous longitudinal angle-irons are supported by covering angle-irons. The angle-irons on the inner edges of the longitudinals are worked in short lengths between the transverse frames, and are double throughout the double bottom. The working of these angle-irons in short lengths is necessitated by the fact that all the longitudinals run up close to the inner bottom, and the continuous transverse frames pass through scores cut in their upper edges. Though a loss of longitudinal strength is thus involved, it still leaves an ample remainder. The detailed arrangements of angle-irons, butt-strap, &c., of one of the lower longitudinals are shown in Fig. 83. The large holes *a, a*, shown in the sketch, are 2 feet by

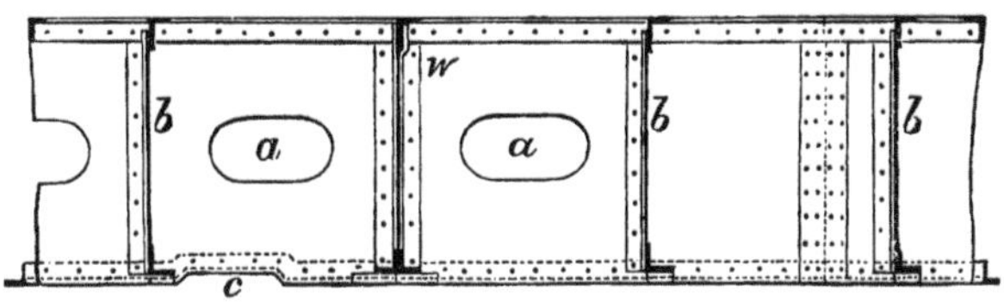

Fig. 83.

1 foot, and are cut for the purpose of lightening the plate, while at the same time they serve as man-holes, which afford communication between the various parts of the space included between the two bottoms. These holes are so arranged as to give as great sectional strength of plate in wake of them as there is along the weakest line across the longitudinal. In the case illustrated by the sketch, this weakest line extends from the bottom of the score cut for the continuous transverse frame, down through a line of rivet-holes at the side of a watertight frame, *w*. It may be interesting to examine the relation between the strength along the weakest line and the strength in wake of the hole, so as to ascertain whether or not uniformity of strength has been, as nearly as

possible, preserved. The score for the continuous angle-iron is $5\frac{1}{2}$ inches deep, and there are eight $\frac{7}{8}$-inch rivet-holes in the breadth of the plate. The total breadth along the supposed line of fracture equals 50 inches,* and hence the effective length is reduced to $37\frac{1}{2}$ inches, and the effective sectional area $=37\frac{1}{2}$ inches by $\frac{1}{2}$ inch$=18\frac{3}{4}$ square inches. If the plate breaks across the lightening hole, there are two rivet-holes in the line of fracture, one $\frac{3}{4}$ inch and one $\frac{7}{8}$ inch, and thus the effective breadth of plate is reduced from 49 inches to $35\frac{3}{8}$ inches, and the effective sectional area $=35\frac{3}{8}$ by $\frac{1}{2}=17\frac{11}{16}$ square inches. But in order that the plate may separate, the angle-irons on the inner edge must also break in this case. These angle-irons are double, $3\times3\times\frac{1}{2}$ inches, and there are two $\frac{3}{4}$-inch rivet-holes in the transverse section of each angle-iron. Their united effective sectional area, therefore, equals $2\times(5\frac{1}{2}-1\frac{1}{2})\times\frac{1}{2}=4$ square inches. By the angle-irons, therefore, the total effective sectional area in wake of the hole cut for lightening is brought up to $21\frac{11}{16}$ square inches, and it has been shown above that along the weakest line the longitudinal has an effective area of $18\frac{3}{4}$ square inches. Hence the proportionate strengths along the two supposed lines of fracture are as 300 : 347, or as 6 : 7 nearly, the margin of strength being left in wake of the large hole.

In order to complete the investigation of the approximation to uniformity of strength in the longitudinal, it is necessary to determine the relation between the strains required to open the butt by shearing the rivets on one side of the butt, or tearing the butt-straps asunder at their weakest section, and the strength of the longitudinal at its weakest section. First suppose the butt-straps to be broken down through the row of rivet-holes nearest to the butt, and the angle-irons on the inner edge to be broken across. The sectional area of the straps is reduced by the $\frac{3}{4}$-inch rivet-holes, and as there are ten rivets in the strap on one side of the plate, which is 46 inches deep, and nine rivets in the strap on the other side, which is $42\frac{1}{2}$ inches deep, the effective sectional area becomes $\frac{5}{16}\{(46-7\frac{1}{2})+(42\frac{1}{2}-6\frac{3}{4})\}=23\frac{13}{64}$ square inches. Adding 4 square inches for the angle-irons on the inner edge, we obtain

* The length of the line of fracture is one inch more than the total breadth of the longitudinal, on account of the diagonal direction of the line joining the bottom of the score with the adjacent rivet-hole.

the total effective area of straps and angle-irons $= 27\frac{13}{64}$ square inches. And since the least effective area of the longitudinal has been found to equal $18\frac{3}{4}$ square inches, the proportion of the strains required to produce fracture in these two ways is as $18\frac{3}{4} : 27\frac{13}{64}$, that is, the strengths are in the ratio of 2 : 3 roughly. If the rivets in the butt-strap on one side of the butt shear, and the angle-irons on the inner edge of the longitudinal break, the butt can separate. In this case eighteen rivets $\frac{3}{4}$ inch in diameter will have to shear twice, since there are double butt-straps, and two rivets of the same size in the flange of the outer angle-iron on the longitudinal will have to shear once, thus requiring a force $= 18 \times 18$ tons $+ 2 \times 10$ tons $= 344$ tons, if we suppose the shearing strength of a $\frac{3}{4}$-inch rivet to be 10 tons for a single shear, and 18 tons for a double shear. Also by punching holes in plates or angle-irons the strength is found to be reduced, and a fair value of the strength per square inch after punching is about 18 tons. Taking this, we have as the strength of the angle-irons on the inner edge 4×18 tons $= 72$ tons, and thus find the total strain required to open the butt in the supposed manner $= 344 + 72 = 416$ tons. Now the effective sectional area of the weakest section of the longitudinal $= 18\frac{3}{4}$ square inches, and taking 18 tons per square inch as the breaking strain, we find that the strain required to break the longitudinal at this section $= 18 \times 18\frac{3}{4}$ tons $= 337$ tons. Hence we have:—Strength of plate : Shearing strength of rivets, &c. : : 337 : 416 : : 6 : $7\frac{1}{2}$ nearly. We thus obtain finally the following approximate proportions between the various breaking strengths of the longitudinal and its butt fastenings. Taking the strength of the longitudinal at the side of a watertight frame as 6, that in wake of the hole cut for lightening will be 7, that along a row of rivet-holes in the butt-strap will be 9, and the shearing strength of the rivets on one side of the butt will be $7\frac{1}{2}$. In all these calculations the continuous angle-iron on the outer edge of the longitudinal is not taken any account of, the longitudinal plate and the short angle-irons only being considered. The results obtained above would, of course, be modified, if the strength of the continuous bar was added in the calculation of the various breaking strains.

It will be remarked in Fig. 83, that in the frame space where the butt of the longitudinal comes, no holes are taken out. The bracket frames, *b*, *b*, are secured to the longitudinal by single angle-irons, which are worked on the opposite side of the brackets to that on which the short frame angle-irons are worked. The watertight frame, *w*, has double staple angle-irons on its end. Where a longitudinal crosses a butt-strap of the bottom plating, as at the point marked *c* in the sketch, it is notched up, and the angle-iron on its edge bent so as to pass over the strap. All the longitudinals between the keel and the longitudinal at the foot of the wing-passage bulkhead are similarly lightened and riveted, but No. 3 is made watertight at all the joints where the continuous frames pass through it, by means of angle-irons similar to those marked *a* in Fig. 84, the

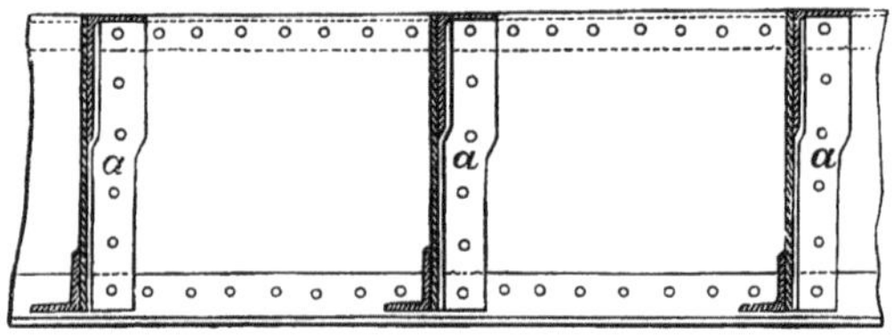

Fig. 84.

edges of which are caulked; the lower edge of the plate-frame is also caulked on the opposite side from that on which the angle-iron, *a*, is worked. It will be seen in Plate 4 that the armour-shelf in this ship is formed and worked in a similar manner to the shelf of the 'Northumberland,' which has been fully described. The skin-plating behind armour is worked in two thicknesses of $\frac{3}{4}$-inch plate, and longitudinal frames or girders are worked outside it, as shown in the section in Plate 4, their dimensions being $9\frac{1}{2}$ by $3\frac{1}{2}$ by $\frac{1}{2}$ inches, their spacing nearly the same as the vertical frame spacing, and their disposition such that one is situated at about 12 inches from the edge of each strake of armour.

The inner bottom extends throughout two-thirds of the ship's length, and, by means of the wing-passage bulkhead on each side, is continued up to the under side of the main deck. The plating on the floor is $\frac{1}{2}$ inch thick, and is worked flush on the upper side, with single-riveted edge strips, and treble-riveted butt-straps worked on the under side. The vertical plating forming the wing-passage bulkheads is $\frac{7}{16}$ inch between the main and lower decks, and $\frac{1}{2}$ inch below the lower deck, and is stiffened by 4 by $3\frac{1}{2}$ by $\frac{5}{8}$ inches angle-irons, placed 2 feet apart. The remarks previously made on the advantages, as regards strength and safety, resulting from the wing-passage bulkheads fitted in the 'Warrior' and 'Northumberland,' apply with equal force to the arrangements of the 'Bellerophon.'

The transverse framing in the double bottom is for the most part similar to that shown in Plate 4, and is made up of the continuous angle-irons, which extend from gunwale to gunwale; of the $5\frac{1}{2}$ by 4 by $\frac{9}{16}$ inches frame angle-irons worked in short lengths between the longitudinals; and of bracket plates $\frac{7}{16}$ inch thick, formed and riveted as shown, and connected to the longitudinals by short angle-irons, $3\frac{1}{2}$ by 3 by $\frac{7}{16}$ inches, worked on the opposite sides of the bracket-plates to those on which the frame angle-irons are secured. At intervals of about 20 feet, watertight frames are worked, and are formed of $\frac{1}{2}$-inch plates fitted between the longitudinals and the inner and outer bottoms, and supported on the upper edge by the continuous angle-irons, and at the bottom and ends by double staple angle-irons.

By these arrangements of the transverse framing we dispense with one-half the frames which would be required if the framing were similar to that of the 'Northumberland,' and at the same time the strength of the structure as a whole is considerably increased. The transverse strength, which is derived from the watertight bulkheads and the plate frames, is distributed over the intermediate spaces by the strong longitudinal frames; and by means of the bracket-frame arrangement the rigidity of the bottom is ensured, thus supplying a feature of the construction, the importance of which was illustrated at the commencement of this work. By means of these arrangements the work of building is also much simplified, and consequently can be performed more quickly and cheaply. One instance of this is found in the fact that on all except watertight frames staple angle-irons are dispensed with, and so the cost of forging is saved; while the frame angle-irons, and the short connecting pieces on the longitudinals, being worked on opposite sides of the brackets, do not require to be joggled over one another, or to have their ends accurately fitted, and thus the cost of moulding and working is reduced.

Between the armour-shelf and the next longitudinal each transverse frame is completed by a plate frame $\frac{7}{16}$ inch thick, which is lightened, as shown in the section, at all except the bulkhead frames, and has its frame angle-iron turned in under the shelf plate and riveted through it. These plate frames are 2 feet apart, and by their use provision is made for the vertical strains which are always experienced by this portion of the framing of a ship, and which are so much greater in a ship which, like the 'Bellerophon,' carries a great weight of armour on her sides.

By means of the various watertight frames before described, both longitudinal and transverse, the space included between the inside and the bottom plating is divided into a series of cells or compartments. The watertight vertical keelson and No. 3 longitudinal on each side, form the boundaries of two large longitudinal compartments; and the wing-passage bulkhead and No. 3 longitudinal, form those of another compartment on each side. Each of these compartments is subdivided by the transverse watertight frames, so that in case of accident happening to any part of the ship's bottom, the water would only have access to a limited space. Watertight manholes are fitted to each compartment to give admission for painting, repairs, &c.

In wake of the armour the frames are formed of 10 by $3\frac{1}{2}$ by $\frac{1}{2}$ inches reversed angle-irons, with double angle-irons $3\frac{1}{2}$ by $3\frac{1}{2}$ by $\frac{5}{8}$ inches on the outer edge. These frames are 2 feet apart, and the transverse flanges of the reversed angle-irons are tapered from 10 inches at the lower edge of armour, to $5\frac{1}{2}$ inches at the foot of the wing-passage bulkhead. Between the armour shelf and No. 2 longitudinal each of these frames is connected with the plate frames described above. The alternate frames only are continued below No. 3 longitudinal, the intermediate frames ending there. The continuous transverse angle-irons of the framing in the double bottom are scarphed to the lower ends of the alternate frames, the scarphs of adjacent frames being carefully shifted so as to bring the ending of the continuous angle-irons alternately 3 feet above and below No. 3 longitudinal. An illustration of the arrangements of the scarphs and butts of the continuous angle-irons which were adopted in the 'Hercules,' will be found in Fig. 94, page 131, and will be described further on.

Before and abaft the double bottom, the arrangements of the framing below the armour shelf are slightly different from those of the framing throughout the double bottom. The spacing of the transverse frames, and the whole arrangement of the framing in wake of armour are, however, identical with those amidships. Outside the double bottom the longitudinal frames are reduced in depth so as to allow the continuous transverse angle-irons to pass above their inner edges, and in consequence of the latter not scoring down, single angle-irons on the inner edge of the longitudinals are made continuous, thus adding considerably to the longitudinal strength. The continuous transverse angle-irons are reduced in size to $3\frac{1}{2}$ by 4 by $\frac{7}{16}$ inches, and the frame angle-irons are reduced to $3\frac{1}{2}$ by 4 by $\frac{9}{16}$ inches. The brackets are replaced by plate frames considerably lightened. At the bow, No. 2 longitudinal is made to work into the watertight flat below the lower deck. This flat extends back to the watertight bulkhead which forms the foremost boundary of the double bottom, and ends forward at the cant frame marked A in Fig. 85, being connected to the transverse watertight plating B by double angle-irons. The plating of the flat is $\frac{3}{8}$ inch thick, and is worked flush on the platform beams, the butt-straps and edge strips being worked above it. The platform so formed supports the store rooms, and in addition adds greatly to the safety of the ship by forming the space below it into a watertight compartment, and thus serving one of the purposes for which the double bottom is fitted in the midship part of the ship. It may be remarked here, that the weight of material which would be required to complete the inner skin, from its present ending forward to the stem, would be much greater than that employed in the construction of the flat; while the fineness of the ship forward renders the size of the space enclosed by the flat comparatively small, and thus removes the objection which might reasonably be made to a large fore compartment. Admission is gained to the space below the flat by means of a watertight trunk which extends up to the main deck, and encloses the scuttle in the flat. It will readily be seen, that in case the bow is broken through, and the lower compartment is filled, the water cannot find its way out upon the lower deck. This is a patented improvement of Mr. Charles Lungley, and was, by his permission, introduced into the construction of the 'Bellerophon.' The arrangement forms part of a plan proposed by Mr. Lungley for making iron ships unsinkable, which will be found described in detail in the *Transactions* of the Institution of Naval Architects for 1861.

The framing of the bow is shown in detail in Figs. 85 and 86, which afford an illustration of the simplification of work, and the increase of strength resulting from the adoption of longitudinal framing. The profile view in Fig. 85 shows the manner in which the longitudinals are run forward and secured to the stem, and the plan in Fig. 86 gives, in full detail, the arrangements by which the fore ends of the longitudinals are converted into powerful breasthooks, by means of small flats of iron plate worked across the ship, and stiffened on their after ends by angle-irons. The

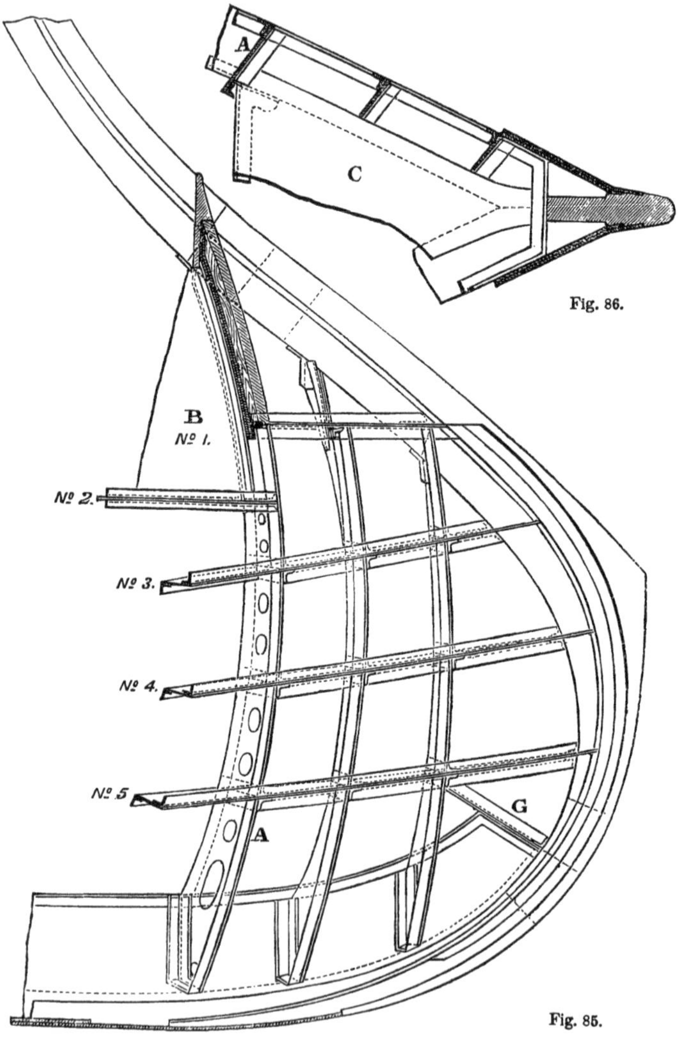

Fig. 86.

Fig. 85.

armour shelf is decreased in width forward on account of the reduced thickness of the armour plating and wood backing, and its fore end is secured to the stem by the double angle-irons on the inner edge of the shelf. The manner in which No. 2 longitudinal is ended has been already described, but it may be added here that at its fore end the angle-irons on the outer edge are doubled as shown in Fig. 85. All the other longitudinals which extend forward to the stem are finished similarly to each other, and it will therefore suffice to describe the arrangements of No. 3, which are shown in plan in Fig. 86. On the fore side of the frame A short pieces of doubling angle-iron are worked on the outer edge of the longitudinal, between the vertical frames. The cant frames on the fore side of A are reduced in depth, in order to allow the plate C to lap on the longitudinal and be riveted to it. The fore part of C is arranged so as to form an edge strip to the joint of the fore ends of the longitudinals on opposite sides. The double angle-irons on the fore end are tap-riveted to the stem, and their rivets, together with the riveting of the outside plating to the longitudinal angle-irons, complete the connection of the breasthook with the bow. It will be remarked that only five of the six longitudinal frames shown on the midship section in Plate 4 are carried right forward. The explanation of this arrangement is that the girth of the ship decreases so rapidly toward the extremities as to render it inconvenient to carry the longitudinal next the keel beyond the double bottom, at the boundaries of which it is therefore terminated both forward and aft. The vertical keelson is connected with the stem in a manner similar to that previously described, and its fore end is secured to a diagonal breasthook, G. A few of the foremost vertical frames in this ship are canted, as shown in the sketches in Figs. 85 and 86, and have their heads and heels secured to the stem and the vertical keelson plate respectively.

The stern framing of this ship is illustrated in Fig. 87. Before the stuffing-box bulkhead the general character of the framing is similar to that already described, special provisions being made, however, for the shaft bearers, &c. Abaft the stuffing-box bulkhead the framing is specially designed to resist the strains caused by the action of the screw propeller, and prevent the local vibration so destructive to the fastenings of an iron ship, and at the same time to accomplish this object without costly forgings or difficult workmanship. Some reference has already been made to the character of the framing between the stern post and the aftermost bulkhead, but, for convenience, it may be better to give a complete description here. The bulkhead is 14 feet before the

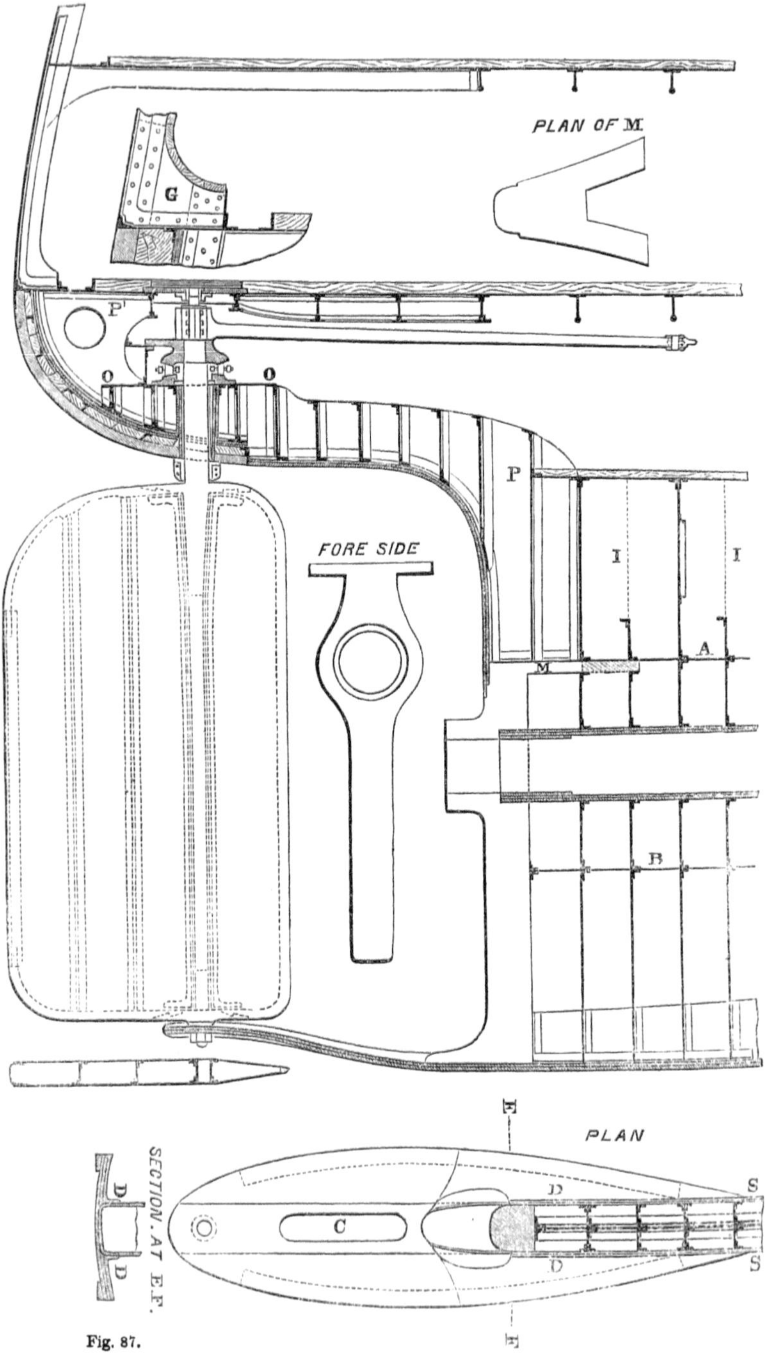

Fig. 87.

post, and is formed of $\frac{5}{8}$-inch plating worked watertight. Its upper edge is secured to the watertight plating on the after part of the lower deck, and thus the after end of the hold is converted into a watertight compartment. All the longitudinals (except the armour shelf and the longitudinal next below it) which extend beyond the double bottom, are stopped at the bulkhead; but the two upper longitudinals on each side are continued beyond the bulkhead, the armour shelf extending around the stern, and No. 2 longitudinal being ended at a transverse frame just before the post. Between the bulkhead and the stern post the longitudinal strength is kept up by the horizontal continuous flats marked A and B in the profile view, and by the wrought-iron tube which takes the engineer's shaft tube; all of which are strongly connected to the post and to the bulkhead. The flat A is constructed of $\frac{3}{4}$-inch plates, lightened by large holes; B is formed of $\frac{5}{8}$-inch plates, and is also lightened; and the tube is made up of two thicknesses of $1\frac{1}{4}$-inch plating, flush riveted. It will be seen on reference to the profile in Fig. 87 that there are three thicknesses of plating at the ends of the tube. The inner thickness there worked forms the bearing on which the engineer's tube fits, and by this means the length of the interior of the iron tube which has to be turned out accurately is very much reduced. The transverse framing between the bulkhead and the stern post is formed of plates $\frac{1}{2}$ inch thick, spaced 2 feet apart and, lightened considerably. This framing, up to the lower deck, is composed of three distinct parts, as will be seen on reference to the sketch. One set of plate frames extends from the keel up to the flat, B; a second extends from B to A, and is pierced with holes for lightening above and below the hole through which the tube passes; and a third set completes the transverse framing up to the lower deck, and is alternately made up of transverse plates considerably lightened in the central part, and of intermediate frames, I, I, formed of plates and angle-irons of moderate scantlings. The large holes cut in the central part of the alternate plate frames above the flat A, in addition to lightening the frames, serve to give easy access to the after part, as do also those in the remainder of the transverse and longitudinal framing. The frames behind armour in this part of the ship terminate in a foot at the lower deck, and are secured to the deck plating; in other respects they are similar to those previously described.

The 'Bellerophon' being constructed with a central battery

and a belt of armour at the water-line throughout her length, it is necessary to introduce a lighter system of framing above the armour plating, and the details of this framing, although it is common to all the unprotected portions of the ship, may, with propriety and convenience, be given here, as they are illustrated in the profile. The frames are formed of 7 by $\frac{7}{16}$ inches plates, with 4 by $3\frac{1}{2}$ by $\frac{9}{16}$ angle-irons on their outer edge, spaced 4 feet apart. The lower ends of these angle-irons are turned in upon the main-deck plating, the feet thus formed being strengthened by $\frac{1}{2}$-inch bracket plates, which are riveted to the plate frames and to the feet of the angle-irons. In order to connect these light frames with the frames behind armour, the transverse flanges of the latter are run up through the stringer plate, and riveted to the bracket plates as shown by G in Fig. 87, which sketch gives an enlarged view of the lower part of one of these frames. Between these frames intermediate stiffeners are fitted, formed of 4 by $3\frac{1}{2}$ by $\frac{9}{16}$ inches angle-irons, having the lower ends turned in on the stringer plate, and riveted to it.

Having completed the description of the framing between the stuffing-box bulkhead and the stern post, we have now to notice the framing on the after side of the stern post. It will be seen from the profile that a plate marked P, $\frac{5}{8}$ inch thick, is worked vertically at the middle line, and taken up on the double thickness of plating at the counter, and in wake of armour, by two 5 by 4 by $\frac{5}{8}$ inches angle-irons; while its lower end is secured to the after part of the flat A, and its foremost edge to the upper plate of the after transverse frame, which is increased in thickness to $\frac{5}{8}$-inch in order to receive the fastenings. In wake of the rudder-hole the plate P is bent so as to form one side of the hole, while another plate of equal thickness forms the other side, and the secure connection of the two is ensured by an internal wrought-iron tube, which is strongly riveted to them both. The upper part of the plate P is secured to the main-deck plating at P′, and it thus forms the aftermost frame of the ship from the main deck to the top of the stern post, and at the same time, with the two thicknesses of plating at the counter and behind armour, keeps up the character of the middle-line arrangements. A considerable number of the vertical frames of the stern of the ship are canted, and are heeled against and connected with the plate P. The method of forming the armour recess at the extreme after part of the ship is shown in the profile, and it will be seen that the armour-plating on the stern entirely protects the rudder-head and steering apparatus. A full description of the peculiar arrangements of the rudder, &c., adopted in this ship will be given hereafter.

Reverting now to the general framing of the 'Bellerophon,' the following brief account of the manner in which it was proceeded with may prove interesting. The two midship pieces of the outer flat keel-plate having been flanged by the smiths were lined to length and breadth, the rivet-holes were marked and drilled, and the edges and butts planed; when these operations had been completed, the plates were placed in position on the blocks. While this was being performed, the corresponding piece of the inner flat keel-plate was flanged, and, when the fixing of the outer flat keel-plates had been completed, was laid in place and fitted, the rivet-holes being marked so as to correspond with those on the outer plates. It was then drilled and planed, and secured by screws to the two outer plates, the three forming a starting length from which to work towards each end of the ship. In doing this the same order was adhered to, namely, first an outer plate and then the inner plate that butted on it. During this time the vertical keel had been planed to width, &c., the rivet-holes had been punched, and the scores for the continuous transverse angle-irons cut out. When a sufficient number of pieces had been prepared and fixed, a piece of the keel angle-iron was fitted in place, and the rivet-holes having been marked, it was taken away and drilled, after which it was replaced and riveted up. The rivets which pass through the frames were omitted in the keel angle-irons until after the former had been fitted. Before fitting the frames the joints of the keel and keel angle-irons were carefully caulked, and when this was completed the keel was ready to receive a tier of bracket-frames, which were got into position and secured by screws. A rib-band having the stations of the frames marked on it was next secured at about 3 inches from the head of the frames and shored. The lowest longitudinal was then screwed upon them, and another tier of frames fixed on it; and so on until the longitudinal framing was completed. As previously stated, great care was taken to make watertight the longitudinal which was at the foot of the wing-passage bulkhead. Before this work was done, the transverse framing behind armour was hoisted in and secured, and while this was proceeding, the frame, continuous, and connecting angle-irons were fitted, the holes marked and punched, and the riveting up of the framing completed. Full details of all these operations will be found in chap. 20.

The bracket-frame system of construction before described for the 'Bellerophon,' has been carried out also in the 'Hercules,' 'Monarch,' 'Captain,' 'King William,' and other ships. The details of the framing of the first two of these ships are, with a few slight modifications, identical with those of the 'Bellerophon;' and the general arrangements of the framing in the 'Captain' are similar to those of the 'Bellerophon.' It may, however, be interesting to examine the differences which exist between the framing of the 'Bellerophon' and that of the ships which have succeeded her; and at the same time to give some details of the construction of these ships which have not been fully illustrated for the 'Bellerophon.'

In the 'Hercules' the lower longitudinal frames are about 4 inches less in depth than those of the 'Bellerophon,' but the thicknesses of the plates and the mode of lightening them are the same. The angle-irons on the inner edge of the longitudinals in the 'Hercules' are single throughout the double bottom instead of being double as in the 'Bellerophon.' A perspective view of the midship framing of the 'Hercules' is given in Plate 5, and it will be seen from it that the vertical portion of the plating of the inner bottom is closer to the ship's side than the corresponding plating in the 'Bellerophon.' This alteration, together with the increase in the thickness of the plating to $\frac{3}{4}$ inch, and the adoption of the vertical stiffeners formed of 7 by 3 by $\frac{1}{2}$ inches angle-irons, with a $3\frac{1}{2}$ by $3\frac{1}{2}$ by $\frac{1}{2}$ inches angle-iron on the outer edge, is made in order to increase the resisting power of the side; and this is still further added to by filling the space between the vertical plating and the ship's side with teak backing as far down as the lower edge of armour. The watertight longitudinal in the 'Hercules' is that next below the armour-shelf, and the arrangements of this longitudinal are shown in plan in Fig. 88. The frames marked *i*, *i*, in this sketch are the intermediates which end upon the longitudinal, the alternate frames only being run through and connected with the continuous angle-irons. It will be seen on reference to the sketch that there is an angle-iron, *a*, worked at the heel of each plate-frame, and joggled out over the transverse flange of the reversed angle-iron. In order, however, to avoid the necessity of caulking the whole width of the

Plate 5

The "Hercules"

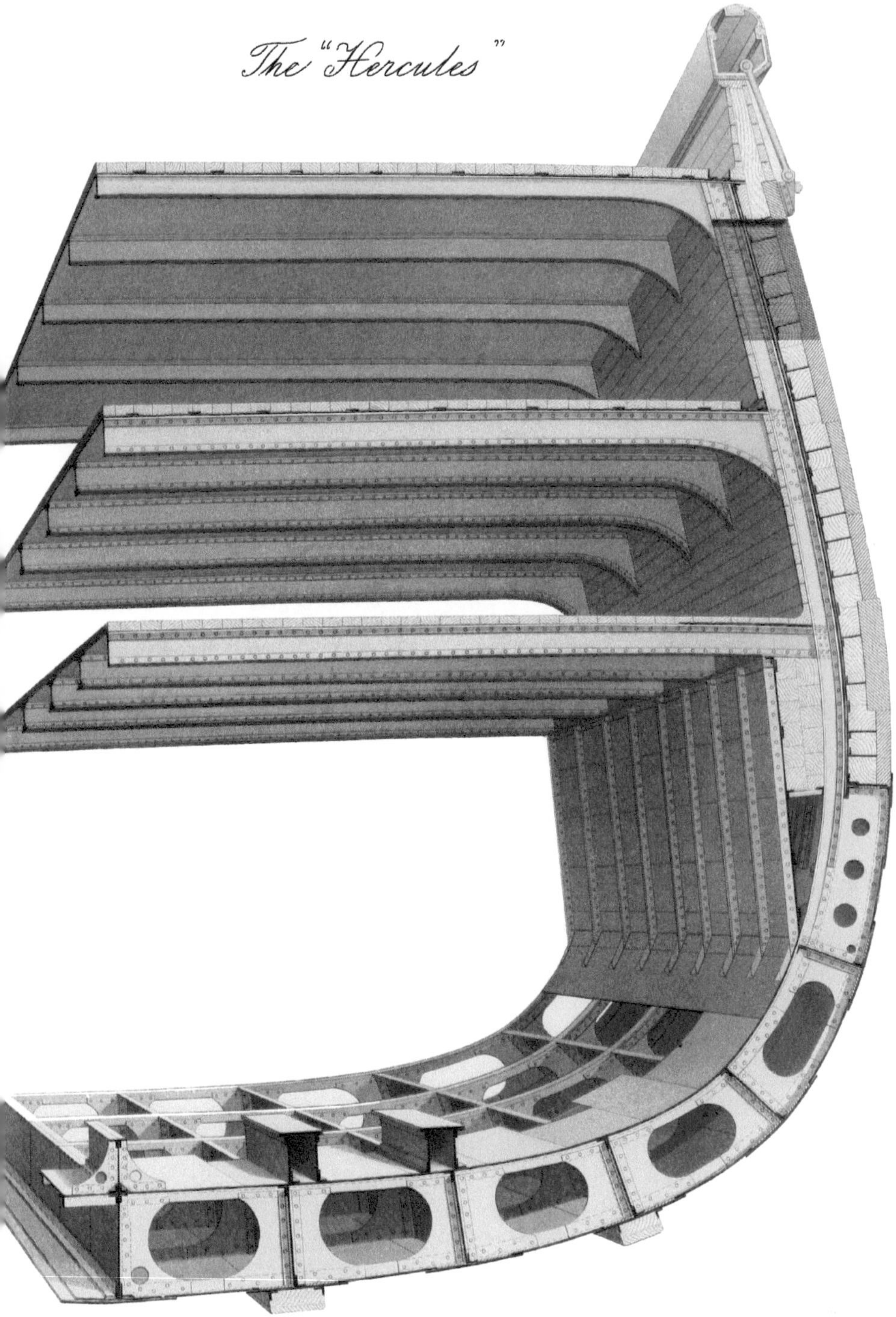

Drawn by J. Maxton. John Murray, Albemarle Street, Novr 1868. Engraved by J. W. Lowry

plate-frame at its lower edge, as was done in the 'Bellerophon,' a short piece of angle-iron marked *b* is worked on the opposite side of

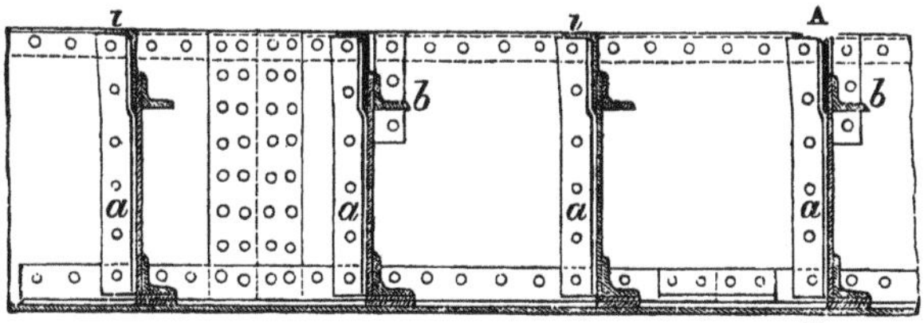

Fig. 88.

the continuous frame, and in order to allow the angle-iron *b* to fit directly against the continuous frame a rectangular piece is cut out of the corner of the plate-frame. The length of *b* is such as to take a rivet in its outer end, by which it is secured to the angle-iron *a*, and so a good caulk can be made. The angle-irons *a* and *b* are worked and caulked before the plate-frames are put in place. One of the officers engaged in the construction of the 'Hercules' says with regard to this arrangement that with ordinary care in the workmanship it is impossible for the water to run down the frame, whereas in the 'Achilles' and 'Bellerophon' it was very difficult to make the wing-passages absolutely watertight.

The lower longitudinals of the 'Captain' are considerably less in depth than those of the 'Bellerophon,' but the thickness of the plates is very nearly the same for the corresponding longitudinals. The watertight longitudinal in the 'Captain' is similarly situated with that in the 'Bellerophon,' but its arrangements differ considerably. The 10 by $3\frac{1}{2}$ by $\frac{1}{2}$ inches reversed frames in wake of armour keep their full width of transverse flange down to the longitudinal, while, in order to reduce the dimensions of the scores in the longitudinal, the transverse flanges are there reduced, at once, to 7 inches, thus leaving a shoulder of 3 inches resting on the longitudinal. On the bracket frames the frame angle-irons are single, their upper ends being turned in under the longitudinal. A straight piece of angle-iron is worked at the heel of each bracket-plate and extends right across the longitudinal, corresponding to the angle-iron marked *a* in Fig. 88, but worked on the opposite side of the plate-frame. In order to make the scores for the continuous frames watertight, a short piece of angle-iron is worked against each frame, and is of sufficient length to be joggled in against the bracket-plate and to take a rivet through it. This short angle-iron is worked on the opposite side of the continuous frame to that on which the short angle-irons marked *b* in Fig. 88 are worked in the 'Hercules,' and it is consequently necessary to caulk the heels of the bracket-plates in the 'Captain' from the outer end of the short angle-irons out to the side. On the bulkhead frames the frame and reversed angle-irons are both double, only one of the reversed bars being continuous, and the scores in the longitudinals being made watertight in a similar manner to the preceding.

It has already been stated that the armour-shelf in the later ships is differently formed from that of the 'Bellerophon,' which it will be remembered is formed of a large flanged plate specially manufactured for the purpose. In the 'Hercules' the arrangements shown in section in Fig. 89 are adopted. The longitudinal plate forming the shelf is 16 by $\frac{7}{8}$ inches, with double angle-irons 5 by 5 by $\frac{5}{8}$ inches on the inner edge connecting it to the plating behind armour, and a single angle-iron of the same size connecting it to the upper strake of the bottom plating. All the rivets in these angle-irons are $1\frac{1}{8}$-inch, and are countersunk in the upper surface of the shelf-plate. The butt-straps are worked below the shelf-plate and only extend between the edges of the longitudinal angle-irons; they are double-chain riveted. The edges of the plate and angle-irons are carefully caulked, and a covering plate 13 inches wide is worked over the seam at the lower edge of armour, being double tap-riveted to the armour, and through-riveted to the bottom plating. This covering plate has been omitted in ships of more recent design as unnecessary.

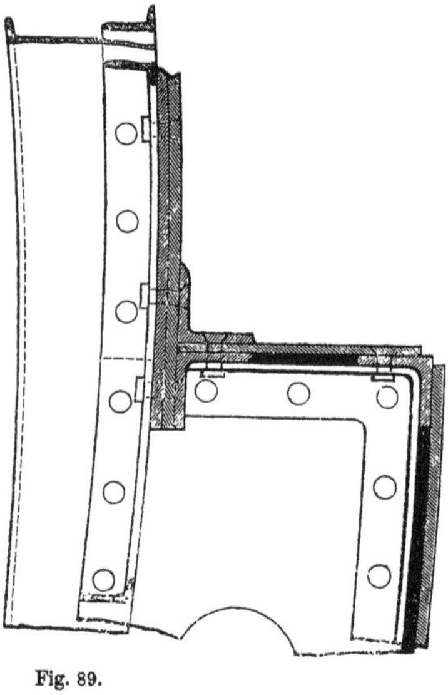

Fig. 89.

The method of fitting the armour-shelf of the 'Captain,' building by Messrs. Laird, is given in Fig. 90, and differs from that of the 'Hercules' in having the outer part of the shelf formed by a deep-flanged angle-iron, which is secured to the longitudinal plate forming the inner part of the shelf by a double-riveted edge-strip, worked underneath. The outer angle-iron of the 'Hercules' arrangement is thus dispensed with, while the upper strake of bottom plating has its edge double riveted to the vertical flange of the angle-iron, instead of being single riveted as in the 'Hercules.' There is, however, an increase of weight over the 'Hercules' plan, and the caulking and any re-caulking that may become necessary, are not quite so well provided for. If the edge of the upper strake of bottom plating did not come up to the lower edge of armour, but stopped about $\frac{3}{4}$ inch below it, the inner edge could then be caulked on the vertical flange of the angle-iron, in case of leakage, without removing the armour. This arrangement has been adopted in the 'Invincible' class of ship, without the covering plate, but of course an open groove is thus left, between the armour and the bottom plating, which has to be cemented. The inside plate of the shelf of the 'Captain' is 10 by $\frac{3}{4}$ inches and the double angle-irons are 5 by 5 by $\frac{3}{4}$ inches; the plates being worked in 48-feet lengths welded in the middle and having treble riveted butt-straps, and the angle-irons being worked in 36-feet lengths and having double-riveted covering angle-irons. The angle-iron forming the outer part of the shelf is 8 by 6 by $\frac{3}{4}$ inches, and is worked in lengths of 48 feet, the butts being secured by treble-riveted straps. At the butts of this angle-iron the plate, which connects the two parts of the shelf to each other, is widened and turned down so as to form a butt-strap. The plate thus made to serve as butt-strap is worked in 16-feet lengths, and has treble-riveted straps. It is evident that where so many longitudinal plates and angle-irons are combined to form this shelf, great care is required in disposing the butts, and the lengths of plates and angle-irons used are as great as to render a very good disposition possible.

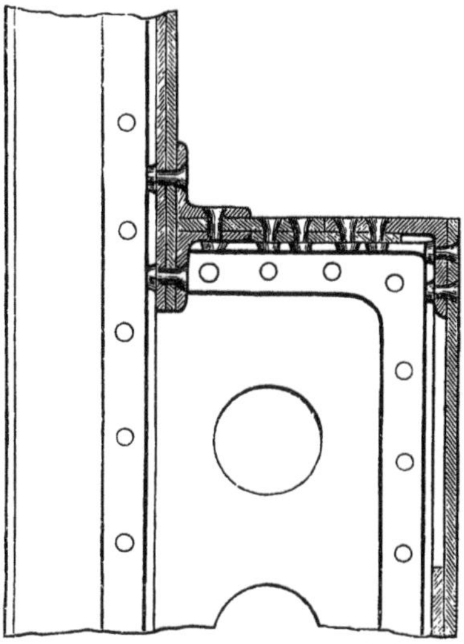

Fig. 90.

Allusion has been made to the armour-shelf arrangements of the 'Invincible' class of ship, and the details of these arrangements are illustrated in Figs. 91 and 92, of which the latter is on a smaller scale than the former. Before proceeding to notice the particulars of this armour-shelf, it may be well to state that in these ships there are no wing-passage bulkheads, but the upper part of the double bottom is made of a considerable depth, as shown in Fig. 92. By this means the capacity of the hold is made greater than it would be if there were wing-passages, and this is a matter of importance in comparatively small vessels which are very highly powered. At the same time, by increasing the space between the two bottoms near the water-line, protection is obtained from the probability of injury by ramming, and by having the upper part of the inner bottom nearly vertical it is made to act as a girder which gives the same kind of strength to the structure as is usually obtained from a wing-passage bulkhead. The upper longitudinal forming the top of the double bottom is made up of two plates, one of which is very wide, and the other forms the armour-shelf. These two plates are connected by a lap-joint as shown in Fig. 91, and the inner edge of the shelf-plate is attached to the skin-plating behind armour by a single continuous angle-iron, the fastenings in the horizontal flange of this angle-iron being made to serve as fastenings in the lap of the shelf-plate and the longitudinal. The arrangement of the outer edge of the shelf-plate has been previously described.

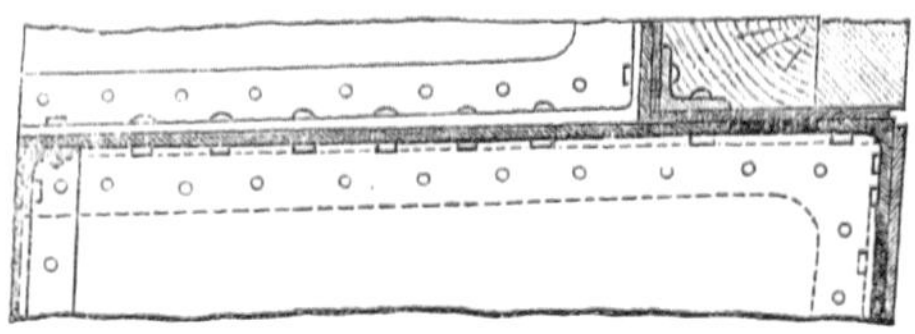

Fig. 91.

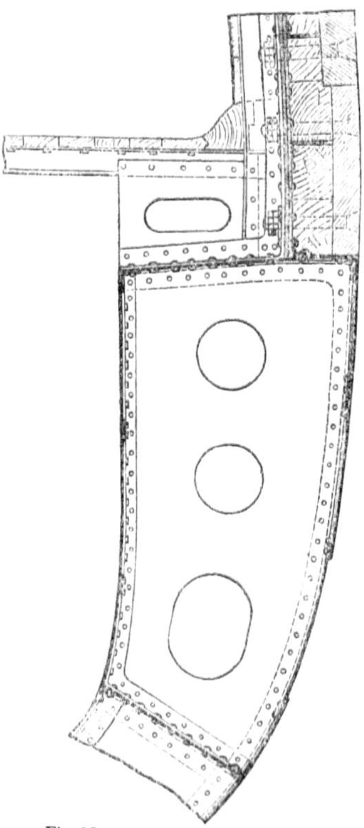

Fig. 92.

In the 'Hercules,' as in the 'Bellerophon,' all the longitudinals except the lowest one are continued forward to the stem. No. 6 longitudinal, in both ships, ends at the boundary of the double bottom. The ends of the longitudinals are connected with the plate-frames by single angle-irons, and the apparently slight character of this arrangement is justified by the considerations that it would be perfectly useless, under any circumstances, to make the connection between the longitudinal and the transverse frame very much stronger than the longitudinal frame itself is at its weakest section; and that in this case the transverse frame could not transmit to the ship a great longitudinal stress. A very similar arrangement is followed at the endings of the longitudinals which are stopped at the stuffing-box bulkheads, only as the angle-iron on the inner edge of the longitudinal is there continuous, it is sometimes turned down against the bulkhead and riveted to it.

The longitudinals of the 'Captain' do not run forward to the stem, and most of them are ended at transverse frames. Here, as in the case before described, the angle-iron on the inner edge is continuous as well as that on the outer edge, and both of these continuous angle-irons are turned down against the transverse frame and riveted to it. Where the longitudinals run well forward, those on opposite sides are joined by a horizontal plate, and the connection of the fore ends is made by continuing the angle-iron around the outer and foremost edges, and riveting it to the transverse plate-frame at which the longitudinal stops. The fine part of the bow is thus greatly strengthened, and the two sides are firmly united. The general rule observed in determining the ending of the longitudinals in ships such as the 'Captain,' of which the extreme forward part is framed transversely, is that the longitudinal shall never be brought closer than 2 feet to the armour-shelf, so as to allow the transverse frames to be run down that distance below the shelf.

Allusion has previously been made to the care which was taken in disposing the butts of the longitudinal frames of the 'Bellerophon,' in order to preserve, as nearly as possible, the continuity of longitudinal strength. A sketch showing the disposition of these butts which was made in the 'Hercules,' is given in Fig. 93. In this sketch the various longitudinal frames are supposed to be laid on one plane, with the outer edge of one coinciding with the inner edge of the frame next higher in number; thus No. 1, or shelf plate, has its outer edge well with the inner edge of No. 2, and so on. The corresponding stations are kept well with each

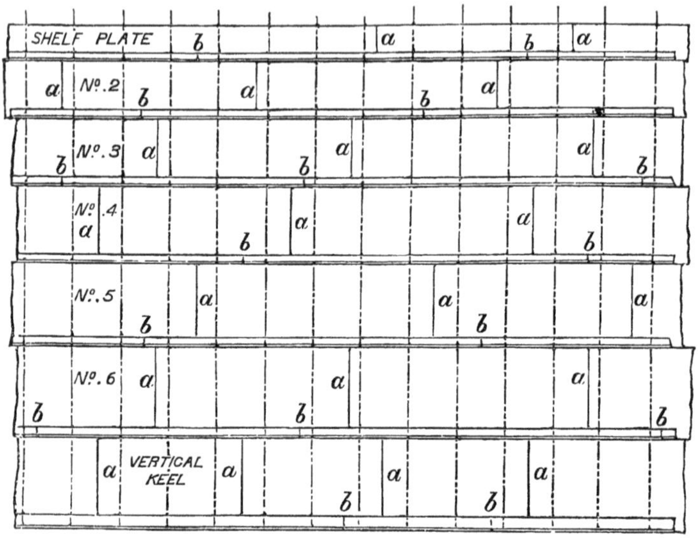

Fig. 93.

other on all the longitudinals, and are shown in the sketch by the dotted straight lines. These stations are 4 feet apart. The butts of the plates are marked *a*, and those of the angle-irons, *b*. The vertical keel is worked in 12-feet lengths; but all the other longitudinals are in 16 and 20-feet lengths, the longer plates being used to give good shift. The butt-straps are double and double-chain riveted. The continuous angle-irons on the outer edges of the longitudinals are worked in 28-feet lengths, the butts being connected by covering angle-irons, double riveted. On reference to Fig. 93, it will be seen that there are never less than two longitudinals between consecutive butts, either of plates or angle-irons, in the same frame space of 4 feet; and in several spaces there are only one or two butts. The butt *b* of an angle-iron is never nearer than 4 feet to the butt *a* of the plate on which the angle-iron is worked, except on the vertical keel, where the distance between *a* and *b* is about 3 feet. By means of this disposition of butts there is no transverse section of the longitudinal framework, where the strength is very much less than that of any other section. In the 'Captain' the longitudinal plates are worked in 24-feet lengths, and the continuous angle-irons in 48-feet lengths. On account of the great lengths of the plates and angle-irons the shift of butts obtained is a very good one, and with the exception of the vertical keel and No. 3 longitudinal, no two butts of longitudinal plates come in the same transverse section, while in no case are two butts of the angle-irons *b* placed in the same frame space. The butts of adjacent longitudinals generally have a shift of 8 feet, and are in no case nearer than 4 feet; and the butts of the angle-irons are never nearer than 4 feet to the butts of the longitudinal plate on which they are worked.

The transverse framing throughout the double bottom of the 'Hercules' is identical in character with that of the 'Bellerophon,' as will be seen on reference to Plate 5. The plates at the water-tight frames of the 'Hercules' are, however, $\frac{7}{16}$ inch thick, instead of $\frac{1}{2}$ inch, as in the 'Bellerophon,' and the continuous transverse angle-irons are there reduced to $3\frac{1}{2}$ by $3\frac{1}{2}$ by $\frac{7}{16}$ inches, while the staple angle-irons on the outer edges and ends are single. Sketches, showing the detailed arrangements of this ship at a bracket and a watertight-frame, have already been given in Figs. 37 and 38, p. 37, and will fully illustrate the preceding remarks. Before and abaft the double-bottom, the transverse framing below the second longitudinal is similar to that of the 'Bellerophon,' before described, and the dimensions of the plates and angle-irons are almost identical with those previously given. In wake of armour the frames of the two ships

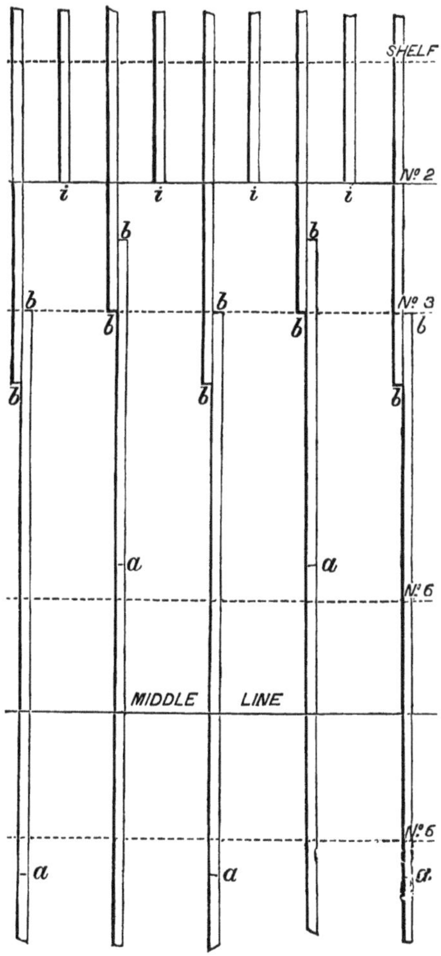

Fig. 94.

are similarly arranged and formed; but the double angle-irons on the outer edges of the 10-inch reversed frames are reduced to $3\frac{1}{2}$ by $3\frac{1}{2}$ by $\frac{1}{2}$ inches in the 'Hercules.' The positions of the scarphs of the alternate frames behind armour with the continuous transverse angle-irons, and of the butts of the continuous angle-irons of the 'Hercules,' are shown in Fig. 94. This sketch is an inside view of an expansion of the frames, and in it the butts are marked *a*, and the scarphs *b b*. The lowest longitudinal, No. 6, is about 6 feet 4 inches from the vertical keel, and the butts *a* of the continuous angle-irons are 18 inches further out, and are placed alternately on opposite sides of the middle line. A covering angle-iron is worked over each butt to keep up the transverse strength. The scarphs *b b* are 3 feet long, and are alternately above and below No. 3 longitudinal. The riveting of the scarphs is shown in Plate 5. The intermediate frames behind armour end upon No. 2 longitudinal, as shown at *i*, *i*, in Fig. 94. This arrangement of butts and scarphs is almost identical with that of the 'Bellerophon.'

It will be remembered that in illustrating the armour-shelf of the vessels of the 'Invincible' class, attention was called to the fact that wing-passages are not formed in these ships, but that the double bottom is made to serve the same purposes. On again referring to Fig. 92 it will be seen that the transverse framing of these vessels above No. 2 longitudinal differs greatly from that of the 'Bellerophon.' Below No. 2 longitudinal, the transverse frames are identical in their character with those of the 'Bellerophon,' being made up of continuous angle-irons, of frame and connecting angle-irons, and of bracket-plates. At No. 2 longitudinal the continuous angle-irons are ended, and forged angle-irons are fitted on the edges of the plate-frames. The lower ends of the outer angle-irons on the frames behind armour are turned in upon the top of the double bottom and riveted to it, and plate-brackets are fitted and riveted to the frames and beams in order to strengthen the connection of the various parts of the framing, and prevent any working at the junction of the protected part with the double bottom. It will be obvious from the preceding remarks that the transverse connection of the frames from gunwale to gunwale, which is obtained in the 'Bellerophon,' is not given by this plan, as there is no direct connection between the continuous transverse-frames, ending at No. 2 longitudinal, and the frames behind armour. But the indirect connection, made by means of the inner bottom and the frame angle-irons between No. 2 longitudinal and the top of the double bottom, gives ample transverse strength, and the advantages previously enumerated as accruing to the plan, are secured without injuriously affecting the strength of the structure.

The framing of the unprotected parts of the 'Hercules' differs from that of the 'Bellerophon' in having frame angle-irons, 7 by $3\frac{1}{2}$ by $\frac{7}{16}$ inches at intervals of 4 feet, instead of frames formed of plates and angle-irons. No bracket-plates are used at the feet of these frames; but the transverse flanges of the alternate frames behind armour are run up through the stringer and directly connected with the angle-iron frames. The whole of the arrangement is thus made simpler and less expensive than the corresponding framing of the 'Bellerophon.'

The general disposition and arrangements of the bow-framing of the 'Hercules' are similar to those previously described for the 'Bellerophon,' and the same precautions are taken to ensure the strength and safety of the bow in case the ship should be used as a ram. Between the stuffing-box bulkhead and the stern post the transverse plate-frames are arranged and lightened in this ship similarly to the corresponding frames of the 'Bellerophon,' only in this ship they are bounded by the flat below the lower deck (corresponding to the flat marked A in Fig. 87), which is made watertight. Above this flat a bracket arrangement of framing is adopted, and so a saving of weight is effected as compared with the corresponding framing of the 'Bellerophon.' The increase in weight of the flat A, required in order to make it watertight, is comparatively small, and it bounds that part of the stern most liable to injury from accidents happening to the propelling apparatus, and forms a smaller compartment than that at the stern of the 'Bellerophon.' In consequence of the flat A being made watertight, the lower deck abaft the stuffing-box bulkhead is not plated over in this vessel, but the lower-deck stringer is continued around the stern. The stern framing abaft the post is of a similar character to that described for the 'Bellerophon,' the principal difference in the arrangements of the two ships being that only a few of the stern frames are canted in the 'Hercules.' The rudder arrangements of the 'Hercules' are of a very novel character, and will be fully described further on.

Before concluding our remarks on the bracket-frame system, it may be interesting to add, that the introduction of this system of construction, so far from adding to the weight of the hull, diminishes it in the proportion shown in the following list, in which it is supposed that the skin-plating behind the armour is $\frac{9}{16}$-inch in every case, and that the armour and wood backing are removed:—

	Weight per 100 feet of length in Tons.
Achilles	1333
Agincourt	1234
Black Prince	1303
Hector	1170
Resistance	1253
Bellerophon	1123
Penelope	680

The first five ships have single bottoms with partial inner plating, and the last two have double bottoms on the new system. It should be added, that the great difference between the weight for the 'Penelope' and the weights for the other ships is largely due to the use of steel in place of iron, and to the fact of her being a comparatively shallow ship.

CHAPTER VIII.

DECK FRAMING AND PILLARING.

THE important part which the decks of a ship play, not only in forming platforms, but in keeping the sides at a fixed distance, and preventing change in either their longitudinal or their transverse form, was fully recognized by wooden shipbuilders, who endeavoured by every means to make a firm connection between the beam ends and the ship's side, to prevent change in the angle which the beam made with the side, and to ensure a good arrangement of pillars underneath the beams, thus greatly contributing to the strength and rigidity of the ship. Nor does the iron shipbuilder neglect any of the precautions formerly practised in wooden ships; but in carrying them into execution he has on the one hand greater difficulties to encounter, and on the other greater facilities to aid him. His difficulties are greater because the unstiffened shell of an iron ship is much more flexible than that of a wooden ship, and unless great care is taken, the fastenings will be injured by the working which results from that flexibility; and he has greater facilities, inasmuch as the connection of the beams with the side and the provision of longitudinal strength are easily made without having recourse to the elaborate arrangements of a wooden ship. The beams of an iron ship are usually spaced so that each shall come on a frame, and be directly attached to it; and the knee which is usually formed on the beam-end is made of sufficient depth to effectually resist the tendency to change of angle between the side and the beam end which results from the strains to which the vessel is exposed. This, which is the principal object of a beam knee, is much more effectually accomplished in an iron than in a wooden ship, owing to the readiness with which the end of a plate-beam is formed into a knee of considerable transverse depth, and to the superiority of riveted iron work to bolt work in wood, especially where, as in this case, the rivets can be placed so much nearer to each other than the bolts. And, further, an iron ship is usually provided with transverse bulkheads, and has in them an amount of strength well adapted to prevent change in the transverse curvature of the sides as high up as the bulkheads extend. In most iron ships longitudinal stringer-plates are worked on the beam ends, and the advantages resulting from the arrangement are very great, and are recognised by all shipbuilders. These plates act as horizontal knees to each beam, and do useful work even if they run out no further than the inside of the frames, and are not fastened to them. If, however, the stringers are attached by means of longitudinal angle-irons to the reversed frames, they add still more to the resistance to change in the longitudinal form of the ship; and if, as is now usual, they are run out between the frames and connected with the outside plating, the greatest possible amount of stiffness is secured. The advantages, in respect of longitudinal strength, which result from the use of stringers will be more fully considered hereafter.

For some time after the introduction of iron shipbuilding the deck beams were made of wood. An instance of this is given in Fig. 72, page 77, which is the section of the 'Recruit' before alluded to when describing the details of her framing. It will be seen that the arrangements of shelf, waterway, clamp, and beam-knees of the upper deck, are exactly similar to those of a wooden ship, except that the lower end of the knee is turned in and screw-bolted to the longitudinal stringer worked between decks.

Another instance of the use of wood beams in an iron ship is found in the iron troop-ship 'Megæra.' Here also the arrangements of the beam-knee, &c., are similar to those of a wooden ship. The knee is secured by screw-bolts, of which the nuts are hove up on the reversed angle-iron. A longitudinal angle-iron is worked above the beams and against the frames, being riveted to the latter and bolted to the former.

The connections between the wood beam and the ship's side in both these cases are altogether disproportionate in strength to the usual mode of connecting iron beams; and the use of a yielding material, such as wood, for beams in an iron ship, is in itself highly objectionable, on account of the working which must, in some degree at least, ensue, and which must eventually injure the fastenings.

Although the employment of iron beams in iron ships has now become almost universal, there are still cases in which wood beams are fitted. Mr. Scott Russell in his work on 'Naval Architecture' gives an instance of an ocean screw steamship constructed for the Australian mail and passenger service in 1852, in which the beams were of wood, and were secured to the frames by bracket-plates, riveted to the frames and bolted to the beams. A similar mode of connecting wooden beams with iron frames has also been adopted in many of the iron-built American monitors. Mr. Scott Russell states that, in his opinion, a wooden deck with iron beams makes a weaker ship than a wooden deck with wooden beams, and adds that in order to give the wooden deck the support which wooden beams afford, and at the same time to preserve the transverse strength given by iron beams he has adopted two different arrangements. One plan consists in putting in iron beams at intervals in order to supply transverse strength, and fitting wooden beams between them to assist the wooden deck. In the other plan the beams have been formed of vertical plates with a timber beam bolted on each side of them. He strongly states his preference, however, for an iron deck and iron beams, and advocates their adoption on the grounds of the increase of structural strength, and the simplicity of the deck framing thus rendered possible.

Iron beams were first employed in steamships above the engines and boilers, and their superior qualities in respect of strength and durability, together with the facilities presented for firmly connecting them with the sides of the ship, soon led to their general adoption. According to calculations made by M. Dupuy de Lôme, and given in his Report, the weights of iron and wood beams of equal strength (for the sections of iron beams then in use) are in the proportion of ·65 : 1, or 1·2 : 1. This latter proportion, which makes the iron beam heavier than the wood beam is deduced from a very defective form of section, and in all the other cases given by the author the iron beam is the lighter. It should be observed, however, that these calculations are based, for the most part, upon theoretical considerations, and that in his work on 'Iron Shipbuilding' Mr. Fairbairn states, as the result of experiment, that as regards the strengths with equal weights for beams or frames it is in favour of oak; but goes on to say that, on account of the superior fastenings in the hull of an iron ship the number of beams can be considerably decreased, and thus a great reduction of weight be effected, while at the same time the strength of the ship as a whole is considerably greater than could be possibly attained in a wooden ship.

The forms of section which have been adopted for iron beams are very various, and are illustrated in Fig. 95. In some light vessels, and in the framing of mere platforms in larger vessels, the section marked *a* has been used, the broader flange of the angle-iron being vertical. In some ships two angle-irons set back to

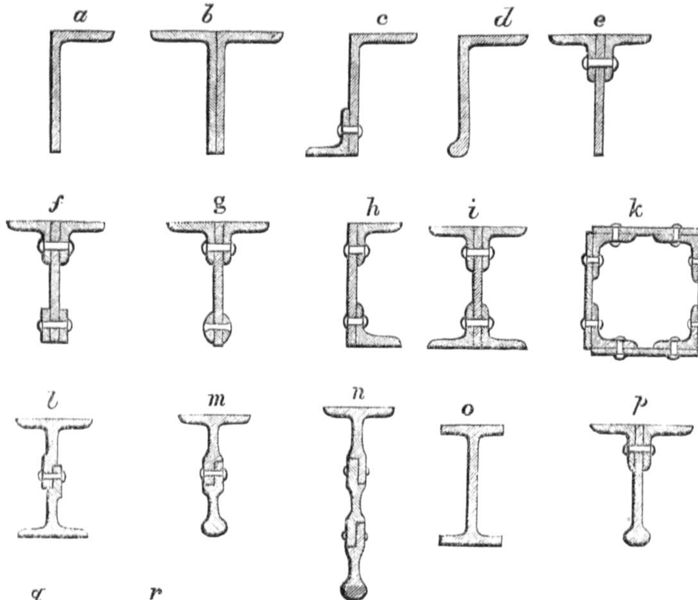

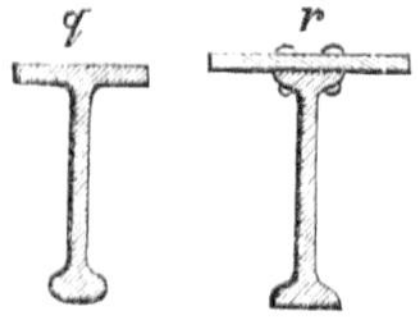

Fig. 95.

back, as shown by *b*, and riveted together, have been used to form the beams; and in other vessels the simple angle-iron arrangement given above at *a*, has been modified by riveting another angle-iron on the lower edge, as shown in section by *c*. In the construction of the composite gunboats for H.M. Service, a bulb angle-iron has been used for the deck beam, of the section illustrated by *d*. In many of the earlier iron ships the beams were formed of a vertical plate with double angle-irons on the upper edge, similar to that shown in section by *e*, and this arrangement has been modified by riveting to the lower edge strips of plate or half-round iron, in the manner shown by *f* and *g*. In some ships the beams have been formed in the manner shown by *h*, and this was the section adopted on some of the decks of the 'Great Eastern.' One of the forms of section now in most common use for large beams is that given at *i*, and it is very generally employed for lower-deck and hold beams of large ships. The hold beams here alluded to are fitted in large merchant ships, when it would be inconvenient to have a complete lower deck, in order to give transverse strength, and very frequently the sectional form adopted, instead of the last-mentioned, is that given in *k*, which is known as a box beam. This form of section is also very commonly used for paddle-beams in steamships, and has been employed in the deck-framing of turret-ships underneath the turrets. T-iron bars are sometimes used for beams, and in some cases beams have been formed of two T bars riveted together, as shown by *l*. The sections marked *m* and *n* are known as the "Butterley Neutral Axis Riveted Beams." The latter of these two sections has not, we believe, been applied in shipbuilding. The objection to beams formed in this manner is that the material required to form the scarphing parts is placed in such a manner as to be almost ineffective as regards strengthening the section. They are therefore heavy in proportion to their efficiency. In other ships I-iron beams have been used, the sectional form given in *o* being arrived at either by rolling it in one, or welding two T-irons together. The section now in most general use for the upper and main-deck beams is that marked *p*, which is composed of a bulb-plate, with two angle-irons on the upper edge. Another form now commonly employed, especially for upper-deck beams, is known as the "Butterley Patent Welded Beam," its section being that shown by *q*. The bulb-half of this beam is rolled separately from the upper or T-half and the two are welded together along the neutral axis of the beam. A similar section has, however, recently been rolled in one, and been brought into use. A new sectional form for beams which has been patented by Mr. Phillips is shown by *r*. It will be seen that the beam is made of an I-shaped girder rolled in one and having a wide plate riveted to the upper flange. This plate is intended to give lateral stiffness to the beam, and prevent it yielding in any but a vertical direction. It appears from experiments made with Mr. Kirkaldy's testing machine that this section is considerably stronger than that marked *i* for equal weights of material per foot of length. The section is proposed, principally, for girder work in connection with civil architecture, but its application to the beams of ships has also been recommended. It may be remarked, however, that in the framing of a deck, the carlings and other longitudinal pieces resist the tendency to bending sideways or buckling, which the top plate of Mr. Phillips' section is intended to prevent; and in ships which have iron decks, the deck plating acts as an immensely strong top flange, which effectually resists any buckling in the vertical webs of the beams.

Mr. Fairbairn states as the result of his experiments on the strength of beams that the I-shaped is the strongest when the flanges are properly proportioned. This section is very difficult to roll in one, and cannot, at present, be obtained of sufficient dimensions for very large beams, unless it is welded along the neutral axis. There seems little doubt, however, as the power of rolling machinery is being increased so greatly, but that ere long manufacturers will be able to supply the largest beams required of this section rolled in one, and at the present time beams of considerable dimensions of I-shaped section, and of the Butterley pattern, are thus made.

Having given most of the sections of beams which have been employed in iron ships, it may not be out of place to consider briefly the principles which should regulate the proportions and forms of the sections. The object to be aimed at in determining on the section to be adopted, is that the maximum strength of the amount of material in the section shall be as nearly as possible attained. The investigations of those who have treated on the "Strength of Materials" show that in order to arrive at this result two things are necessary; the first, that the moment of inertia of the section shall be made as great as possible, consistently with the retention of the sectional strength required to resist the vertical shearing strains resulting from the load; and the second, that the section shall be of uniform strength, so as to ensure that the top and bottom of the section shall begin to yield simultaneously under a breaking load, the one giving way to a compressive and the other to a tensile strain. In order to meet the first requirement the sections of beams have been made up of a vertical web, and of either top and bottom flanges, or only a top flange. The varied arrangements of the webs and flanges have already been described. In a deck beam it is imperative that there shall be a top flange in order to receive the fastenings of the deck planking; but in many cases the bottom flange has either been wanting altogether, as in the sections *a*, *b*, and *e* in Fig. 95, or has been imperfectly supplied, as in the sections *f* and *g*. In all the other cases given in the preceding sketches, the bottom flange of the beam is formed of a flange or bulb rolled on the web-plate, or of angle-irons riveted to it. By these arrangements of web and flanges the material in the latter is placed at the greatest possible distance from the centre of gravity of the section, and is thus made most effective in resisting the bending moment. The ordinary assumption made in proportioning the material in the web and flanges, is that the web bears the shearing strains, and the flanges resist the bending moment. This is a very good approximation in the case of beams with top and bottom flanges such as those whose sections are shown by *l*, *o*, *p*, and *q*, in Fig. 95. But in the cases where the web has a considerable sectional area, as compared with the areas of the flanges, it is necessary to take into account the resistance to bending offered by the material in the web; and this is still more necessary when the beam has no lower flange. In order to meet the second requirement, and arrive at an uniformly strong section, the usual method, for beams with top and bottom flanges, is to make the sectional areas of the flanges inversely proportional to the resistance offered to rupture by compression and extension respectively, or, in other words, inversely proportional to the *moduli of rupture* by compression and extension respectively. In wrought-iron beams the area of the bottom flange is generally about five-sixths that of the top flange, and the flanges together have an area about equal to that of the web. Mr. Rankine states that for a T-iron beam of uniformly strong section, the web should be of three times the sectional area of the upper flange, and this proportion is also applicable to the sections marked *a*, *b*, and *e*.

Beams are generally of uniform depth throughout their length, except at the ends, where the knees are considerably deeper than the beam amidships. It may be supposed that the principle of uniformity of strength might, with advantage, be applied to the longitudinal sections of beams. If it were applied here the depth of the web would be made proportional to the bending moment at every section, and the beams would consequently require to be decreased in depth from the middle toward the ends. But in practice, as has already been stated, the beams are considerably deepened at the ends in order to make a firm connection with the side; and, consequently, if the beam were decreased in depth from the middle outwards as far as the knee, the weakest section would be immediately adjacent to the strongest. It may be added to the above that a beam in a ship has to act not merely as a girder which supports a load, but as a strut or tie between the two sides, and as part of the ship's frame, and consequently it would not be sufficient to determine its form and dimensions simply from considering it as a girder only. On the whole, therefore, we may conclude that the general practice of keeping the beams of uniform depth, is preferable to attempting to vary their depth according to a bending moment the amount of which is not known, and which varies considerably under different circumstances. The usual practice among shipbuilders is to go upon some successful example in designing the beams of a new ship, instead of acting on partial and theoretical considerations simply.

Lloyd's rule with respect to the form and depth of beams is as follows:—"Beam-plates to be in depth one-quarter of an inch for "every foot in length of the midship-beams, and to be in thick- "ness one-sixteenth of an inch for every inch in depth of the said "beams, and to be made of H iron, T bulb-iron, or bulb-plate "with double angle-irons riveted on the upper edge; the two "sides of each of these angle-irons to be not less in breadth than "three-fourths the depth of beam-plate, and to be in thickness one- "sixteenth of an inch for every inch of the two sides of the angle- "iron; or the beams may be composed of any other approved "form of beam-iron of equal strength. Where beams below the "upper or middle deck (including orlop beams) have no deck laid "upon them, the angle-irons on their upper edges are required to "be of the dimensions of the angle-irons of the reverse frames." The Liverpool Rules require that the beams shall be formed of bulbed iron with strongly bulbed lower edge, with double angle-irons on the top edge, or of bulbed T iron, or of any other approved form, and their regulation as to depth is almost identical with that of Lloyd's Rules.

The beams of the earlier iron ships were usually formed of plates and angle-irons. In the 'Dover' the beams were formed as shown by *e* in Fig. 95, and in the 'Birkenhead,' as shown by *i*. As the lengths of the beam-plates required were greater than could then be rolled in one, it was usual to join the several lengths of a beam-plate either by a lap or butt-joint. The latter was most common, and double butt-straps were usually fitted, the angle-irons on the edges being joggled over them. In some ships the lengths of plate were welded together instead of being lapped or butt-strapped. As the lengths in which plates can be rolled have been so greatly increased since the commencement of iron shipbuilding, it is now usual to have the plates of the made beams of most ships in one length, and the angle-irons are also rolled in one length in most cases. In large ships, however, where the breadth is great, the plates of the made beams are still made up of two or more lengths welded to each other, and in some cases the angle-irons on the edges are also made up of two lengths connected by cross-welding. The particulars of the manufacture of the made beams of the 'Northumberland' will illustrate the foregoing remarks. The longest beams on the main and lower decks are made up of four lengths, and the others of three lengths of plate, and in all cases the angle-irons on the edges are in two lengths. The arrangement of the butts of plate and angle-irons in a three-piece beam is given in Fig. 96, and the arrangement of butts in a four-piece beam in Fig. 97. In both the sketches the butts *a*, *a*, are those of plates, and

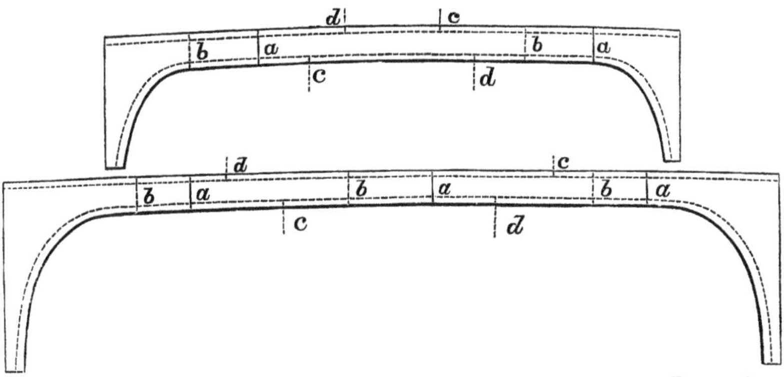

Figs. 96, 97.

those dotted and marked *b*, *b*, are the butts of plates of the adjacent beams on each side of the first. The butts of the angle-irons on one side of the beam-plate are marked *c*, *c*, and those on the other side of the same beam-plate are dotted and marked *d*, *d*. It will be seen that care is taken to shift the butts, so as to prevent any unavoidable weakness at the weld from occurring near to a similar place of weakness. The plates forming the beam-arms are 5 feet long on one end and 7 feet on the other end of the same beam, and a shift of 2 feet is given to the welds of adjacent beams by bringing the long and short arms alternately on the same side of the ship. The iron for these beams was rolled on the premises, and it was considered more economical to roll plates which would admit of the two arms of a beam being punched out of one plate, than to form the knees in the usual manner. A plan showing the way in which the beam-arms were lined on the plate is given in Fig. 98. When the lining-out had been completed the beam-arms were punched out, and the jagged edges were trimmed off after the beam had been put in place. The surplus pieces of plate were used for making welds and forming knees to the Butterley beams. The process of welding the beam-plates together was conducted in the following manner:—The butts of the plates were each snaped away with the hammer and slightly upset, and the plates were fixed on low carriages or trollies, and secured in the proper position for welding by a clamping arrangement, which allowed the welding to be conveniently performed. The carriages on which the beam-plates were fixed ran on a portable railway, and the beam was thus easily conveyed from the furnace to the anvil. The blast-furnace was also portable, and was constructed of iron with a lining of fire-bricks, and supplied with coke through a rectangular orifice in the top. The parts requiring to be heated were laid over the opening in the top of the furnace, the adjacent parts of the beam-plates being kept cool by a covering of fire-brick and wet loam. At the same time a piece of square iron was heated in an adjoining fire ready to be placed in the lips of the scarphs, so as to weld the two together. The latter operation was performed by hand, the workmen employed being two smiths, two hammermen, and two helpers. It should also be stated that each piece of plate after being heated was bent to its curve on the iron slabs used for bending angle-iron frames. The angle-iron for these beams was first straightened, then welded to the required lengths, and the parts which fitted against the beam-arms were bent on the slabs. While this was being done, the beam-plates were finally shaped by the beam-mould, and the rivet-holes on both edges were punched. The angle-irons were then temporarily secured to the beam-plates and the rivet-holes marked, after which they were removed to the punching machine, and had the holes punched. They were then brought back and riveted up on the beam-plates. The mode of welding the lengths of angle-iron for these beams was identical with that described above, only two heats were required

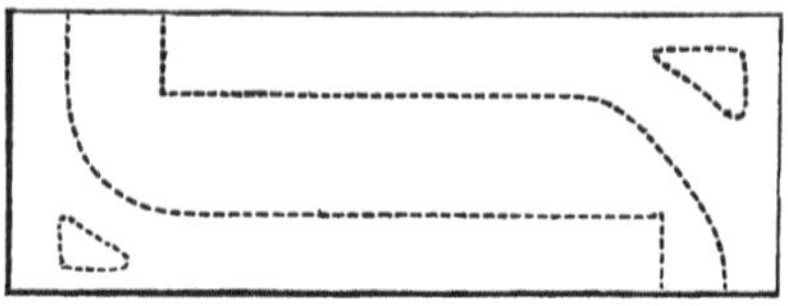

Fig. 98.

in this case, as each flange of the angle-iron was welded separately.

In the new Indian troop-ship 'Euphrates' the central part of the beam-plate is rolled in one, and the beam-knee ends are welded on to the central piece. The fibre of the beam-knees is in the same direction as that of the central piece of plate. The angle-irons are rolled in two lengths, the positions of the welds being interchanged on alternate beams, the lower becoming upper and *vice versâ*. In the made beams of the 'Serapis,' another of the Indian troop-ships, each beam-plate is made up of two lengths, the welds on alternate beams having a shift of 14 feet 6 inches, and the angle-irons are rolled in one length.

Having thus illustrated the manufacture of made beams, we proceed to describe briefly the mode of manufacture of the patent welded beam of the Butterley pattern. It has already been stated that the T part and the bulb part of this beam are rolled separately and afterwards welded along the neutral axis; the following details of the mode of welding are taken from the specification of Mr. Alleyne's patent, dated 8th December, 1859. The edges which are to be welded are introduced into the grooves of an H-shaped piece of iron, which the patentee calls a "glut." The parts to be welded are then heated by means of small cupola furnaces, and, when the welding heat has been attained, the joint is completed either by hammering or by passing it under the rolls. A portion of the glut may be burnt during the operation, but the edges of the T and bulb-irons, as well as the inside parts of the glut itself, are protected from oxidation. The specification goes on to state that, although the form of the glut may be varied from that described above, yet the same principle must be carried out in making the weld, and the edges which are to be welded must be embraced by the glut.

The bulb-iron now so generally used for deck-beams is rolled in one, and the angle-irons on the upper edge are worked after the bulb-iron has been bent to the round-up. The mode of manufacture of H-iron beams has already been alluded to, and the mode of welding the two T-iron bars which usually compose a beam of this section is similar to that described for the Butterley beams.

The different modes of bending beams to their round-up are worthy of remark. On the Mersey the beams are bent to their round-up and straightened cold in a screw-press worked by hand;

on the Clyde they are usually bent cold by Rennie's patent beam-bending machine; on the Tyne also the beams are bent cold; but in some yards the beams are made moderately hot, and bent to their round-up on the slabs used for bending frames, and in other cases the beams are heated and then placed upon blocks laid to the required curve, upon which they are allowed to settle, care being taken to prevent them from twisting during the operation.

The superior convenience and strength of attachment of the beams to the side obtained by the use of beam-knees, have led to their almost universal adoption. There are two modes of forming beam-knees which are now in general use. The first and most common method is that illustrated by Fig. 99. After the beam has been bent to the round-up, the end is split up for a short distance, and the lower part is turned down to form the outline of the knee. A piece of plate, marked *a*, is then welded in so as to fill the space thus left, and it is shaded in the sketch in order to distinguish its outline.

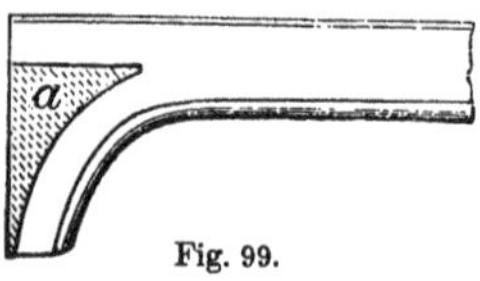

Fig. 99.

When Butterley beams are used, the welding of the two parts is generally discontinued at such a distance from the end as to allow the beam-end to be turned down to form the knee in the manner above described. Sometimes the plate *a* does not exactly fill the space at the beam-end, but leaves a triangular hole in the beam-end, as shown in the upper and main-deck beams of the 'Warrior,' in Plate 3. The second method of forming the beam-

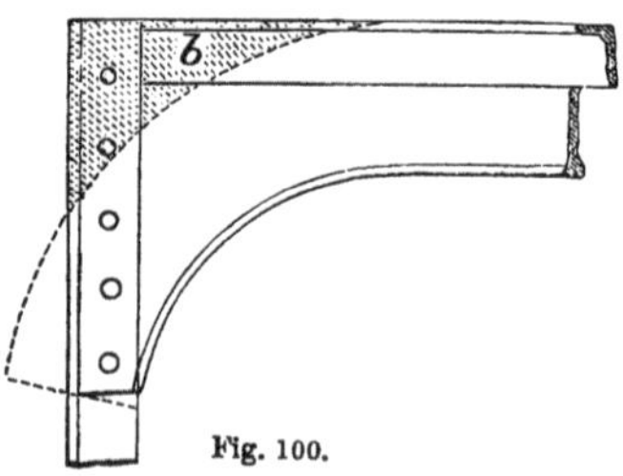

Fig. 100.

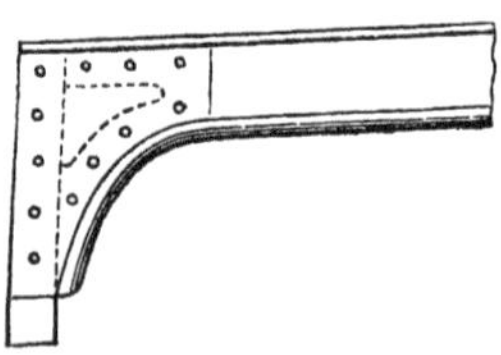

Fig. 101.

knee is used on the Clyde when the beam has no upper flanges rolled on it. The beam-end is itself turned down in order to form the lower part of the knee, and a piece of plate, marked *b* in Fig. 100, is welded on to form the upper corner. A third mode of forming beam-knees sometimes practised on the Tyne is illustrated in Fig. 101. It consists in turning down the lower part of the beam-end, as in the method first described, and riveting on a piece of plate to the side of the beam-web, in order to preserve the form of the knee. This method is heavier and less neat than the method generally followed of welding in a piece of plate. At Messrs. Harland and Wolff's yard, at Belfast, the knees to beams are usually formed by welding on a piece of plate below the beam-web proper, and when, as is most common, bulb beams are employed, the bulb is cut away from the lower edge as far in as the weld comes.

In Fig. 102 there is given a sketch of the mode of slinging the beam, and the arrangement of the anvils, which are adopted at Messrs. Laird's works, in order to readily perform the welding of the knees. The sketch needs no explanation; but it may be remarked that the fire in which the beam-ends are heated is placed so as to allow the beam to be transferred from it to the anvils

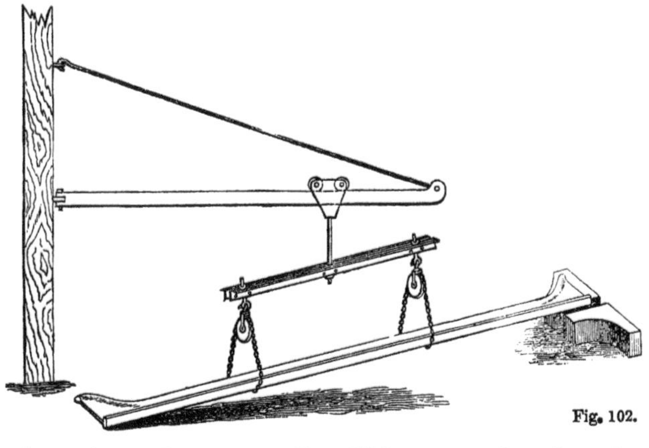

Fig. 102.

while slung from the crane. It will be seen, also, that the arrangements of the crane and slings are such as to allow the beam to be easily handled, and that by the use of two anvils, shaped and placed as shown, the welding of both sides of the beam-knee is very readily accomplished.

The angle-irons on the edges of beams are taken account of and worked in the manner described for the made beams of the 'Northumberland,' and the holes in beam-plates and angle-irons are usually punched. It is a very common practice in many ship-building yards to use the steam riveting-machine for riveting the beam angle-irons to the plates; but on the Tyne the riveting is performed by hand, the reason given for this course being that when the machine is used the beams have their round-up increased and have to be again taken to the bending machine and brought to their correct form after the riveting is completed.

Coming now to the illustration of the different means which have been adopted for connecting the beam-ends with the ship's side, we commence with a description of the arrangements of the 'Birkenhead,' which are fully shown in the enlarged views of the beam ends given in Fig. 71, page 75. The connections of the beams on the main and upper decks were very similar. A horizontal shelf-plate was worked below the beam-ends, and secured to short pieces of reversed angle-iron, and a stringer-plate and angle-iron were worked on the upper side of the beams. It will be seen from the sketches that, on account of the frames having been ended below the upper-deck shelf-plate, both shelf and stringer-plate on the upper deck were extended out to the skin-plating and secured to it, instead of ending inside the frames as on the main deck. On the lower deck only a shelf-plate was worked, the stringer-plate being omitted. These arrangements differ from those in general use in more modern ships, in having no beam-knees, nor any direct connection between the beams and the frames; in many instances where the plan was adopted, the beams were not stationed at frames.

The beams of H.M.S. 'Vulcan,' built in 1846, were formed and connected as shown in Figs. 103 and 104. In wake of the upper deck beam-ends a longitudinal clamp-plate, 30 by ½ inch, was worked on the inside of the frames. The beam-end was formed into a triangular knee, which was connected with the clamp-plate by double angle-irons. A stringer-plate and angle-iron were worked on the beam-ends and connected with the upper edge of the clamp-plate. With this arrangement also, there was no necessity for stationing every beam at a frame. The lower-deck beams of this ship were stationed at frames and connected to them by bracket-plates, in the manner shown in Fig. 104. This mode of connection was extensively practised previously to

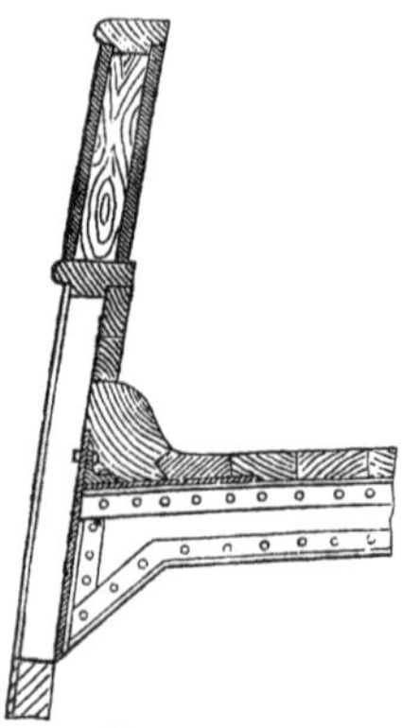

Fig. 103.

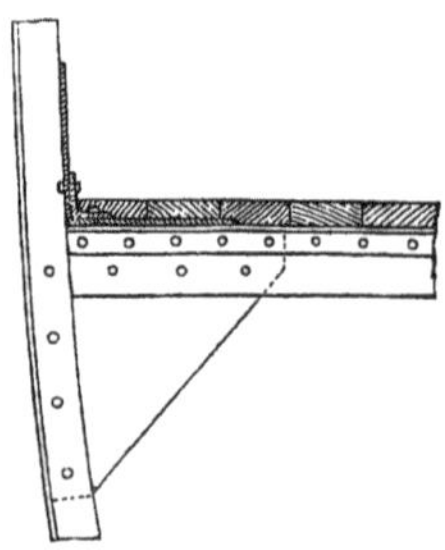

Fig. 104.

the introduction of the present mode of forming beam-knees, and is still sometimes adopted.

Having thus illustrated some of the principal means of connection which were adopted in a few comparatively old iron ships, we pass to the consideration of the arrangements of beam-ends now in common use. The employment of beam-knees has now become almost universal, and the only exceptions worthy of remark are those in which bracket-plates, or shelf-plates and brackets, are used. The latter arrangement is sometimes adopted for lower-deck and hold-beams, and an illustration of its details is given in Fig. 105. The beam-plate is run into the bosom of the frame and riveted to it, and a plate-shelf is worked under the beams and secured to the outside plating by intercostal angle-irons.

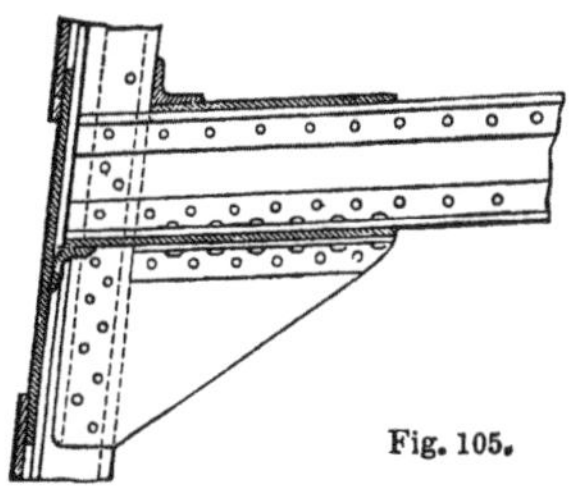

Fig. 105.

The connection is further strengthened by a bracket-plate being worked below the shelf and directly under the beam, so that the rivets connecting the angle-iron on the upper edge of the bracket with the shelf, pass up through the angle-iron on the lower edge of the beam. A direct connection is thus made between the bracket-plate and the beam. The connection of the beam-end is completed by a stringer worked in the ordinary manner. In the vessels of the 'Invincible' class the battery deck is higher than the main deck before and abaft the battery, the difference in height being equal to the depth of the battery beams at the side. This arrangement is made in order to obtain a good height of the port-sill above the water-level, and at the same time to make the height and weight of the belt of armour-plating as little as possible. This break in the deck would cause a great reduction in the longitudinal strength, unless the stringer were continued along the battery beams,* and hence the stringer is worked below, and the battery beams have no knees, bracket-plates being worked below the stringer-plate, in order to strengthen the connection. The whole arrangement is very similar to that shown in Fig. 105, except that there is no stringer-plate upon the beam-ends, a gutter waterway only being worked.

* It will be noticed also from the 'Bellerophon's' section in Plate 4, that the main-deck beams have no knees, and that a stringer-plate is worked beneath them. The explanation of this arrangement is identical with that given above.

Another special arrangement of the beam-ends is illustrated in Fig. 106 and represents the connection sometimes adopted for the arms of beams, which form the sides of the principal hatchways. The sketch is taken from the 'Queen,' a vessel of which the details of the framing have already been described. The beam-plate is run out to the side and attached to the side of the frame. A plate bracket or knee, marked A, is fitted below the beam and extends as far down as the next deck, its foot being connected with the stringer-plate. The plate A is connected with the beam-plate by a double-riveted strap, and the outer edge of the plate overlaps the frame and is riveted to it. The beam angle-irons on the lower edge are continued along the inner edge of the plate A by short lengths of angle-iron, which are butted at *b*, *b*, the butts being secured by double-riveted covering angle-irons. This very strong connection of beam to side is made in order to provide against the loss of transverse strength consequent on the fact that throughout the length of the hatchway there is no complete transverse tie.

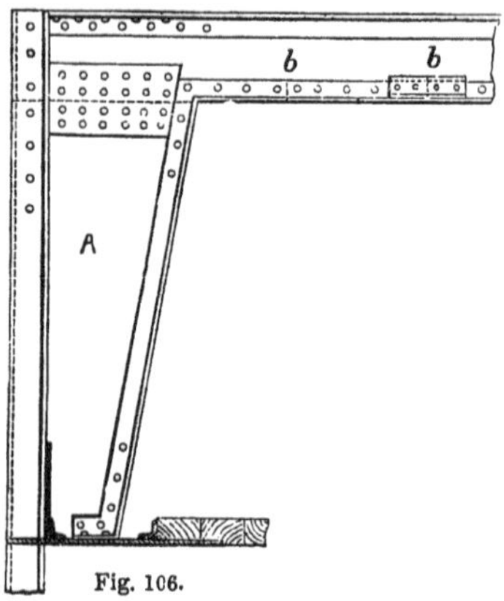

Fig. 106.

Lloyd's Rule with respect to the connection of the beams with the ship's side, is as follows:—"All beams are to be well and "efficiently connected or riveted to the frames, with bracket-ends "or knee-plates; each arm of the knee-plates at ends of beams not "to be less in length than twice and a half "the depth of the beams, and to be in thick-"ness equal to the beams." The Liverpool Rules agree with Lloyd's as to the depth of the beam-knees. The usual mode of connecting a beam-knee with a frame in a vessel of moderate dimensions, in which the frames are of comparatively small moulding, is illustrated in Fig. 107. In the sketch a reversed frame is worked upon the frame in wake of the beam-ends, and the knee is run into the bosom of the frame and ended just inside the fore and aft flange. Of the double angle-irons on the upper edge one is continued out

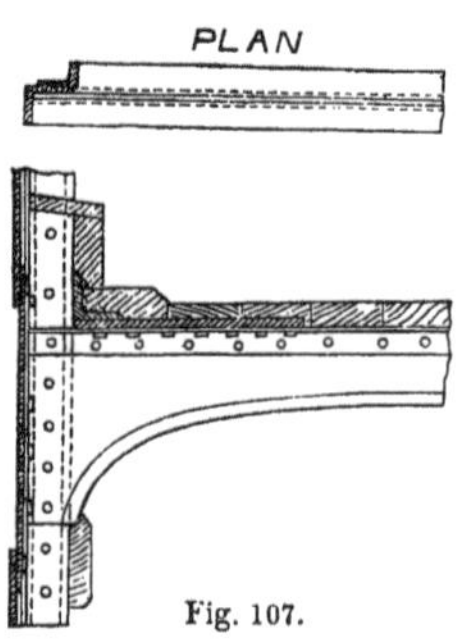

Fig. 107.

nearly to the flange of the frame angle-iron and the other is stopped against the reversed frame, as shown in the plan. The beam-knee is secured to the frame by a single line of rivets in this case, but in some ships where the moulding of the frames is greater, there are two or three lines of rivets, generally placed zigzag. An instance of this is given on the main deck of the 'Hercules' in Plate 5, which illustrates the connection of the beams of the more recent iron-clad frigates with the 10-inch frames in wake of armour. In these vessels the beam-knee and one of the upper flanges are run out as far as the inner edge of the $3\frac{1}{2}$ by $3\frac{1}{2}$ by $\frac{1}{2}$ inches angle-iron and fitted against it, the other flange is stopped against the reversed frame, and the knee is secured to the frame by two rows of rivets placed zigzag. On the upper deck of most merchant vessels the beams are fitted against the outside of the transverse flanges of the frame angle-irons, instead of being run into the bosom of the frames as is done on the middle and lower decks. This difference is made on account of the reversed frame ending just above the middle deck of a three-decked ship, and the lower deck of a two-decked ship, and consequently the outer edge of the knee can be either fitted against or brought just clear of the outside plating. In the unprotected parts of the iron-clads where there are no reversed frames the beams are run in and secured in a similar manner.

In the preceding remarks on the attachment of the beam-end to the frames, it has been assumed that both beams and frames are transverse. This is the usual case met with by the iron ship-builder; but there are some instances in which a few of the frames at the extremities of an iron ship are canted, and the common practice in such cases is to flange the beam-knees so as to make them fit against the sides of the frames, the beam itself being kept transverse. The objection to this practice is that the strength of the beam, considered as a transverse strut or tie, is somewhat reduced. In some of the iron-clads a large number of the frames at the bow and stern are canted, and the deck-framing is arranged diagonally. This is the case in the vessels of the 'Northumberland' class, in which the last transverse beam is that to which the head of the fore stern post is secured. Abaft this beam the deck framing is made up of diagonal half-beams and a middle-line carling which extends between the stern-posts. The half-beams stand in the planes of the cant frames and are consequently nearly at right angles to the side. Their inner ends are secured to the aftermost beam, and to the middle-line carling. Their outer ends are secured to the frames by beam-knees in a similar manner to that previously described. The transverse strength of a deck thus framed is, of course, less than that of a deck framed transversely, but in this case the diagonal arrangement has the advantage of giving great support to the armour-plated stern in case it is struck by heavy projectiles. In some of the later iron-clads, the forward and after parts of the decks are also framed diagonally. In the 'Bellerophon,' the diagonal half-beams have their inner ends attached to two fore and aft carlings, one on each side of the middle line, which extend from the aftermost beam to the stern. Between the fore and aft carlings athwartship carlings are worked in order to complete the connection of the two sides of the ship.

In vessels built on the longitudinal system of Mr. Scott Russell, the deck framing consists, for the most part, of longitudinal girders running from one transverse bulkhead to the next, and of longitudinal stiffeners worked between the girders underneath the deck plating. The only transverse framing usually adopted in the decks of these vessels is formed by continuations of the partial bulkheads, and by the upper edges of the watertight bulkheads. In the 'Great Eastern' the upper deck is a cellular iron structure of which the framing consists of longitudinal girders. The details of the construction of this deck will be described hereafter. The main-deck beams of this ship are placed transversely, as are also the beams of the decks below the main deck. Where the transverse beams of these decks come on partial bulkheads, they are secured to the upper parts of the bulkheads, no beam-knees being formed on their ends. Between the partial bulkheads the beams have the usual form of knee and are secured to the skin-plating by double angle-irons.

Returning to the consideration of the usual mode of deck-framing, the question of the arrangement of the beams is one which requires attention. Lloyd's Rules state that upper-deck beams in vessels with one or two tiers of beams, and the upper and middle deck beams in vessels with three tiers of beams, are to be fastened to alternate frames. The arrangement of the lower-deck or hold beams, as settled by the Rules, is regulated by the depth in hold. Thus, for vessels of 12 feet and under 13 feet depth of hold, or where the tonnage exceeds 200 tons, the hold-beams are not to be further apart than every eighth frame; for vessels of 13 feet depth and under 15 feet, the hold-beams are to be fastened to every fourth frame; for vessels of 15 feet depth and under 18 feet, the hold or lower-deck beams are to be fastened to every second and fourth frame alternately; and for vessels of 18 feet depth and above, the hold or lower-deck beams are to be fastened to every alternate frame. The Rules also provide for the use of orlop-beams fastened to every sixth frame, in two-decked vessels where the depth from the upper side of the upper-deck beams to the top of floor-plates exceeds 24 feet, and in three-decked ships where the depth from the upper side of the middle-deck beams to the top of floor-plates exceeds 24 feet, and where the depth from the under side of the lower-deck beams exceeds 15 feet. A depth of 25 feet is allowed instead of 24 feet for flush-decked ships in which the depth of the lower hold does not exceed 16 feet. These orlop-deck beams are required to have stringer-plates and angle-irons worked on their ends, of equal strength with those on the lower-deck beams. When the spaces between beams exceed two frame-spaces a knee or bracket plate is to be riveted to alternate frames and to the under side of the stringer-plate. It is also stipulated that where deviations are made from the foregoing Rules in wake of engine-rooms or hatchways, or where no deck is intended to be laid, a sketch, showing the proposed arrangements which are to be substituted, is to be forwarded for the consideration of the Committee.

The Liverpool Rules agree with Lloyd's in the arrangement of the upper and middle deck beams, and give very nearly the same directions with respect to lower-deck or hold and orlop beams. The differences of depth in hold required by the two Rules, for the same arrangement of lower-deck and orlop beams, are comparatively small, as will be seen on reference to the Appendix.

In the iron-clad frigates of the Royal Navy the deck-beams are, in most cases, secured to alternate frames; and as the frame-space in the later iron-clads is 2 feet, the space between the beams is generally 4 feet. In some ships, however, where the weight of the armament is very great, the beams in the battery-decks are only 2 feet apart.

In wake of hatchways and mast-holes, of which the length generally exceeds the usual space between the beams, the deck-framing is completed by half-beams, of which the outer ends are secured to the frames in a similar manner to the beam-ends, and the inner ends to the fore and aft carlings forming the framing of the hatchways or mast-holes. At the larger hatchways over engines, boilers, and the cargo-holds of merchant-ships, there are several half-beams abreast of the hatchway, and the carlings are made very strong in order to secure the inner ends. At these parts of the decks the transverse strength is considerably reduced, and various means of compensating for this loss of strength have been adopted. One has already been given in Fig. 106, which shows the arrangement made in the 'Queen,' built by Messrs. Laird. In some ships the depth of the hatch-beams has been increased, and by this means an increase of strength has been obtained; while in many vessels transverse carlings have been fitted so as to form continuations of the half-beams across the hatch, and are fixed in place after the engines, boilers, or cargo have been put in or stowed. This latter arrangement is specially suited for engine and boiler hatchways which do not require to be used frequently. The transverse carlings are generally secured to the fore and aft carlings by screw-bolts, so as to be readily removed. In some ships the fore and aft carlings are omitted at the sides of the hatchways over engines and boilers, the inner ends of the half-beams being simply connected with the deck tie-plates; and when the hatch is to be closed, middle pieces are fitted to each pair of half-beams, being connected with them by vertical scarphs, fastened with nut and screw bolts.

In the larger number of shipbuilding yards the lengths of beams and the forms of the knees are given out from the mould-loft, but on the Tyne it is customary to take the lengths from the ship. On the Clyde it is the usual practice to rivet the beams to the frames before hoisting them into place in vessels which do not exceed from 500 to 600 tons, and for ships up to 1600 tons this course is frequently adopted, but for larger vessels it is customary to proceed in the manner commonly followed, and get in the beams after the frames have been put in place and set fair. On the Mersey a strake of outside plating is worked at or near the beam-ends, and the ship is shored on it before the beams are put in place; but in many yards the ship is faired by means of ribands and cross-spalls only before the beams are fitted. In many cases the holes for the rivets in beam-knees are punched in the frames before they are hoisted into place. These holes are easily set off, as the heights of the decks are marked on the frames and the depths of the knees are known. When the lengths of the beams are taken from the ship the stations of the holes are marked on the mould, and transferred from it to the beam, and the holes are punched or drilled before the beams are put in place. When the beams are worked to the lengths given from the mould-loft, the holes in the beam-knees are generally punched or drilled before the beams are put in, and the holes in the frames are drilled, or punched with a machine known as a *bear*, after the beams are fixed in place. Illustrations of the usual modes of arranging the fastenings in beam-knees have already been given, and it may be added here that care should always be taken not to weaken the frame by two rivet-holes placed abreast each other or near together. It sometimes happens that the beams require to be straightened after being put in place and secured, and a sketch showing the machine used for this purpose is given in Fig. 108. The side view shows how the arms of the clamp are formed so as to clutch the bulb of the beam, and the plan illustrates the manner in which the screw is worked in the clamp. The end of the screw is furnished with a solid head formed so as to fit against the bulb, and when the machine has been fixed at any part of the beam where it has been bulged sideways by an accident, the beam is straightened very quickly by heaving up the screw.

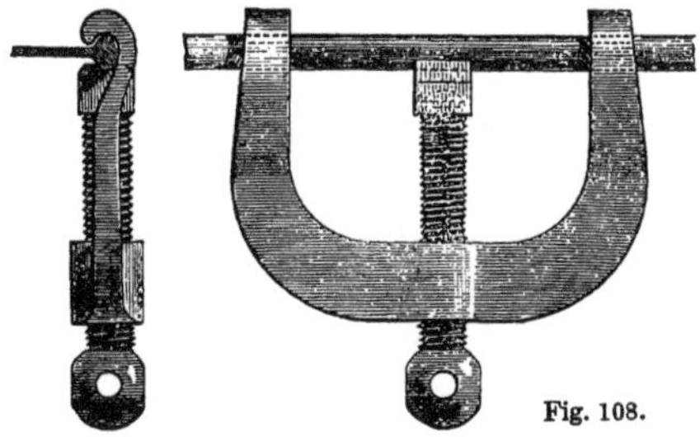

Fig. 108.

The pillars to the beams of an iron ship are a very important means of increasing her structural strength, not merely by acting as vertical struts, but, when properly secured at the heads and heels, by forming vertical ties. Mr. Rankine states, in 'Shipbuilding Theoretical and Practical,' that the use of pillars at the middle line increases the strength of the beams by about one-half, and points out the increase in the resistance to transverse bending obtained by connecting the beams with the floors. The great advantages gained by placing the beams of the different decks directly over each other, in respect both of pillaring, and of facilities for framing hatchways by which ready access is given to lower decks and holds, have led to the very general adoption of this arrangement. Lloyd's Rules require that the beams shall be placed over each other and pillared where practicable. The Liverpool Rules state that stanchions are to be fixed to every beam amidships for one-third of the vessel's length, and to alternate beams forward and aft; and require that hold-stanchions shall have two 1-inch rivets through palms upon their heads and heels. In most parts of a ship it is possible to place the pillars directly above one another, and so form a strut which extends from the floors to the upper deck. When this direct vertical support cannot be kept up, on account of the interior arrangements of the ship, there is still an indirect support derived from the thrust of the pillars which heel on a deck being transmitted by it to the pillars beneath. The principal use of pillars is to transmit the loads on the decks to the floors, keelsons, &c.; and many ship-builders, regarding this as the only use, firmly connect the heads of the pillars to the beams, and are contented with leaving the heels unsecured against upward tensile strains. It cannot, however, be doubted that if the pillars are to assist in resisting the strains which often result from a ship's motion in a seaway, it is requisite that they shall be so secured at the heels as to form ties. The special importance of securing the heels of pillars in war-ships carrying heavy guns has been repeatedly demonstrated. One instance occurred on board the 'Viper,' where the heels of the pillars were not securely fastened, and were consequently drawn out of their sockets by the deck being slightly lifted by the violent concussion caused by the explosion of her guns. The pillars in most common use in ships of the mercantile marine are formed of solid bars, but in the ships of the Royal Navy the pillars are formed of wrought-iron tubes welded in solid at the heads and heels. This latter form is preferable, as it possesses considerable lateral stiffness, whereas the solid bar can be easily bent sideways by a horizontal blow. The facilities offered in an iron ship for connecting the pillars to the beams and floors are much greater than those in a wooden ship. In considering the various modes of securing the heads and heels of pillars, the connections of the pillars between decks will be first illustrated. A very common mode of securing the heads of pillars to bulb-beams is that given in Fig. 109. It will be seen that the head of the pillar is formed into a palm on one side of the beam, and can thus give direct support to the beam and be securely fastened to it. It may be added that the holes for the bolts which secure the pillar-head should be placed as near as possible to the neutral axis of the

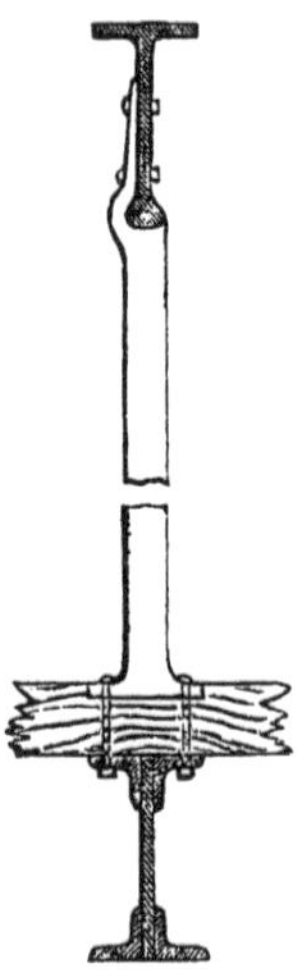

Fig. 109.

beam, in order to preserve the strength. When made beams are used, a horizontal palm is formed on the pillar-head, and is bolted to the horizontal flanges of the angle-irons on the lower edge of the beam. This arrangement is illustrated in Fig. 110, and a similar mode of connection is adopted when H-iron beams are used. There

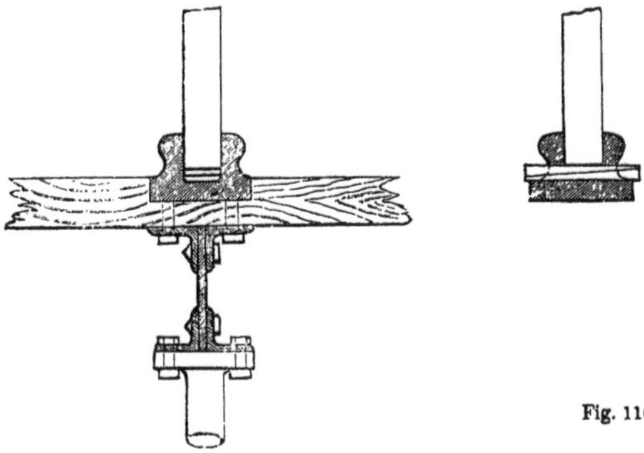

Fig. 110

is a greater variety in the modes of securing the heels of pillars between decks than there is in the fastenings of their heads. In one arrangement, illustrated in Fig. 110, the heel of the pillar merely fits into an iron shoe or socket which is bolted to the deck, and wedges are driven in through slots in the sides of the shoe in order to set up the pillar. Another very common mode of securing the heels of pillars is that given in Fig. 109. Horizontal palms are formed on the pillars and are bolted to the deck; and, where possible, the bolts are made to pass through the beam-flanges underneath, in the manner shown in the sketch. This mode of fitting gives good security to the heel, and allows the pillar to be easily removed, when required. In many ships, instead of heeling the pillars on the deck planking, they are brought down on the beams and secured, the planking being fitted around them. In wake of the capstan-bars the pillars are made to turn up, in order to give room to work the bars, and the arrangement usually followed in H.M.'s Service is illustrated in Fig. 111. The palm is formed separately from the pillar, and its lower end is arranged as a hinge. An iron shoe is bolted on the deck, and is slightly wedge-shaped, as is seen from the transverse view given in the sketch. When the capstan is not in use the pillar is let down, and the heel is forced into the iron shoe, which by its wedge-shape sets the pillar up to its proper height. In order to support the deck in the neighbourhood of the capstan, when it is required to turn up the pillars to work the capstan-bars, screw-pillars are used in some ships, and, being similar in their arrangements to the ordinary "screw-jack," they can be transferred to any part of the deck. In other ships additional hinged pillars are fitted to the beams just outside the sweep of the capstan-bars, and are usually kept turned up, but when the capstan is in work they are let down and support the deck above.

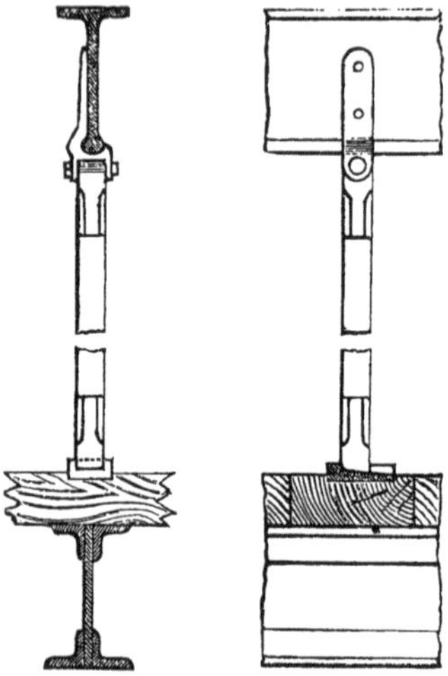

Fig. 111.

Coming now to the consideration of the modes of securing the heads and heels of hold stanchions or pillars, the connections with the lower-deck beams hardly require notice, as they are identical in character with the arrangements already described for pillars between decks. The heels of the stanchions are formed differently in different ships in order to connect them with the keelsons or hold-stringers. When the ship has a flat keelson-plate, the stanchions have horizontal palms on their lower ends, and are bolted through the keelson-plate and angle-irons. When a centre-plate or a side-bar keel is run up to form the keelson, and is taken up by angle-irons on each side, the heels are generally secured by bolting them through the vertical flanges. When an angle-iron keelson is employed without a centre plate, the connection of the heel is similar to the preceding; and when a bulb-iron keelson is worked it is usual to secure the stanchions in a manner similar to that shown for a pillar-head in Fig. 109.

Mention has already been made of the difficulty sometimes experienced in efficiently pillaring beams, and in a steam-ship the engine and boiler rooms are the parts of the ship where this difficulty most frequently occurs. In order to support those parts of the deck-framing which cannot be directly pillared, wrought-iron girders of I-shaped section formed of plates and angle-irons are often worked below the beams, and are ended under beams at which pillars can be fitted. Similar girders are also worked under those parts of a deck which have to sustain any special load, in order to distribute it, as for instance in wake of mast-steps, &c.

CHAPTER IX.

DECK STRINGERS AND PLATING.

COMING now to the consideration of deck stringer-plates, we may first observe that the aspect of a ship regarded as a girder having top and bottom flanges and an intervening web, is, although familiar, not sufficiently borne in mind by shipbuilders, who often fail in properly adjusting the proportionate strength of the flanges. Thus in an iron ship the bottom flange, formed by the keel, keelsons, and bottom plating, is usually enormously strong, both in tension and compression; but the upper flange is often neglected, and in some ships it is merely a flange of wood. Both Lloyd's and the Liverpool Rules provide that stringer and tie plates shall be worked on the upper decks of all ships, but no mention is made of iron decks, the adoption of which has been so strongly advocated by Mr. Fairbairn, Mr. Scott Russell, and others, and which have now become almost universal in her Majesty's service. It seems probable, however, that the great disproportion between the strength of the top and bottom of an iron ship as commonly built, and the evident want of economy of material involved, will lead to the introduction into general use of, at least, partial iron decks. At present, in cases where great longitudinal strength is required (as for instance in the 'Queen,' of which ship's framing a view is given in Plate 1) a partial iron deck is adopted. In the 'Queen' a belt of plating $\frac{5}{8}$-inch thick is worked on the upper-deck beam-ends, and extends out to a distance of 7 feet from the side, and upon its outer part a stringer-plate $\frac{5}{8}$ inch thick and 2 feet wide is worked and stiffened by three longitudinal angle-irons. A longitudinal tie-plate $\frac{5}{8}$ inch thick and 3 feet wide stiffened by a double angle-iron stringer, is worked at 6 feet from the middle line. Thus on the upper deck of this ship, of which the extreme breadth is 41 feet, there is a partial iron deck, of which the total width is 20 feet, and this partial deck is stiffened by the plate and angle-iron stringers above described. The doubling of the sheer-strake and the other strakes of the upper part of the outside plating shown on the section, together with the other arrangements of stringers, &c., form a very efficient top-flange in this ship. Another illustration of the use of partial iron decks is found in the steamships built at the Thames Iron Works for the "Compagnie générale transatlantique," from the designs of M. Forquenot. In these vessels there is a partial iron deck, on the upper and main decks, which extends in as far as the sides of the boiler-hatch, and the upper flange is still further strengthened by box-stringers on these two decks.

Iron upper decks have also been worked in the very long fine ships built by Messrs. Harland and Wolff, of Belfast, in some of which ships the length is between 10 and 11 times the beam. Thus, for instance, in the 'Istrian,' 'Iberian,' and 'Illyrian,' of which the breadth extreme is 37 feet, the length of keel 390 feet, and the depth in hold 29 feet 3 inches, there is an iron upper deck, and the stringers on the middle-deck beam-ends are 5 feet broad, while there are tie-plates 22 inches broad on each side of the hatchways on the middle deck.

In the greater number of iron ships, however, the only partial iron deck which is fitted is composed of the stringer-plates, and of fore and aft and diagonal tie-plates. These are of service in adding strength to the top of the girder formed by the ship, and in preventing change in the longitudinal form. In a wooden ship all the deck planking tends to prevent this change, by resisting alteration in the angles between the beams and the various strakes of planking; but in an iron ship the wooden deck has very little power to resist this change, since the nature of the materials and fastenings is such as to always allow some motion, and would probably admit of enough motion to injure the fastenings of the ship's side before they themselves came effectually into play. Mention has previously been made of the fact, that the stringer-plates and angle-irons on the beam-ends act as horizontal knees to the beams, and their efficiency in this respect is increased by the practice, now introduced into common use, of working upon the stringer-plates continuous angle-irons, which serve both as stiffeners to the stringers and as gutter waterways at the side. The tie-plates, usually worked on the various tiers of beams, are arranged so as to have two placed longitudinally, one on each side of the hatchways, and the remainder placed diagonally and running from side to side between the hatchways. These plates serve to prevent the racking forces which are brought into play when the ship is heeled over, or lies across a series of waves. For, though the decks and bottom of an iron ship form in most cases the top and bottom of the girder, there are positions, as we have seen, in which a ship is placed when at sea, where they become the web of the girder, and the sides of the ship form the upper and lower flanges. And between these extreme positions there are intermediate ones, in which considerable forces are acting, and tending to produce change in the longitudinal form, and consequently to rack the deck-framing. It may be doubted, however, whether the disposition of material in the form of fore and aft and diagonal tie-plates is as good as it would be if the material were disposed in additional stringer-plate. In ships which have no iron upper deck, the side plates, and especially the sheer-strakes, form an important part of the top of the girder, and are so recognised in both sets of Rules, where additional longitudinal strength is obtained when required, either by increasing the thickness of the sheer-strake, or doubling it for the whole or part of its length, and by increasing the thickness of the stringer-plates. In such vessels, therefore, if the tie-plates were dispensed with, and the iron thus saved put into additional thickness of stringer-plate, it would be directly connected with the sheer-strakes instead of lying many feet away from them, and would, as stringer, tend to resist the first changes of form, while as tie-plate certain alterations must take place in the curvature of the beams before it can aid the sheer-strakes. By securing the stringer to the sheer-strake, the material in both is also rendered much more efficient to resist compressive strains. There is, in fact, no part of an iron ship of more consequence than the stringer-plates; and the due proportioning of plates, butt-straps, and rivets in their construction requires most careful attention.

Lloyd's Rules with respect to stringers and tie-plates are as follow:—" All vessels to have stringer-plates upon the ends of " each tier of beams. Those upon the ends of upper-deck " beams in vessels with one or two decks or tiers of beams, and on " ends of middle-deck beams in vessels with three decks or tiers " of beams, to be in width one inch for every seven feet of the " vessel's entire length, for half her length amidships, and from " thence to the ends of the vessel they may be gradually reduced " to three-fourths the width amidships—in no case, however, is the " width to be less than 18 inches amidships. The stringer-plates " are to be fitted home and riveted to the outside plating at all "upper decks, and at the middle deck in vessels having three "decks, with angle-iron of the dimensions given in Table G: the "middle-deck stringer-plates to have an additional angle-iron "extending all fore and aft inside of the frames, riveted to the "reversed angle-iron on the frames, and to the stringer-plate. "Stringer-plates on ends of beams below the upper deck in vessels "with two decks, or below the middle deck in vessels with three "decks, may be reduced in width to three-fourths of the midship "breadth above named; this breadth is to be extended all fore and "aft, and to have an angle-iron of the dimensions given in Table G, "extending all fore and aft, riveted to the reversed angle-iron of "the frames and to the stringer-plates. In cases where no deck "is laid and the width of stringer-plate on ends of hold-beams is "objected to, it may be reduced, provided such reduction be fully "compensated for. The objectionable practice of cutting through "the stringer-plates for the admission of wood rough-tree stan- "chions will not be allowed."

"All vessels to have tie-plates, ranging all fore and aft upon "each side of the hatchways on each tier of beams, and in addition "thereto the beams of the upper and middle decks, in three-decked "or spar-decked ships, and of the upper deck in vessels of one or "two decks must have the tie-plates fitted from side to side "diagonally, whenever the arrangements of the deck will admit "of them; the tie-plates are to be in width once and a half the "depth of the beams and of the thickness required for stringer- "plates, and to be well riveted to each other and to the beams, "deck-hooks, and transoms; and all butts to be properly shifted. "Upon hold-beams where no deck is laid, or where tie-plates would "interfere with stowage of cargo, an angle-iron of the dimensions "given for angle-iron on beam stringers, placed at middle line, ex- "tending fore and aft wherever practicable, and well riveted to all "beams, deck-hooks and transoms, will be admitted in lieu thereof."

Until 1867 the Liverpool Rules on this subject were as follow:— Stringer-plates are to be laid upon the ends of each tier of beams, and riveted thereto through both beam angles. Main-deck stringer-plates may be reduced in width one-fourth at ends of vessels, and one-sixteenth in thickness; this reduction to begin at one-fifth the length of the vessel from each end. Stringer-plates on upper deck, and in vessels with three decks on main and upper decks, to be fitted and riveted to shell-plates with angles as per table for keelsons.

All stringer-plates are to extend fore and aft, where practicable, and not to be stopped at bulkheads. If desired, stringers on orlop-deck beams may be diminished in width not exceeding one-third, if proportionately increased in thickness. Angle-iron on gunwale-stringer not to be butted at scuppers, but to be formed around them, or if butted to be otherwise strengthened. Poop and forecastle stringers may be one-third lighter than lower-deck stringers. Tie-plates to be laid upon each tier of beams alongside of hatches, and on main deck with double angle-irons on upper side, ranging all fore and aft; to be riveted to both angles of the beams, and riveted at ends of vessel to the stringer-plates. In vessels under 600 tons, and on orlop-deck beams where no deck is laid, two angle-irons, back to back, each side of hatchways, same size as for keelsons, to be riveted through and through, and to the beams, may be substituted for tie-plates. The Rules at present enforced are given in the Appendix.

In the Liverpool Rules the breadth and thickness of the stringer and tie plates are given in Table No. 5, which will be found in the Appendix, as will also the Table G, referred to in Lloyd's Rules. By a comparison of these rules and tables, it will be seen that for small ships the Liverpool Rules require a much wider stringer than Lloyd's, and that the disproportion in the widths required decreases as the size of the vessel increases, until the sizes approximate to equality in the largest ships. The thickness required by both rules is nearly the same. The stringer angle-irons required by Lloyd's are about the same size for the smaller ships, and of greater size for the larger ships than are required by the Liverpool Rules. According to Lloyd's Rule, the principal deck-stringers may be reduced at the extremities to three-quarters of the midship breadth, the tapering being begun at one-fourth of the vessel's length from each end; but the Liverpool Rules allow the main-deck stringer to be reduced one-sixteenth in thickness at the extremities, in addition to being diminished in breadth to an equal extent with that permitted by Lloyd's, and require the tapering to be commenced at one-fifth the length from each end. The tie-plates required by Lloyd's Rules are narrower than those required by the Liverpool Rules, except for the smaller ships, and the latter Rules require a double angle-iron stringer to be worked upon the main-deck tie-plate. The diagonal tie-plates required by Lloyd's, are not mentioned in the Liverpool Rules.

M 2

A more detailed statement of the differences existing between the two sets of regulations will be found in Chapter 19. The practice of cutting holes in the upper-deck stringer in order to allow the top timbers to pass down, which is forbidden by Lloyd's Rules, was formerly very common, and is described and strongly condemned by M. Dupuy de Lôme, in his Report. As now fitted, the upper-deck stringer-plate is left uninjured except by the holes required for fastenings, and by the scuppers.

The great importance of the preservation of a continuity of longitudinal strength has been previously illustrated, and in the case of deck-stringers this continuity is preserved, in some measure, by butt-strapping the various lengths of plates and angle-irons, and by continuing the stringers through the bulkheads. In a well-built iron ship care is taken also that the butts of the stringer-plate shall give good shift to the butts of the strake of outside plating in wake of the beam-ends; and it will be remembered that in one of the cases of weakness illustrated in the commencement of this work, a want of care in this respect led to very serious results. The arrangement of the butt fastenings of stringer-plates is also a subject requiring careful consideration. If we suppose the stringer-plate under notice to be that on the middle-deck beams of a three-decked ship, the plate has to be scored in between the frames, and at each beam it crosses it is perforated by a number of holes for the purpose of receiving rivets to secure it to the beam. The plate is thus weakened by the sectional area of iron punched out, and by the loss of strength in that which remains, caused by the punching. In order, therefore, to secure breaking lines of equal strength along the lines of the beams, and across the butts, it is necessary to arrange the fastenings so that the tensile strengths at the butt and beam respectively shall be equal. If the fastenings are not so arranged, the butt is made too strong, and labour wasted, or too weak and material sacrificed. It is consequently necessary to make calculations, in order, as far as possible, to ensure uniformity of strength.

In a paper by Mr. Barnaby, in the *Transactions* of the Institution of Naval Architects for 1866, on "Economy of Material in Iron Decks and Stringers," the author proposes a novel mode of lightening stringers and tie-plates without reducing their strength; on the contrary, he states that the tensile power of the plates will be increased when thus lightened. The principle on which this arrangement rests is, that when strains are suddenly applied to the plates, it is necessary to consider not only the number of tons required to break the weakest sections, but the amount which it would stretch before breaking, in other words, the work done in producing rupture. In order, therefore, to make the amount of work done as great as possible, it is necessary to reduce the strength of the plate between the weak sections, at the butts and beams, to the strength at these sections, or even to less than this, in order to obtain long spaces of uniform strength to give elongation. If these long spaces of uniform strength are not provided, and the plate is consequently left with strong parts between the beams, no practical elongation will take place in these strong parts under the action of a sudden strain; but the stretching will be thrown almost entirely on the weak points, and if any one of these is weaker, in any sensible degree, than the rest, it will be confined to that point. The author states that the fact that the strains of greatest magnitude in a ship are sudden makes the principle above stated of no slight importance to naval architects, because by its application the time is increased during which a given force must be applied in order to produce rupture. He illustrates the application of this principle to stringers and tie-plates, by supposing a stringer or tie-plate to cross a series of beams 3 feet 6 inches apart, and to have the strength at each beam reduced to seven-ninths the full strength of the plate, by the holes punched in it to receive the fastenings. If this plate is brought under the action of a steady strain, it is a matter of indifference how many such points of weakness there may be, or how much stronger the material may be which lies between these weak points. For under these circumstances the strength of the tie will be measured by the strength at the weakest place. But when the plate is brought under the action of strains which are sudden in their nature, like the most severe strains in a ship, the principle above stated becomes applicable, and the long spaces of uniform strength required are obtained by cutting holes, 2 feet long and 5 inches broad, in the supposed plate, in all the beam spaces except those in which the butts come. Sketches showing the proposed arrangement of holes and fastenings of the supposed stringer or tie-plate, are given in the *Transactions*, and are well worth careful study.

The various arrangements of deck-stringers which have been, and now are in use, are illustrated in the sketches of framing and beam connections which have been given. In iron ships which have wood beams stringer-plates are not usually worked, but in some ships stringer angle-irons have been worked on the beam-ends. Mr. Scott Russell gives two modes of working stringers on wood beams which are attached to the frames by bracket-plates. For a simple wood beam the stringer-plate is bolted through the beam, and the connection thus made is strengthened by riveting it to a short piece of angle-iron worked on the upper edge of the bracket-plate. When the beam is made up of a central web-plate and two wood beams, a short piece of angle-iron is worked on the upper edge of the web-plate, under the stringer, and is riveted to both. Passing to the illustration of the stringer arrangements of some of the earlier iron ships which have iron beams, we need only refer to the arrangements of the 'Birkenhead' given in Fig. 71, page 75, which were described when speaking of the connection of the beams to the side. On the lower-deck beams of the 'Recruit' the stringer arrangements were of a very singular character, as will be seen on reference to Fig. 112. The stringer plate proper, *a*, is 8 inches wide, and is scored in over the frames and fitted against their flanges, but is not connected to the outside plating. The only fastenings in the stringer consist of two rivets in each beam. In addition to this a longitudinal strip of plating, *b*, is worked on the beams in a position corresponding to that occupied by the binding-strake in a wood ship, and is secured by a single rivet in each beam.

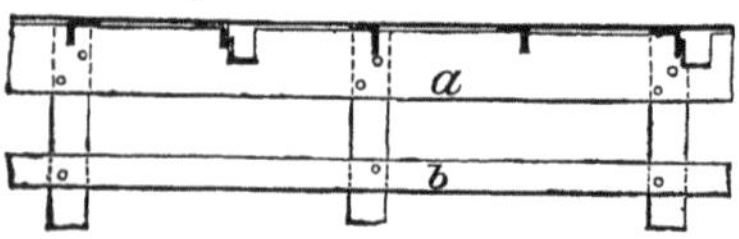

Fig. 112.

The modes of working stringer-plates now in general use are the following:—1st, that in which the stringer-plate is run along inside the frames, and connected to the reversed frames and angle-irons; 2nd, that in which the stringer is directly connected with the outside plating; and 3rd, that in which the stringer-plate is connected to a vertical stringer or clamp-plate worked inside the frames above the beams. The first two modes of working stringers are those most commonly followed, and in some ships the second and third modes are combined. The first mode is generally adopted on the lower-deck and hold beams, and sometimes also on the middle deck. Illustrations of this arrangement are given on the lower and middle decks of the 'China,' as shown in Plate 2. The second mode of working stringers is almost universally adopted on the upper-deck beams, and can, in general, be readily performed as the frames usually end underneath the upper-deck stringer-plate. Illustrations of the common mode of securing the stringer on the upper deck to the outside plating are given in Plates 1 and 4.

Both Lloyd's and the Liverpool Rules require that this mode shall be adopted on the upper deck in all ships, and on the upper and middle decks in three-decked ships. On the middle deck the stringer is usually scored in between the transverse frames, and secured to the outside plating by short pieces of angle-iron. Sectional views of this arrangement are given in Plate 3, and it will be seen from Plate 2 that the 'China's' upper-deck stringer is fitted similarly, on account of the fact that her frames are run up to form the topside framing. In Fig. 113 there is given a plan which illustrates the usual mode of scoring in and securing a stringer-plate. The sketch is taken from the 'Captain,' and represents a part of the lower-deck stringer where it is not required to be made watertight. It will be remarked that the cost of fitting is reduced to a minimum by cutting away the corners of the plate which would come against the frame angle-irons. In those parts of the ship where this stringer has to be made watertight, in order to form a top to the wing-passage, a different mode of fitting is adopted, and is illustrated in plan and section in Fig. 114. Intercostal plates and angle-irons are carefully fitted between the frames, and the plates overlap the continuous stringer-plate sufficiently to allow the continuous stringer angle-irons to be worked along upon the inner edges, as is seen in the section. The fastenings of the stringer angle-irons are thus made to work in as fastenings in the inner edges of the intercostal plates. The edges of the plates and angle-irons in this arrangement can be caulked with facility and the whole made watertight. The stringer-plate is worked in 16-feet lengths, with treble-chain riveted butt-straps;

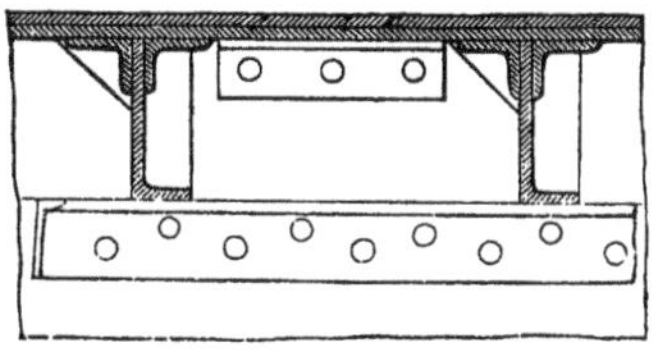

Fig. 113.

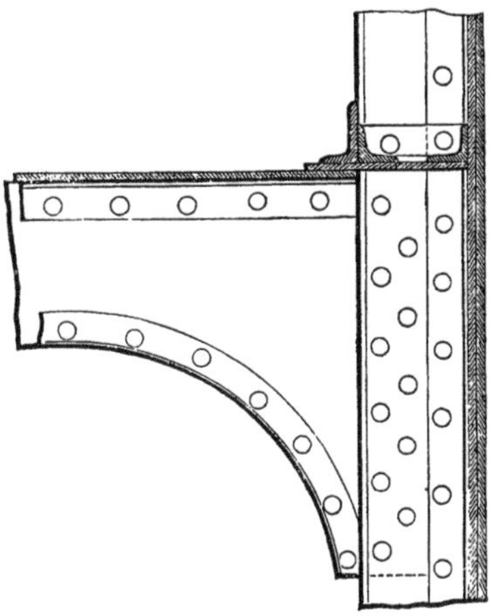

PLAN

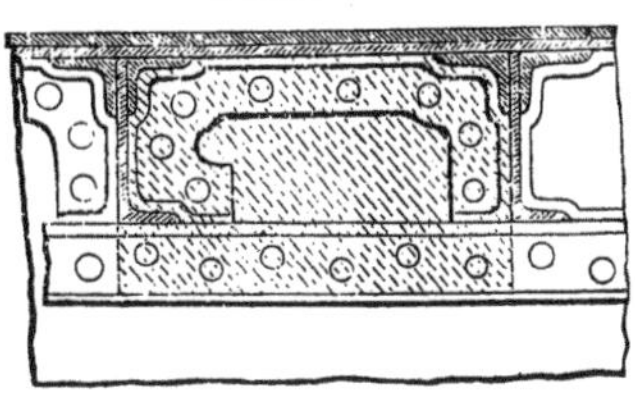

Fig. 114.

and the continuous stringer angle-iron is worked in lengths of 72 feet, formed by welding two 36 feet lengths together, the butts being secured with treble-riveted covering angle-irons. In some ships where the lower or middle-deck stringer has to be made watertight the intercostal plates are worked out beyond the frames so as to admit of their being connected with the stringer-plate by a flush joint and an edge-strip, the latter being worked underneath the stringer, and in short lengths between the beams. This is the arrangement adopted in the 'Hercules,' in which ship also the staple angle-irons connecting the intercostal plates with the frames and skin-plating are worked underneath the plates.

The third mode of fitting stringers is illustrated in Fig. 104, page 148, and the combination of the second and third modes previously referred to is illustrated on the middle and lower decks of the 'Queen' in Plate 1.

In cases where great longitudinal strength is required, box-stringers are formed on the beam-ends, and it will be remembered that the construction of these box-stringers is one of the means by which the defects of longitudinal strength previously described have been remedied. Their easy application in ships of which the stringer arrangements are of the usual character, has rendered the use of box-stringers very general in such cases. The simplest arrangement of box-stringer on the beam-ends is that given in Fig. 115, and it will be seen that the top and front plates, and their connecting angle-irons, are the only additions to the ordinary stringer. No direct connection is made in this case between either the stringer-plate or the top-plate, and the outside plating, but as usually fitted the stringer-plate would be carried out and connected to the plating. On lower decks where the stringer-plate is run along inside the frames, a very efficient box-stringer is formed by the arrangement shown in Fig. 116, which is taken from Lloyd's illustrations. Intercostal flanged-plates are fitted between the frames, and overlap and are riveted to the top-plate of the box-stringer. A direct connection is thus made with the outside plating, without weakening the stringer or top-plate by scoring them in between the frames. When box-stringers

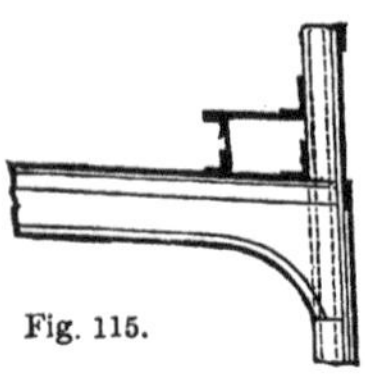

Fig. 115.

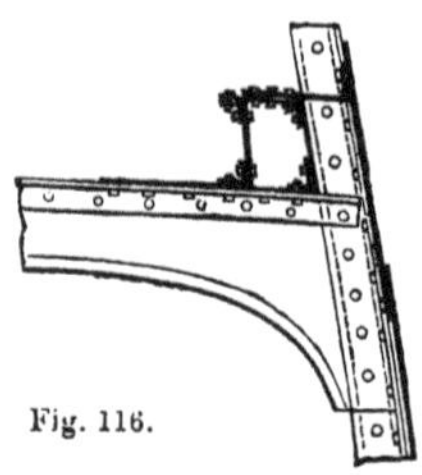

Fig. 116.

are used on the upper-deck beam-ends the arrangements are usually similar to those shown in Fig. 190, page 243, and, as will be seen, are of a very simple character. In the 'Queen' a box-stringer is worked on the amidship part of the lower deck, and formed as shown in Fig. 117, by adding top and front plates and connecting angle-irons, to the stringer arrangements which are continuous throughout the length of the ship.

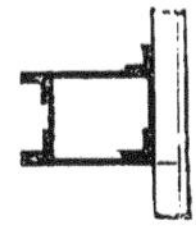

Fig. 117.

A very peculiar arrangement of upper-deck stringer was fitted in the 'Sentinel' (before referred to) before and abaft the cabins. A stringer-plate of the usual form was worked on the beam-ends, and, by means of a bent plate formed and secured as shown in Fig. 118, a kind of cellular arrangement was made. As far as the cabins extended a box-stringer was worked on each side, the fittings being of such a character as to obviate the inconveniences which would have been caused if a box-stringer had been continued throughout the ship's length.

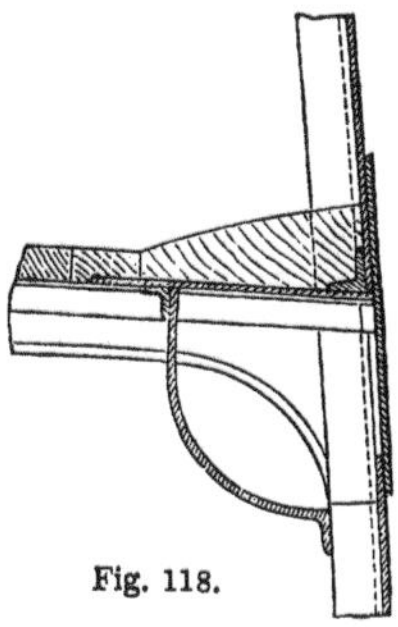

Fig. 118.

In a paper on "Iron-Plated Ships," in the *Transactions* of the Institution of Naval Architects for 1863, Mr. Fairbairn proposes a novel means of strengthening the upper deck. He suggests that six longitudinal cells should be worked under the deck beams, four of them being rectangular in section, and the other two forming a continuous stringer-bracket on each side. The transverse beams are to be H-iron girders, and the upper deck is to be plated over, while the hull of the ship is to be built on the system usually adopted in H.M. Service, which combines the longitudinal and transverse framing. The author states that this arrangement of girders or cells, &c., would make the upper part of the ship of a corresponding strength to that of the lower part, and supports his proposal by the statement that "the cellular form is the only one "calculated to attain the maximum powers of resistance with a "flexible material such as wrought iron, and it has been demon-"strated by direct experiment that nearly one-half the material "is saved by the cellular system, or in other words it would require "double the weight of metal on the deck of a ship to resist a cor-"responding force of compression to that of extension." It will be obvious that the height between decks is usually insufficient to admit of this arrangement.*

At the extremities of an iron ship the stringers on the opposite sides of a deck are usually joined, and the stringer angle-irons continued around the fore and after ends. Angle-iron stiffeners are also worked in many ships around the inner edges of the fore part of the stringers, which are thus converted into breasthooks, as was previously stated.

Having illustrated the different arrangements of deck-stringers, we turn to the consideration of the disposition and fastenings of iron decks. The importance attaching to this subject results from the fact previously stated, viz., that iron upper decks have become almost universal in ships of war built in this country, and that the employment of, at least, partial iron decks is becoming more general for ships of the mercantile marine.

The usual disposition of plating, edge-strips, &c., of an iron deck is shown in Fig. 119, and it will be seen that one strake of

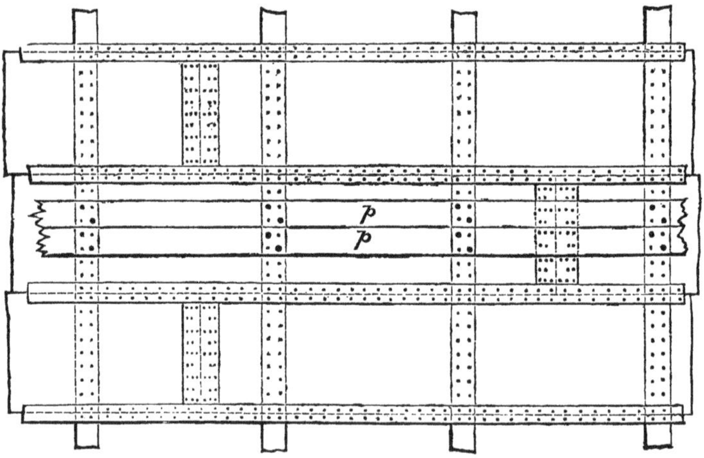

Fig. 119.

plating comes between successive butts in the same beam space. The butt-straps are treble-chain riveted, the alternate rivets being left out in the rows next the butt. The edge-strips are single-riveted, and both edge-strips and butt-straps are worked on the upper side of the deck, the planking being scored down over them.

* Full details and illustrative sketches of this proposal will be found in the *Transactions.*

Double straps are very commonly employed in the butts of deck plating. The holes in the beam flanges serve alternately for the rivets in the iron deck and for the screw bolts in the deck planking. Two strakes of the planking are shown in the sketch and marked *p, p*, while their fastenings to the beam flanges are made larger for distinction. In some ships the deck fastenings are brought out into the spaces between the beams, and this arrangement is preferable to the preceding, as it tends to more uniformity of strength in the plating than can be attained when a very weak line is caused in wake of each beam flange by punching rivet and bolt holes, while the plating between the beams retains its full strength.

The following details of the deck-plating of the 'Warrior' are of interest as illustrating the arrangements of the first iron-built armour-clad frigates of the Royal Navy. The whole surface of the upper deck is covered with $\frac{1}{4}$-inch plates worked flush at the joints, and having double-riveted butt-straps, and single-riveted edge-strips. A tie-plate 24 inches broad is worked on each side of the hatchways, and a stringer-plate 36 inches broad is worked round the side, with diagonal tie-plates 24 inches broad running across the deck. The whole of these plates are $\frac{5}{8}$ inch thick and are worked above the $\frac{1}{4}$-inch plating; the butt-straps are double and treble riveted. The lengths of the plates are not less than 15 feet, and the rivet heads are countersunk on the upper surface. All the joints of the plates are made watertight. The stringer-plate is attached to the sheer-strake by angle-irons, 6 by 4 by $\frac{3}{4}$ inches for 20 feet before and abaft the armour plates, and 5 by $3\frac{1}{2}$ by $\frac{5}{8}$ inches for the remainder of the length of the deck. Short angle-irons 4 by 4 by $\frac{9}{16}$ inches are also worked between the frames on the underside of the stringer and are riveted to the sheer-strake. The main deck is entirely covered with $\frac{1}{4}$-inch plates, and two tie-plates 20 inches broad, with a stringer-plate 4 feet 6 inches broad at the ship's side, are worked upon the $\frac{1}{4}$-inch plating. The thickness of these plates, their lengths, arrangements of butt-straps, &c., are the same as for the upper deck. The plating on this deck is scored out against the outside plating and attached to it by short angle-irons between the frames, both above and below, and secured to the reversed frames by a continuous angle-iron 4 by 4 by $\frac{7}{16}$ inches. On the lower deck, stringer and tie plates are worked, the former being the whole breadth of the wing-passage, and being secured to the outside plating in a similar manner to the main-deck stringer. All these arrangements are illustrated by the section of this ship given in Plate 3.

From the following details of the deck-plating of the vessels of the 'Northumberland' class the reader will see what was considered to be an efficient top to the girder formed by those very long armour-clad ships, plated to the extremities. The upper deck has a tie-plate $\frac{5}{8}$ inch thick and 30 inches wide on each side of the hatchways, the width being decreased gradually to 20 inches at the ends. Between the tie-plates $\frac{5}{16}$-inch plating is worked. At the side a $\frac{5}{8}$-inch stringer-plate is worked of which the width amidships is 5 feet, and forward and aft 4 feet. The deck between the tie-plates and the stringer is covered with $\frac{1}{2}$-inch plating. The butt-straps of the $\frac{5}{8}$ and $\frac{1}{2}$ inch plating are double and treble riveted, and the plates are worked in lengths of 16 and 20 feet. All the plating is worked flush on the beams, and has single-riveted edge-strips. The rivet-holes are countersunk in the upper surface of the plating, and all the joints and butts are made watertight. The stringer plates are attached to the sheer-strakes by angle-irons 6 by 4 by $\frac{3}{4}$ inches, worked above the stringer, and by short angle-irons 4 by 4 by $\frac{9}{16}$ inches between the frames under the stringer. On the main deck a tie-plate 20 inches broad, and a stringer-plate 4 feet broad are fitted on each side, the thickness of both being $\frac{5}{8}$ inch, and the lengths of plates at least 20 feet. The butts are fastened and the rivets countersunk as on the upper deck. The stringer-plates are scored home to the ship's side and united thereto by short angle-irons 4 by 4 by $\frac{9}{16}$ inches, worked above and below the stringer, and secured to the reversed angle-irons by a continuous angle-iron of equal dimensions worked on the upper side. On the lower deck a stringer-plate $\frac{1}{2}$ inch thick and of the breadth of the wing-passage, is fitted, and secured similarly to the main-deck stringer. Tie-plates $\frac{1}{2}$ inch thick and 18 inches broad are worked on each side of the hatchways, and the butt-straps of both stringer and tie plates are double and treble riveted.

The 'Bellerophon's' deck plating is arranged as follows:—On the upper deck, from about 70 feet from the bow, and from 7 feet out on each side of the middle line, the beams are covered with $\frac{1}{2}$-inch steel plates, and on the fore side of this with $\frac{1}{4}$-inch steel-plates. Over the central battery, and in wake of masts and capstans, the 14-feet space at the middle line is filled in with iron plates, as are also the corresponding spaces at the extremities of the ship. The butt-straps to the $\frac{1}{2}$-inch steel-plates are double, each being $\frac{1}{4}$ inch thick, and the straps to the $\frac{1}{4}$-inch plates are single, and of the same thickness as the plates. The butts are treble-chain riveted, with every other rivet left out in the rows nearest the butts. The rivets used in the $\frac{1}{2}$-inch plating are $\frac{5}{8}$-inch, and those in the $\frac{1}{4}$-inch plating, $\frac{1}{2}$-inch. In the single-riveted edge-strips, the pitch of the rivets varies from 3 to $3\frac{1}{2}$ inches. The holes in the steel-plates require to be carefully drilled. The whole surface of the main-deck is covered with $\frac{1}{2}$-inch iron plates, worked flush with edge-strips and butt-straps of the same thickness. The edge-strips are single, and the butts treble riveted, the diameter of the rivets being $\frac{5}{8}$ inch, and the pitch from 4 to $4\frac{1}{2}$ inches. On this deck, as well as on the upper deck, the plating is directly attached to the plating behind armour, but in this case the plating has to be scored in between the frames and secured by short angle-irons. The beams under the central battery of this ship are placed higher than the beams of the main deck before and abaft the battery, the lower side of the battery-beams being well with the upper side of the main-deck beams. The object of this arrangement is to have the port sill high out of the water, and at the same time to make the height and weight of protecting armour before and abaft the battery as small as possible, as explained on page 149. In order to keep up the longitudinal connection on the main deck, the stringer-plates on the beams before and abaft the battery are run along under the battery beams as shown in Plate 4, and the plating in the battery is made to scarph for the length of two beam spaces with the plating on the other parts of the main deck. In consequence of these arrangements no knees are formed on the battery-beams, but the connection with the side is made by vertical partial bulkheads, which are fitted between the main and lower deck stringers, and secured to them and to the frames. It may be added that the main-deck plating in this ship is worked less on account of the structural strength, than on account of the protection against vertical fire which it affords to the parts of the ship before and abaft the battery. On both the main and upper decks, the joints of the plating are made watertight, and the holes in the upper flanges of the beams are occupied alternately by rivets and screw bolts, the former securing the plating, and the latter the deck-planking. On the lower deck a stringer-plate $\frac{1}{2}$ inch thick is worked, the butts being secured by single butt-straps, treble-chain riveted. Tie-plates 18 inches broad and $\frac{1}{2}$ inch thick are worked at the sides of the hatchways, and the butts are fastened similarly to those of the stringers. The part of this deck abaft the stuffing-box bulkhead is entirely covered with $\frac{9}{16}$-inch plating worked watertight, and forming the after end of the hold into a compartment, as previously described.

The arrangements of the 'Hercules'' deck-plating are in most respects similar to those of the 'Bellerophon.' The principal differences are that the 'Hercules' has $\frac{3}{8}$-inch plating on the main-deck outside the battery, instead of $\frac{1}{2}$-inch in the 'Bellerophon,'* and that the butts of plating are double-chain riveted in the 'Hercules.' No tie-plates are worked on the lower-deck beams in

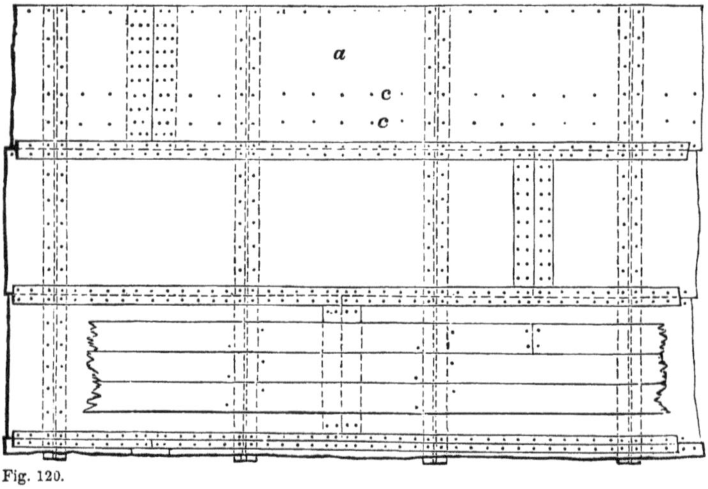

Fig. 120.

this ship, nor is the deck plated over abaft the stuffing-box bulkhead, as the flat below the lower deck is worked watertight instead. A plan of a portion of the upper-deck plating of the 'Hercules' is given in Fig. 120. The plate *a* is the stringer and the row of rivet-holes on its outer edge takes the fastenings of the stringer angle-irons connecting it with the skin-plating. The rows marked *c*, *c* take the fastenings of the angle-irons which form the gutter waterway, of which a section is given in Plate 5. The butts of the

* The height of the main deck of the 'Hercules' is much greater than that of the 'Bellerophon,' and therefore less exposed to injury in a naval engagement.

stringer *a* are treble-chain riveted, and in order to keep the bottom of the gutter waterway flush, the butt-straps are worked below the stringer. Where the rows *c*, *c* cross the butt one rivet is left out in the butt fastening in order to avoid too great a weakening of the angle-irons. The diagonal arrangement of butts is adopted, and there are consequently two strakes between consecutive butts in the same beam space. The edges of the strakes of deck are joined by single-riveted straps, and the butts by double-chain riveted straps. These straps are double to the ½-inch steel plates, and single to the ¼-inch steel plates. In this ship the fastenings of the deck-planking are brought out between the beams as shown for a few strakes in the sketch. This arrangement has the double advantage of rendering the pitch of the holes in wake of the beams considerably greater, and of preventing the deck-plating from "bagging" down, or becoming concave on the upper surface, between the beams, when it is subject to a compressive strain. It will readily be seen that the deck fastenings in this case hold up the plating and form points of support intermediate between the beams.

In low-decked turret-ships, where there is great need of protection against vertical fire, and where, unless special arrangements were made, the upper deck would be very seriously weakened by the large holes required for the turrets, it is usual to work a very strong iron deck. In some ships this deck is made up of layers of plates. A brief description of the arrangements adopted on the main deck of the Italian iron-clad turret-ship 'Affondatore,' will be of interest. In this ship the armour is only carried up to the height of the main deck, and, for the reasons above stated, the plating of that deck is formed of two layers of plates, each 1 inch thick. The edges of the plates of one thickness are placed just in the centre of the strakes of the other thickness, and the butts of the two thicknesses form steps in which the shift for adjacent strakes of the same thickness varies from 5 to 6 feet, the lengths of plates used being about 12 feet. The butt of an underlying strake is placed about midway between the butts of the strakes of the upper thickness which rest upon it. The upper thickness of plating is ended inside the frames, but the lower thickness is run out to the side, and attached to the plating.

In the paper on "Iron Decks and Stringers" previously alluded to, Mr. Barnaby proposes a new mode of plating iron decks, by means of which the tensile strength of the unperforated plates intervening between adjacent butts is to be made equal, or nearly equal, to the strength of the intervening plates together with that of one of the butted plates across a line of rivet-holes on a beam. By this arrangement spaces of uniform strength are formed in which elongation can take place, and thus increase the amount of work requiring to be done to produce rupture. The principle here carried out is identical with that explained when illustrating Mr. Barnaby's proposal with respect to stringers and tie-plates. The deck fastenings are put between the beams, instead of in the beam-flanges, and instead of strapping the butts, spaces are left between them, while three plates intervene between consecutive butts in the same beam space. The length of the intervals which separate the ends of adjacent plates of the same strake, is determined by the number of rivets which can be placed in the edge of the butted plate between the beam and the butt, as there must be sufficient to break the plate across the beam. A short piece of edge-strip is worked on the underside from the beam to the butt, and doubles the shearing strength of the rivets, so that the intervals between the butts will be about one-third of the beam space.* Mr. Barnaby sums up the advantages of the proposed plan as follow :—

1. In the ordinary system one-fifth or one-sixth of the iron is punched away ; by that proposed only one-ninth or one-tenth is punched out. There is from this cause a gain in tensile strength, to which must be added an increase of strength in the iron between the holes. These are together equal to
2. The strength of an iron deck under compression is limited not by the area of section, but by its resistance to buckling between the beams. According to the ordinary mode this is very small, since it is quite free to bend downwards between the beams. But by spacing the deck fastening at intervals of 2 feet instead of 3 feet 6 inches, the tendency to buckling would be reduced. The wooden deck would thus by its own resistance to compression, and by the support it gives to the plates, play a most useful part in compression,

* For full details of the proposal see the *Transactions* of the Institution of Naval Architects for 1866.

although it is powerless against extension when in connection with iron. I therefore conclude that no loss of compressive strength is incurred by the holes in the plates.

3. All the holes for receiving the deck-fastenings may be punched, whereas if the fastenings are in the beam-flanges the holes for them must be drilled either in the plates or in the beams.
4. The expense of cutting, fitting, punching, and riveting butt-straps is avoided. Where the material employed is steel, the gain is more considerable, as all the holes in the butts of the plates and in the straps would require to be drilled.
5. The weight of material omitted at the butts amounts to one-seventh of the whole material employed.
6. There is a gain in strength against injury and rupture by the action of sudden forces, the amount of which is not susceptible of calculation, but which being in proportion to the extent of the spaces of uniform strength which have been introduced is, I think, very considerable.

There can be no doubt that, where the strength of the deck is required for structural purposes only, this is a very economical and good arrangement, although not applicable to decks which are liable to be struck by shot or shell.

The upper deck of the 'Great Eastern' differs in its construction from that of other iron ships. It is a cellular iron structure of which the top and bottom are formed of long plates, worked in two thicknesses, each of ½-inch plate. Sketches are given in Mr. Scott Russell's work on Naval Architecture which illustrate the details of the construction. In the description of the upper deck there given it is stated that the butts are secured by double-riveted butt-straps, and it is added that "the plates themselves being in pairs break "joint and serve as butt-plates to one another, and it is only "where they do not supply this function that butt-plates are "needed."

The employment of wood planking for the decks of iron vessels is almost universal, even when iron decks are laid on the beams. In some cases, however, iron-plates similar to those used in the flats of stoke-holes have been laid on the beams and riveted to them so as to form the surface of the deck. The objectionable features of this arrangement are, the discomfort resulting to the crew, and the fact that the moisture in the interior of the ship rising in the form of vapour becomes condensed under the iron deck, of which the upper surface is acted upon by the external atmosphere, and the water thus produced does injury to both ship and cargo. In a few other cases, as in the 'Scorpion' and 'Wivern' built by Messrs. Laird, iron plates have been laid on the deck-planking. In these ships this was done to protect the deck, but the inconveniences resulting were such as to cause the removal of the iron plates. The objections stated above to an exposed iron deck, do not apply to one on which a wooden deck is laid, as the wood-planking renders it almost unaffected by slight variations of the external temperature. As the employment of wooden decks is thus universal it is necessary to consider the deck-fastenings usually employed, in both iron beams and iron decks. In some ship-building yards the holes for the deck-fastenings are punched in the beam-flanges or angle-irons before the beams are put in place, without taking account of the positions which the strakes of planking will occupy. This practice is a bad one, as it often brings the fastening too near the edge of the strake; and, in order to avoid this evil, it is now the common practice to drill or punch the holes for the deck-fastenings after the beams have been put in, and the disposition of the deck-planking has been made. In most ships each plank is secured by a bolt in each beam-flange. Lloyd's Rules require two bolts in each plank in every beam where the planks exceed six inches in width, one of which may be a short screw-bolt if the planks are not wider than eight inches. The different modes of fastening which have been adopted are illustrated in the following

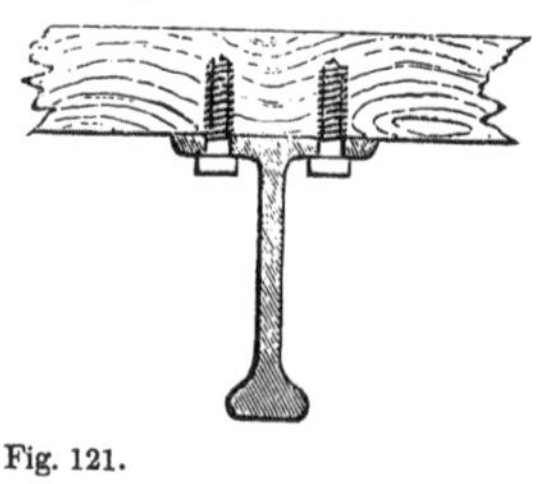

Fig. 121.

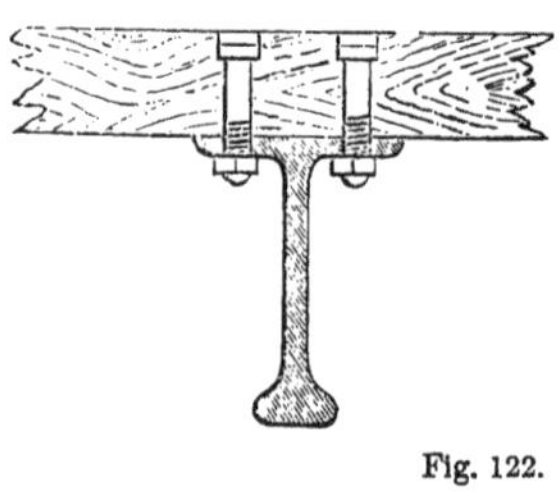

Fig. 122.

sketches. The fastening shown in Fig. 121 consists of a wood screw hove up from beneath and passing into the plank for about three-fourths of the thickness; that illustrated by Fig. 122 is formed by a screw-bolt driven down from above, and secured by a nut hove up underneath the flange or iron deck; that given in Fig. 123

consists of a screw-bolt hove up from beneath, and secured by a nut let down into the plank; and the fourth arrangement in Fig. 124,

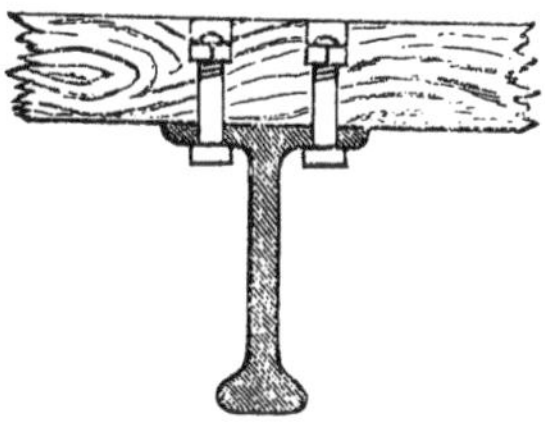

Fig. 123.

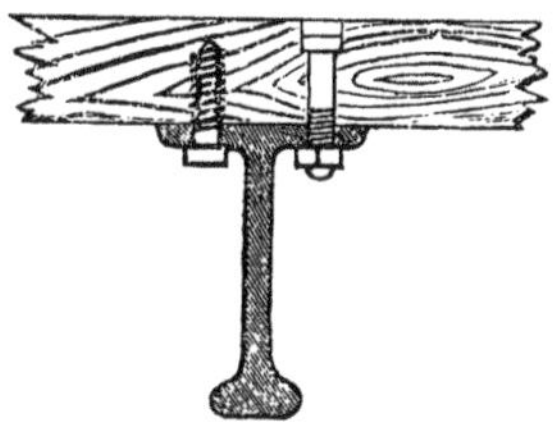

Fig. 124.

which is sometimes made in the decks of merchant ships, is formed by combining the first two fastenings, care being taken that the wood screws on adjacent beams shall fasten opposite edges of a plank. The second plan of fastening is that most usually employed.

An unusual kind of deck-fastening employed by Messrs. Harland and Wolff, the eminent shipbuilders of Belfast, is illustrated in detail in Fig. 125. This is a watertight fastening, and, as will be seen, the iron nut is itself screwed up into the deck-plating, and the screw-bolt hove down into the thread on the inside of the nut. The bolt-head *h* is very wide, and is hove down upon a washer, *w*, so as to ensure a fair bearing, and prevent the bolt-head from being worked into the planking when hove in. The nuts are formed of malleable cast iron, and the cavity below the screw thread is filled with tallow before the bolts are put in. In order to ensure that the bolt-holes in the planks shall be correctly bored, a nut with the bolt-hole bored through the bottom is used. This nut is screwed up into the deck and the auger passed up through it in boring the hole. After the hole has been bored this nut is taken out of place and one of the complete nuts put in its place, after which the deck-bolt is hove in.

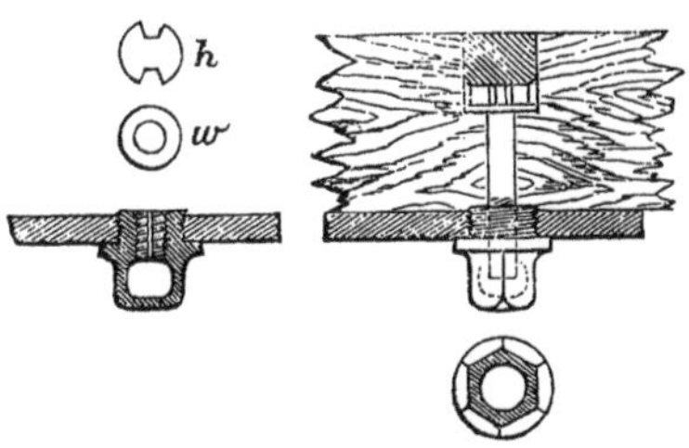

Fig. 125.

The decks of an iron ship are usually caulked in a similar manner to those of a wooden ship; but in some few instances iron tongues have been worked in the edges of the planks so as to form a stop for the caulk. The expense of the latter plan has led to its disuse. Where holes are made in the surface of the deck for bolts or nuts, plugs dipped in paint are driven in to prevent leakage.

CHAPTER X.

OUTSIDE PLATING.

THE skin of a ship is obviously a very important part of her structure; and in an iron ship, where the edges and butts of the plates are strongly connected, it forms a continuous shell well adapted for resisting strains in all directions. But while it is true that a well-built iron ship, of which the proportions, scantlings, and arrangements of framing, fastenings, and butts of plating, have been well considered, forms a structure of almost perfect rigidity, and of immense strength, it is also true that in many ships in which the quantity of material employed is large, a proportionate strength is not obtained, on account of the fact that proper care has not been taken in determining the details of the construction. Among the most important of these details, the arrangements of the plates of the skin and their fastenings are placed by the common consent of shipbuilders. Instances of weakness resulting from a bad shift of butts in the plating and longitudinal framing were given in the commencement of this work. The arrangements of outside plating which have been, and are at present in use, will now be more fully considered and described; and, as in the preceding pages the older methods have been first given, so here the plating of the earlier iron ships will be first illustrated.

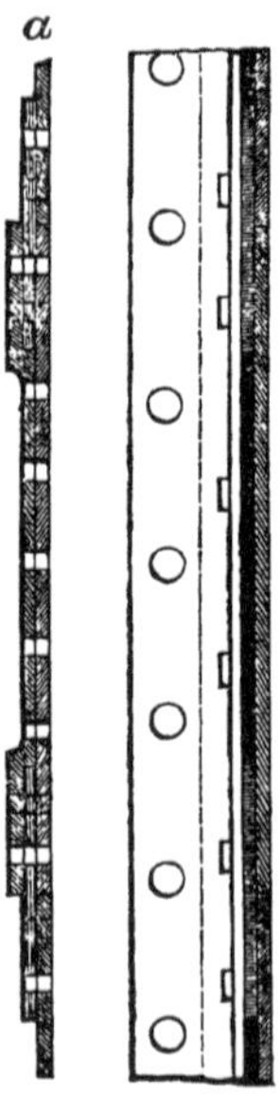

Fig. 126.

The oldest method of plating is that shown in section by Fig. 126, and further illustrated by the part section of the 'Dover,' given in Fig. 70, p. 74. The edges of the strakes of plating were fitted against one another, and the flush-joints thus formed were covered by internal edge-strips. These strips were worked in as great lengths as possible, and their butts connected either by short butt-straps, or by thinning down the ends and riveting through the overlap. In this, as in all succeeding arrangements, the butts of the plates were flush-jointed, and secured by internal butt-straps. These straps were in some ships worked between the edge-strips, and in other ships were joggled over the strips so as to extend the whole width of the plate, and take the rivets forming the edge-fastening, as shown by the section marked *a* in Fig. 126. The latter method was superior to the former, both as respects strength and the facilities it afforded for caulking. On account of the edge-strips being worked inside the plates, liners had to be fitted at each frame. These liners were of the same thickness as the edge-strips, and, usually, were as wide as the frames. The edge-riveting adopted was sometimes single, and sometimes double; all the rivets were countersunk on the outside of the plating.*

The greater simplicity of the *clinker* arrangement of plating, shown in Fig. 127, led to its general adoption in preference to the arrangement above described. In the clinker-built ships the plates of adjacent strakes were lapped over each other, and the edge-riveting passed through both thicknesses. The liners between the plates and the frames were wedge-shaped, and had to be specially prepared. In some vessels, instead of fitting these plate-liners, mere plate-washers were put in between the plating and the frames in wake of the rivet-holes, and thus the plates were only supported at these points and at their edges. This latter practice, however, was strongly condemned by the greater number of shipbuilders. In addition to the expense of properly fitting the liners of a clinker-built ship, there was the disadvantage of having the vertical strains borne by the rivets, instead of by the plate-edges as in the preceding arrangement. But, on the other hand, the clinker plan of plating had the advantage of only requiring in the edges one-half the rivets which were necessary when internal joint-strips were employed; and the edges of the plates required much less care and precision in fitting. The cost of materials and workmanship was thus reduced, and by means of the lap-joint the caulking of the edges was more easily performed. For these reasons the clinker arrangement was prevalent for some years, until it was superseded by the now almost universal mode of plating. In some of the earlier iron ships the above-named advantages of the flush and clinker plans of plating led to the com-

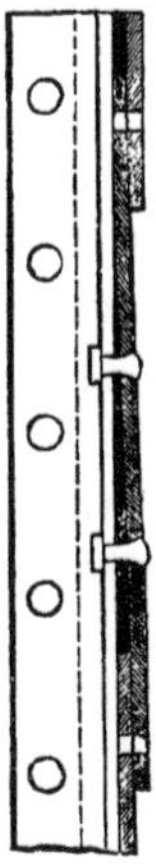

Fig. 127.

* It may be of interest to state that this plan of plating has recently been adopted in the construction of some iron ships building at the Palmer Company's yard at Jarrow-on-Tyne.

bination of the two methods. Thus in the 'Dover' and 'Megæra,' the plating was worked flush on the broadside where there were considerable vertical strains, and clinker-fashion from the keel out to the turn of the bilge where these strains were inconsiderable.

The usual lengths of plates formerly employed were from 7 to 8 feet, and their breadth amidships 2 feet. The shift of butts generally adopted was that shown in Fig. 133, p. 189, where there is one strake between every two butts in the same vertical line. The thickness of the plates used varied, of course, with the dimensions of the vessel, and the common rule formerly in use was to make the thickness of plating in similar vessels proportional either to a single dimension in each, or to the cube roots of the products of the three principal dimensions in each. The thickness of the plating near the keel was usually greatest, and that of the plating between wind and water least; the variation in some cases being as great as from 1 inch to $\frac{3}{8}$ inch. The plates used were very light as compared with those now employed. The processes of fitting the joints and butts of the plates were extremely rough, and were usually completed by the projecting parts being hammered down by the workman, the joint being made watertight with the caulking-tool after the plates were riveted. In some cases the Scotch shipbuilders did not take even this trouble to make close joints, but drove in strips of iron, and caulked the edges to them, in the plating above the water-line. An instance of want of care in this respect is found in the case of the 'John Garrow,' an iron sailing-ship of 555 tons, built at Aberdeen in 1838, which, after a single voyage to Bombay was thoroughly repaired at Liverpool by Mr. Grantham. Among other defects named by the arbitrators appointed to determine whether the contract between the builders and the proprietors had been properly carried out were the following:—"That the outside seams "or joints of the plates were very large and filled in with wood "and iron cement, and that it was necessary that these joints "should be cleared out and caulked in the usual manner. The "spaces between the plates and frames ought to have been filled "with wedge-shaped liners, instead of cement as they were." This employment of cement instead of making close joints is stated by M. Dupuy de Lôme to have been very general at Glasgow and Greenock when he visited those places, but it was from the first opposed by most shipbuilders, and has fallen into disuse.

Having briefly illustrated the modes of plating which have been adopted, we turn to the consideration of the plan now in general use. This plan is illustrated in Fig. 128. It was introduced, we believe, simultaneously and independently by Mr. Scott Russell and Mr. J. R. Napier. In this arrangement each alternate strake is worked directly on the frames, and the intermediate strakes form an outer layer, each strake of which overlaps the edges of the two adjoining strakes of the inner layer. The strakes worked on the frames are termed *sunken* or *inside* strakes, and those of the outer layer *raised* or *outside* strakes. It will be evident that while this plan has the same advantages in respect of riveting and caulking as are possessed by the clinker arrangement, it has the additional advantage of requiring liners to one-half the strakes only, and these liners are all of parallel thickness, instead of being wedge-shaped as in the clinker plan of plating. Hence it follows that the modern plan more effectively combines the shell of a ship with the frame, and does this at a less cost for workmanship in preparing the liners than would be necessary if the clinker plan were adopted. The butts of adjacent plates in the same strake are connected by internal butt-straps, and the edge-riveting is sometimes double, and sometimes single, while in some ships one part is double-riveted and the rest single. The subject of riveting will be fully treated of further on.

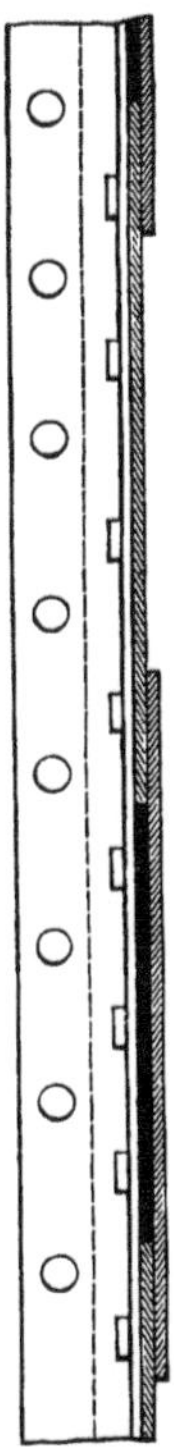

Fig. 128.

Another plan of plating is illustrated in section in Fig. 129. A patent was taken out by Mr. Seaton in 1852 for applying this arrangement to the bottom-plating of ships from the keel up to the water-line, and in 1856 Mr. Lamb, of Southampton, patented the application of this arrangement to outside plating in general. All the plates are worked directly on the frames, liners being altogether dispensed with, and the flush longitudinal joints are covered by external edge-strips. In the specification of Mr. Lamb's patent it is stated that these strips may either be confined to the breadth sufficient to take the rivets forming the edge-fastenings, or they may be worked as a complete outer skin, thus forming a flush bottom within and without. The arrangement illustrated by Fig. 129 is that commonly known as Lamb's patent. This plan has the advantages, before spoken of, which result from the fact that flush horizontal

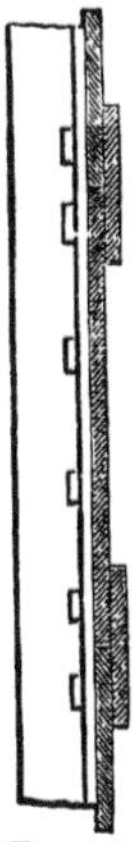

Fig. 129.

seams will sustain the greatest vertical force, and take its shearing effect from the rivets. But it also possesses the disadvantages which are caused by the number of rivets in the edge-fastening being double the number required for an ordinary lap-joint of equal strength. It may be further assumed, as warranted by general experience, that in a ship plated in the usual manner the strength of the side is amply sufficient to resist vertical strains; and that the reduction in vertical strength made by the usual plan of plating is compensated for by the increased economy, and the strength to resist disturbance in the relative positions of the plates, which result. Experience shows also that strip-iron is deficient in strength both when sheared and when in bars, and this together with the width, and consequently weight, required for the strips, constitute objections to the plan. It has, however, been employed in the construction of a few vessels, and has been highly spoken of by some shipbuilders. Mr. Grantham says, "I "consider this plan gives the greatest amount of strength with the "same amount of iron of any system of jointing yet proposed; "and I should like to see a large vessel so built." The plating of the large unarmoured iron frigate 'Inconstant' is, for a particular purpose, worked flush with external edge-strips, the strips being made thick enough to receive the fastenings of wood-planking worked upon the outside as a sheathing. In the earlier iron-clads the skin-plating in wake of armour is worked flush with external edge-strips, as will be seen on reference to the section of the 'Warrior' in Plate 3. In the 'Bellerophon,' and other of the later iron-clad frigates, the skin-plating in wake of armour is worked flush, and made up of two thicknesses. The arrangements adopted in the 'Bellerophon' and the 'Hercules' are given in section in Plates 4 and 5. The outer thickness of skin-plating has single-riveted edge-strips on the outside, but there are no edge-strips to the inner thickness. The widths of the strakes of plating are so determined as to admit of one row of rivets in the edges of the inside thickness being worked in as fastenings in the longitudinal girders. In the vessels of the 'Invincible' class now building, the two thicknesses of skin-plating behind armour have no edge-strips or butt-straps, but are single-riveted to each other at both edges and butts. This plan reduces the tensile strength of the two thicknesses to a certain extent, but this is not so objectionable as it may at first appear, in consequence of the fact that the double thickness of plating is introduced in order to increase the resisting power of the side, and is not required for structural strength. It should be added that the employment of two thicknesses of plating has the advantage of requiring much smaller rivets than would be necessary if the plating were in one thickness, and the frames are consequently much less weakened.

The unprotected parts of the later iron-clads above the armour-belts are flush-plated, the edge-strips being worked inside. In order, however, to avoid the use of liners, all the plates are worked directly upon the frames, and the edge-strips are worked in short lengths between the frames. In wake of the channels doubling plates are worked between the frames to receive the fastenings of the chain-plates, and consequently the edge-strips are there dispensed with. This plan of plating has not been adopted in any other parts of the iron-clads, the bottom-plating of which vessels is in all cases worked on the usual plan of inside and outside strakes.

Another mode of plating, which has been proposed by Mr. Daft, differs from all the preceding arrangements in not making close joints at the butts of the plates. The system on which the butts and joints are formed grew out of a desire to apply zinc-sheathing directly to iron ships. Longitudinal strips of plating are worked inside the plates in order to take the edge-riveting, and the plates are separated at both edges and butts so as to form grooves. Strips of teak are afterwards fitted into these grooves, and the zinc sheathing is fastened to the teak. But independently of the facilities for sheathing furnished by this method, the proposer claims for it the advantages both of the flush and lap jointed systems of plating. He supports this claim by the statement that, by means of the separation of the edges and butts of adjacent plates, all the joints are virtually lap-joints, which require less care in fitting, and can be better caulked than butt-joints; while by means of the teak strips filling up the grooves the ship's bottom is made a flush-surface. Full details of the system and of the advantages claimed for it will be found in the *Transactions* of the Institution of Naval Architects for 1866. The objections made to its adoption are, that there is an increase in the weight of material employed, and in the labour of punching, riveting, &c., as compared with the usual systems of plating.

Coming now to the illustration of the disposition of the butts and edges of outside plating, we shall assume that the system of plating with alternate inside and outside strakes is the only one taken into consideration. The usual mode of procedure in arranging the plating of an iron ship is as follows:—The breadths of the plates are set off on the drawing of the midship section, and then either a model of one side of the ship or an expansion drawing is prepared, on which to set off the edges and butts of the plates. The expansion drawing is necessarily inaccurate, on account of the fact that the surface of a ship is what is known in geometry as an *undevelopable* surface, that is, it cannot be truly flattened out on a plane surface. Consequently the lines drawn upon the expansion drawing are only rude approximations to the true forms of the edges. On the model, however, the form of every plate is truly shown, and fair edges can be more readily obtained. These reasons have led to the general adoption of the model for the purpose of disposing the plating, though in some cases the expansion drawing is still used for this purpose. The breadths of plates having been transferred to the model, or to the expansion drawing, lines are drawn through the points thus obtained, to represent the edges of the plates, and the butts are disposed of.

In some ships all the strakes, varied in breadth so as to give fair lines, are run throughout the whole length; but in many vessels the number of strakes is reduced at the bow and stern, by working some of them as *stealers*, i. e., stopping them short of the stem and stern-post, and working the stealer and an adjacent strake into one. The manner in which stealers are worked is illustrated in the succeeding sketches. The first plan, illustrated in Fig. 130,

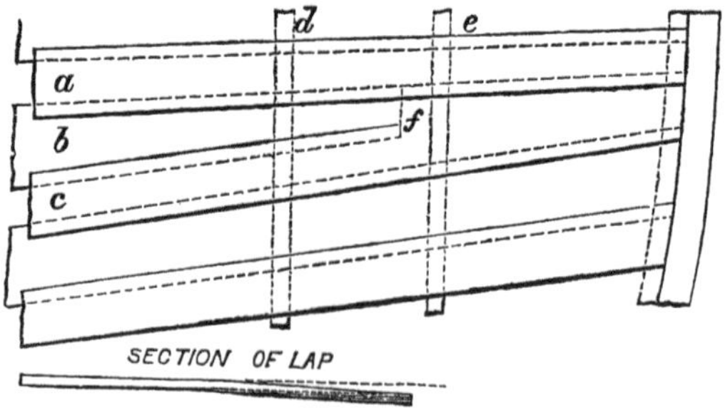

Fig. 130.

was adopted in the 'Achilles.' The strake of plating marked *b* was first worked, and before it was put in place the lower edge, from the frame *d* to the butt *f* was twisted outward through a distance equal to the thickness of the plating, and was planed straight through for the breadth of the lap as shown in dotted lines in the horizontal section of the lap given in the sketch. The strake *c* was then put on and the foremost plate was shaped so as to complete the strakes *b* and *c* from the butt *f* to the stem. On the fore side of the butt *f* the plate *c* was twisted in on the upper edge so as to underlap the lower edge of the strake *a*, which was then put in place. The butt *f* was about 15 inches wide, and was placed at such a distance from the frame *e* as to admit of a treble-riveted butt-strap, of which the after edge was well with the end of the snape in the lower edge of the plate *b*. The butt *f* and the chased part of the edge of *b* were caulked similarly to a butt-joint, but the lap of the plates *a* and *c* was caulked in the usual manner. On the frame *e* and the other frames between it and the stem, tapered liners had to be fitted under the strake *c*, which was there worked clinker-fashion on account of the strake *b* having been butted.

The next two sketches are illustrations of different methods of arranging the butts of stealers, which have been adopted in the 'Captain.' The first of these in Fig. 131 shows the plan followed

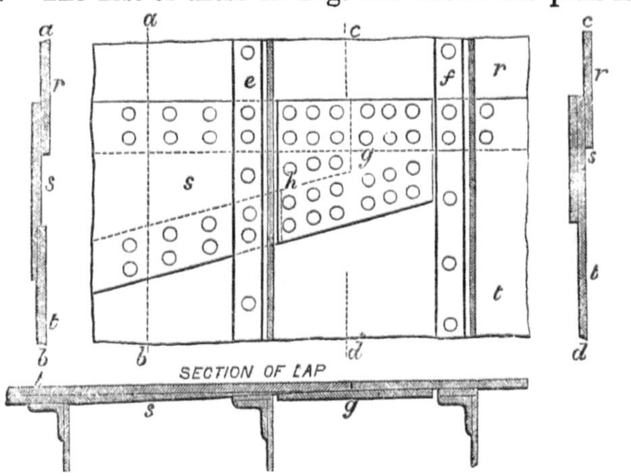

Fig. 131.

when the stealer is an inside strake. From the section at *a b* it will be seen that the lower edge of the stealer *s* is planed away so as to allow the upper edge of the strake *t* to chase in. The rabbet or chasing thus formed is snaped away as shown by the horizontal section through the lap of the stealer. It ends at the dotted line marked *h*, just before the frame *e*. Beyond the line *h* the stealer is reduced in width by the breadth of the lap, and the strakes *s* and *t* are flush-jointed as far forward as the butt *g*. From the section at *c d* and the plan, it will be seen that the butt-strap to the stealer serves also to secure the flush-joint of the plates *s* and *t*.

The second arrangement of the butt of a stealer given in Fig. 132 is that adopted where the stealer is an outside strake. Here

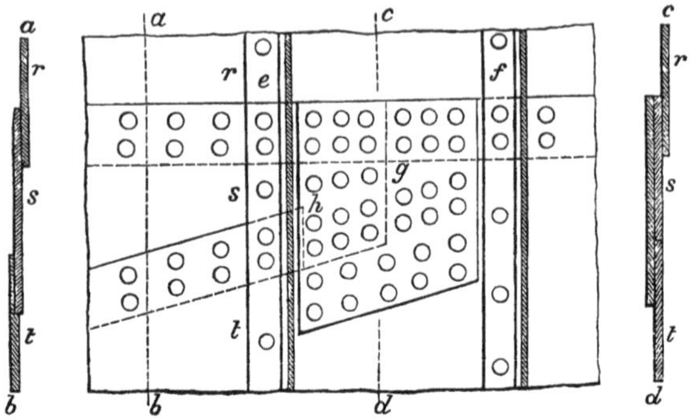

Fig. 132.

the upper edge of the strake *t* is chased away for some distance abaft the frame *e*, as shown in the section at *a b*. The chasing ends at the dotted line marked *h*, between which and the butt *g* the strakes *s* and *t* are flush-jointed, and are secured by the butt-strap in a similar manner to that described above.

Near the extremities of the ship the usual plan is to thin away the laps of the plates, so that the inside and outside strakes may chase into each other, and form a flush surface. In some ships the same object is attained by working the plates flush-jointed for a short distance, and securing the edges by internal strips. In the later iron-clad frigates of the Royal Navy the outside bottom plating for about 40 feet from the bow has been doubled, in order to take the wear of the anchors and cables, and to increase the strength for ramming.

The butts of outside plating on one side of a ship are generally opposite those on the other side. The exceptions, in most cases, are the butts of the garboards, and those of the plating of the fine parts of the vessel, where the two sides are within a few feet or inches of each other. In arranging a shift of butts in which the butts of the garboard on one side are not opposite those on the other side, it is not possible to alter the shift so as to have the remaining butts alike on both sides, and at the same time to get as good an arrangement where the alteration is made as at all other parts of the ship, with plates of the same length. On this account, and in order to avoid the use of longer plates, the arrangement of butts made for the garboards and adjoining strakes is sometimes carried throughout the ship, thus making the whole of the butts on one side fall in different frame spaces from those which they occupy on the other side.

In arranging a shift of butts care has to be taken to adjust it so as to suit the positions of the scuttles and ports, and, in fact, these go far towards determining the positions of some of the joints and butts of the topside plating, and fixing the lengths and breadths of the plates. In all well-built ships the butts of deck-stringers, and of internal plating and angle-irons, are shifted as far as possible from the adjacent butts of outside plating. The importance of this was illustrated in the case previously given (p. 14), where a ship with a large quantity of material in her hull, broke down through a series of butts of plating, stringers, &c.

The butts of outside plating are generally placed midway between the frames, and their usual arrangements are shown in the following sketches. The disposition of butts given in Fig. 133 is

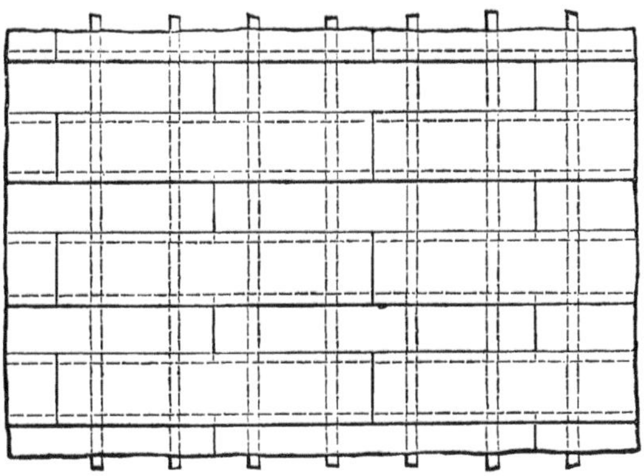

Fig. 133.

that which was formerly almost universally adopted. It is known as the *brick* arrangement, and with the short plates formerly in use no better arrangement could be made. It is still adopted in the construction of many iron ships. The butts of alternate strakes are in one vertical line, each butt being placed at the middle of the length of the plates above and below it. It may be remarked, in passing, that this arrangement brings the butts of the inside strakes on one set of vertical lines, and those of the outside strakes on another set. Some builders, in order to make all the strakes appear of equal breadths, work the inside strakes broader than the outside strakes by twice the width of the lap. Others prefer having the inside and outside strakes of equal breadths. The cost of labour and materials is the same for both arrangements, but, presuming the butts to be places of weakness even when strapped, the latter arrangement is the stronger of the two, although the difference is not great. For, if the plates of the inside strakes are all wider than those of the outside strakes, the ship will not be so strong through a line of butts of the inside strakes as through one of the lines of butts of the outside strakes. But if the strakes are of equal width the strength at the two lines of butts will be equal. And it is evident that the ship will be stronger through any line of butts in the latter arrangement, than through a corresponding line of butts of the inside strakes in the former arrangement.

The next disposition of butts given in Fig. 134 is known as the

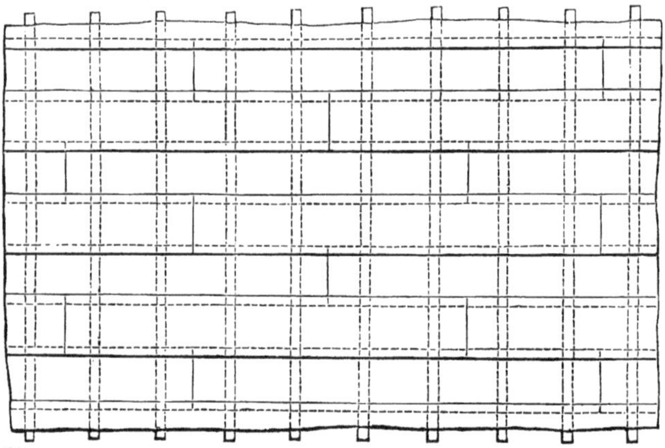

Fig. 134.

diagonal arrangement, and is now in common use on the Mersey, Clyde, and Tyne, and in the construction of the iron ships of the Royal Navy. There are always two strakes between consecutive butts in the same frame space, and the butts of adjacent strakes are never nearer than two frame spaces. It will also be remarked that the successive butts in the same frame space are alternately those of inside and outside strakes.

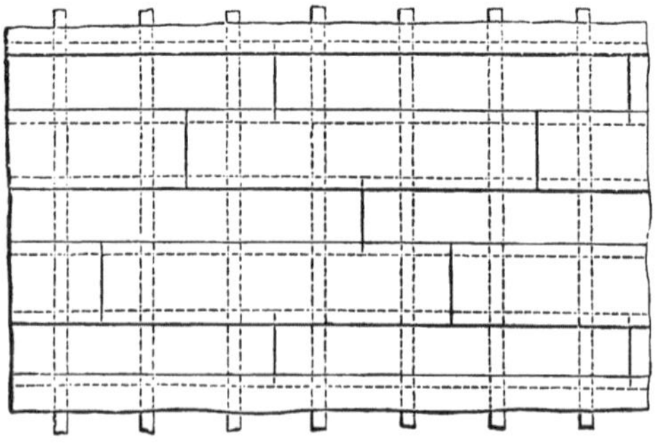

Fig. 135.

The disposition of butts illustrated by Fig. 135 is adopted in the outside plating of the 'Inconstant,' the arrangements of which have been previously described. The sketch is, however, drawn to represent the ordinary mode of plating. There are three passing strakes between consecutive butts in the same vertical line, and the lengths of plate used extend over four frame spaces. In some instances the butts of successive strakes are only one frame space apart; but as this is 3 feet 6 inches in this ship, the shift obtained very nearly agrees with that required by Lloyd's Rules, which state that a shift of two frame spaces must be obtained, and fix the spacing of the frames at from 21 to 24 inches.

The fourth disposition of butts shown in Fig. 136 has four

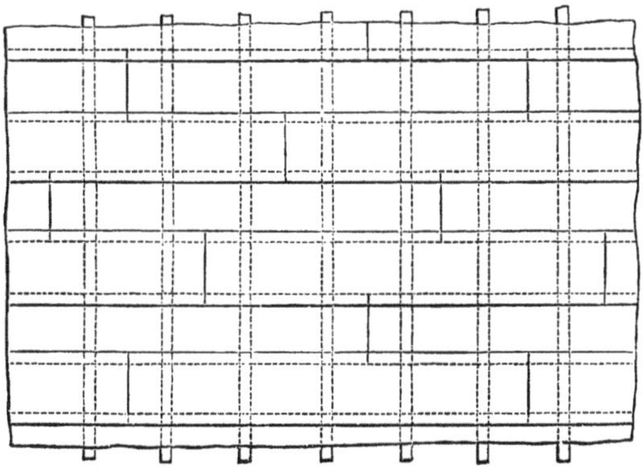

Fig. 136.

passing strakes between consecutive butts in the same vertical line. Each plate is five frame spaces in length and the butts of adjacent strakes are never nearer than two frame spaces. In this, as in the second disposition, successive butts in the same vertical line are alternately those of inside and outside strakes.

The sketch in Fig. 137 illustrates the arrangements of the butts and edges of the armour plates, and of the two thicknesses of skin-plating behind armour, of the 'Hercules.' The butts of armour plates are marked *a*, those of the outer thickness of skin-plating *b*, and those of the inner thickness *c*. The lines *s*, *s* are the stations of the frames behind armour, and are 2 feet apart. The armour plates are 16 feet in length, and their butts are arranged brick fashion, and placed directly upon the stations. The skin-plating is in 12-feet lengths, the diagonal arrangement being adopted for the butts of each thickness. It will be seen from the sketch that the butts *c* of the inner thickness come in the frame spaces between those in which the butts *b* of adjacent strakes of the outer thickness are placed, and *vice versâ.* The butts *a*, *a* of the armour are in most instances placed on the stations midway between the

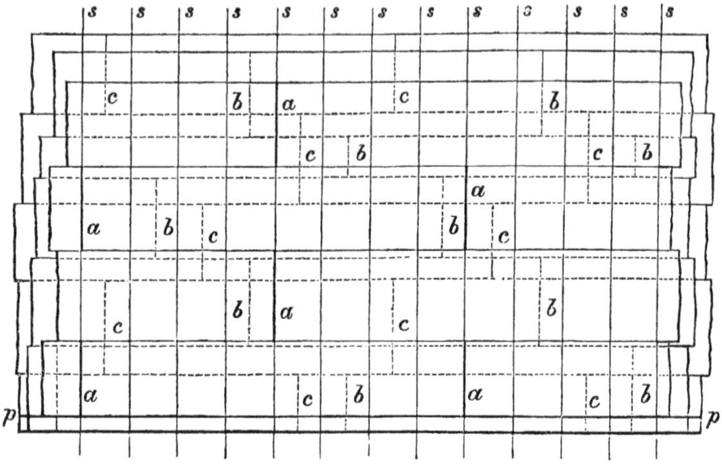

Fig. 137.

butts *b* and *c* of the underlying strakes of skin-plating. It may be remarked here that in some of the iron-clads the butts of the skin-plating behind armour are placed upon the frames, instead of being midway between them as is usual.

Lloyd's Rules require that all plates, except the fore and after hoods, shall be at least five frame spaces in length, and that no butts of outside plating shall be nearer to each other than two frame spaces. These requirements are illustrated by Fig. 136. The Liverpool Rules stipulate that all butts of garboard strakes, shell-plating, stringers, and scarphs of keels, shall be two frame spaces apart; that butts in the garboards must not be opposite each other; and that butts of the upper-deck stringer-plates must not be nearer than three feet to butts of the sheer strake.

The lengths of bottom plates are generally determined so as to give a good arrangement of butts, and the breadths are modified by the girth of the frames and weight of the plate, but are usually about one-fourth or one-fifth of the lengths. The use of longer plates tends to make a ship stronger and lighter, but at the same time more expensive, as the price per ton increases with the weight. The minimum length allowed by Lloyd's would average about 9 feet; on the Tyne the lengths of plate in general use vary from 8 feet to 10 feet 6 inches; on the Mersey and Clyde they are about 10 feet. In the 'Great Eastern' all the bottom plates are

10 feet long and 33 inches wide. In the ships of the Royal Navy the plates are usually in 12-feet lengths, although in some cases 16-feet lengths are introduced to give a good shift, and in a few instances greater lengths have been employed. In the 'Resistance,' for example, there are a few 1-inch plates, 22 feet long, and 3 feet 6 inches wide; in the 'Captain' the skin-plating behind armour is in 16-feet lengths.

The preceding remarks have been limited to the *arrangement* of outside plating; there remains another subject of very great importance to be considered, viz., the modes of fitting and securing the plates. Lloyd's Rules state that "all plates are to be well-"fitted and secured to the frames and to each other; the butts to "be closely fitted by planing or otherwise, and to be united by "butt-straps of not less than the same thickness as the plates, and "of sufficient breadth for riveting, and to be fitted with the fibre of "the iron in the same direction as the fibre of the plates to which "they are riveted." The Liverpool Rules are identical with Lloyd's in their specification for butt-straps. Their requirement with respect to the fitting of butts is fuller than Lloyd's, and is as follows:—"Butts to be closely fitted either by planing or "jumping; when jumped the ridge formed by jumping is to be "chiselled off the inside, in order that the butt-straps may fit closely. "The ridge outside to be hammered into the seam." Both rules require that the rivet-holes shall be punched from the faying surfaces, and countersunk through the outer plating, in order to preserve a comparatively flush surface.

Before giving a detailed description of the modes of securing outside plating, it may be interesting to give a brief general outline of the process of plating. After the frames have been put in place, faired, and fixed, the lines for edges of the inside strakes of plating are got in, and marked upon them. The plates of the inside strakes are then prepared. If the plates are light they are put in place, and the rivet-holes in the frames, and the positions of the edges and butts, are marked on them. If the plates are heavier than can be conveniently handled, the account for them is taken by means of a mould or template of wood or iron, and the holes and stations are transferred from the mould to the plates. The plates are then lined, and the edge and butt fastenings having been set off, and the fastenings to frames having been marked, the holes are punched, the plate edges sheared, and their butts either jumped or planed. These operations having been completed the plates are fixed in place by means of screw-bolts, and the butt-straps are fitted, marked, and punched. In the meantime, as soon as two inside strakes have been worked, the plates for the outside strake which overlaps them are prepared. These plates are taken account of in a similar manner to the plates of the inside strakes, only the edge fastenings have to be marked on the plate or mould when it is first put up in place. The edge fastening requires great care in its marking, transference, and punching, and by far the greater number of what are termed *blind*, or *half-blind*, holes are found in the edges. The plates are put in place, secured, and have their butt-straps fitted, similarly to the inside strakes. When the plates have been temporarily secured, the plate liners on the frames are prepared and fixed in position. It is usual, in parts of the ship where there is moderate curvature, to bend the plates to an approximation to the form required, by means of rollers. In general, the fitting, marking, and fixing of the outside plating are performed by a party of workmen known as *platers*, assisted by a number of labourers or helpers, and while these proceed with their work other workmen are engaged in riveting up and caulking the plating which has been fixed. A more detailed description of the operations will be found in chap. 20.

The care in fitting the butt-joints of outside plating, which is now enjoined by both Lloyd's and the Liverpool Rules, has not been generally taken in ships of the mercantile marine. In a paper published in the *Transactions* of the Institution of Naval Architects, for 1862, Mr. Grantham speaks as follows on the subject:—" To make the ends a tolerable fit, a hammer is generally " used, knocking down the higher parts and throwing up a burr on " one edge; but as this operation is very imperfect, the plates in " reality only touch at points; and even in what appears to the " eye a good joint, the plates often only touch at two or three " points. In other cases where still less care is observed, the joints " will touch at one corner of the plates only, and be entirely separated from all the rest. These defects are quite visible at first, " but when the butt-strap is put on, the light can no longer be " seen through the joints, and thus observation is partially prevented. When the men are told of this, the reply is that the riveting " will stretch the plates and close the joint; but though this cannot " be relied upon to any extent to compensate for the width too " frequently left, yet all questions are set at rest by the caulking

" tool knocking down the burr which has before been raised, and " closing the plates on the outer part of the joint. This system " makes the joint watertight, but is so entirely defective in strength " that no one ought to be satisfied with it. All parts not really " in contact, however slight the separation may appear, are deprived " of the great additional strength which well butted plates must " afford." The remedy for this bad workmanship is found in the use of planing and slotting machines, which are now employed in many private yards. In the ships of the Royal Navy the butts of plates are always planed and accurately fitted.

The mode of punching rivet-holes in bottom plating is a matter requiring great attention. The importance of care in punching holes is not confined to bottom-plating, but is also essential to good work throughout a ship; still, as the connection of the plates forming the shell of an iron ship is, by the common consent of shipbuilders, regarded as one of the most important features of her construction, we may, with convenience and propriety, here give a slight sketch of the process of punching as usually performed. After the positions of the holes have been marked on a plate, it is taken to the press or punching-machine and held in position by several men, who also shift the plate at intervals so as to bring the stations of the holes successively under the punch. The form of punch usually employed is shown by *a* in Fig. 138, and is flat-ended. When a centre-punch is used to mark the stations of the holes for punching, a pointer is sometimes formed at the centre of the end of the punch, as shown by *b* in Fig. 138, in order to feel for the punctures and to ensure accuracy. It is also found that punches distress the iron less when the ends are formed as shown by *c* in Fig. 138, instead of being flat. It is usual to have the holes $\frac{1}{16}$ inch larger than the rivets, in order to allow for their expansion when heated; it is evident, however, that the difference between the diameters of the holes and the rivets should vary with the size of the rivet. Mr. Fairbairn states that, for ordinary work, the proportion of the diameter of the punch to that of the hole in the die varies from 1 : 1·15 to 1 : 1·2. By this means the holes when punched are

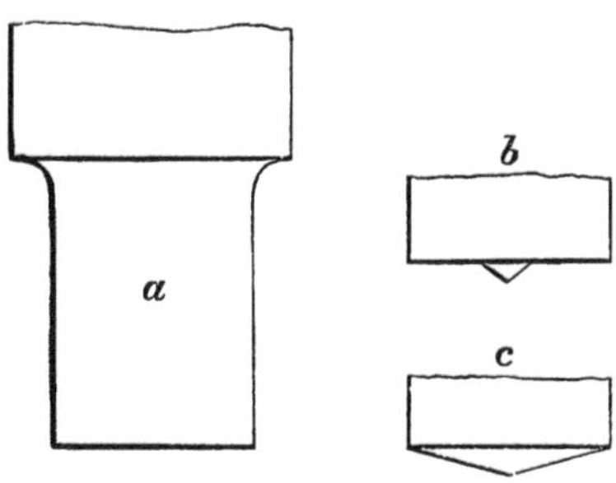

Fig. 138.

made slightly conical, as shown in Fig. 139. It is the fact of this slight countersink being obtained by punching which makes punching from the faying surfaces such an important matter.

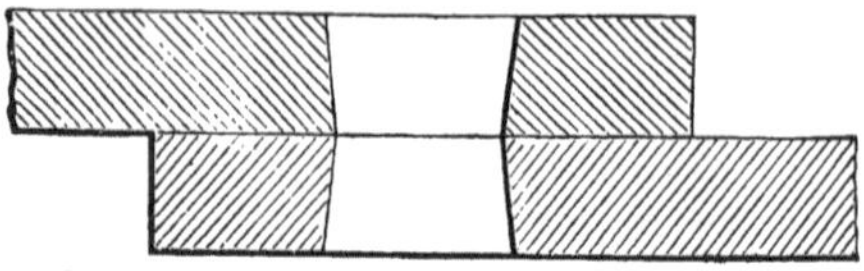
Fig. 139.

Some shipbuilders do not carry out any fixed plan in punching plates, and others punch all the outside plating from the inside, so that the holes in the edges of inside strakes are not punched from the faying surfaces. In order to carry out the regulations of both Lloyd's and the Liverpool Rules, the holes in the frame angle-irons should be punched from the outside, and the corresponding holes in the plating should be punched from the inside. The holes for edge-fastenings should be punched from the inside of the outside strakes, and from the outside of the inside strakes; the holes for butt-fastenings should be punched from the inside of all the plates, and from the faying surface of the butt-straps. The holes are countersunk in the outer surface of the plating, in order to allow the rivets to be knocked down into them, and make the plating flush, or nearly so. Both Lloyd's and the Liverpool Rules require that the countersinking shall extend through the whole thickness of the plate.

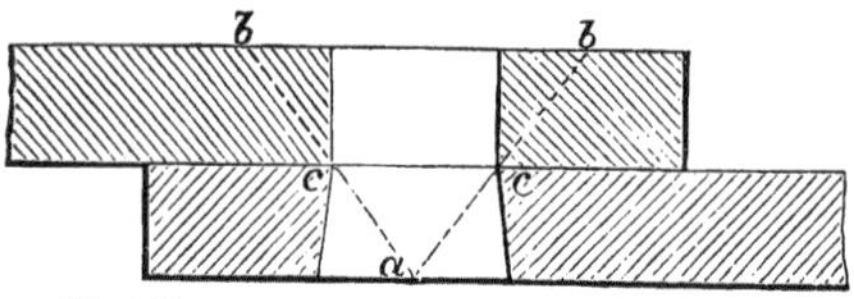

Fig. 140.

The rule which is sometimes employed for guiding the countersink is illustrated by Fig. 140. The centre *a* of the hole on the inner surface of the plating is joined with the boundaries *c*, *c*, at the common surface of the plates, and the lines *a c* produced give the taper, *c b*, of the countersink. It is a very common practice, however, to leave a small shoulder of about $\frac{1}{16}$ or $\frac{1}{8}$ inch at *c*, as shown in Fig. 141, instead of counter-

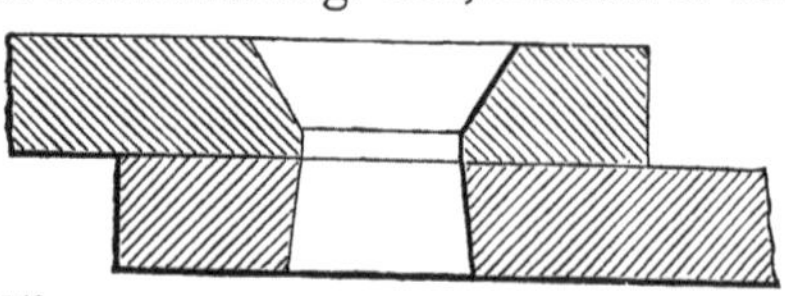
Fig. 141.

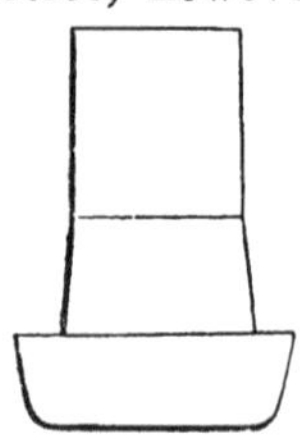
Fig. 142.

sinking quite through. The last sketch shows the form of the hole before the rivet is put in, and the form of rivet now in common use is illustrated by Fig. 142. It will be seen that under the head of the rivet there is a slightly conical part, which fills the countersink made by punching the holes in the inner plate. In knocking down the rivet the large countersink in the outer plate is also filled so that the rivet-shank then has the form of two truncated cones with their smaller ends joined. This form has the great advantages of completely filling the hole when it is riveted up, and of drawing the plates close together and making a tight joint. In many instances, where the rivet-heads have been worn away by the corrosive action of bilge-water, the rivets have by this means been kept in place, and the plates held together. The importance of the last-named advantage will appear when it is stated that in some ships the entire heads of many of the rivets of the bottom have been worn down in less than five years, and it has been necessary to re-rivet the greater part of the bottom. Lloyd's Rules require that the points of the rivets shall be round or convex, and not be below the surface of the plating. The Liverpool Rules require the rivet points to be perfectly fair with the surface of the plating. The practice of shipbuilders also differs on this point. On the Mersey the points of the rivets are made flush, but on the Clyde and Tyne the usual custom is to have the points flush above the light water-line, and about $\frac{1}{16}$ inch convex below it. The heads of the rivets are generally *laid-up*, that is, are made close to the surface, against which they fit by a few heavy blows given by the workman, known as "the holder-up;" this is required by the Liverpool Rules only.

In the preceding description of the manner i nwhich holes are generally punched and countersunk, and rivets formed and riveted up, it has been supposed that when two plates which are to be riveted are put together, the corresponding holes are coincident and have a common centre. This is what is aimed at in all well-built ships, and is required by both Lloyd's and the Liverpool Rules. But in practice it often happens that holes are not coincident, and either occupy a position similar to that shown in Fig. 143, when they are said to be half-blind, or are even more eccentric and are nearly blind altogether. In rough imperfect work this fault is of very common occurrence; and even in cases where care is taken in marking, transferring, and punching the holes, it frequently occurs. Nor can this be wondered at when it is remem-

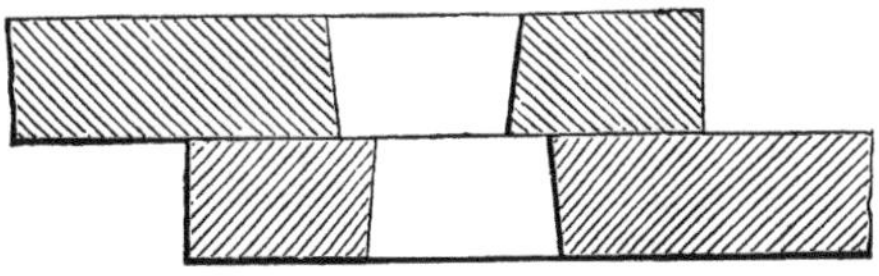

Fig. 143.

bered that the plate is held and shifted by manual labour while the holes are being punched, and that, consequently, slight deviations from the true positions of the holes are almost unavoidable. But, on the other hand, it cannot be doubted that the worst faults in punching are caused by the gross carelessness of workmen. In order to avoid this source of error and to ensure correct holes, two proposals have been made. The first of these is to perform the punching by means of a self-regulating machine, such as the Jacquard machine of the late Mr. Roberts;* and the second is to drill the holes. To the first proposal there is the great objection, that although such a machine is suitable for punching a great number of plates of the same pattern, in plating an iron ship, where the shapes of the plates and the positions of the holes differ widely, the machine could not be applied. To the second proposal it has been objected that there is a difficulty in fixing the drill in the exact position required to make the hole true, and that drilled holes are not always perfect, nor so well suited to rivets as to bolts. Mr. Fairbairn says, on this point, "according to our judgment, " drilled holes are not always perfect, and are never so sound, nor " yet so secure for rivets as those which come from the punch; and " for this reason, that in punching a hole through an iron plate it " is not the same as a drilled hole, exactly cylindrical, parallel, or " smooth, but the frustum of a cone, and hence follows the superiority of the joint as more easily adjusted, and more closely " incorporated with the plates."

In practice, when the holes are badly punched the workman drives in a steel drift-punch, of which the end is tapered and the centre is nearly parallel, and the plate is thus forced and torn and the holes enlarged, so that the rivet passes obliquely through the plate and is very imperfectly riveted up. If the rivet is very hot and the hole not very irregular, it may be filled when the rivet is knocked down; but if the hole is much distorted the rivet will not fill it, and when put under strain the rivet becomes loose. This process of drifting the holes cannot be too strongly condemned, as it considerably reduces the effective strength of the iron, especially in the edges and butts. The frequency of its being required

* The machine here alluded to was used in the construction of the Conway and Britannia tubular bridges, and a full description of it is given by Mr. Clark in his work, 'The Britannia and Conway Tubular Bridges.' He states that by means of it 3168 holes, each $1\frac{1}{8}$ inch in diameter, were punched per hour in the $\frac{1}{2}$-inch plates forming the bottom of the tube.

has led some shipbuilders to adopt and advocate the practice of drilling the holes in outside plating.*

---

* Since this was written the following letter has appeared in Mr. Colburn's valuable journal, *Engineering*:—

"TO THE EDITOR OF 'ENGINEERING.'

"SIR,—It has often appeared inexplicable to me that whilst engineers now-a-days so generally insist on the importance of drilling instead of punching the rivet-holes in girder and boiler work, the subject has not received the same amount of attention at the hands of our naval architects.

"Some years ago I chanced to see a large Messageries Impériales' steamer which had struck upon a submerged rock in the Mediterranean, the impetus of the vessel being so great that she had not been brought to a stand until pretty nearly evenly balanced on the rock.

"We are taught that a sea-going ship may be considered in the light of an immense girder, which, under varying circumstances, may be called upon to carry a load in the middle whilst supported at the ends, or *vice versâ*; but it was evident that either the rules upon which this vessel had been constructed, or the workmanship, or both, were at fault. I am disposed to think *both*, for no sooner did she begin to fill than she parted amidships, and settled down fore and aft in deep water (I hope I am correct, or at least understandable, in my nautical phraseology). The rents, of course, commenced at the bulwarks, and gradually extended downwards, choosing, however, *the lines of rivet-holes*. No doubt this would have been the case under any circumstances, and that it is impossible by any system of riveting, except, perhaps, thickening the edges of the plates, to preserve the full strength of it. The result of the experiments of Mr. Fairbairn and other well-known engineers would appear to prove that we cannot, in any appreciable degree, avoid the loss of strength occasioned by diminishing the sectional area of the plate; but, although they show that this loss is further unnecessarily increased by *forcing* out the portions of iron by means of a punch, instead of *cutting* them out with a drill, I am not aware that there have been any satisfactory experiments made to test the effects of the subsequent operation of drifting, which all punched holes must undergo.

"The manager of one of the largest shipbuilding yards in the north told me the other day that it was their practice to punch the rivet-holes an eighth of an inch less than they were intended ultimately to be, and to make the two holes (in a manner) correspond, by forcing through a steel drift the size of the rivets. One advantage attends this plan, that is, *it tests the quality of the iron*. None but a tough strong iron, with a hole anything like within its own diameter of the edge of the plate or bar, would stand such fearful treatment as this. But whilst there may be no crack that even the sharp eye of the inspector can detect, who shall say how many or how few more blows would suffice to produce one, how nearly the strength and tenacity of the iron are exhausted, or how fit it is to fulfil its proper functions as a portion of the vessel?

"Taking into consideration the *inevitable* loss of strength occasioned by the diminished sectional area, the *avoidable* loss caused by punching, and, as I believe, the *still greater* loss which the drifting process entails, is it safe to assume in many cases that there is 30 per cent. of the original strength of the bar or plate left? To remedy the two last-named evils, I would suggest the practicability of drilling and countersinking at one operation the rivet-holes after the plates are placed in position on the sides of the vessel, simultaneously through skin and framing. This could be accomplished by a portable multiple drilling-machine, having its own engine attached, of course supplied with steam by means of a flexible or jointed steam-pipe.

"The construction of such a machine with six, eight, or ten drills of readily variable pitch 'presents no mechanical difficulty.' It would not be an expensive

We now turn to the arrangement of the rivets which attach the plates to the frames and to each other. Lloyd's Rules give eight diameters as the distance apart of the rivets in the frames, and the Liverpool Rules give the same distance apart (8 diameters) from centre to centre of the rivets. In bulkhead frames of merchant-ships, and in watertight frames of ships of the Royal Navy, the pitch of the rivets is reduced to from five to six diameters. In the laps of the plating the usual pitch of the rivets is from four to five diameters. Lloyd's Rule with respect to the edge-riveting is, "that all vessels are to have all edges or " horizontal joints of outside plating double-riveted from the keel " to the upper part of bilges, all fore and aft; but vessels of 700 tons " and above, intended for the highest grade, are to have all edges " or horizontal joints of outside plating double-riveted throughout." The Liverpool Rules require all vessels to be double-riveted in bottom, bilges, and sheer strake, and all vessels above 600 tons to be double-riveted throughout. Lloyd's stipulate that all butts of outside plating are to be double-riveted, and the Liverpool Rules require butts to be double or treble-riveted.

The arrangement of rivets in the laps and butts of outside plating is a much disputed subject. Some shipbuilders, among whom is Mr. Grantham, approve of double-riveting for butts, but doubt its utility for horizontal joints; others think with Mr. Fairbairn that double-riveting is sufficiently strong in the longitudinal joints, but comparatively weak in the butts. The simplest mode of connection of both edges and butts is that made by single-riveting. This is not allowed by the Rules for butts, but it has been adopted by Mr. Scott Russell in the construction of the 'Annette' and other longitudinally framed vessels. He justifies the course adopted by the statements that the widths of plates used were greater, and that the longitudinal frames being con-

---

machine, and could be easily applied to most parts of the shell of the vessel, whilst an ordinary radial drill, also with its own engine, could be applied to the more inaccessible parts.

"The plan might be somewhat more costly than punching, but so is drilled girder work, and the stability of a bridge is not of more importance than the strength of a ship.

"If this suggestion has not been made before, I shall be glad if you think it worth publishing, and subscribe myself,

"Yours, &c.,

"G. HUTCHINSON.

"Skerne Iron Works, Darlington, Sept. 21, 1867."

tinuous and firmly riveted to the plates, formed efficient ties across the butts, and made a stronger union than any other arrangement. Of course, if the longitudinal strength is ample for the purpose with this arrangement, no objection can be offered to it, and it must be added that in these cases the strength of the bottom plates is said to be really limited by the resistance to buckling, the tensile strength being excessive. It may be remarked, however, generally, that while it is perfectly true that the longitudinal frames thus succour the butts, it is also true that the longitudinal strength of the ship's framing is reduced proportionately to the help thus afforded; or, in other words, since the single-riveted butt is of itself confessedly weaker than the other parts of a plate, it follows that in order to bring up the strength of the butt to that of the other parts of the plate, the effective strength of the longitudinal frame crossing the butt must be proportionately reduced. Hence it is evident that in a longitudinally-framed ship, if the full strength of both bottom-plates and longitudinals is requisite, there are the alternative courses of either giving surplus strength, and therefore weight, to the frames in order that they may succour the single-riveted butts, or of more efficiently riveting the butt-straps. But it is not only in ships which are thus framed tha the butts of outside plating have been single-riveted, for, in many vessels which are transversely framed, this course has been also followed without adopting a shift of butts which would have supplied the requisite strength. Allusion has already been made to this practice when illustrating some of the weaknesses of iron ships, and its effect in reducing longitudinal strength has been pointed out. An illustration of a single-riveted lap-joint is given in Fig. 144. Objection has been made to single-riveting as applied to the laps of thick plates, on account of the supposed want of closing power which it presents; but this is answered by the statements that the laps are made close and secured by screw-bolts before the riveting is begun, and that in single-riveting the rivets are placed closer.

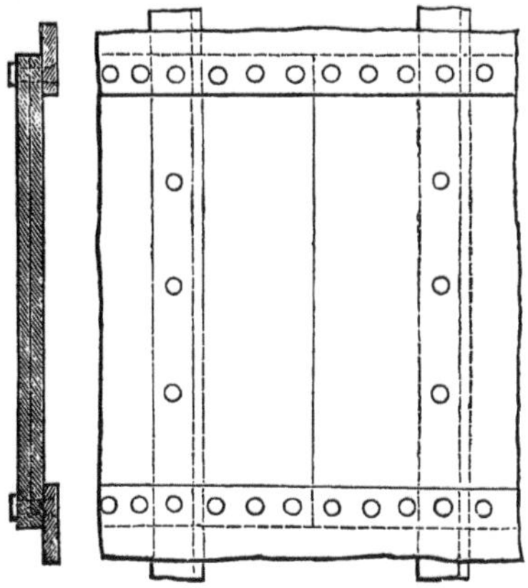

Fig. 144.

When double-edge riveting is adopted, it may be either *zigzag*, as shown in Fig. 145, or *chain*, as shown in Fig. 146. The first of these arrangements requires that each butt shall pass through the centre of a rivet in each lap, as will be seen on reference to the sketch. This cannot be well avoided, for if the butt were placed between two rivet-holes in the laps they would be too close to the butt to make good work. The rivets in the butt are not very

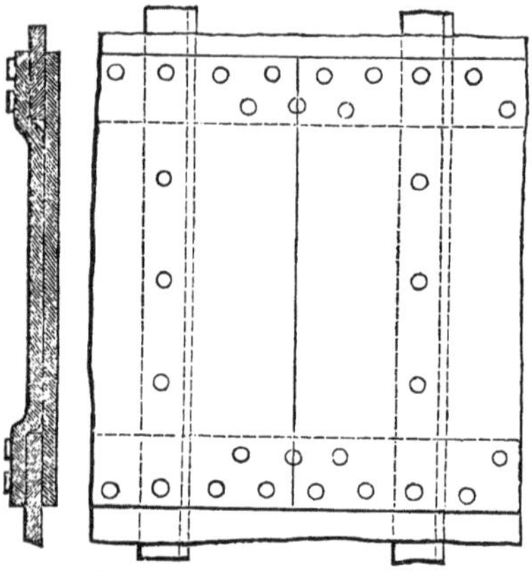

Fig. 145.

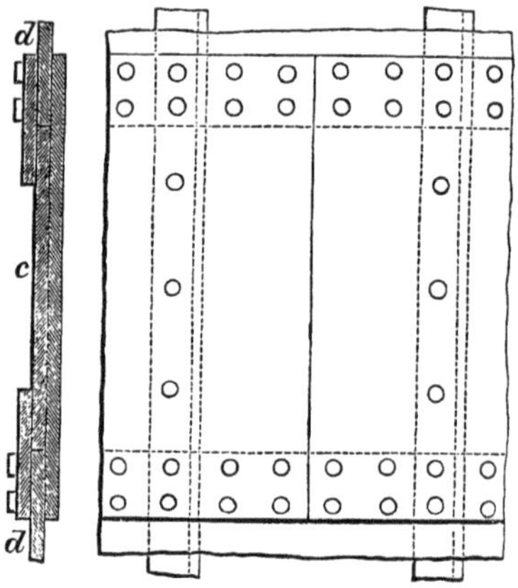

Fig. 146.

objectionable, for, though they do not help to unite the two plates, they may keep the butt from opening under a transverse strain. An objection to the zigzag system is that a long space is left clear at every frame in one row of rivets, as is shown in the sketch in Fig. 145. In some ships, however, in order to avoid this evil, two rivets have been put in each frame, and while the continuity of the edge-riveting has thus been kept up, the frames have, of course, been weakened.

Chain-riveting was proposed by Mr. Fairbairn, and largely adopted in the construction of the Britannia Bridge. It is generally regarded as a stronger mode of connection than the zigzag system described above, and this opinion is supported by the experiments made by Mr. Fairbairn, and by those conducted by Mr. Mumford, in 1857, under the direction of Lloyd's Committee. In the Report made by the latter gentleman it is stated that the breaking strains of plates $13\frac{1}{4}$ by $\frac{3}{8}$ inches, which were connected by double-chain and zigzag riveting respectively, were 50 and 42 tons. On the other hand, in his history of the Britannia and Conway bridges, Mr. Clark states that zigzag or diagonal riveting was proved to be stronger than chain-riveting by the experiments made in order to show the great importance of friction between the surfaces of the connected plates. Lloyd's Rules do not state which system of riveting must be followed, but in the illustrative sketches accompanying the Rules the edges and butts are chain-riveted. The Liverpool Rules require that all double-riveting shall be arranged chain-fashion. When the edges are double-chain riveted, as shown in Fig. 146, the butts of outside plating can be placed well clear of the rivets in the laps, and the pitch of the rivets can generally be so regulated as to leave no vacant spaces at the frames. These facts, together with the supposed stronger character of the connection, have led to the extended use of chain-riveting. Zigzag riveting is, however, still very extensively employed in the ships of the mercantile marine, especially in the laps of outside plating. The breadth of lap required by Lloyd's Rules is $3\frac{1}{4}$ diameters for single-riveting, and $5\frac{1}{2}$ diameters for double-riveting. The Liverpool Rules do not state the lap necessary for single-riveting, and agree with Lloyd's in their requirement for double-riveting. Lloyd's also require that the rivets shall not be nearer to the butts or edges of plating than a space not less than their own diameter. It follows from this last regulation that in a double-riveted lap the distance between the rows of rivets would equal one and a half times the diameter of the rivet. In many shipbuilding yards, however, it is customary to have the breadth of lap for double-chain riveting 6 diameters instead of $5\frac{1}{2}$ diameters as required by the Rules. The rivet-holes are then placed at $1\frac{1}{6}$ diameter from the edges, and the distance between the rows is $1\frac{2}{3}$ diameter. For single-riveted laps the breadth in many cases is increased to $3\frac{1}{2}$ diameters.

The butts of outside plating are secured by butt-straps, which are in most ships double or treble riveted. Before proceeding to illustrate the various arrangements of butt-fastenings, it will be necessary to consider the various modes of fitting butt-straps. At the butt of an inside strake the strap usually extends the whole width of the strake, as shown in section in Fig. 144, p. 201, and takes the rivets in the laps on both edges. For the butts of outside strakes the arrangement illustrated in section by Fig. 145 was formerly in general use. It will be seen that the ends of the strap are joggled over the edges of the adjacent inside strakes so as to take the edge-riveting. This arrangement was superseded by the plan shown in Fig. 146, where the butt-strap *c* is fitted between the edges of the inside strakes, and covering plates, *d, d,* are worked as shown, and take the edge-riveting and a few rivets through the strap *c*. This plan was adopted in some of the first iron-clads of the Royal Navy, but it has since been displaced by a simpler and lighter arrangement, which consists in dispensing with the covering plates *d, d,* and sometimes having the butt-strap *c* a little thicker in order to give sufficient strength. In the case of outside strakes where the frames are spaced a moderate distance apart, it is often well to make the butt-strap and liners to the adjacent frames in one piece, especially in parts where the strain is great, such as the plating near the screw-propeller, and that forward in a ship with a ram-bow, &c. Lloyd's Rules require this arrangement to be carried out at the butts of sheer strakes, which are made outside strakes in order to allow the butt-straps to be thus fitted.

Butt-straps are usually of the same thickness as the plates they connect. In a letter appended to the Report of Mr. Mumford's experiments, Lloyd's surveyors for the Port of London state that butt-straps of the same thickness as the plates are not equal to double-riveting, and should in all cases (except where broad liners are used behind bulkheads) be $\frac{1}{8}$ inch thicker than the plates. In some of the later iron-clads, however, the straps to the butts of inside strakes are $\frac{1}{16}$ inch thinner than the plates they connect, and the straps to the butts of outside strakes are the same thickness as the plates; the latter arrangement is consequent on the fact that the straps are fitted between the edges of the adjacent inside strakes, and are 10 or 11 inches narrower than the plates they connect. It will hereafter be seen, on examining the case of the 'Hercules,' that these thicknesses are sufficient when the riveting arrangements are properly carried out.

It will be remembered that both Lloyd's and the Liverpool Rules require that the fibre of the iron in the butt-straps shall be in the same direction as that of the plates. The importance of this provision has been proved by experience, and follows from the fact that wrought iron is considerably stronger lengthwise of a plate than crosswise, the difference in tensile strength in the two directions amounting to between 3 and 4 tons per sq. inch of sectional area. It was formerly very common to roll the iron for butt-straps in a long bar of the required width and thickness, and then cut off the straps to the length required. This plan is, of course, forbidden by the provision of the Rules with regard to the direction of the fibre, and the straps are now cut from plates having the fibre in the direction of the breadth of the straps.

The rivets of butt-fastenings, like those connecting the plate-edges, are placed either zigzag or chain-fashion. The usual arrangement of butt-strap with double zigzag riveting is given in Fig. 147. This plan of butt-fastening is allowed by Lloyd's for all ships, and has been adopted in the 'Warrior,' 'Black Prince,' and others of the first iron-clads of the Royal Navy. In some of the larger iron-clads built since the 'Warrior,' and in some iron-cased ships for foreign Governments which have been built in this country, the butts of outside plating have been treble-zigzag riveted as shown in Fig. 148. This plan is heavier and more costly than common double-chain riveting, and is stronger than double-chain riveting only in cases where increased rivet-power is required.

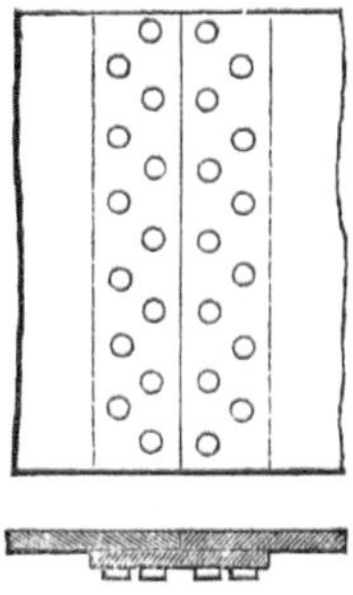

Fig. 147.

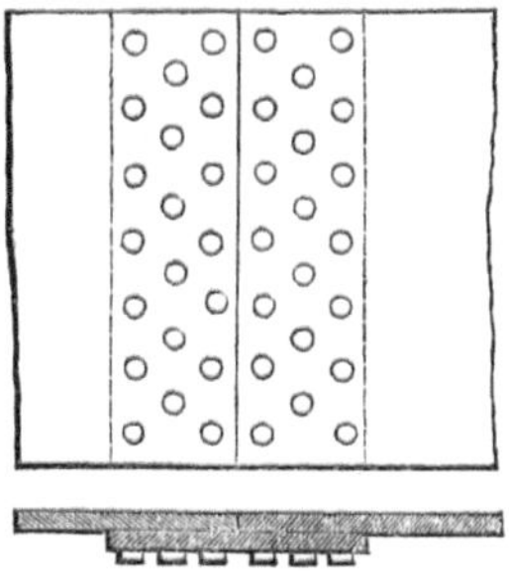

Fig. 148.

The double chain-riveted butt-fastening illustrated in Fig. 149 is now very extensively used in the construction of merchant-vessels, and has been adopted in some of the more recent iron-clads of the Royal Navy. Sketches showing the butt-fastenings employed in the 'Hercules' are given in Fig. 153, and will be further alluded to hereafter. In the 'Bellerophon' and 'Captain' the butts are secured, as shown in Fig. 150, and the arrangement of rivets there shown is that usually referred to as treble-chain riveting. In a paper on 'The Strength of Iron Ships' in the *Transactions* of the Institution of Naval Architects for 1860, Mr. Fairbairn advocated the adoption of quadruple-chain riveted butt-straps, and stated that the arrange-

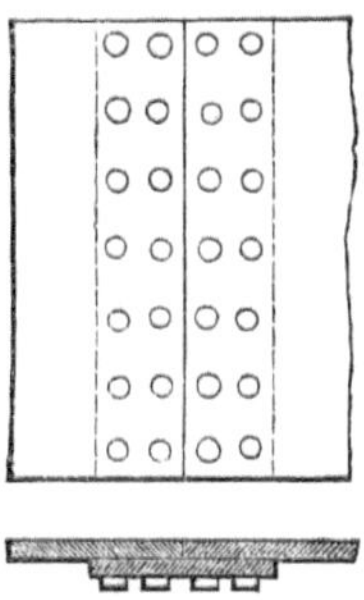

Fig. 149.

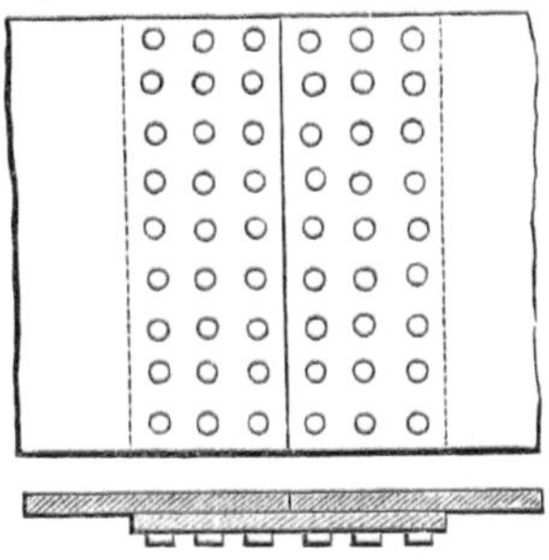

Fig. 150.

ment would add 20 per cent. to the strength of the ship, while the expense and weight would be but slightly increased. In the very long fine ships built by Messrs. Harland and Wolff, of Belfast, the butts of the sheer-strakes are quadruple-chain riveted for about half the length amidships. Another proposed arrangement of a

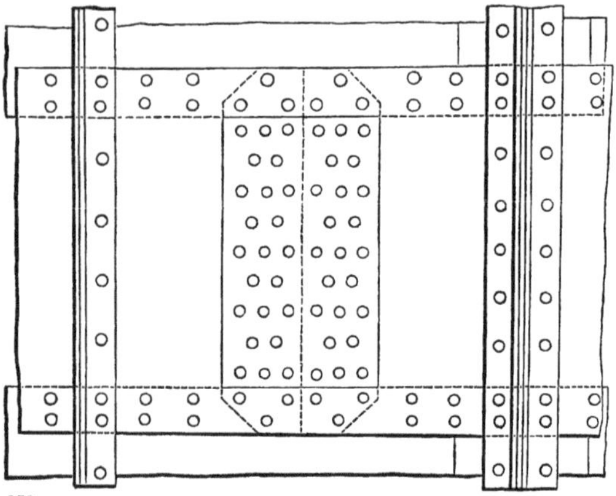

Fig. 151.

treble-chain riveted butt is shown in Fig. 151. The arrangement of butt-fastening given in Fig. 152 was proposed by Mr. Fairbairn in the paper previously referred to. This plan of fastening has been largely adopted in the 'Northumberland' in places where a very efficient connection is required, such as the sheer strakes, strakes under ports, garboards, &c. It will be remarked that the number of rivets employed is nearly identical with that required for a double-chain riveted butt-strap, but that the alternate rows are placed so as to form treble-chain riveting. The intermediate rivets near the butt are introduced in order to allow of the joint being effectually caulked. It should be noticed that this arrangement of rivets makes it necessary to have the

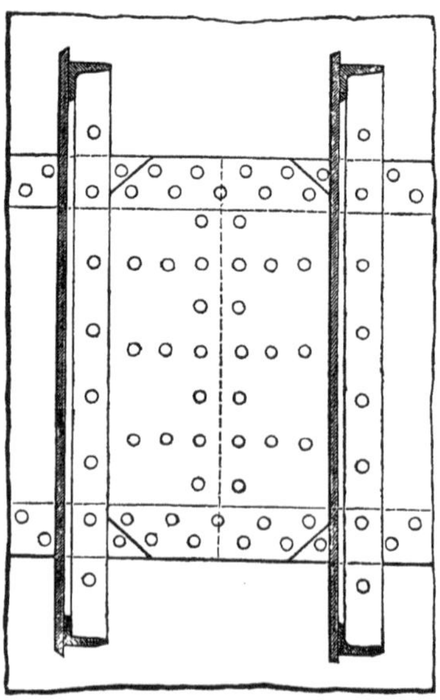

Fig. 152.

strap thicker than the plate, in order to secure equal strengths along their respective breaking lines. The sketch in Fig. 153 shows an inside view of a portion of the bottom plating of the 'Hercules,' with the arrangement of the rivets in the butts, edges,

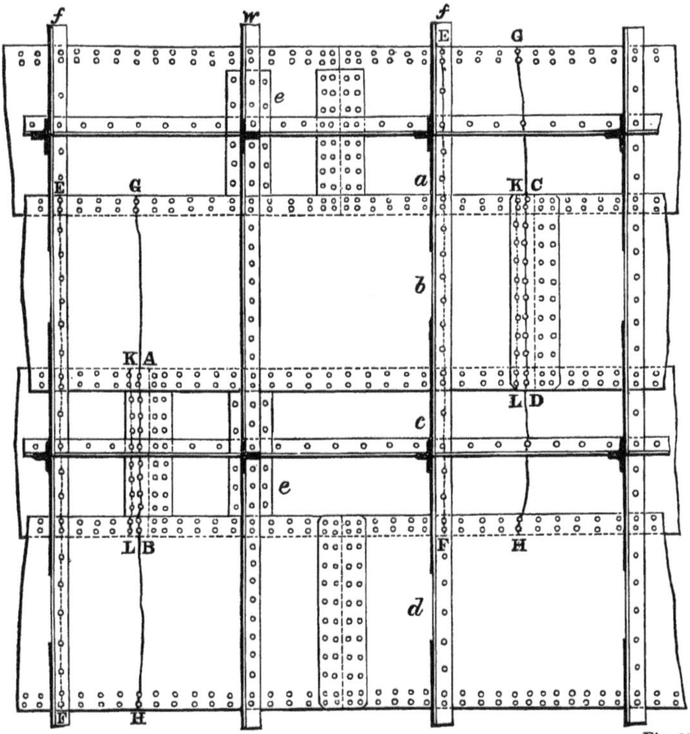

Fig. 153.

and frames. We shall revert to this sketch hereafter, in order to calculate the breaking strengths of the plating and butt-straps, and to investigate the relations between the strengths of the butt-fastenings and the plating.

For double-riveted butt-straps Lloyd's Rules require the same breadth of lap as for double-riveted edges, so that the whole breadth of the strap equals 11 diameters of the rivet. The Liverpool Rules require 12 diameters for the breadth of a double-riveted strap, and in H.M's Service the breadth employed is from 11 to $11\frac{1}{2}$ diameters. Neither of the Rules gives any regulation as to the breadth of treble-riveted butt-straps. In H.M.'s Service the usual practice in fitting treble-riveted straps has been to allow the same distance between the rows of rivets as is allowed by Lloyd's for double-riveting, viz. $1\frac{1}{2}$ diameter. The whole breadth of the strap has thus equalled 16 diameters; but in some later ships, in order to keep the outer rows of rivets further from the plates and strap, the breadth of the strap has been increased to $16\frac{1}{2}$ diameters; in the 'Captain' the butt-straps are $17\frac{1}{2}$ diameters in breadth.

What rule should be followed in determining the thickness of outside plating, is a very difficult and much disputed question. Lloyd's Rules proceed upon the basis of the tonnage of vessels, and the Liverpool Rules upon the depth in hold. Another rule given in Professor Rankine's 'Shipbuilding Theoretical and Practical,' and said to have been deduced by Mr. J. R. Napier from the practical working of a great number of iron ships, is as follows:—

$$\text{Thickness of skin in inches} = \frac{\text{Displacement in tons} \times \text{Length in feet}}{800 \times \text{Breadth in feet} \times \text{Depth in feet}}$$

This expression is obtained from the consideration that the mean thickness of the skin should be proportional directly to the load displacement and length, and inversely to the breadth and depth. The constant in the denominator is determined from several examples of well-built ships. In practice, however, the usual mode of settling this point, as to the thickness of the skin-plating, is by means of the experience gained in preceding ships. As previously stated, the plating of the earlier iron ships was much lighter than that now employed; but it is of interest to know that some of those earlier vessels were in good condition after twenty years' work. The employment of thicker plates for the skin of iron ships has many manifest advantages, of which the chief are that the increase of strength due to increased thickness of plating is far more rapid than the increase of weight, and consequently of displacement, thus caused; and that greater durability is also ensured, while local strength is added to the bottom, enabling it to resist more effectually a blow caused by striking a rock, or any hard-pointed substance. The great liability to injury from this cause which is experienced by iron ships, has led to the adoption of double bottoms in many vessels. In a valuable paper published in the *Transactions* of the Institution of Naval Architects for 1864, the late Mr. Letty made the following statements, which are very interesting as illustrating the various arrangements of plating which are possible for a given total weight. "If we increase the thickness "of the plates and keep the total weight of plating the same, we "must reduce the breadths of the laps and the weight of the butt-"covers. By taking an average length and breadth of plate "(10 feet by 2 feet 6 inches) it appears that the weight of "hull-plating will be the same in the three following arrange-"ments:—

"I. $\frac{3}{4}$-inch plating, laps double riveted, butts quadruple "chain-riveted as recommended by Mr. Fairbairn, with "a distance of twice the diameter of the rivet between "the rows of rivet-holes in the butts.

"II. Bare $\frac{7}{8}$-inch plating (·82 inch) double-riveted laps and "butts as required by Lloyd's.

"III. Full 1-inch plating (1·09 inch), single-riveted laps and "butts, or a double bottom of $\frac{5}{8}$ inch full, and $\frac{3}{8}$ inch."

The thicknesses of plating required by Lloyd's and the Liverpool Rules respectively will be found in the Tables in the Appendix. It appears on comparison that the Liverpool Rules require, in most cases, thinner plating than Lloyd's. In the Liverpool Rules the plating from the garboard to the sheer strake of the two higher classes of ships is required to be of an uniform thickness; the garboard must be $\frac{1}{16}$ inch, and the sheer strake $\frac{1}{8}$ inch thicker than the remainder of the plating. According to Lloyd's Table the space between the garboard and the sheer strake is divided into four parts; the general rule observed with respect to the plating of these parts being, that the thickness must be $\frac{1}{16}$ inch less than that of the plating of the part below it; the garboard is $\frac{1}{16}$ inch, and the sheer strake $\frac{1}{8}$ inch thicker than the adjacent strakes of plating. According to theoretical investigations, a good distribution of material would be made, if the outside plating between the upper part of the floor and the neutral axis of the ship for transverse strains, were made one-third less thick than the plating of the bottom and topsides. But this would be an arrangement specially suited to the position usually occupied by a ship when floating in still water; whereas, when at sea, her position, and consequently the position of the parts forming the top and bottom of the girder, are constantly changing; this constitutes the prime argument of those who advocate the adoption of plating of uniform thickness from keel to gunwale. The common practice of iron shipbuilders is in accordance with the arrangement required by Lloyd's, or only differs in some small degree from that arrangement.

It is obvious that the strains experienced by the extremities of an iron ship are less intense than those borne by the parts of the structure nearer the middle of the length. On this account it is usual to reduce the thickness of the outside plating toward the bow and stern. The theoretical law for the decrease in thickness cannot be adopted in practice, as the local strength of the plating of the extremities would be so greatly reduced as to endanger the structure. Rules are given both by Lloyd's and the Liverpool Underwriters to regulate the decrease of thickness. Lloyd's Rules state that "in vessels under 1200 tons, the plating may be "reduced $\frac{1}{16}$ inch forward and aft, for a distance not exceeding one "quarter of the length of the vessel from each end, below the "upper edge of main sheer-strake, down to a perpendicular height "from upper side of keel of three-fifths the internal depth of hold; "and in ships of 1200 tons and upwards a reduction of two-sixteenths "will be allowed; the plates next abaft and next afore the quarter "length of the vessel to be of an intermediate or graduated thickness "between that required in midships, and the reduction allowed at "the ends. In screw propelled vessels, however, no reduction is to "be made in the plating at the after end below the lower part of "the rudder-trunk." The Liverpool Rules allow a reduction of one-sixth of the total thickness forward and aft on all outside plating from the thickness required amidships, the reduction commencing at one-fifth of the vessel's length from each end. In the 'Warrior' the plating is of the midship thickness for a length of about 250 feet and before and abaft is reduced $\frac{1}{16}$ inch. In the 'Hercules' and other of the more recent iron-clads, the reduction in thickness has been commenced before and abaft a length of about 100 feet amidships. Mr. Fairbairn is of opinion that the reduction in thickness should never exceed one-third of the midship thickness.

Great objection has been made to the thickness of plating still required by Lloyd's Rules at the bow and stern below the water line, and it has been stated that thick plates are not wanted, and only cause a waste of material without adding to the safety of the ship. In answer to this objection, it has been urged that the friction of the water at the bows is so considerable as to cause a more rapid decrease in the thickness of the plating there than amidships; and that the wear of the anchors and cables on the plating necessitates the keeping up of the thickness. The last regulation of Lloyd's Rule, with respect to the plating on the sterns of screw-steamers, is due to the facts that in wake of the screw-shafts, internal hooks and other strengthenings cannot be fitted, and that in some vessels while the plating aft has been thinner the rivets have become loose and the plating has required to be replaced.

The following account of the plating of ships which have been previously mentioned may be interesting as illustrating the practice of shipbuilders in various periods of the progress of iron shipbuilding. The 'Birkenhead' of 1400 tons was plated as follows:—Up to 4-feet water-line, $\frac{5}{8}$-inch plates lapped and double riveted; above the 4-feet line $\frac{9}{16}$-inch plates, lapped and single riveted up to the 9-feet water-line; and above this line worked flush with internal edge-strips. The butts were double-riveted, and the edge-strips and butt-straps were $\frac{1}{16}$ inch thicker than the plates. The 'Megæra,' of 1395 tons, had her plating arranged as follows:—All the plates were $\frac{1}{2}$ inch thick, except the strakes in wake of the lower-deck beams which were $\frac{9}{16}$ inch, and those second out from the keel which were $\frac{5}{8}$ inch. Up to the strake next above the floor-head the plating was worked clinker-fashion; but above this height it was flush-jointed with internal edge-strips. All laps and butts were double-riveted. The 'Himalaya's' plating was worked in alternate inside and outside strakes as far up as the wales; these strakes together with the topside plating were worked flush. The ship's burden is 3453 tons, and the thicknesses of plating used are as follow:—

	Amidships.	Forward and Aft.
	inch.	inch.
Garboard-strakes	$1\frac{1}{16}$	$\frac{7}{8}$
8 next ,, ...	$\frac{13}{16}$	$\frac{11}{16}$
5 ,, ,, ...	$\frac{11}{16}$	$\frac{9}{16}$
Wales (2 strakes)	$\frac{13}{16}$	$\frac{5}{8}$
Topsides	$\frac{1}{2}$	$\frac{7}{16}$
Sheer strake	$\frac{5}{8}$	$\frac{9}{16}$

The details of the plating of the steam-ships 'Queen' and 'China' have been given previously, and may be considered as good examples of the practice of mercantile shipbuilders.

The outside plating and skin-plating behind armour of the 'Warrior,' of 6039 tons' burden, were arranged, as shown in the following table :—

	For a length of about 250 ft. amidships.	Forward and Aft.
	inch.	inch.
Middle-line or keel-strakes { Upper ..	$1\frac{1}{8}$	$1\frac{1}{8}$
Middle-line or keel-strakes { Lower ..	$1\frac{1}{4}$	$1\frac{1}{4}$
Next two strakes	$1\frac{1}{8}$	1
Next strake	1	$\frac{7}{8}$
Thence to 14-ft. water-line	$\frac{7}{8}$	$\frac{13}{16}$
From 14-ft. water-line to port-sill, except behind armour	$\frac{13}{16}$	$\frac{12}{16}$
Plating behind armour : lower-strake	$\frac{5}{8}$	..
,, ,, remainder	$\frac{9}{16}$	..
Strake under ports (doubled)	$\frac{9}{16}$	..
Between the ports forward and aft	..	$\frac{9}{16}$
Sheer strake (worked in two thicknesses) ..	$\frac{5}{8}$	$\frac{9}{16}$

The thicknesses of the outside plating of the 'Hercules' of 5226 tons burden are as follow :—

	For a length of about 100 ft. amidships.	Forward and Aft.
	inch.	inch.
Middle-line or keel-strakes { Upper ..	1	1
Middle-line or keel-strakes { Lower ..	$1\frac{1}{8}$	$1\frac{1}{8}$
Next strake	1	$\frac{7}{8}$
Next four strakes	$\frac{7}{8}$	$\frac{13}{16}$
The remainder	$\frac{13}{16}$	$\frac{3}{4}$

It must be remembered that an inner bottom $\frac{1}{2}$ inch thick extends for 216 feet amidships in this ship, so that the strength and safety of the ship are considerably increased. As previously stated the bottom plating forward is doubled for about 40 feet abaft the stem. The skin-plating behind armour is worked in two thicknesses, each being $\frac{3}{4}$ inch thick amidships, and a little less forward and aft. The plating of the unprotected parts above the armour belt is $\frac{1}{2}$ inch thick. The details of the modes of working both of the last-named assemblages of plating have been previously explained.

CHAPTER XI.

BULKHEADS.

THERE is, probably, no subject on which iron shipbuilders are more generally agreed, than on the desirability, both as regards safety and structural strength, of the employment of watertight bulkheads. These were first introduced, we believe, by the late Charles Wye Williams, of Liverpool, who is entitled to the foremost place among the introducers of iron ocean-steamers, and, in fact, contributed in a pre-eminent manner to the introduction of ocean steam-navigation. The increased safety resulting from the adoption of these bulkheads proceeds from the fact that a leak or fire in any compartment can in most cases be prevented from affecting the other compartments. The increase in structural strength caused by the use of bulkheads has already been alluded to when speaking of the various systems of framing, and of the connection of the beam-end with the side. Watertight bulkheads in iron vessels are almost always placed transversely, but there are, in many instances, longitudinal watertight divisions also. In many steam ships the longitudinal bulkheads enclosing the coal bunkers are made watertight, and thus form subdivisions in the compartments bounded by the transverse bulkheads. A very common practice in the earlier steam ships was to have an arched passage between the engines and boilers, the plating in the arched bulkhead being watertight. The shaft-passage bulkheads of many screw-steamers are watertight, and form the sides of a longitudinal compartment extending from the engine-room to the stuffing-box bulkheads. It will be remembered also that in the 'Birkenhead,' notwithstanding her disastrous loss, longitudinal bulkheads were fitted, as shown in Fig. 71, p. 75; and in the armour-clad frigates of the Royal Navy the wing-passage bulkheads form longitudinal divisions of the hold, as previously explained, while advantage is taken of the subdivisions formed by the bulkheads of magazines, shell-rooms, chain-lockers, shaft-passages, and passages between engines and boilers, all of which are made watertight. The 'Great Eastern' is, however, the ship in which the greatest prominence is given to longitudinal bulkheads. In this vessel there are two longitudinal bulkheads extending up to the upper deck, and running about half the length of the ship, being so placed as to form the sides of the engine and boiler spaces, &c. In general, however, the watertight bulkheads are, as before stated, vertical and transverse. Lloyd's Rules require that the transverse watertight bulkheads of a steamer shall be placed so that, in addition to the engine-room bulkheads, there may be two bulkheads built at a reasonable distance from the ends. In sailing-ships the foremost or collision bulkhead only is required. The Liverpool Rules are almost identical with Lloyd's. In the earlier iron sailing-ships it was usual to have at least three watertight bulkheads; two being placed at some distance from the extremities, and one in the mid-ship part of the vessel. In steamers there were generally four bulkheads; two toward the extremities, and two enclosing the engine space. An instance of the arrangements of the bulkheads in a comparatively old iron steam-ship is found in the 'Birkenhead,' just referred to. There were six transverse bulkheads, which extended up to the upper deck. The engine-room was divided into five watertight compartments, and the coal bunkers were enclosed by two longitudinal bulkheads. Between the engines and boilers there was a bulkhead which had a large arched opening in it, that served for communication and ventilation. In the fore and main holds two vertical longitudinal bulkheads were fitted, and extended up to the orlop deck, as shown in Fig. 71, p. 75. The greater interest attaches to this description of the bulkhead arrangements of the 'Birkenhead,' as she was lost at sea, and went down rapidly. The explanation of this occurrence is found in the fact that some of the compartments had been opened up in order to afford communication between them.

In iron steam-ships the common practice now is to place the engines and boilers in watertight compartments. In some vessels the engines and boilers are in one compartment; but in others, and especially where the power is large, the boilers are placed in a separate compartment from the engines; and in some paddle-wheel steamers the boilers are divided, one-half being placed in a compartment before the engines, and the other half in a compartment abaft them. All the arrangements, together with the disposition of the other watertight bulkheads in the fore and after holds of merchant vessels, depend entirely on the shipowner and the shipbuilder, and are made to conform to the purposes for which the ship is to be employed. The collision bulkhead forward, required by both Lloyd's and the Liverpool Rules, is fitted in almost all iron ships, and experience has fully justified the importance attached to it by the Rules. In many instances where the bows of vessels have been broken through by accidental collision, the ships have been kept afloat by means of this bulkhead. A case in point is found in the 'Samphire,' a steam-packet running between Calais and Dover, which had her bow stove in by collision with another vessel, but was kept afloat by her bulkheads, and brought into harbour. In the collision between the 'Haswell' and 'Bruiser,' the former ship was saved in a similar manner, and brought into port, although the plating on the port-bow was greatly injured, and in several places completely fractured. The bulkhead usually placed a few feet before the stern post of a screw-steamer, has also been of great service in many cases where the water has entered the after compartment through accident to the stern or its fittings, rudder-braces, &c. By means of the small compartments thus formed at the bow and stern, those parts of the hull most liable to injury may be penetrated or otherwise injured, and yet the trim or speed of the ship may be very little affected by the quantity of water which enters.

In the iron-clad frigates of the Royal Navy the transverse bulkheads have been arranged so as to enclose the engine and boiler-rooms, the magazines, and fore and after holds; in all cases the aftermost bulkhead has been placed at about 12 or 14 feet before the post, in order to take the stuffing-box arrangement of the engineer's shaft tube, and the foremost bulkhead has been moderately close to the stem. In the later frigates, such as the 'Bellerophon' and 'Hercules,' which have an inner bottom, there are nine transverse watertight bulkheads. One of these bounds the double bottom forward, another forms the stuffing-box bulkhead, and, of the remainder, two enclose the boiler space, and a third divides it into two compartments, while two enclose the engine-room, and two bound the compartment in which the after magazines are placed.

Mr. Scott Russell gives it as his opinion that a ship should have at least as many watertight compartments as there are breadths in her length, and states that for the most part the watertight bulkheads in a ship are not sufficiently numerous. No arbitrary rule, however, can be satisfactorily adopted; much must depend upon the internal arrangements of the vessel, and especially upon the height to which the bulkheads rise above the water-line. The importance of having numerous watertight compartments has been repeatedly shown by ships thus constructed having been saved, and by vessels with too few compartments having been lost. When bulkheads are introduced not merely as a source of structural strength, but as a means of giving safety to the ship, it is requisite that they should divide the ship in such a manner that if one of the compartments were injured and filled the vessel would still float. It has been objected to the division of the hold into numerous compartments by transverse bulkheads, that the stowage of the ship's cargo is thereby greatly interfered with. In order to avoid this inconvenience, and yet preserve the safety of the ship, Mr. Lungley has proposed a plan for dividing the hold into compartments by means of watertight decks or flats, and for obtaining admission to these compartments by means of watertight trunks which extend up above the water-level. It will be remembered that an illustration of this plan has already been given, as applied in the 'Bellerophon' between the fore end of the double bottom and the stem.

We now come to the illustration of the construction and modes of securing transverse watertight bulkheads. The plating of the bulkheads is in most cases lap-jointed both at the edges and butts and single riveted; but in some instances the plating has been worked flush, and the edges and butts have been secured by single-riveted strips. The *brick* arrangement of butts shown in Fig. 133, p. 189, is commonly adopted, but in some of the bulkheads of the later iron-clads the *diagonal* arrangement shown in Fig. 134, p. 190, has been followed. It may be remarked, in passing, that while the latter disposition of butts is the stronger of the two, the former disposition gives ample strength to resist extension in the plating of the bulkhead. In order to stiffen the plating, angle-iron or T-iron bars are worked, and in most cases are placed vertically. Lloyd's Rules require the bulkheads to be supported vertically by angle-irons not exceeding two feet six inches apart. The Liverpool Rules state that bulkheads are to be stayed on both sides with angle-iron bars four feet apart, one set being vertical and the other horizontal. It has been previously remarked, that in a well built iron-ship change in the transverse form is almost entirely prevented by the bulkheads. This statement, however, implies that the bulkheads are so constructed as to be themselves rigid; but it is a fact that in some ships, where the bulkheads have not been properly stiffened, they have buckled when the vessel has been afloat and become fair when she has been docked. The arrangement of the stiffening bars in both horizontal and vertical directions is in perfect accordance with that which theory suggests as the most effective for resisting change of form, as it consists of a network of bars at right angles to each other connected by the web formed by the plating. This mode of stiffening is now largely adopted, and is especially suited for the bulkheads bounding the engine and boiler spaces and cargo-holds in large ships, where the transverse strength of the decks is considerably reduced by the large hatchways required for getting in the machinery, boilers, and cargo. It should be mentioned, however, that when vertical stiffeners only are worked, the decks and platforms act as horizontal stiffeners and add greatly to the strength of the bulkhead.

Before passing on to the illustration of the different modes of working bulkhead plating and stiffeners, it may be well to add that some shipbuilders have favoured the proposal to place the plates with their greatest length vertical, instead of horizontal as is almost always the case. This arrangement has been carried out in many merchant ships, being very commonly followed on the Clyde and at Belfast, and in a few cases in vessels of the Royal Navy. In the latter ships the bulkheads have had to sustain great vertical pressures from heavy weights directly above them, and the stiffeners have been formed of T-iron bars which also served as seam-strips to the vertical joints of the plating. An illustration of a bulkhead with the plates placed with their greatest length vertical is given in Fig. 156, p. 221, and will be again referred to hereafter.

A sketch is given in Fig. 154, which illustrates the manner in which bulkheads are sometimes fitted in small vessels. The ship in which this arrangement is adopted is 23 feet 6 inches broad, and 11 feet in depth of hold. The plating is $\frac{1}{4}$ inch thick, and is stiffened by vertical angle-irons $2\frac{1}{2}$ by $2\frac{1}{2}$ by $\frac{5}{16}$ inches, spaced 33 inches apart. The plating is lap-jointed both at the vertical and horizontal joints, and the rivets in the laps are $\frac{1}{2}$ inch in diameter.

The usual mode of constructing the transverse watertight bulkheads of the larger ships of the mercantile marine is shown in Fig. 155, which is taken from the 'Paramatta' of 3000 tons, built at the Thames Iron Works for the Royal West India Mail Co. The details of this bulkhead were given by Mr. Mackrow in a paper read before the Institution of Naval Architects in 1863. In this case also the plating is connected by lap-joints both at the edges and butts. One unusual feature in this bulkhead is that the plates at the middle are $\frac{5}{8}$-inch, and those at the sides $\frac{9}{16}$-inch, instead of being uniformly thick as is generally the case. There are two sets of stiffening bars formed of 4 by 3 by $\frac{3}{8}$ angle-irons spaced 30 inches apart, one set being horizontal and the other vertical, worked on opposite sides of the bulkhead.

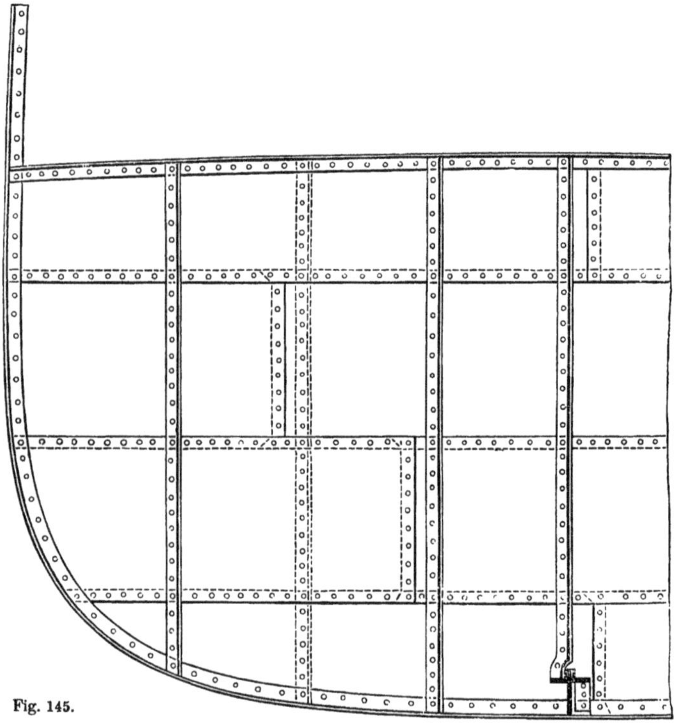

Fig. 145.

One objection to the lap-jointed bulkhead is that liners are requisite under the stiffeners, and in order to dispense with liners some builders have had the stiffening-bars worked in the forge so as to fit them directly upon the plating; but the cost of workmanship causes this plan to be seldom adopted. Another feature of lap-jointed bulkheads requiring notice is that at those parts where the horizontal and vertical laps cross each other there are three thicknesses of plating, and in order to preserve the uniformity of the work, and to caulk the joints of the plating, it is necessary to hammer out the corners of the butt-lap to a thin edge, so that they may overlie each other and form but one thickness. It has also been stated that in lap-jointed bulkheads the rivets in the edges are very liable to be sheared when the vessel receives

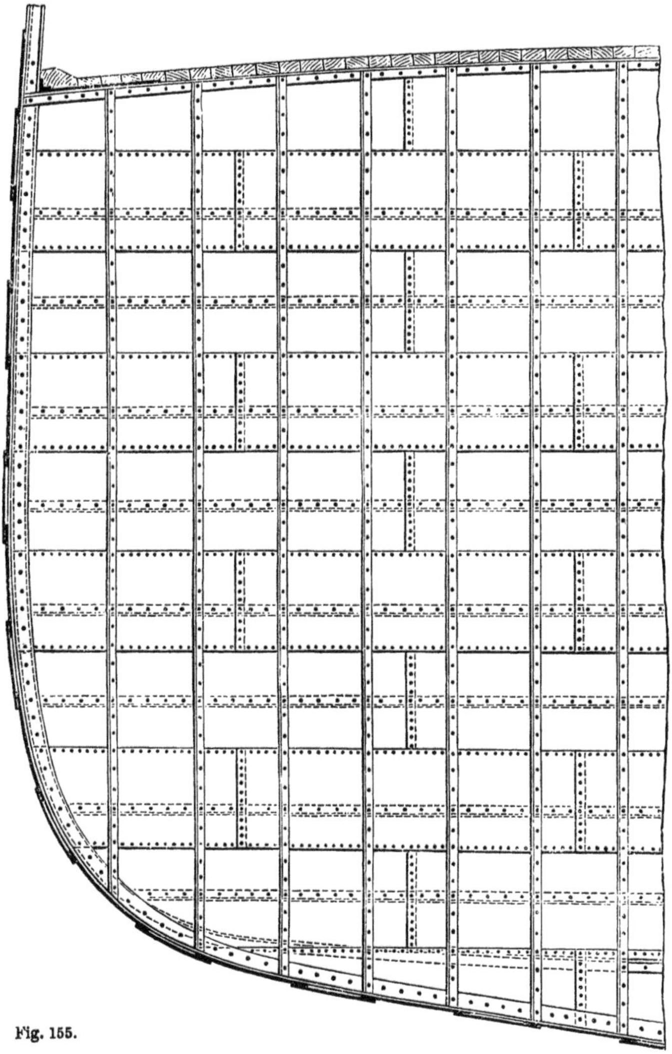

Fig. 155.

a very severe shock by grounding, or otherwise. In the case of the mail steamer 'Tyne,' which was wrecked in the Channel, this shearing did take place in the riveting of the engine-room bulkhead, which was directly over the part where she settled. It may be remarked, however, that the arguments used in favour of lapped joints in the outside plating, apply also to the case of bulkheads, and that in respect of the number of rivets, and the cost of workmanship, lap-jointed bulkheads are superior to flush-jointed bulkheads, while the latter have the advantage as regards strength.

In the paper previously referred to, Mr. Mackrow proposes a new plan of arranging the plating and stiffeners of bulkheads. The plating is worked flush or jump-jointed, and supported at the vertical and horizontal joints by T-irons $4\frac{1}{2}$ by $4\frac{1}{2}$ by $\frac{5}{8}$ inches. The vertical T-irons are on one side and the horizontal on the other side of the plating, and, in addition to stiffening the bulkhead, serve as seam-strips. Between successive vertical T-irons an angle-iron stiffener is worked. The advantages claimed for this arrangement are, the prevention of the rivets being sheared off as they were in the 'Tyne,' the dispensing with liners behind the stiffeners, and the consequent economy of material in the bulkhead. The author states that a bulkhead built on his plan would be about $2\frac{3}{8}$ per cent. lighter than the 'Paramatta's' bulkhead previously described.

It may be added here, that in the middle line bulkhead forward of vessels of the 'Northumberland' class, the plating is worked flush with horizontal edge-strips and vertical butt straps. The stiffening bars to this bulkhead are vertical, and in consequence of the use of edge-strips it is necessary to employ liners in wake of the stiffeners. In the aftermost transverse bulkheads of the iron-clad frigates the plating is worked in a similar manner, but there are no stiffening bars. It must be recollected, however, that these bulkheads are effectually stiffened by means of the horizontal iron flats and the wrought-iron shaft-tube on the aft side, and by the longitudinal frames and shaft-passage bulkheads on the fore side.

In the transverse bulkheads of the later vessels of the Royal Navy the arrangements differ from all the preceding plans. In order to illustrate the system followed, we will take as an example one of the midship bulkheads of the iron-clad frigate 'Hercules.' The plating is $\frac{9}{16}$ inch thick, with the exception of the upper strakes, which are $\frac{1}{2}$ inch; it is worked flush with T-iron bars $4\frac{1}{2}$ by $2\frac{1}{4}$ by $\frac{1}{2}$ inches as horizontal seam-strips, and single-riveted plate butt-straps. Vertical angle-iron stiffeners 3 by $3\frac{1}{2}$ by $\frac{9}{16}$ inches are worked at intervals of four feet on the opposite side of the bulkhead to that on which the T-bars are fixed. This arrangement is identical with Mr. Mackrow's plan as far as the horizontal T-iron stiffening bars are concerned, but differs from it in having plate butt-straps and a different arrangement of vertical stiffeners. The bulkhead proper is bounded by the inner bottom and the vertical wing-passage bulkheads; but, in order to complete it between these bulkheads and the frames, a partial bulkhead is fitted, the outer edge of which laps upon and is riveted to the 10-inch frame.

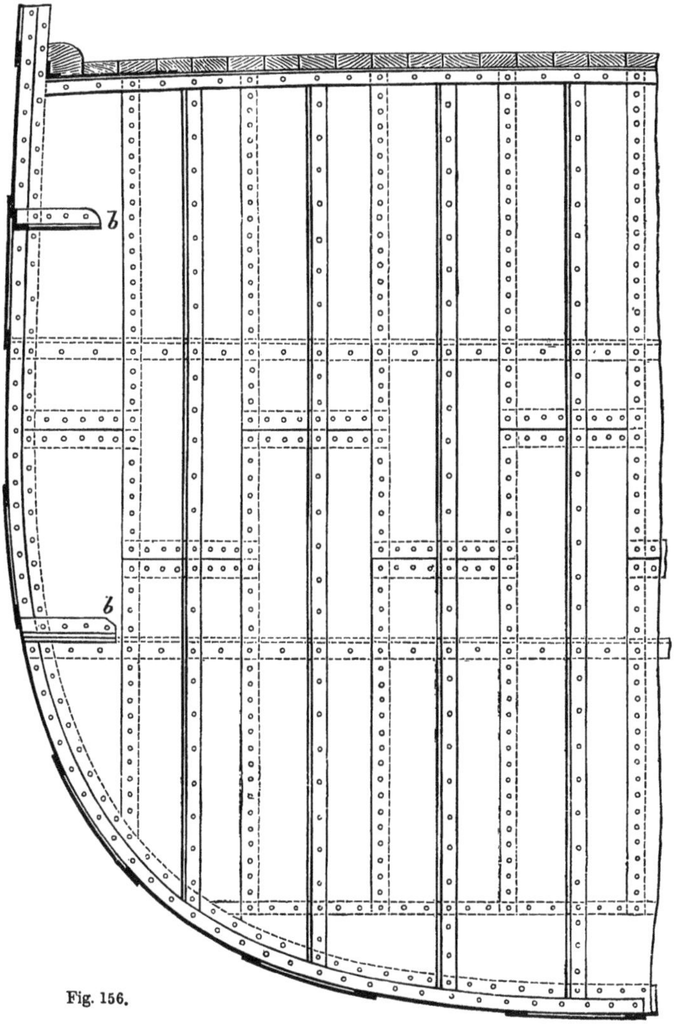

Fig. 156.

The sketch in Fig. 156 affords an illustration of a bulkhead in which the plates are placed with their greatest length vertical, and gives an example of the mode of construction practised in some vessels belonging to the French mercantile marine. It is taken from the *Traité Pratique de Construction Navale* by M. de Freminville. The vertical joints of the plating are lapped, and the horizontal joints worked flush, and connected by plate butt-straps worked on the opposite side to the vertical stiffening bars. By this means the vertical stiffeners are worked directly on the plating, and no liners are required. Two horizontal angle-iron stiffeners are also shown in the sketch; these are placed at the height of the decks and platforms in most cases. It will be remarked, from the horizontal section of the bulkhead given in Fig. 157, that the plating is arranged clinker-fashion, and that consequently tapered liners are required behind the horizontal stiffeners. M. de Freminville states that in some instances the angle-irons have been worked in the forge in order to fit directly upon the plating, but that the expense of workmanship has caused this plan to be but seldom followed.

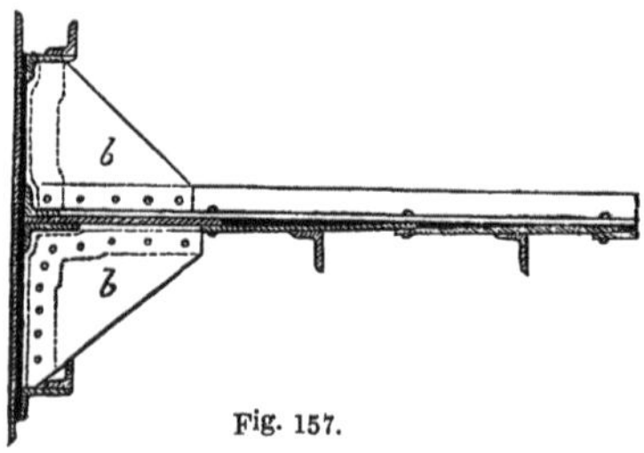

Fig. 157.

In the paper previously alluded to, Mr. Mackrow expresses the opinion that if the plates were worked flush with their greatest length vertical, and T-iron bars were employed both as vertical seam-strips and stiffeners, there would be a considerable advantage with regard to weight; while for vessels of 1000 tons and under, the plates might be worked in one length, and for larger vessels in two lengths connected by a butt-strap. It has been previously stated that bulkheads constructed in this manner have been fitted in some vessels of the Royal Navy, in places where there was a special need of vertical strength. It will be obvious that in bulkheads built on this plan the shearing strength of the rivets in the vertical joints and the horizontal ties at the heights of the decks form the only provision for resistance to horizontal extension, but this would, no doubt, ordinarily be ample to resist all probable strains.

Having thus considered the modes of plating and stiffening practised in the construction of bulkheads, we turn to the illustration of the different means employed for making a secure and watertight connection between them and the ship's side. In some of the earlier ships the bulkheads were secured to the outside plating by a single angle-iron, as shown in Fig. 158. In order to make watertight work the pitch of the rivets in this angle-iron was made very small, thus reducing the effective sectional area of the bottom-plating in wake of the bulkhead, and causing great weakness. Many ships have been lost by their being broken down through the line of rivet-holes at a bulkhead when exposed to great longitudinal strains, the builders having, in their over-anxiety to make a watertight division, neglected the corresponding reduction in strength. Some builders, in avoiding this fault of construction, have secured the bulkheads so slightly to the skin-plating, that they have either burst when under the pressure of water in the compartment, or the rivets have been broken or made loose and a leak has been caused. Another mode of attaching the bulkhead to the plating of the bottom, which was very commonly employed, is given in Fig. 159. Two frame angle-irons were worked on the edge of the bulkhead, and the usual spacing of the rivets in both flanges was $4\frac{1}{2}$ diameters. In some ships, in order to dispense with liners between these frame angle-irons and the plating, the angle-irons were worked in the forge, so as to fit directly upon the plates and be caulked to them. This plan was not usually adopted, and in many cases the joints were made watertight by forcing in cement.

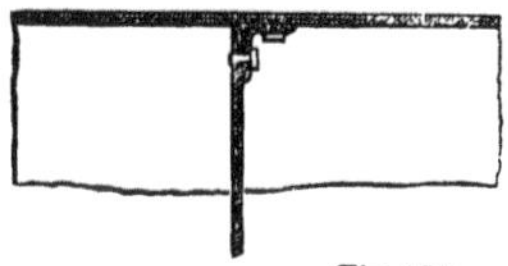

Fig. 158.

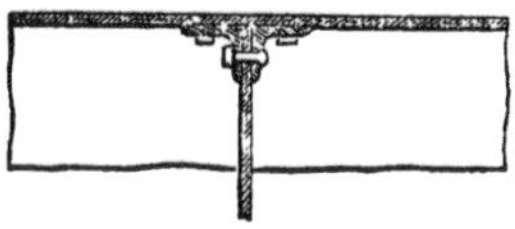

Fig. 159.

These two modes of connecting bulkheads with the side are still practised with some modifications. Both Lloyd's and the Liverpool Rules require that the bulkheads shall be either fitted between two frames and be well riveted through them, or be attached only to one frame and have horizontal bracket or knee-plates riveted to the bulkhead and to the outside plating. A sketch showing the latter mode of connection is given in Fig. 160. From the elevation it will be seen that the fore and aft flange of the bulkhead frame is considerably wider than that of the other frame angle-irons. This difference is made in order to allow the riveting through the bottom plating to be zigzagged, and thus increase the distance between the holes in the two lines of rivets. The plates marked *b*, *b* are the horizontal brackets or knees previously referred to, which are fitted on alternate sides of the bulkhead, as shown in the elevation. The importance attaching to the employment of these brackets has been already alluded to when illustrating some of the weaknesses of iron ships. By this means the strength of the bulkhead is distributed over the adjacent parts, and the evil effects are prevented which would ensue if the bulkheads formed a series of rigid divisions in the structure, between which the hull was comparatively flexible, and might work to such an extent as to loosen or break the rivets in the neighbourhood of the bulkheads and cause leaks. The longitudinal keelsons and stringers in the hold of a vessel, and the stringer-plates and angle-irons on the various decks, also act as very efficient distributors of the strength of the bulkhead over the intervening parts of the bottom. In the ships built on the longitudinal system of Mr. Scott Russell, the transverse and partial bulkheads are relied upon to supply the transverse strength of the parts of the structure lying between them, and it has been proved by experience that the longitudinal frames distribute the strength of the bulkheads. But it has also been proved that the necessity exists in vessels built on the transverse system for the employment of longitudinal plates or brackets at bulkheads in order to avoid the localization of their strength and the consequent injury to the hull. For this reason, as well as on account of the very efficient aid they give in connecting the bulkheads with the plating, bracket-plates are required by the Rules, and generally employed by shipbuilders. It is worthy of note that the brackets are usually placed near the centre of the strakes of outside plating on which they come.

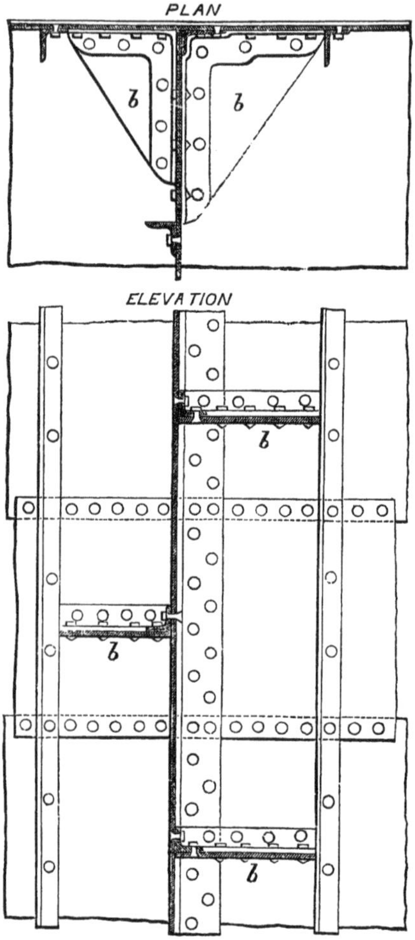

Fig. 160.

Although brackets are not required by the Rules in cases where the bulkheads are attached to double angle-iron frames, they are often fitted, especially in large ships. An instance of this is given in Figs. 156 and 157, pages 221, 222, where the brackets are marked *b*, *b*. It will be remarked that one of the double frame angle-irons has its transverse flange considerably deeper than that of the other, and takes an inner row of rivets. In the elevation of the bulkhead it will be noticed that the brackets *b*, *b* are near the edges of the strakes of outside plating. This is contrary to the usual practice of English shipbuilders, and to the requirement of Lloyd's Rules. It seems probable, however, that the lower bracket is placed where it is, in order to take hold of the horizontal stiffener to the bulkhead, and that the upper bracket is situated as in the sketch in order to be as nearly as possible half-way between the deck and the horizontal stiffener next below it.

In ships built on Mr. Scott Russell's longitudinal system, a belt of plating is worked on the inside of the longitudinal frames in wake of the bulkhead, and connected with the bulkhead by means of a single angle-iron. The watertight division is completed between this belt of plating and the outside plating by means of plate-frames fitted between the longitudinals. The vertical angle-iron stiffeners on the bulkhead are connected with the longitudinal framing of both the hull and the deck. Mr. Russell thinks this connection a very desirable feature, and recommends that in ships which are framed transversely the bulkhead stiffeners should be connected to some longitudinal framing, such as keelsons or hold-stringers.

In vessels which are constructed with a double bottom, the bulkheads are usually attached to the inner skin by either single or double angle-irons, according to the size of the ship. In the 'Hercules,' for example, the angle-irons connecting the bulkhead with the inner skin are double. In the 'Captain,' instead of directly attaching the bulkhead plating to the inner bottom, it is lapped upon and riveted to the edge of a partial bulkhead or deep frame, formed of plates and angle-irons and fitted against the inner skin. In these ships the watertight divisions are completed out to the outside plating by means of watertight plate-frames fitted between the inner and outer skins. Before and abaft the double bottom, the bulkheads are simply lapped upon, and riveted to, the continuous reversed angle-irons.

In the 'Great Eastern' the transverse bulkheads are attached to the inner skin, but all the watertight divisions of the hold thus formed are not completed between the two skins by plate-frames. Mr. Scott Russell was led to adopt this course by the considerations that the transverse strength of the ship was amply provided for, and that it was unnecessary to subdivide the double bottom to so great an extent as would have been done if all the bulkheads had been completed.

Before proceeding to notice a few of the proposed methods of attaching transverse bulkheads to the ship's side, it may be proper to call attention to a very important feature in the ordinary bulkhead connections, viz., the means employed for strengthening the outside plating in wake of the bulkheads. Mention has already been made of the fact that the strength of the plating is seriously reduced by the small pitch of the rivets in the bulkhead frames rendered necessary in order to make watertight work. This pitch should be about six diameters in the fore and aft flanges of the bulkhead frames, and four diameters in the transverse flanges. In order to strengthen the section of the plating thus weakened, it is the practice of many builders to increase the breadth of the liners between the outside strakes of the bottom plating and the bulkhead frames, so that they may act as straps across the lines of rivet-holes. Both Lloyd's and the Liverpool Rules require that the liners shall extend in one piece from the fore side of the frame before the bulkhead to the aft side of the frame abaft the bulkhead, and this is the general arrangement followed by the principal merchant-shipbuilders. It will be readily understood from the horizontal section given in Fig. 157, page 222, where the liner is drawn in black for the sake of distinction. In the iron-built vessels of the Royal Navy it is also usual to increase the breadth of the liners at the bulkhead frames, but not to as great an extent as is customary in vessels of the mercantile marine. In the 'Hercules' the liners overlap the transverse frames about 3 inches, and take a row of rivets on each side of the frame. These liners are marked *e*, *e* in Fig. 153, page 207. In the new Indian troop-ships the liners at the bulkheads are made broad enough to take two rows of rivets on each side of the frame angle-irons.

Coming now to the illustration of some proposed modes of securing bulkheads, we first call attention to the arrangement shown in Fig. 161. This was patented by Mr. Hodgson in 1859, and consists in working the bulkhead into two parts near the side, and making them diverge so that their flanged outer edges may be attached to the reversed angle-irons on two successive frames, and thus make watertight joints.

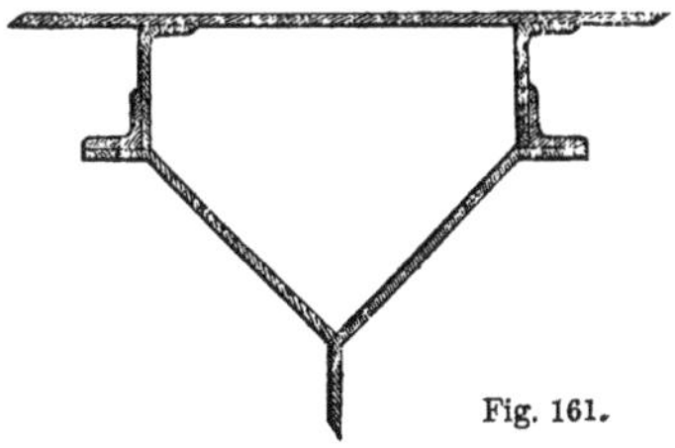

Fig. 161.

The second plan, shown in Fig. 162, was patented by Mr. Rae in 1860. It will be seen, on reference to the sketch, that the bulkhead frame is formed of a broad flanged T-iron, of which the arms are sufficiently broad to allow the rivets in the bottom plating to be placed zigzag fashion. The athwartship flange of the T-iron is rolled three times the usual thickness, and is grooved on the inner edge. The bulkhead plating is worked flush and fitted into this groove, its fastening consisting of through-rivets, as shown in the sketch. The proposer also patents the use, in wake of bulkheads, of broad-plate liners inside the bottom plating, combined with corresponding plates on the outside, the latter being thinned away at the fore and after edges in order to offer as little resistance as possible to the ship's progress.

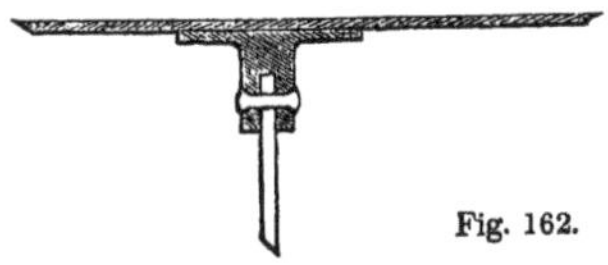

Fig. 162.

The third mode of connecting bulkheads with the ship's side was patented by Mr. Ash in 1860, and is illustrated in Fig. 163. A belt of plating is worked inside the ship and secured to the reversed angle-irons on two successive frames. The bulkhead is then attached to this belt of plating by a single frame. This plan greatly resembles the arrangement of the bulkhead connections made by Mr. Scott Russell in ships built on his longitudinal system, and having no double bottom.

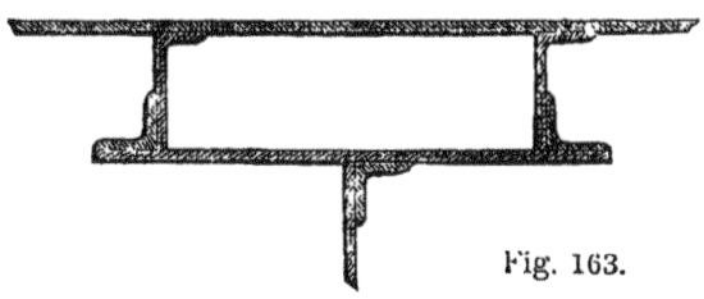

Fig. 163.

In all these proposals the aim is to avoid weakening the side plating in wake of the bulkheads, and to prevent the strength of the bulkhead from being too much localized. They are all more expensive and less simple than the ordinary plans of attaching bulkheads, and, while both in the first and third plans the direct connection of the bulkheads with the outside plating is avoided, it must be evident that this is done at the expense of forming spaces between the frames on each side of the bulkheads which are inaccessible for painting and cleaning, and where, consequently, oxidation is likely to proceed very rapidly. It may be added, however, that the idea of connecting the bulkheads with an internal belt of plating is favoured by some eminent shipbuilders, although it has not been generally practised.

The upper edges of transverse bulkheads are usually brought up to a deck above the load water-line, and secured by means of single or double angle-irons. Lloyd's Rules require all bulkheads, except the aftermost in steam ships, to be carried up to the upper deck in two-decked vessels, and to the middle deck in vessels with three decks. The aftermost, or stuffing-box bulkhead is allowed to be brought above the load water-line and secured to a watertight flat which extends around the after part of the vessel, and renders the lower after body a watertight compartment. In many cases, however, the bulkheads of merchant ships do not extend to the height above the load water-line which is requisite for safety; and in some instances they have actually been stopped below the load line, thus rendering them perfectly useless as respects safety, although, of course, they are of use in giving structural strength. In the ships of the Royal Navy the bulkheads are usually brought up to the main deck and connected with the deck plating, but some of the bulkheads toward the bow and stern are, in some ships, ended underneath the watertight platforms.

The great importance attaching to the preservation of the continuity of the longitudinal pieces of the framing has led to the bulkheads being pierced by the keelsons and stringers, and very great care is needed to make the bulkheads watertight around the longitudinal framing. Watertightness is usually secured by working angle-irons around the keelsons or stringers in such a manner as to allow the caulking of the joints to be readily performed. The arrangements of a box, I-shaped, and bulb-iron keelson, and of stringers formed of a plate and two angle-irons, or of double angle-irons set back to back, are illustrated by Figs. 164-168. It will be observed that in all these arrangements the edges of the angle-irons can be caulked at all the joints. It has been previously remarked (when describing the watertight work in connection with the longitudinal frames of the 'Northumberland' which is illustrated in Figs. 79, 80, 81, pages 103 and 105), that a stop-water, formed of canvas steeped in paint or of some other material, must be fitted between the continuous plates and angle-irons in order to prevent the water from passing from one compartment to another.

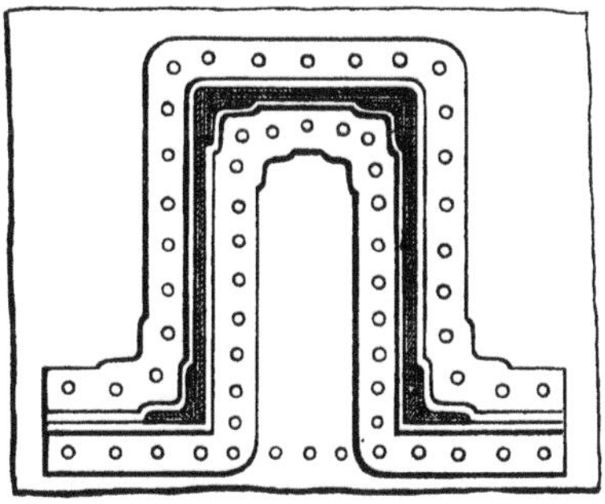

Fig. 164.

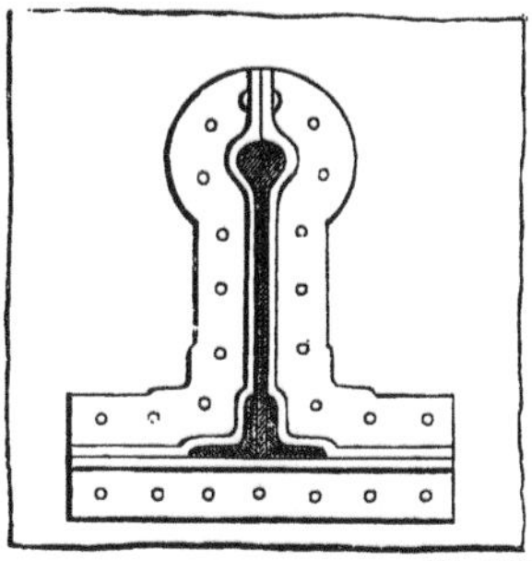

Fig. 166.

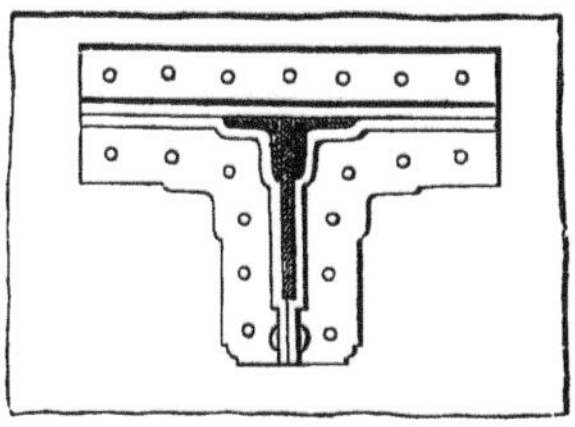

Fig. 167.

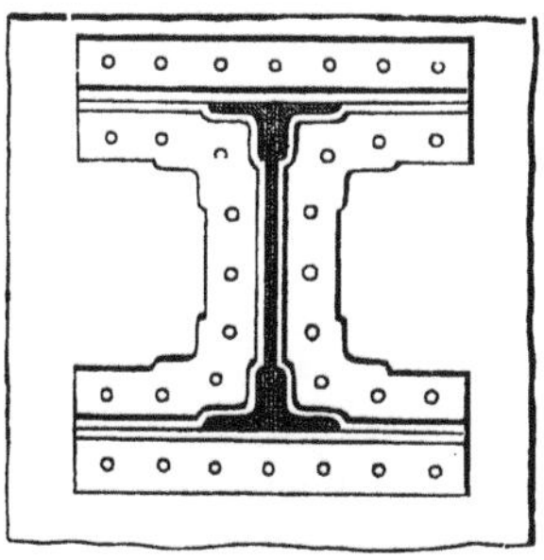

Fig. 165.

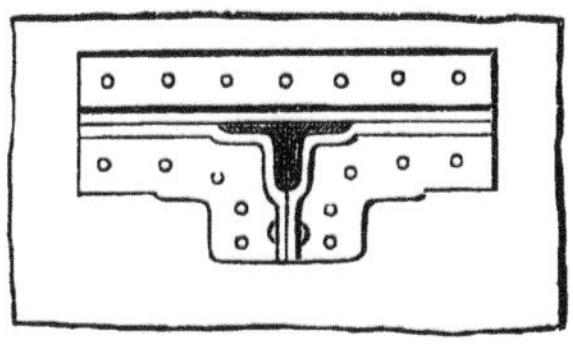

Fig. 168.

For the sake of convenience, the reasons before given for the use of these stop-waters may be repeated here. Unless a stop-water were fitted in each of the joints of the continuous plates and angle-irons in the neighbourhood of each bulkhead, it would be necessary to caulk the edges of the angle-irons throughout the ship. For if this were not done, the water might enter at some distance from the bulkhead, and, running along between the plate and angle-iron, find its way into the adjacent compartment. When stop-waters are fitted, the passage of water in the manner described is effectually prevented. It has been found also that a thick coat of paint will act very efficiently as a stop-water, and thus canvas may be dispensed with.

Before leaving this subject of watertight work it may be well to call attention to the manner in which the wing-passage bulkheads of the iron-clads are made watertight where the deck beams pass through them. The sketches given in Figs. 169 and 170 are illustrative of the arrangements followed in the 'Northumberland'

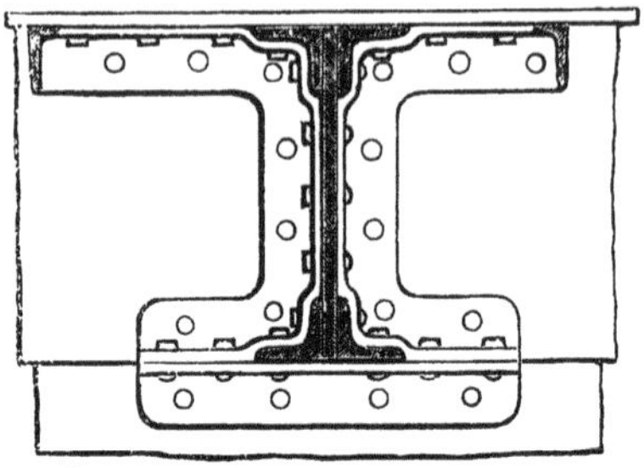

Fig. 169.

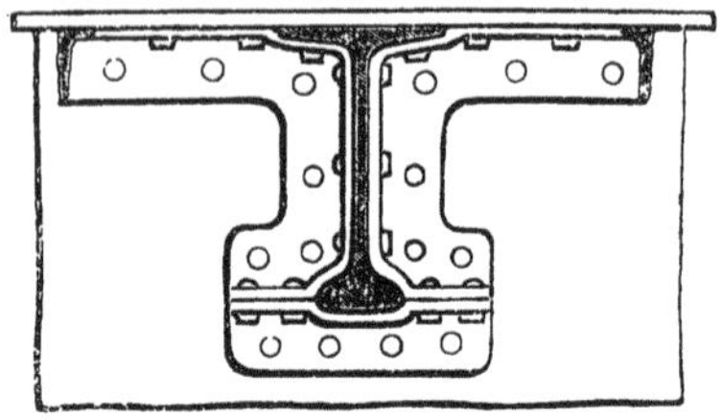

Fig. 170.

for I-shaped and Butterley beams, respectively. The staple angle-irons between the beams are turned out horizontally at their lower ends, and a short piece of angle-iron is worked beneath in order to allow the edges to be caulked. In some cases the half-beams of the Butterley pattern are made watertight as shown in Fig. 171. Here the lower ends of the staple angle-irons are forged so that they clasp the bulb of the half-beam and take a rivet through the vertical flanges, which are brought together below the half-beam.

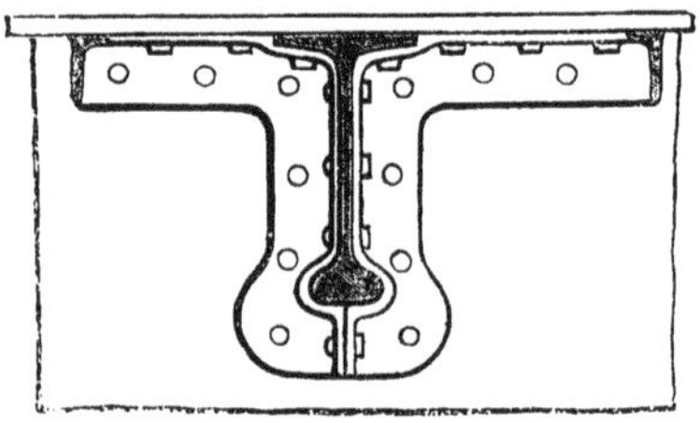

Fig. 171.

Watertight doors are fitted to bulkheads in order to afford ready communication between the different compartments under ordinary circumstances, and to allow the compartments to be completely separated in case of an accident occurring in any one of them. In the hold it is especially desirable that there should be direct communication between the compartments containing the engines and boilers, and large watertight doors are generally fitted at the openings in the bulkheads made for the purpose of affording this communication. In most merchant steamers the boilers are placed in the compartment next to that in which the engines are situated; but in many of the ships of the Royal Navy there is an intermediate compartment between the engine and boiler rooms, and, in order to connect the two, a watertight passage is constructed and watertight doors are fitted at each end of the passage. In Figs. 172 and 173 sketches are given which show the details of one of these watertight doors as fitted in H.M.S. 'Minotaur.' The door is formed of a wrought-iron plate 1 inch thick, and slides in a frame of wrought iron which is fastened to the bulkhead. This frame is rabbeted, as shown in the sections, at both the top and bottom, and also at one end. From the horizontal section through B, in Fig. 172, it will be seen that the part of the rabbet in which the door rests when it is open, is of parallel width, but that the part of the rabbet occupied by the door when it is closed (as in the sketch) is wedge-shaped. As the plate forming the door is of parallel thickness it is necessary to work wedge-shaped bearings on the top and bottom edges in order that the door may fit the rabbets tightly when it is closed. The shape of these bearings will be seen from the vertical section through C, in Fig. 173, and it will be evident that the section adopted for the bottom bearing is such that the weight of the door is made to assist in making the rabbet watertight. In order to make a watertight joint at the edge of the door which is not buried in a rabbet when the door is closed, a bearing is fitted which has a wedge-shaped horizontal section, as shown in the sections at A and B, in Fig. 172, and this bearing is pressed tightly against a corresponding wedge-shaped bearing fastened on the bulkhead. The bearings which are fitted on the edges of the door are all

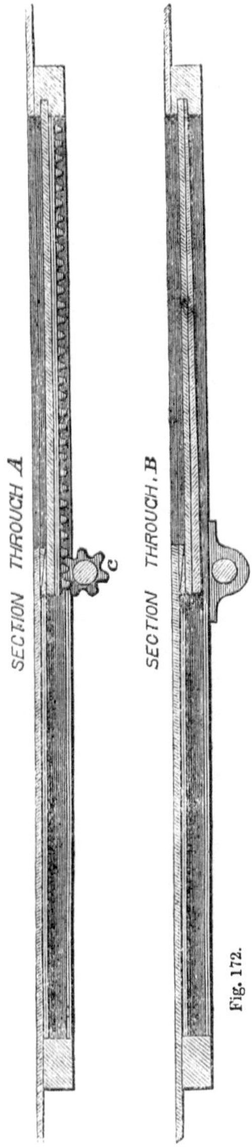

Fig. 172.

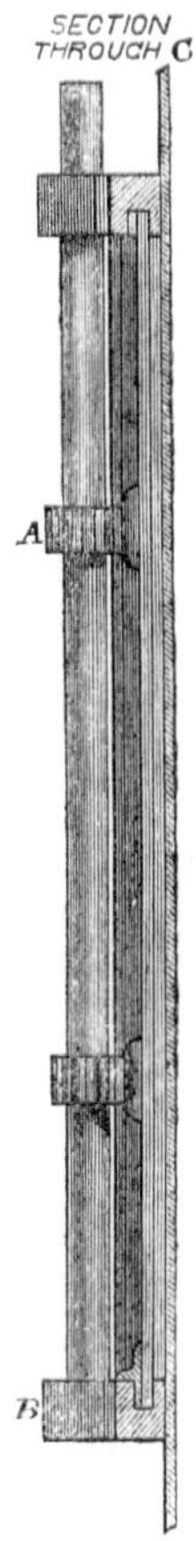

Fig. 173.

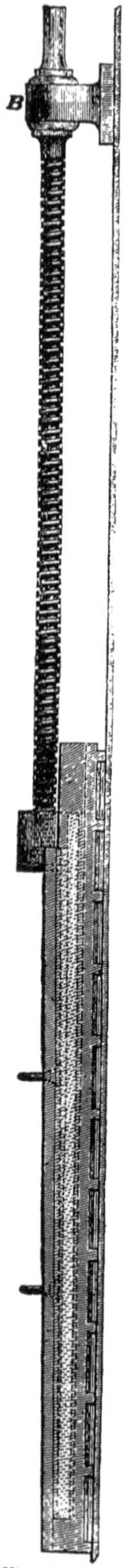

Fig. 174.

brass castings. Motion is given to the door, when it is required to be opened or closed, by means of the vertical rod C, which extends up to the main deck and is worked from thence, so that even when the hold is partially filled with water the door can be closed. The vertical rod carries two pinions which gear with the iron racks attached to the door, as shown in Fig. 173, and by this means the door can be moved horizontally. It may be remarked here that the principle of making both door and slide of a wedge-shaped section, and securing water-tightness by the use of brass bearings, is almost universally adopted in fitting watertight doors. The arrangements for opening and closing the door described above are also those generally adopted, although, instead of moving the door horizontally as in this case, the motion is in many instances vertical. An illustration of the latter arrangement is given in Fig. 174, which is taken from the watertight doors fitted in H.M.S. 'Penelope.' The sketch shows a side view of the door when closed. Here the vertical shaft extends from the upper part of the door up to the main deck. The lower end of the shaft has cut in it a coarse-threaded screw which works in the large metal nut that is fastened to the upper part of the door. As any vertical motion of the shaft is prevented by the bearing B, it is obvious that when the shaft is turned from the main deck the door must move vertically. In this case the door and frame are of cast iron, and the vertical wedge-shaped section is given to the edges of the door in casting. Metal bearings are worked on both sides of the door, on the vertical and the lower edges, in order to make watertight joints with the rabbets of the frame.

In vessels where the bulkheads extend up to the main deck it is very desirable that there should be direct communication between the various parts of the lower deck. For this purpose light watertight doors are fitted in the ships of the Royal Navy, similar to that shown by the horizontal section in Fig. 175, which is taken from the 'Bellerophon.' The door is formed of a wrought-iron plate stiffened by strips on the edges and under the middle hinge, and when closed fits tightly into the rabbets of a wrought-iron frame which is fastened on the bulkhead.

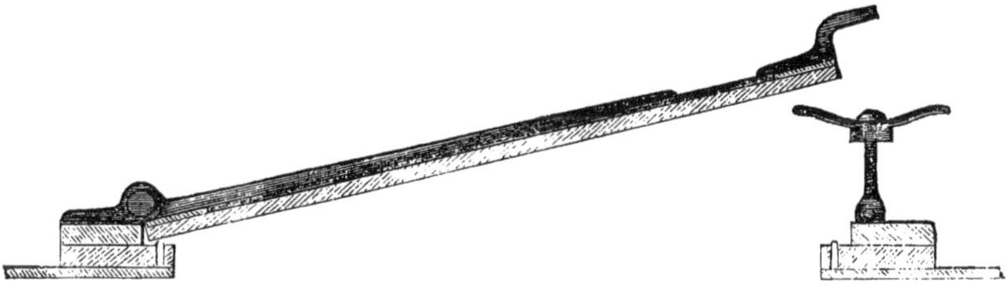

Fig. 175.

The door is hung to the frame by three hinges, similar to that of which the details are given in the section. Beads of india-rubber are fitted in the rabbets of the frame, and project above the surface of the rabbet. When the door is to be closed, the clamp-screws (or "butterfly nuts"), which are hinged to the frame, are turned back from the doorway, and when the door has been brought into the rabbet they are turned up into the forks of the lugs on the edge of the door and are hove up. By this means the door is pressed tightly against the india-rubber beading, and the rabbet is made watertight There are no brass bearings fitted on these doors, and, as they are above the lower deck, it is obvious that there would be ample time for the doors being closed and secured by manual labour in case of any compartment being injured.

In the iron-clad frigates of the Royal Navy admission is obtained to the wing passages from the lower deck through openings

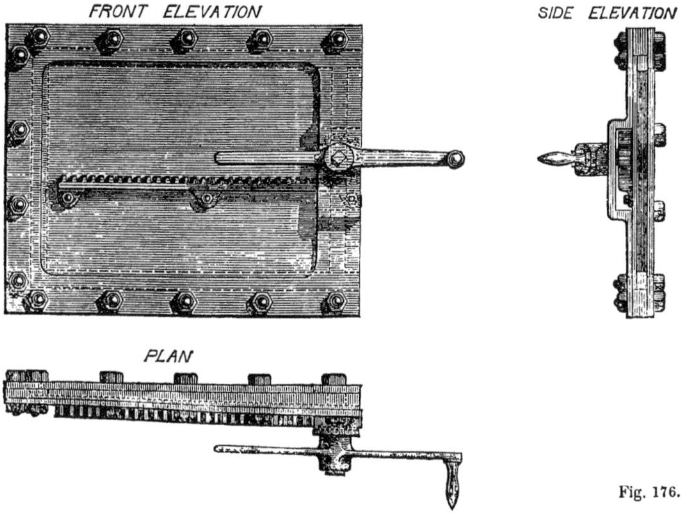

Fig. 176.

which can be closed by small watertight doors. The sketches given in Fig. 176 show the details of one of these doors as fitted in the 'Minotaur.' The doors and frames are of wrought iron, and the bearings on all the edges are of brass. The door is moved horizontally by means of a pinion fixed at one end of the frame, which gears with a rack bolted on the door. The details of the horizontal grooves are shown in the plan, and it will be seen that the wedge-shaped section is given to the edges of the door by means of metal bearings, as before described.

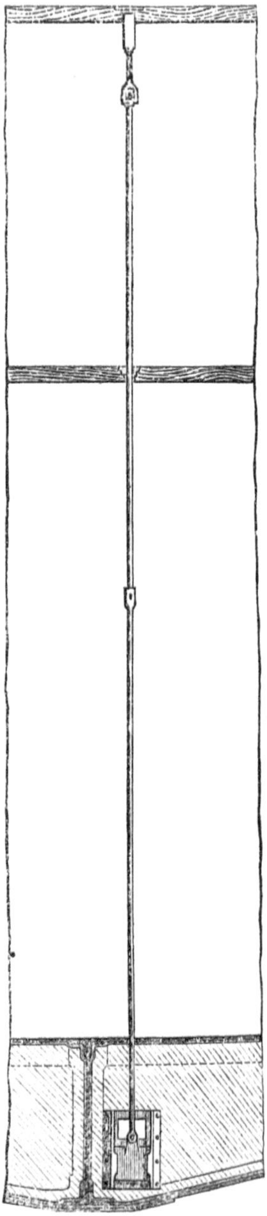

Fig. 177.

Having thus illustrated various modes of fitting and working watertight doors, it may be well to give the particulars of a sluice-valve before leaving the consideration of the fittings of watertight bulkheads. Sluice-valves are usually fitted in order to allow the water to pass from one compartment to another when required. This is absolutely necessary when each compartment is not supplied with the means of drainage, and is required by Lloyd's Rules under such circumstances. In the ships of the Royal Navy which have a double bottom, sluice-valves are fitted to the watertight frames in the double bottom. The example chosen to show the mode of fitting and working sluice-valves is taken from one of H.M. ships, and is illustrated by Fig. 177. It will be seen, from the sketch, that the valve is placed as low down as possible, and is bolted to the watertight frame, while, by means of a vertical connecting-rod it can be worked from the main deck. Where the rod passes through the lower deck a stuffing-box is fitted around it, and its upper end is secured by means of a swivel joint to the lower end of a cup-screw which is let down into the main deck. When it is required to open or close the valve, a key is employed with a mortice on the lower end which fits upon the head of the screw. By means of the swivel joint mentioned above, the connecting-rod does not turn with the screw to which its upper end is secured, and when the screw is turned the rod moves upwards and opens the valve. It may be added that when the valve is open, the end of the screw is above the level of the main deck, and, consequently, it can hardly fail to be noticed; by this means the probability of leaving the valve open by accident is much reduced. When the valve is closed a cover is hove down upon the top of the screw-cup, which is thus kept free from dirt and in working order. Similar arrangements are usually made at the upper ends of the vertical shafts by which watertight doors are moved, so that it may at once be evident when the doors are open. Mr. Roberts, of Millwall, has patented an arrangement by means of which the position which the door occupies is indicated by a tell-tale placed on the deck from which the door is worked. The tell-tale consists of a hand travelling round a dial, the motion of the hand being communicated from the vertical shaft by means of a train of mechanism.

The modes of testing the watertight work in bulkheads and other divisions in the hold are worth a passing notice. One very common method has been previously referred to, viz., the use of a fire-engine and hose, by means of which a stream of water at a high velocity is brought upon the part to be tested; this plan is especially suited for watertight doors in bulkheads, and side scuttles. In some cases the compartments are filled up to the height of the top of the floors, and the remainder of the work tested with the fire engine. In H. M. Service watertight compartments in the hold are usually tested either by filling them entirely with water, or only up to the height of the load water-line, according to circumstances. Watertight flats, such as crowns to magazines, platforms, &c., are usually tested with a depth of from 9 to 12 inches of water, which is allowed to remain for about two days, in order to ascertain the amount of leakage. Double bottoms are generally tested by a pressure equal to that due to the height of the load water-line above them, which is, of course, as great a pressure as they can ever have to bear. The great importance attaching to the ensuring of watertightness in bulkheads, &c., is so obvious as to need no enforcement, although it often happens that imperfect workmanship escapes detection unless carefully tested.

All the preceding remarks have had reference to watertight bulkheads; but there are cases in which it would be inconvenient or impossible to introduce complete transverse bulkheads, and yet by the use of partial bulkheads sufficient structural strength may be ensured. It will be remembered that the use of these partial bulkheads forms part of the longitudinal system of framing practised by Mr. Scott Russell. In the ships of the Royal Navy, built on the bracket-frame system, the watertight plate-frames at intervals of about 20 feet form partial bulkheads. In some of the iron-clads partial bulkheads have been fitted in the wing passages and placed between the complete bulkheads. In wake of engines and boilers also, when deep transverse bearers are fitted, they may be regarded as acting as partial bulkheads and adding considerably to the transverse strength of this part of a ship. In many vessels built on the transverse system deep plate-frames have been worked at intervals in order to give the requisite strength, and at the same time allow the convenient stowage of the cargo. In ships which have to carry timber or any other cargo requiring great length for its stowage, these deep frames are particularly suitable. The usual plan is to have a deep beam-plate in the same transverse plane as the plate-frame, and to connect the two by means of brackets at the ends of the beam-plate. The various lengths of plate in the deep frames are connected by lap or butt joints, and frame and reversed angle-irons are worked on the outer and inner edges respectively.

CHAPTER XII.

TOPSIDES.

THE topsides of iron vessels were at first almost universally constructed of wood, and Mr. Laird was, we believe, the only English builder, who in 1842, was constructing vessels with iron topsides. An example of the mode of fitting wooden topsides adopted in some of the earlier ships is given in Fig. 178, which is taken from the 'Dover.' The roughtree stanchions are of wood, and are run down by the side of the frames, and bolted through the outside plating. It will also be noticed that the wooden gunwale, or covering board, is fitted around the stanchions, and bolted to them and to the beams. The upper edge of the outside plating is rabbeted into the gunwale, and the topside above the gunwale is similar to that of a wooden ship. There is no stringer-plate on the beam-end, and consequently the watertightness of the gunwale depends entirely on the fitting and caulking of the stanchions and covering-board.

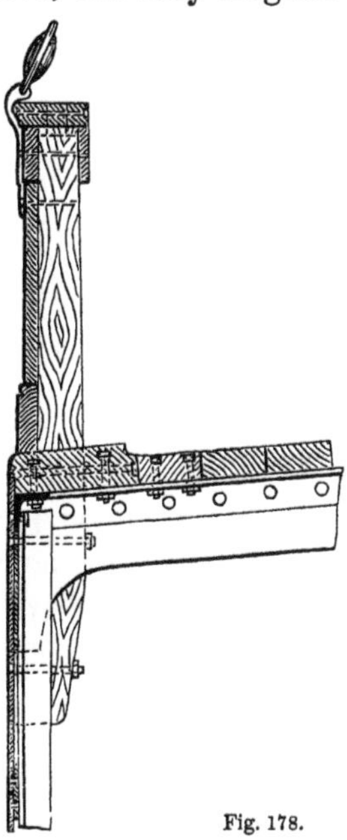

Fig. 178.

In most ships where deck stringer-plates were fitted, the practice was to cut holes in the stringers in order to allow the timber stanchions to be continued down by the side of the frames and secured by bolts through the plating and the frames. It will, no doubt, be remembered that this practice has been previously alluded to, when illustrating the modes of fitting deck-stringers, and the serious reduction which is thus made in the longitudinal strength has been pointed out. Lloyd's Rules expressly forbid this arrangement of the stringers and stanchions.

The topsides of the 'Birkenhead,' built by Mr. Laird, are

shown in Fig. 71, p. 75. In order to avoid cutting through the upper-deck shelf and stringer-plates, the transverse frames are stopped below the shelf, and the side plating only is continued up for a short distance. The upper edge of the plating is stiffened by a continuous angle-iron, which also serves to receive the fastenings of the wooden planksheer. The waterway is worked directly against the side plating, and bolted through it and the stringer-plate, and by this means the low iron topsides are considerably stiffened. In most of the ships which have been constructed with topsides similar to the preceding, portable iron stanchions and guard-rails have been fitted upon the planksheer.

The topside arrangements of the 'Recruit' have been before alluded to, and are of a very unusual character, as will be seen from Fig. 72, p. 77. It will be remarked that the outside plating is continued up to the rough-tree rail, and that inside the ship the topsides are planked similarly to those of a wooden ship. The iron frames extend up to the rough-tree rail, and top timbers are bolted to the sides of alternate frames, the outside plating on the topsides being riveted to the iron frames, and the inside planking being bolted to the top timbers.

Attention has been called to the 'Megæra' as an example of the once common practice of fitting wooden beams in iron ships. In this vessel the frames and outside plating are continued up for a short distance above the upper deck, and a gunwale or covering board is worked upon the upper ends. It is customary to complete the topsides above this gunwale by, what is termed, a top-gallant bulwark formed of wooden berthing and stanchions. The berthing or thin planking on the bulwarks is usually rabbeted or tongued.

The last example we shall give of a comparatively old topside arrangement is taken from the 'Vulcan,' and is illustrated by Fig. 103, p. 148. The frames and plating are ended similarly to those of the 'Megæra,' and the top-gallant bulwark is shown in the sketch. In describing the beam connections mention has been made of the vertical clamp-plate worked inside the frames, and it will be seen that by this arrangement a very strong connection is made between the stringer and the frames, and the frames can be run up to form the topsides without cutting the stringer.

Coming now to the illustration of the topsides of more modern iron vessels, we will take first those cases where the topsides are either wholly or partially of iron.* In Plate 2 is seen the topside of the steam-ship 'China,' built by Messrs. Napier. It will be observed that the frames and side plating are continued up to form the bulwark for a height of 4 feet above the deck, and that a light top-gallant bulwark of wood, about 2 ft. high, is worked above them. When this plan of fitting topsides is adopted, it is usual to stop the alternate frames below the upper-deck stringer. As it would be very expensive to fit the stringer-plate and angle-irons accurately between the frames which are run up, it is a very common practice in merchant ships to depend upon the fitting and caulking of the covering board for the water-tightness of the gunwale. A very similar arrangement to the preceding is shown in Fig. 179. The principal differences consist in dispensing with the top-gallant bulwark, substituting a simple rough-tree rail, and working a vertical clamp-plate and a gutter waterway on the beam-ends. In cases of weakness, additional strength is often given to the upper part of a ship by means of rail and clamp-plates worked upon the topside frames. In some vessels with gutter waterways the spaces between the frames, enclosed by the outside plating and the angle-iron worked inside the frames, are filled up with cement, in order to secure watertightness, and to prevent the water from lodging.

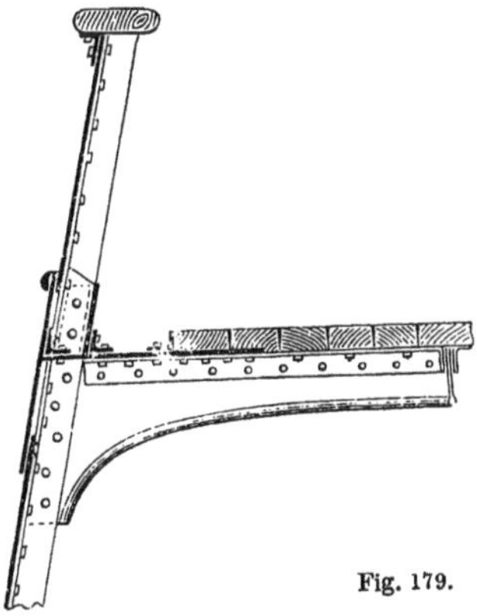
Fig. 179.

It is generally considered that greater simplicity and strength are obtained by ending the frames below the upper-deck stringer, and an illustration is given in Plate 1 of a common mode of forming the topsides of vessels in which this arrangement is carried out. The topside plating forms a continuation of the side plating, and is supported by iron stanchions. Above the iron topsides a light wooden bulwark is fitted. The total height of the bulwark is about 5 feet 6 inches, and the height of the iron topsides 4 feet. It may be added that the arrangement here shown is modified, in some cases, by fitting a simple rail instead of a top-gallant bulwark, and by varying the form of the iron stanchions. An illus-

* Several of the following sketches are taken from the very admirable illustrations which accompany Lloyd's Rules.

tration of this statement is given in Fig. 180, and needs no further description. In many ships which have iron topsides, light angle-iron or T-iron frames are employed instead of stanchions formed of bar-iron. The illustration of this arrangement, given in Fig. 181, is taken from the 'Bellerophon.' In this ship the topsides are supported by light

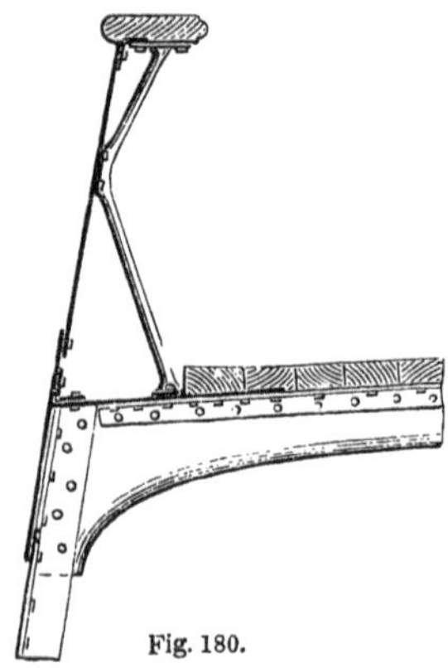
Fig. 180.

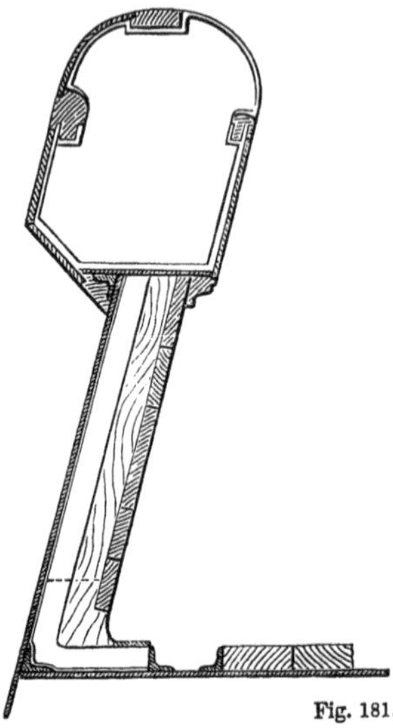
Fig. 181.

angle-iron frames of which the lower ends are turned in on the deck plating. It will be remarked that wooden stanchions are also fitted in order to receive the fastenings of the inside planking.* A plate-rail is fitted upon the top of the frames and supports the hammock berthing. The addition of the inside planking and wooden stanchions to the ordinary iron topsides was made on account of the great facilities thus obtained for fixing cleats, &c., on any part of the topsides where they might be required; a matter of great importance in a fully rigged ship of the size of the 'Bellerophon.'

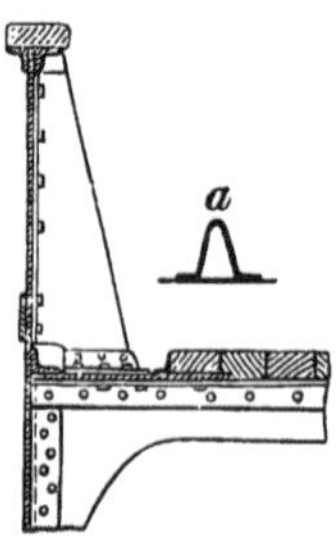

Fig. 182.

Another mode of fitting topsides is given in Fig. 182. The plating of the topsides is in this case supported by light frames formed of plates bent to the sectional form shown by *a* in the sketch. The heels of these frames are secured to the stringer-plate by bent angle-irons fitted around them. Double angle-irons are worked on the upper edge of the plating, and receive the fastenings of the wooden rail.

* These stanchions and the internal topside planking have been omitted in the section given in Plate 4.

Another arrangement of iron topsides is illustrated in Fig. 183. The transverse frames are run up about 2 feet above the upper deck, and a horizontal gunwale plate is worked upon them. Above the gunwale plate the bulwarks are completed by light plating, supported by bar-iron stanchions similar to those before described.

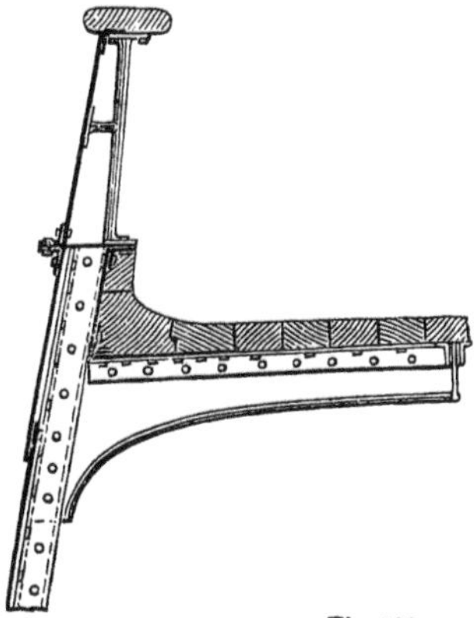

Fig. 183.

In some vessels which have poops and forecastles, the bulwarks in wake of them are constructed of a rounded form at the gunwale. This plan is recognized by Lloyd's Rules, which provide that the beams of the poop or forecastle may be of plain angle-iron, of a size not less than that of the main frames, and that the angle-iron beams must be properly riveted to every alternate main frame with a scarph not less than 4 feet in length. An illustration of this kind of gunwale is given in Fig. 184, which is taken from the 'Sentinel,' a ship of which the details of the construction have been previously given. The plating is continued up over the gunwale, and a stringer angle-iron is worked at the beginning of the round-down in order to form a finish to the deck planking. Light guard rails and stanchions are usually fitted upon the poops and forecastles.

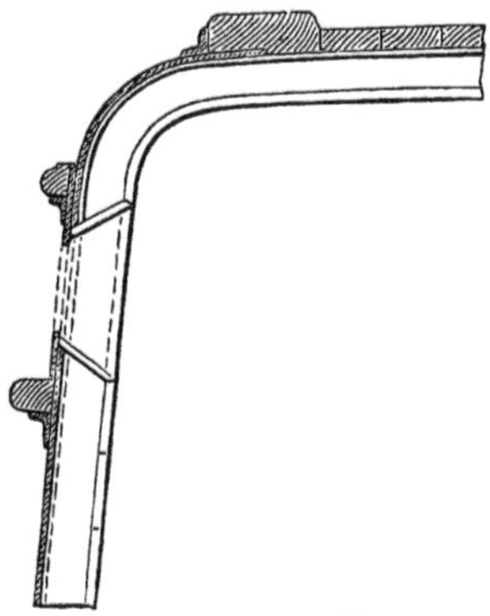

Fig. 184.

We next turn to the consideration of various arrangements of wooden topsides now commonly employed. As before stated, the practice of cutting the stringer-plate, in order to allow the wooden stanchions to be passed down through it, is now forbidden by Lloyd's; and the general custom of shipbuilders is to leave the stringer unpierced, except by the rivet-holes and the scuppers. Various plans have been adopted in order to secure the heels of the topside stanchions. One very common mode is that shown in Fig. 185. A wooden covering board or gunwale is worked on the stringer and forms the waterway. The heels of the stanchions are let down into the gunwale and through-bolted. The arrangements of the rails and outside planking or berthing are similar to those

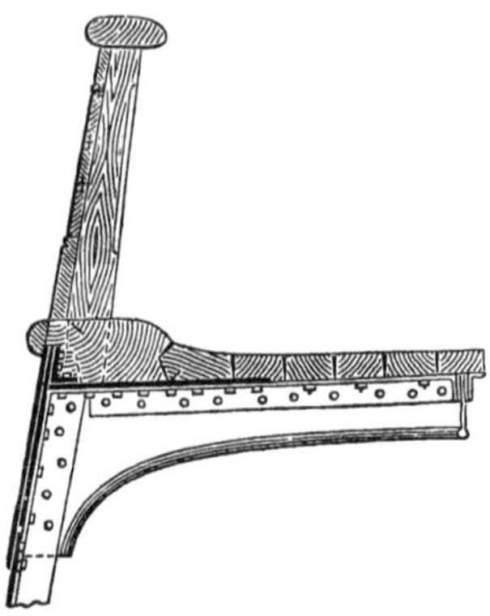

Fig. 185.

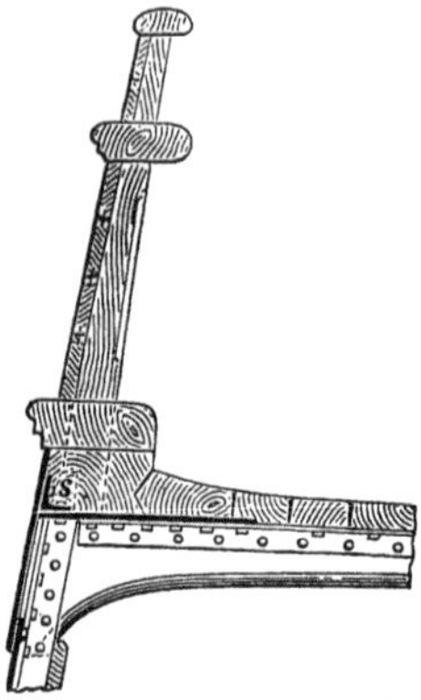

Fig. 186.

of a wooden ship's bulwarks. In some vessels the foregoing plan is slightly varied by working an additional covering board upon the wooden waterway, as shown in Fig. 186, and by having an additional rail.

A very common plan of securing the rough-tree stanchions is to extend the sheer strake up above the stringer-plate, and to work a vertical clamp-plate parallel to the sheer strake, and at a distance from it equal to the moulding of the heels of the stanchions. An illustration of this is given in Fig. 187. The stanchions step in between the sheer strake and the parallel plate, and are through-bolted. The spaces between the heels of the stanchions are filled in within wooden chocks in most instances. In the case shown in the sketch, a wood waterway is worked; but in some ships

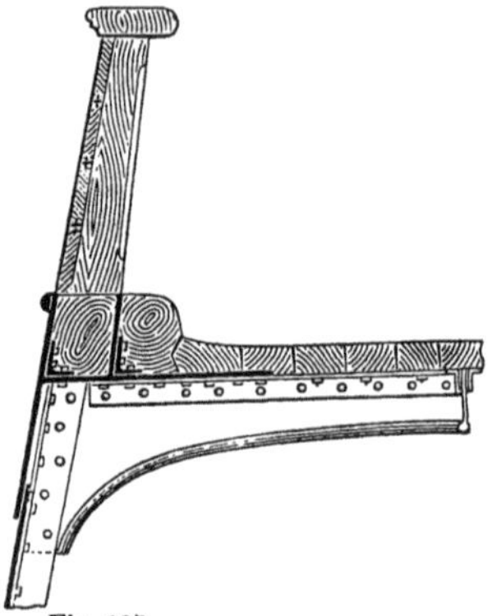

Fig. 187.

a covering board is also fitted upon the waterway, and in others both waterway and covering board are omitted and a gutter waterway formed, as shown in Fig. 188. A slightly modified form of this arrangement is given in Fig. 189, where the chocks between the heels of the stanchions and the waterway are not worked up to the full height of the sheer strake, and the covering board is supported by an angle-iron on the outer edge, and on the inner edge by a shallow strake of spirketing. Another modification of the preceding arrangement consists in working a deep angle-iron

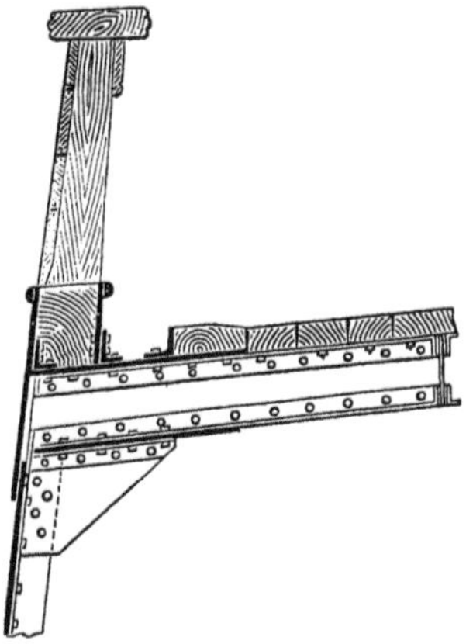

Fig. 188.

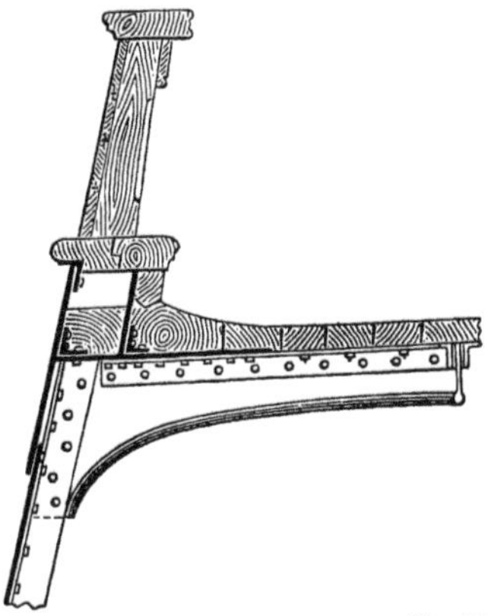

Fig. 189.

parallel to the sheer strake, instead of having a plate and angle-iron.

When box-stringers are fitted on the upper deck, wooden topsides are sometimes adopted, and an arrangement is given in Fig. 190, which will illustrate the manner in which the topsides and stringer are combined. As it would cause a serious reduction in the strength of the stringer if the top plate were pierced, the heels of the stanchions are fitted into iron sockets riveted to the stringer. The remainder of the arrangements of the bulwark require no description.

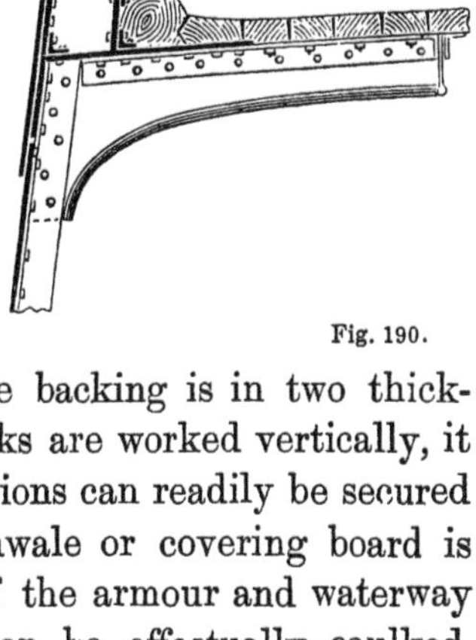

Fig. 190.

In the 'Warrior' the topsides are of wood, and are fitted as shown in Plate 3. The only matter requiring remark, is the mode of securing the heels of the wooden stanchions. As the backing is in two thicknesses, and in the outer layer the planks are worked vertically, it is evident that the heels of the stanchions can readily be secured between the vertical planks. The gunwale or covering board is very wide, and since the upper edges of the armour and waterway are rabbeted into the gunwale they can be effectually caulked, and the passage of water down between the teak planks in the backing can be prevented.

In the 'Hercules' there is only one thickness of backing, and the topside stanchions are let down, dowelled, and bolted, as shown in section in Fig. 191. As it would be a very expensive and wasteful process to cut away the upper strake of backing in order to let down the stanchions, another plan is adopted. Thin strakes of planking are worked between the skin plating and the stanchions, and the spaces between the heels of the stanchions are filled in solid with short vertical timbers extending up to the height of the top of the spirketing. Outside the protected portion of the ship the stanchions are let down into, and dowelled and bolted to a wooden gunwale worked upon the deck plating. It will be remarked that in this case, as well as in that illustrated by Fig. 188, gutter waterways are fitted in combination with wooden topsides. It has been previously pointed out in how great a measure these waterways add to the strength of the ship; and it may be added here, that they are also most efficient in clearing water from the decks, and are very highly spoken of by practical seamen.

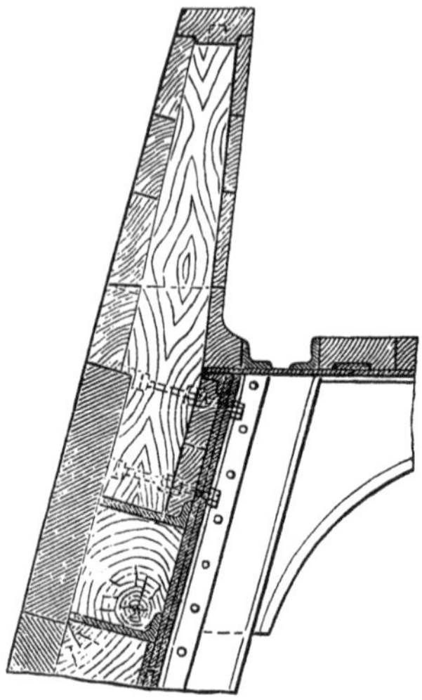

Fig. 191.

In concluding this chapter, it may be well to call attention to the topsides of turret ships, which are usually fitted so that they can be turned down when required. Of these topsides two examples are given in Figs. 192 and 193, which are taken from H. M. S. 'Scorpion,' built by Messrs. Laird, and the Italian ironclad 'Affondatore,' built at the Millwall Company's Works. In the 'Scorpion' the bulwarks are made in lengths of about 8 feet, in order that they may be lifted easily, and each length is secured by means of two hinged stanchions similar to that shown in the section in Fig. 192. When in place, the inside part of the stanchion is secured by the pin marked *a*, which is removed when the topsides are to be turned down, and they then move upon the hinge marked *b*. It will be seen that the lower part of the bulwark is covered with thin iron plating $\frac{1}{2}$ inch thick, stiffened by angle-irons along the edges, and the upper part is formed of a top-gallant bulwark of wood. In the 'Affondatore' the bulwarks are in lengths of 8 feet 9 inches, each length being secured by three hinges. The bulwarks are lower than those of the 'Scorpion,' and are made up of light plating ($\frac{3}{16}$-inch), with a plain rail on the upper part, and stiffening angle-irons on the edges. The hinges are differently formed and secured from those of the 'Scorpion,' as will be seen from Fig. 193. When in place the bulwarks are secured by stays similar to *s*, and T-iron and angle-iron stiffeners are worked upon the bulwark plating in order to stiffen it in wake of the stays. In these ships it is usual to place

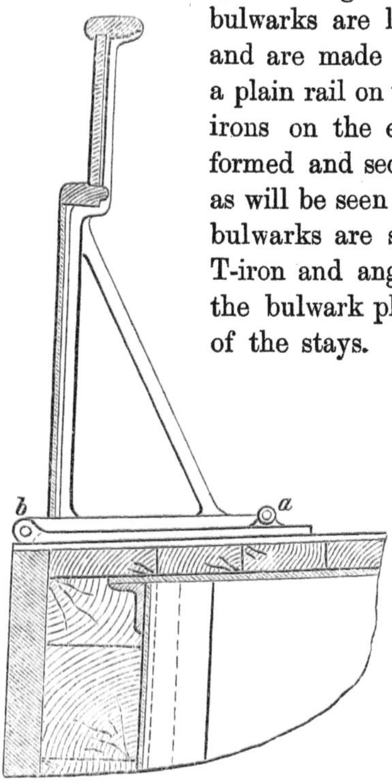

Fig. 192.

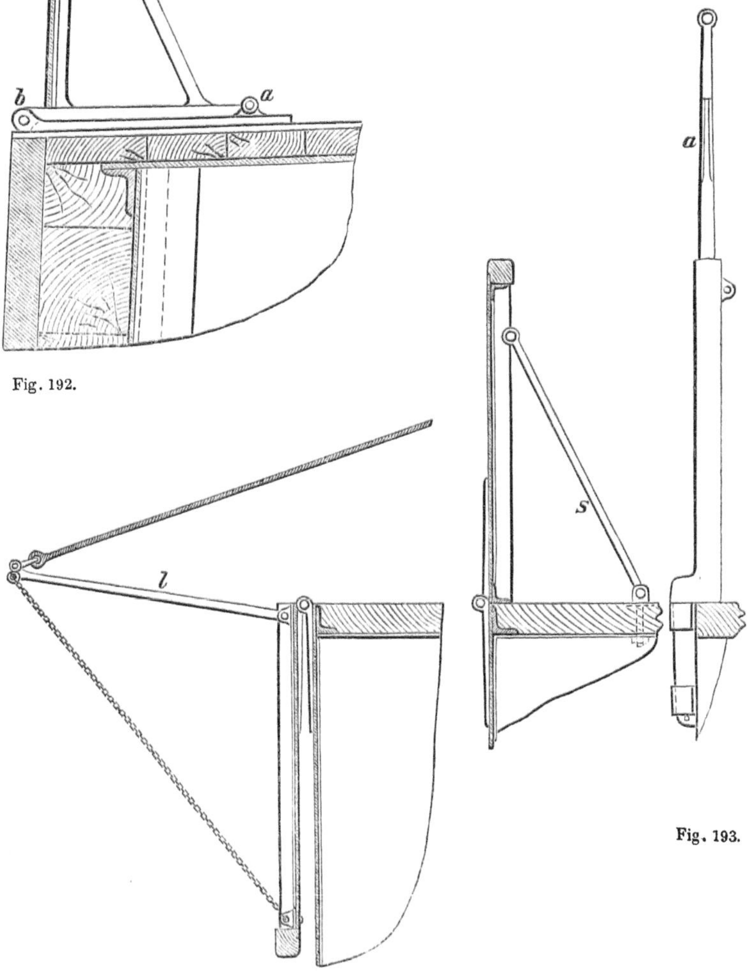

Fig. 193.

the awning stanchions in such positions as to bring them between two lengths of the bulwarks. A side view of one of these stanchions, as fitted in the 'Affondatore,' is marked *a* in Fig. 193, and will serve as an example of the ordinary method of securing them. In these low-decked vessels it is very essential to have the means of speedily clearing the water from the deck; for this purpose flap ports are fitted in the bulwarks, and are so arranged as to open outwards under pressure, while they prevent the entrance of the water. It will be obvious that when the bulwarks were turned down against the side some difficulty would be experienced in raising them again, unless special provision were made for that purpose. In Fig. 193 there is given a separate sketch showing the plan adopted in the 'Affondatore.' The bulwark is shown turned down, and it will be seen that in order to give a good lead to the rope by which the men pull it up, the lever *l* is fitted, its outer end being connected with the upper part of the topsides by a chain. By this means the bulwarks can be brought up to the position in which the men can take hold of the lever *l*, and then the operation can be easily completed. There are two levers to each length of the bulwarks. On reference to the sketch in Fig. 192, it will be evident that the foot of the stanchion in the 'Scorpion' will serve the same purpose as the lever *l* in the 'Affondatore,' as the men can take hold of it to raise the bulwark.

CHAPTER XIII.

RUDDERS.

IN this chapter we propose to give a brief sketch of the modes of forming and fitting the rudders of iron ships, without entering on a discussion of the relative advantages of the various proposals which have been made for increasing the steering power of ships, which is a subject that scarcely falls within the compass of a work that, like the present, is limited to practical information. In the course of the following remarks it will be necessary, however, to point out some of the reasons which have led to the adoption of novel forms of rudders, and to notice the results of their trials.

In the early iron ships the main pieces of the rudders were formed of hollow iron plates, the space between the plates being filled in with fir or some light wood. A good illustration of this is found in Fig. 194, which gives a section of the 'Dover's' rudder. The main piece, or front, and the sides of the rudder were formed by a bent plate, and the space between the side-plating was filled in with fir secured by through-bolts on the after edge. The rudder-head was hollow and cylindrical for about one-third of the depth of the rudder, and down to the part where the breadth of the rudder began to increase rapidly, it was welded. The rudder was hung to three braces, riveted to the hollow-plate stern-post, by means of three similar braces riveted to the bent plate forming the front of the rudder, and was kept in place by a long pintle-bolt which passed down through all the braces. The upper end of this bolt was supported on a shoulder formed on the inside of the upper part of the rudder-head, and was easily got at from the upper deck. When the rudder was to be unshipped, the bolt was withdrawn. In the sketch the pintle-bolt is shown in section, and it will be seen that its centre is coincident with that of the fore part of the rudder, so that although the front of the rudder was straight, the size of the rudder-hole was only required to be as much larger than the dia-

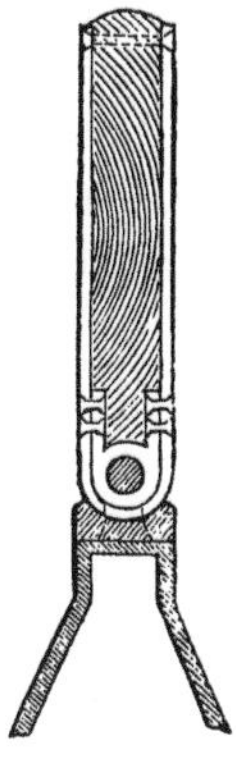

Fig. 194.

meter of the rudder-head as would give room for shipping and unshipping the rudder. The casing of the rudder-hole was made of thin plating worked watertight, and extending from the counter up to the deck above.

M. Dupuy de Lôme gives an account of a mode of forming the rudders of small vessels which was practised at Glasgow at the time of his visit in 1842. This method is interesting on account of its simplicity, as will appear from the following description given in the Report:—" They pass through all the braces " a bar of round iron, and rest its lower end upon the keel. They " then rivet on the sides of the bar the plates which form the " rudder. In case of these rudders being damaged, the ships must " be docked, or the whole of the stern post is spoiled by stripping " off the bottom plating. If it becomes necessary to shift or " change the rudder, it must be taken to pieces in place, in the " same way as it was constructed, and on this account the method " is not suited to vessels which are liable to be long absent from " port."

When the hollow-plate stern posts were displaced by the solid bar posts, the great diminution made in the siding of the posts necessitated a corresponding reduction in the siding of the rudders, and so led to the introduction of a solid forging for the main piece. This arrangement had the great advantage of giving the necessary resistance to torsion with much smaller dimensions than would have been required for a hollow-plate main piece. In addition, the pintles were forged in one with the main piece, and the frame of the rudder was either scarphed or welded together, and then plated over, the space between the side-plating being left empty. It was then thought unnecessary to fill in the interior of the rudder with fir, as the siding was so small as to render the increase of weight very trifling in case the rudder was by accident filled with water.

We next come to consider the mode of forming rudders now generally followed. The main piece or front of the rudder is, almost invariably, a solid forging, the pintles being forged in one with it, and in most cases the back of the rudder is also formed by a solid iron frame which is welded to the main piece. In some cases, however, the body of the rudder has been made of a single plate, and the rudder-head and pintles have been separately forged and riveted to the plate-rudder. Lloyd's Rules state that the main piece of the rudder is to be made of the best hammered iron, and the plating to be carefully stayed and riveted. The Liverpool Rules simply require that rudder-frames shall be forged solid. In both cases dimensions are given for the rudder-heads and heels of various classes of ships. The rudder of a large iron ship is now usually made in the following manner:—The main piece is made in one forging from the head to the heel, lugs in the rough being brought on to receive the other portions of the frame. The forging of the main piece is so formed as to admit of the pintles being worked out of the solid, and they are afterwards shaped out under a slotting-machine and completed by hand with chisel and file. A turning-machine has been sometimes used for the purpose of finishing the pintles, and machinery may be advantageously employed also for boring the holes in the braces, and taking out the gulleting in the back of the rudder-post. The back of the rudder is formed in a separate forging, and is connected with the main piece by horizontal stays. The stays and back of the rudder are welded to the main piece by means of the lugs left on it for the purpose, and thus the rudder-frame is made into one solid forging. The back and the heel of the rudder are usually tapered considerably from the dimensions of the head, and the sides are made nearly in one plane, and covered with thin plating worked flush. The front, back, and stays of the rudder-frame are perforated through and through with holes for the riveting of the side-plates, and the edges of the plates are connected by internal edge-strips worked between the stays. After account has been taken of the rivet-holes in the rudder-frame, and the edge-strips have been fitted in place and the holes for the edge-riveting punched, the plating of each side is taken off, the holes for the fastenings in the rudder-frame are drilled, and the edge-rivets are put in. When this work has been completed, the plating is replaced on the frame, and the through-riveting is proceeded with. The edges of the plating are caulked in the ordinary manner, and the whole is supposed to be perfectly watertight. There is, however, a difficulty in ensuring this at first, and when it is remembered that the rudder is very liable to be struck, it will appear that the chances are greatly against the inside of the rudder being free from water. With a view to ensure the exclusion of the water, rudders are frequently filled in between the frame and stays with some light wood (generally Dantzic fir) and then covered by the side-plates in the usual way.

The sketch in Fig. 195 will fully illustrate the preceding description. It is taken from the rudder of an armour-plated frigate of the 'Northumberland' class. The diameter of the rudder-head is $11\frac{1}{2}$ inches, and the siding of the back tapers from $11\frac{1}{2}$ inches at the upper part to 5 inches at the heel. Between the front and back of the rudder there are four horizontal stays, marked *s*, and upon the upper and lower stays stop-cleats, marked *c*, are riveted, and serve to prevent the rudder from being put over past a certain angle, or being driven beyond it by a sudden blow. It may be remarked here that the number of cross-stays similar to *s* varies with the size of the rudder, and in vessels of moderate dimensions they are entirely dispensed with. In very small vessels the back piece also is omitted and the plates are secured to the main piece, their after ends being brought together and riveted. It will be noticed that the pintles are forged in one with the main piece, and that their centre is in a line with the centre of the rudder-head, as is usually the case in iron rudders. The pintle at the heel fits into a socket in the after end of the keel-piece, and the other pintles fit the braces forged on the rudder-post. The pintle next below the upper one and that next above the lower

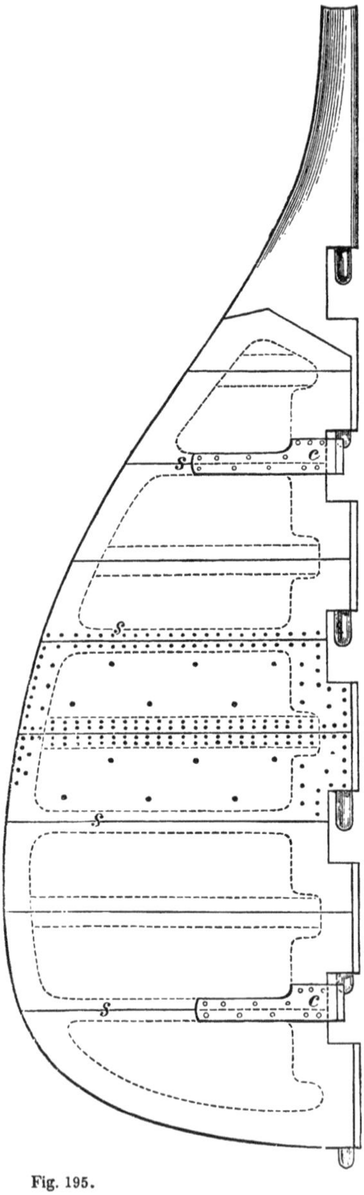

Fig. 195.

one are considerably shorter than the other pintles, and they have steel pins screwed into their lower convex surface. These steel-pointed pintles bear on corresponding steel pins fitted in the braces, when the rudder is in place, and the weight of the rudder being taken by these convex surfaces of steel, the friction is reduced to a very small amount, and the rudder is made to turn readily. The importance of this arrangement will appear when it is considered that the total weight of this rudder is 15 tons, of which weight the frame makes up 12 tons and the plating, fillings, &c., the remaining 3 tons. For a considerably smaller and lighter rudder one of these steel-pointed pintles would be considered sufficient. Another advantage attaching to this arrangement, which also deserves notice, is that with it there need be little trouble taken in bringing the lower part of every pintle to bear accurately on the upper part of every brace, as is usual in a wood ship. If there are two sets of pins it is sufficient to bring them accurately into contact, and if there is but one the only care required is to give sufficient play between the several pintles and braces to ensure the rudder turning entirely on the pins. The sketch in Fig. 195 shows the arrangement of a portion of the riveting of the $\frac{3}{8}$-inch plating forming the sides of the rudder. The space between the frame and plating is filled with Dantzic fir.

In some cases, instead of hanging the rudder in the manner described above, the following plan is adopted. The recesses in the front piece are made only of the same length as the pintles, and the usual kind of brace is replaced by iron straps, which are bent so that they may fit around the pintles and have their fore ends riveted to the stern-post. The rudder has to be in place before these straps can be riveted; and when once fixed, the rudder cannot be unshipped without cutting out the rivets in the straps. These inconveniences form grave objections to the plan, but it is frequently adopted in merchant ships, and is cheaper than the usual method.

Heel-ropes are usually fitted to large iron rudders, and in order to receive them iron tubes are worked through the thin plating, their ends being turned back outside the plates and beaten down or clenched. One other feature of the ordinary mode of fitting iron rudders requires a passing notice. The axis of the rudder is usually at an equal distance from the back of the post throughout its length, and as the rudder generally tapers considerably from the head to the heel the consequence is that when the rudder is put hard over, a hollow is found on one side of it and a projection on the other, both of which tend to seriously obstruct the action of the rudder. This evil is, however, entirely removed by making the distance of the axis of the rudder from the back of the post correspond to the siding of the rudder at every point of its length, thus following the practice of the wooden shipbuilder.

Balanced rudders have in several instances been fitted to iron ships. The screw steam-ship 'Great Britain,' built at Bristol, was originally supplied with a rudder of this kind, the head of the after post above the bearing for the after end of the screw-shaft being rounded in order to serve as the axis of rotation of the rudder. Of the total area of the rudder one-third was before the axis and two-thirds abaft it. When the ship was wrecked in Dundrum Bay the rudder was knocked away, and when the repairs were performed a solid rudder-post was fitted, and the ordinary form of rudder was adopted. Within the last few years balanced rudders have been more frequently employed, both in ships with single and twin screw-propellers. This adoption of balanced rudders has been consequent on the recognition of the great advantages they possess as compared with ordinary rudders in respect both of the increased area of rudder-surface thus obtainable, and the ease with which they can be put over to very large angles. As instances of twin-screw ships with balanced rudders we may refer to the American monitors, and the ships of the 'Invincible' class in our own Navy. In these vessels, which have a single dead-wood, the rudder-head is secured at the ship's counter, and the heel is kept in place by a large pintle which steps into a socket in the after end of the keel. The keel is prolonged for a few feet abaft the post in order to receive the pintle in the rudder-heel, and the fore edge of the rudder clears the after side of the body-post by about 2 feet. As examples of single-screw ships with balanced rudders, we may mention H.M.'s ships 'Bellerophon,' 'Monarch,' 'Inconstant,' and 'Hercules,' and the Prussian iron-clad 'King William.' The rudder arrangements of the 'Bellerophon' and 'King William' are of an almost identical character, and it will suffice, therefore, to illustrate those of the former ship; the arrangements adopted in the 'Hercules' are very different, and will be fully described hereafter. In the Bellerophon' the rudder is constructed as shown in side view and section in Fig. 87, p. 120. The front, upper, and lower edges of the rudder-frame are formed by a solid forging, and the after edge by a bent plate as shown in the section. The rudder-frame is completed by four vertical frames formed of plates and angle-irons, the two foremost frames being placed close together and at equal distances on each side of the axis of the rudder. By means of this arrangement of the rudder-frame its weight is considerably diminished, and its cost considerably reduced, from what would be rendered necessary if the frame were one solid forging; while the rudder is made as strong as usual on those parts which have to resist torsion, or are most exposed to violent blows. The plating on the sides of the rudder is $\frac{7}{16}$ inch thick, and the inside of the rudder is filled with a mixture of cork cuttings and Hay's glue, which is used in order to keep out the water. The rudder-heel is steadied by a pintle 8 inches in diameter, having a large nut hove up on its lower end below the keel-plates. The rudder-head is a solid forging of circular section of which the diameter is 12 inches. Its lower end steps into the socket formed by the two foremost vertical frames in the rudder and the side-plating, and is run down about 4 feet 6 inches into the rudder. The upper part of the pintle-forging at the rudder-heel also extends up into the socket about 2 feet, and it will be seen from the side view of the rudder in Fig. 87, that the forgings forming the pintle and rudder-head are both secured to the forged rudder-frame by shoulders having a hook-scarph at each end. The siding of the rudder is tapered considerably before the axis, as will be seen from the section, and thus its retarding effect on the ship is considerably reduced. It may be added that in this case, as in all modern balanced rudders, the proportion of the area of the rudder before the axis to that abaft it is about 1 to 2. One other feature requiring notice is the casting fitted between the upper edge of the rudder and the counter, in order to prevent the rudder from being driven upward by striking the ground, and so injuring the inboard arrangements. This casting is shown in Fig. 87. In order to allow the rudder to be unshipped a transverse slot is cut in the casting, the width of the slot being a little greater than the siding of the rudder. When the rudder is placed in an athwartship position, it can be lifted up into this slot and the heel-pintle withdrawn from the socket in the keel-plates. In order to allow the rudder-heel to be carried aft to clear the keel-plates the fore side of the casing of the rudder-hole is inclined as shown in the sketch, and when the rudder is in place a metal casing is put into the rudder-hole which just allows the rudder-head room to work. At the upper end of the casing a stuffing-box arrangement is fitted to prevent the passage of water inboard. It will be remarked, from the sketch of the stern in Fig. 87, that the upper end of the rudder-head is steadied by an angle-iron socket worked underneath the main deck, and that a scuttle is made in the deck to allow the rudder-head to pass up when the rudder is being shipped or unshipped.

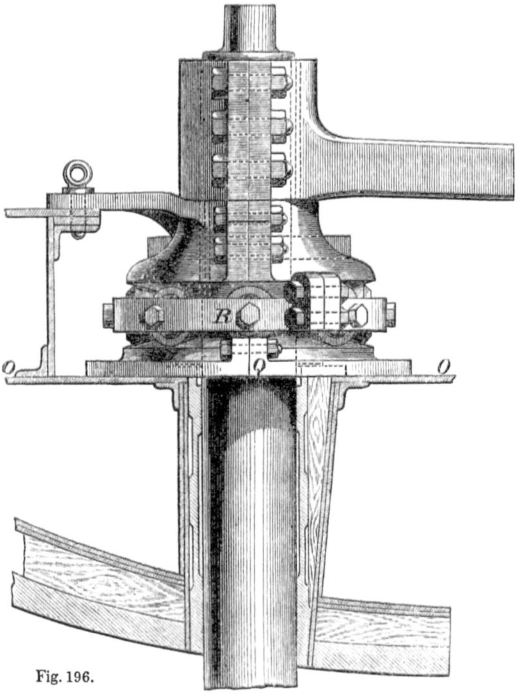

Fig. 196.

We now come to notice the arrangements made for supporting and working the rudder. As previously stated, the weight of the rudder is taken inboard, and the pintle at the heel is merely intended to steady it. The four aftermost vertical frames support a horizontal platform, marked O O in Fig. 87, on which the weight of the rudder is taken. Upon this platform is fixed an arrangement of friction rollers which is shown in the sketch just referred to, and is illustrated on a larger scale in Figs. 196 and 197. From the elevation of the rudder-head given in Fig. 196 it will be seen that a circular forging, Q, is secured to the platform O O, and forms a table, upon the upper bevilled surface of which the friction rollers rest. These rollers are of brass and are conical frusta in shape, being secured in the cone band R as shown in Fig. 197. The rudder-head is of uniform diameter as far up as the upper side of R, but from that point, and throughout the depth of the forging S, the diameter is reduced to 11 inches, thus forming a shoulder of $\frac{1}{2}$ inch at the upper and lower edges of S. By means of this shoulder all the weight of the rudder is transmitted by S to the friction rollers underneath it, so that the working of the rudder is rendered extremely easy. It may be of interest to state that this arrangement has stood the test of actual employment most satisfactorily, and that the officers who have had opportunities of observing it in work have all expressed their entire approval of it. The sketch in Fig. 196 shows the manner in which the rudder can be locked at any angle which may be desired. The after part of the table S is formed in such a manner as to overlap the fore edge of the locking plate, and the usual form of locking pin is employed to keep the rudder fixed. In the sketch one of these pins is shown in position.

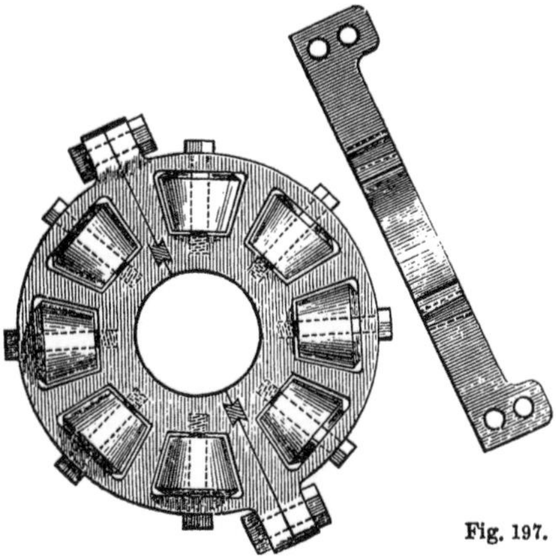

Fig. 197.

In designing the rudder of the 'Hercules' provision has been made to allow it to be used as a balanced rudder when under steam, and to lock the part of the rudder before the axis and steer only with the part abaft the axis, if it should be thought necessary, when the ship is under sail. The change from the preceding arrangement has been made on account of the opinion expressed by the officers who have commanded the 'Bellerophon,' that the great area of the balanced rudder, while adding very greatly to the ship's manœuvring power under steam, tends to destroy the ship's way when she is tacking, and causes her occasionally to miss stays. The details of the arrangements of the rudder of the 'Hercules' are given in side view and sections in Fig. 198. The two parts of the rudder before and abaft the axis are separately built, and are connected by pintles and braces, the braces being forged on the after edge of the fore part and the pintles on the fore edge of the after part. From the sketches it will be remarked that the fore part has its frame formed in one forging, while the after part has its frame made up of a forging and a bent plate, with two vertical frames intermediate between the fore and after edges. The after part is sided uniformly throughout, but the fore part is tapered in siding, as shown in the sections, the reason for this difference being the same as has been given for a similar arrangement in the

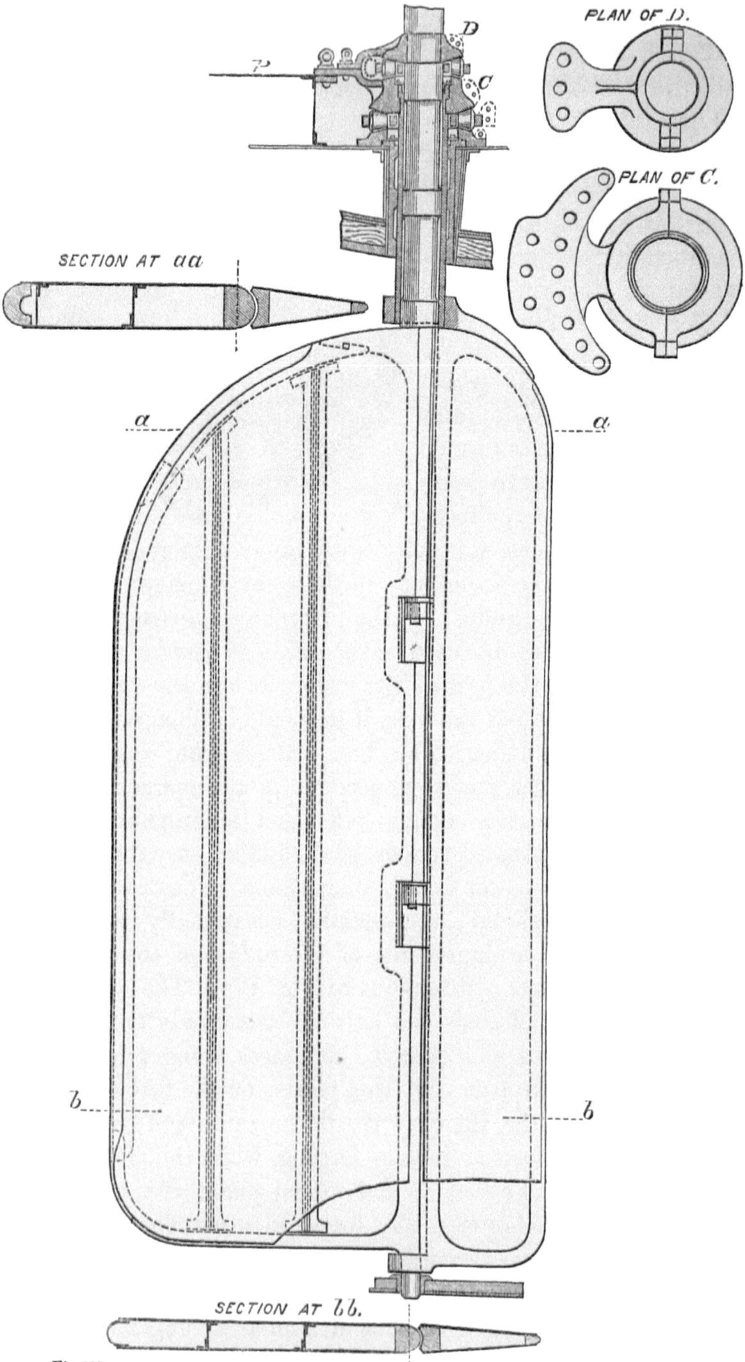

Fig 198.

'Bellerophon.' The heel of the rudder is steadied by the after end of the keel piece, and the lower pintle on the after piece of the rudder passes through a brace welded to the frame of the fore piece, and through a hole in the keel piece. The after piece of the rudder is attached to a solid iron rudder head, very similar in form to that of the 'Bellerophon,' its lower end being connected with the forged frame by hook-scarphs, as shown in the side view in Fig. 198. The fore piece of the rudder is attached to a tubular rudder-head at the lower end of which there is formed an arm that clasps the upper part of the frame. The solid rudder-head attached to the after piece passes up through the tubular rudder-head, and its upper end is steadied by a socket, worked below the main deck, similar to that fitted in the 'Bellerophon.' The side view of the rudder, &c., shows how the weight of the rudder is taken inboard on a platform built on the stern frames, and that in this ship there are two sets of friction rollers. The table marked C which rests upon the lower set of conical rollers is attached to the tubular rudder-head, and its after part is arranged to receive the locking-pins which also pass through the fixed locking-plate P at the stern of the ship. A plan of the table C is given in Fig. 198, and the aftermost row of holes there shown is that which these pins pass through in order to lock the fore part of the rudder in any position that may be desired. The second system of rollers rests upon the table C, and supports a table D which is connected with the solid rudder-head. A plan of D is also given, and it will be seen that its after part is arranged so that it can be locked to the foremost row of holes in C. In the side view the two locking-pins are shown in the positions they occupy when C is locked to D, and D is locked to the fixed locking-plate. From the preceding description it will be obvious that when it is desired to use the rudder as a simple balanced rudder the two parts can be made to move together by locking the tables C and D to each other, and when the rudder has been put over to the desired angle it can be fixed there by putting in the locking-pins in the after row of holes in C. If, on the other hand, it is desired to use only the after part of the rudder, the fore part can be locked in its amidship position, and the after part, being moved by the solid rudder-head which turns within the tubular head, can be locked at any angle by putting in the pins which pass through the holes in the after parts of C and D. It need only be added that provision is made in this case also to prevent the rudder from being driven upward if it strikes the ground, by means of a forging fitted between the counter and the upper part of the rudder; and that the shipping or unshipping is performed in a manner similar to that fully described for the 'Bellerophon.'

Before concluding this chapter it may be well to notice that bow-rudders have been fitted to iron ships, and are often employed in double-ended ships which are intended to run in both directions. The steam ferry boats which cross the Mersey between Liverpool and Birkenhead have a rudder at each end, and the 'Waterwitch,' which is fitted with a hydraulic propeller and is double-ended, has a similar arrangement of rudders. In these and other ships which have bow-rudders it is usual to form the bow-frame in such a manner that the rudder can be placed in a recess, and the rudder is so shaped that when locked in its amidship position it completes the form of the body both longitudinally and vertically. The rudder-head is formed by a solid forging which passes inboard, and the rudder itself is very lightly framed and plated. Proposals have been made to fit bow-rudders in very long ships in conjunction with the ordinary stern rudder, in order to decrease the time required for turning. These proposals have, however, been rejected for the twofold reasons, that great difficulty would be experienced in working a bow and a stern rudder in conjunction, and the fact that a bow-rudder is much less effective than a stern rudder, as it acts upon water moving much slower than that which is driven upon the stern rudder by the screw.

CHAPTER XIV.

IRON MASTS.

ONE of the more recent applications of iron in the equipment of ships has been its employment in the construction of masts and yards. In his book on "Iron Shipbuilding" Mr. Grantham states that an iron mast was placed in one of the City of Dublin Company's steamers about 35 years ago; but it is only within the last ten or fifteen years that iron lower masts have been constructed for large ships. At first there were great objections raised on account of the supposed danger incurred from the impossibility of cutting away an iron mast when the ship's safety seemed to demand such a measure; and it was further urged that in case of shipwreck the means of constructing a raft would be considerably smaller in a ship with iron masts than in one of which the masts were of wood. These objections have, however, been over-ruled by other and more important considerations, and at the present time the lower masts of a very large number of our iron ships, and of many wood ships, are built of iron, while in numerous instances the topmasts, topgallantmasts, and the yards are also made of iron or steel. Before passing to the illustration of the construction of iron lower masts, it may be of interest to give a brief summary of the advantages which are supposed to be gained by their employment. In a vessel of moderate size the large dimensions of a wood lower mast compel the builder to make it up of several pieces, which are coaked and bolted to each other, and bound together by numerous iron hoops. The various pieces used in a large mast are difficult to procure, and even when the greatest care is taken to secure a good combination of the different parts, it is absolutely impossible to prevent the gradual decay of the material. An iron lower mast of the same diameter would be made up of plates, each bent to form an arc of a circle (generally 120 degrees) and connected at the edges and ends by through-riveted lap-joints or covering strips, the structure being usually stiffened by continuous T or angle-irons. No difficulty is experienced in procuring the materials for the mast, however large its dimensions, and no danger is incurred from decay, as the interior of the mast can be readily got at, cleaned, and painted, while the various parts of the structure are well combined. Iron masts have usually been made of the same diameter as the wood masts they have replaced, and it appears that the strength of a well built iron mast is nearly the same as that of a tree of Riga fir of the same dimensions. Iron masts are stated by some authorities to be considerably lighter than wood masts of the same dimensions. Thus Mr. Grantham states that, for vessels of the same tonnage, the weight of the three lower masts and the bowsprit when built of iron, is only two thirds their weight when built of wood; and adds that he is confirmed in this opinion by independent calculations made by Mr. John Vernon. In the ships of the Royal Navy which have iron masts it is found that for the larger masts the weights are nearly identical with those of wooden masts of the same dimensions, but that for the smaller masts those made of iron are rather heavier than those built of wood. The explanation of the difference between Mr. Grantham's estimate and the observed weights of the iron masts in H.M. ships, is found in the fact that a very strong system of stiffeners is adopted in these masts, and consequently their weight is increased.

The relative cost of iron and wood masts is a matter requiring notice. In a paper on "Iron and other masts and spars" in the *Transactions* of the Institution of Naval Architects for 1863, Mr. Lamport states as the result of his experience that the first cost of iron lower masts is greater than that of wood masts in vessels under 700 tons, and that it is only for vessels of at least 1000 tons that the gain on iron masts is a decided one. Mr. Grantham expresses a similar opinion, and remarks that when compared with wood-built masts in large vessels iron masts are rather less expensive. Clyde shipbuilders differ considerably in their opinions on this point, some thinking with Mr. Lawrie that there is a considerable gain in the use of iron masts and spars, and others agreeing with Mr. Lamport. The iron lower masts employed in the ships of the Royal Navy are more expensive than wood masts of the same dimensions. The advantages gained in respect of strength and durability are, however, such as to outweigh any of these considerations of expense.

The Liverpool Rules give a table of dimensions for iron masts, which will be found in the Appendix, in which the diameter, thickness of plating, &c., is stated for masts of various lengths. Lloyd's Rules do not give any regulations or dimensions for iron masts. Before concluding these preliminary remarks it may be added that iron masts are usually made to serve as ventilators to the interior of the vessel, and that in order to remove the objection previously mentioned, several modes of cutting away iron masts have been proposed, and in some cases adopted. The plates used in the construction of lower masts vary from $\frac{1}{4}$ inch to $\frac{5}{8}$ inch in thickness, and their breadth is usually one-third the circumference, although in small masts plates having a breadth of half the circumference, and in large masts of one-fourth the circumference, are sometimes employed. The longitudinal seams of the plating are generally single-riveted, and the butts are double-riveted, except in wake of wedging decks where they are often treble-riveted. Internal stiffeners are fitted throughout the length of many masts, and in some cases horizontal cross-stays are also worked at intervals.

Coming now to the illustration of the various modes of forming iron masts we would first call attention to the section given in Fig. 199, in which there are four plates in the circumference, con-

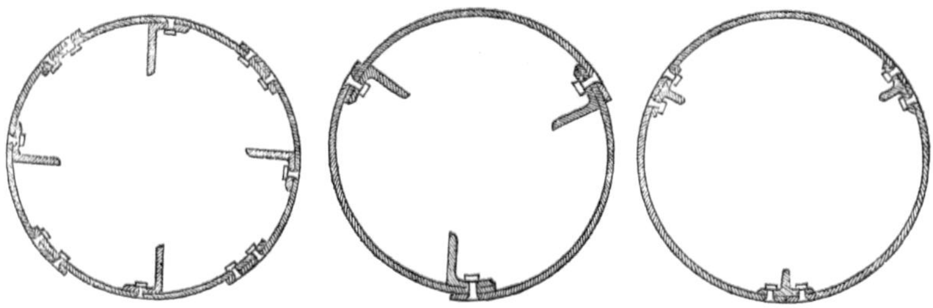

Fig. 199. Fig. 200. Fig. 201.

nected by double-riveted lap-joints, and stiffened by four continuous angle-irons worked upon the centre of each plate. Sometimes instead of having the stiffening bars as in this case of simple angle-irons, they are formed of double angle-irons set back to back, or of T-irons. In the masts of many merchant ships the stiffening bars are dispensed with, and the thickness of the plating somewhat increased in order to give the requisite strength. In other masts the angle-iron stiffeners are placed as shown in the section in Fig. 200, so that the edge riveting shall work in as fastenings in the stiffeners. In masts where the plating is worked flush at the edges it is usual to have the stiffening bars of T-iron, and to place them, as shown in section in Fig. 201, so that they shall serve as edge strips. In order to still further stiffen masts the flanges of the stiffening bars are often connected by braces or horizontal stays, formed of T or angle-iron, or of plate. In addition to adding to the strength of the mast these cross-stays also afford the means of climbing up inside for the purpose of inspecting, cleaning, and painting it. A section showing a mast constructed with T-iron stiffeners and plate stays is given in Fig. 202, which illustrates the arrangements adopted in the masts of the 'Defence' and other of the earlier iron-clads. These plate stays are placed at intervals of from 4 to 6 feet. In some masts the plate stays are arranged as shown in Fig. 203, a bent plate

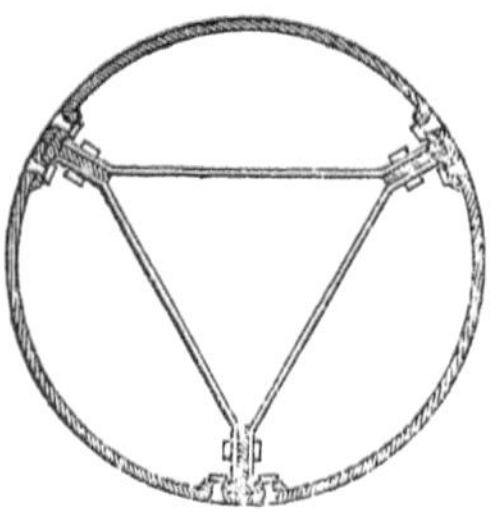

Fig. 202.

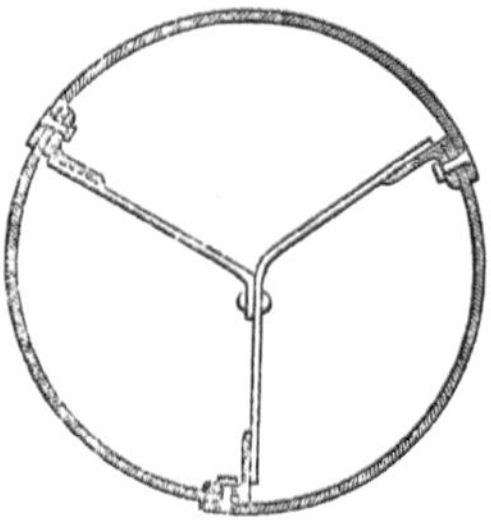

Fig. 203.

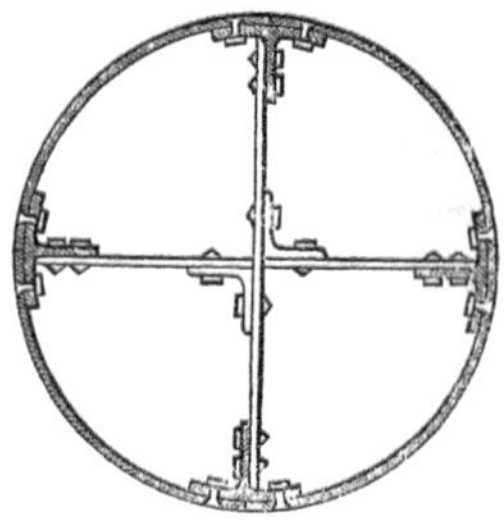

Fig. 204.

forming two parts and a separate plate the remaining part. Mr. Grantham gives a section of a mast, from which Fig. 204 is taken, and in which there are four plates in the circumference, the seams being flush-jointed and the stiffeners formed of T-bars placed as usual upon the seams. The plate cross stays in this instance are placed in diametral planes, being riveted to the flanges of the T-irons, and connected with each other at the centre of the mast by short angle-irons. In the 'Resistance' the masts were constructed as shown in section by Fig. 205. There are four plates in the circumference, the edges being flush-jointed, and covered by T-iron stiffeners. At intervals of 10 feet the stiffening arrangement shown in the sketch is fitted, consisting of a ring of angle-iron worked in short lengths between the T-iron stiffeners and connected by a horizontal stiffening plate *p*.

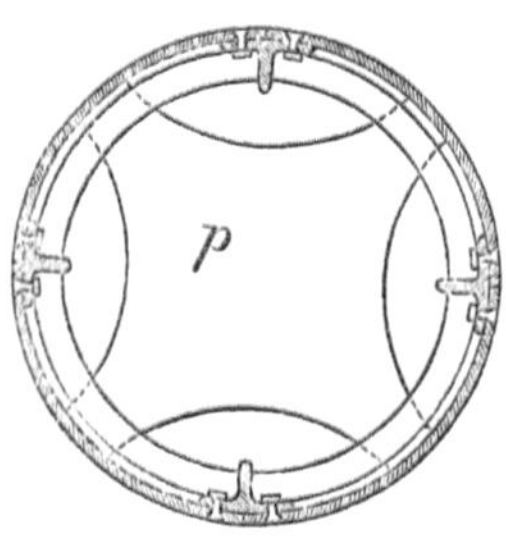

Fig. 205.

In the masts of the 'Hector' a very unusual mode of stiffening was adopted, as will be seen from the section given in Fig. 206. At intervals of 9 feet angle-iron stiffening rings are fitted within the angle-iron stiffeners and connected with them by short reversed angle-irons worked on the stiffeners. In addition to this a bolt is put in through each lap and passed through a cylindrical tube or washer fitted between the plating and the stiffening ring as shown. In the construction of the masts of the 'Bellerophon' and others of the later iron-clads the arrangements have been similar to those shown in Fig. 207. The cross-stays are formed of T-iron, and are placed at intervals of about 6 feet throughout the masts. This mode of forming and stiffening masts may be taken as the illustration of the practice of H.M. Service, and as an example of the dimensions of plates, and stiffeners which are employed, we have given the following particulars of the masts of the 'Bellerophon.' The fore-mast is 33 inchesin diameter and the main-mast 35 inches, both masts being formed of $\frac{7}{16}$ inch plates riveted to three T-irons 6 by 4 by $\frac{1}{2}$ inches, and supported by T-iron cross stays 5 by 3 by $\frac{1}{2}$ inches, at intervals of 6 feet. The mizen-mast is 24 inches in diameter, and is formed of $\frac{3}{8}$ inch plates riveted to three T-iron stiffeners 5 by 4 by $\frac{1}{2}$ inches, and supported at every 6 feet by cross stays of T-iron 5 by $2\frac{1}{2}$ by $\frac{3}{8}$ inches.

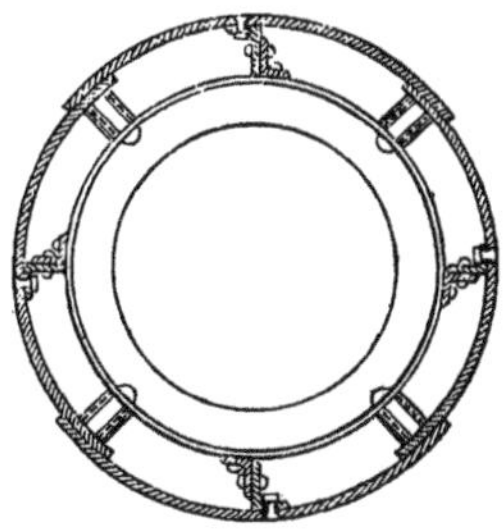

Fig. 206.

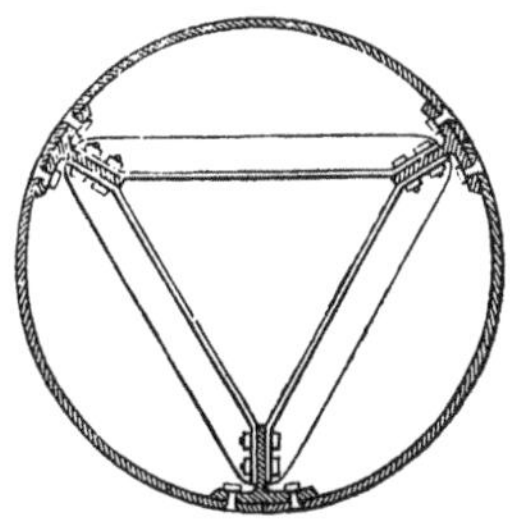

Fig. 207.

In the sea-going turret-ship 'Monarch' where the usual arrangement of shrouds is not adopted, and the masts have, in consequence, less support than is commonly the case, it has been considered desirable to increase the diameters of the masts and the dimensions of the plates and stiffeners of which they are built. In this ship the fore and main-masts are 40 inches in diameter, and are formed of $\frac{1}{2}$ inch plates riveted to three T-irons 6 by 5 by $\frac{5}{8}$ inches, and supported by cross stays of T-iron 5 by 3 by $\frac{1}{2}$ inches at intervals of 6 feet. The mizen-mast is 36 inches in diameter, and is formed of $\frac{3}{8}$ inch plates riveted to three T-irons $5\frac{3}{4}$ by $4\frac{1}{2}$ by $\frac{9}{16}$ inches and supported by cross stays of T-iron 5 by $2\frac{1}{2}$ by $\frac{3}{8}$ inches spaced as in the fore and main-masts.

The plates used in the construction of the masts of the ships of the Navy are of the "*best best*" quality, in lengths of at least 12 feet, their edges being single riveted to the T-irons, and their butts double chain riveted to covering plates worked inside. The T-bars which form the continuous stiffeners are usually welded up into one length, but when this cannot be accomplished the lengths used must be at least 24 feet and butt covers are fitted. In the latter case great care is required in arranging the shifts of butts of the T-bars and plates in the mast, and the evil effects of neglecting this precaution have been shown in some cases where spars which have been constructed with an ample weight of material have given way under severe strains, and the fractures have shown a bad disposition of the butts of plating and stiffeners to be the cause of the accident. Another very important feature in the construction of iron masts is the accurate fitting of the butts. This accuracy is required in all the specifications for iron masts built for H.M. Ships, and is aimed at by most private builders. It will be obvious that too much care cannot be taken in securing this result, when it is considered that so great a thrust is brought upon the mast by the tension of the shrouds, and that when the butts are badly fitted the strain is in some measure thrown upon the rivets, and they are consequently injured. It should be stated, however, that the masts of some merchant ships are lap-jointed at the butts, as well as at the edges of the plates. All the rivets in mast-work are countersunk on the outside of the plates, and chipped fair with the surface.

Iron lower masts are very often of uniform diameter from the heel up to the trestle-trees, but their heads have, in many cases, been made of the same external form as is commonly adopted for wood masts. Mr. Grantham gives two illustrations of this form of mast head, and in both cases the dimensions of the head are considerably less than the diameter of the lower mast from the hounds downward, while upon the shoulders thus formed the trestle-trees rest and are bolted to the head. When this form of mast-head is adopted, the lower part of the head is generally run down some distance into the body of the mast, and firmly connected with the plating and stiffeners. In more recent masts the square section of the head has been given up, and the head forms a continuation of the mast below the hounds, its diameter being slightly reduced by a gradual taper. In Fig. 208 sketches are given showing in plan and elevation the mainmast head of a sloop of war of 1100 tons. This mast is built on the plan illustrated by Fig. 207, and the T-iron stiffeners extend throughout its length. The sketches show the mode of fitting wood trestle-trees to an iron mast with this form of head. As there are no shoulders at the hounds, special provision has to be made for supporting the trestle-trees, and this is accomplished by working a plate and a ring of angle-iron around the mast, and fitting plate-knees *k*, *k*, which correspond with the cheeks usually worked below the trestle-trees of a wood mast. The plan shows very clearly the spread of the knees and the arrangement of the plate and angle-iron below the trestle-trees.

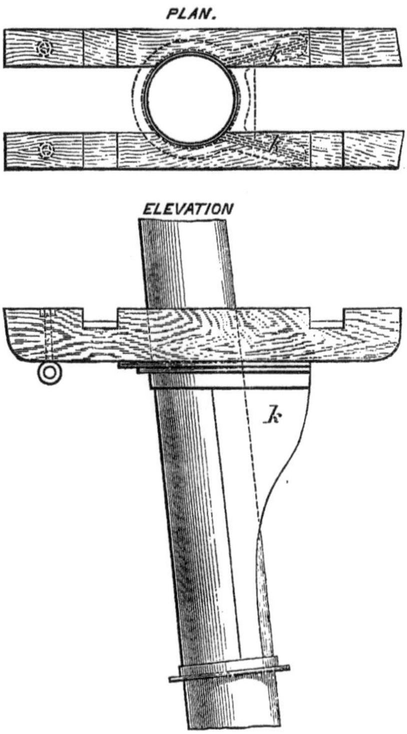

Fig. 208.

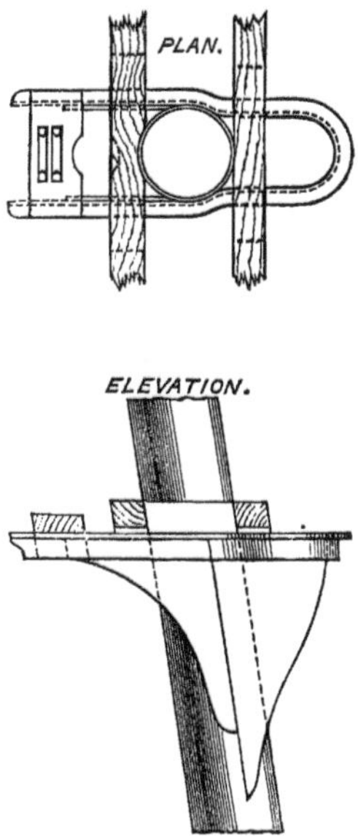

Fig. 209.

Iron trestle-trees are often fitted to iron lower masts, especially in vessels of the mercantile marine, and an illustration of such an arrangement is given in plan and elevation in Fig. 209. From the plan it will be remarked that the trestle-trees are formed of one length of angle-iron, the fore part being curved in such a manner as to fit around the heel of the topmast. The angle-iron trestle-trees are riveted to the sides of the mast, and are supported at the fore and after sides of the mast by plate-knees fitted as shown in the elevation. The wooden cross-trees are in this case fitted directly upon, and bolted to the trestle-trees. In many ships the cross-trees also are of iron. It may be added that these sketches are taken from the mizenmast of the sloop of which the mainmast head is illustrated in Fig. 208, the vessel being barque rigged.

Mr. Lamport remarks with respect to mast-head arrangements that they should be such "as to ensure the pressure from the "heel of the topmast, &c., being as near the centre of the mast as "practicable, and at the same time to allow the drag of the shrouds "to be fairly carried over the plates of each side. These desiderata "are easily attained by carrying up the plate on the aft side of the "mast in a straight line to form the back of the head, riveting a "strong plate over the round of the mast at the hounds to a strong "angle-iron running around the outside, and building upon it the "remaining three sides of the square doubling or head." The section of the head which he proposes would be square on three sides, and on the aft side a circular arc forming a continuation of the curve of the mast. There would be no plate-knees underneath the plate upon which the head is built, corresponding to the cheeks fitted to wood masts.

One other mast-head fitting requires notice, namely, the manner in which the cover is formed which admits of ventilation, but prevents the entrance of wet. Sketches showing the details of one of these covers are given in plan and vertical section in Fig. 210. It will be seen that a ring of angle-iron is worked inside the mast at the upper edge of the plating, and upon the rim thus formed the angle-iron *r* is riveted. The cover is formed of a circular plate with a ring of deep flanged angle-iron riveted to the edge. When in position the cover is supported by the radial frame marked *f* in the plan, the vertical sides of which are secured to the angle-iron *r*. From the vertical section it will be evident that the cover thus formed and supported fulfils the conditions for which it was designed.

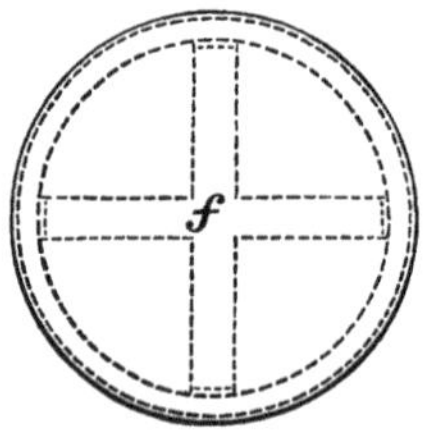

SECTION.

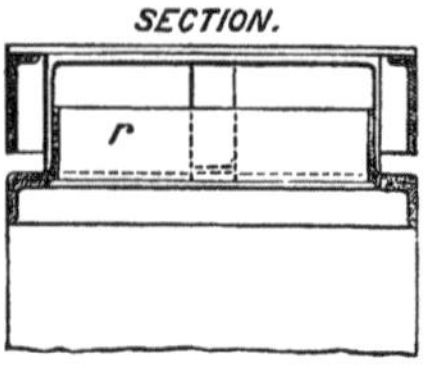

Fig. 210.

The various fittings of an iron lower mast (sling and stay cleats, eyes for braces, halyards, stays, &c.) are of a similar character to those of a wooden mast of the same dimensions, the only differences being due to the fact that the means of attaching the eyes, &c., have to be varied on account of the different material in the mast. Iron caps are now universally fitted to iron lower masts and are secured by being shrunk on. It does not fall within the compass of this work to give the details or dimensions of the various fittings, information with respect to which will be found in works on masting.

Incidental reference has been made to the usual practice of working doubling plates upon the masts in wake of the wedging decks. These plates serve the double purpose of giving additional rigidity in wake of the wedges, and preventing corrosion in the mast-plating itself. The latter feature is especially important on the upper deck, where, most probably, there will always be some moisture which has passed down between or through the wedges, and tends to cause more or less rapid corrosion. By means of the doubling plates the strength of the mast is not affected by the corrosion, and new doubling plates can be readily worked if required. We shall describe the details of the framing of mast holes in the following chapter.

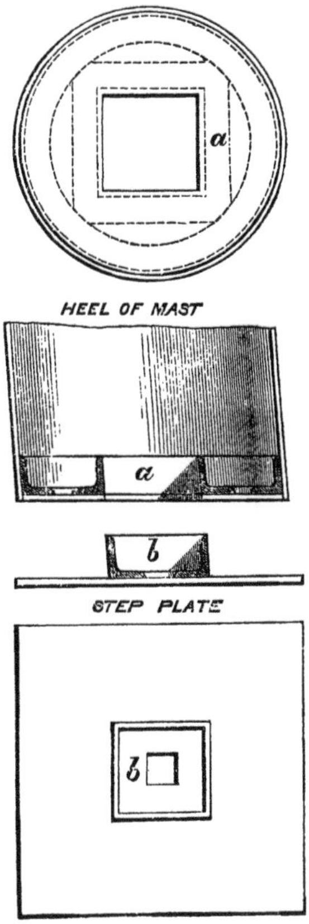

Fig. 211.

The modes of forming the heels of iron lower masts require a brief description. At first the heels were constructed of the same external form as the heels of wood masts, rectangular tenons being formed upon them and fitted into mortices in the top of the mast-steps. This arrangement effectually prevented the masts from rotating when the yards were swung round and the ship under sail, in exactly the same manner as had been done previously for wood masts. Now, however, the heels of iron masts are usually formed in an entirely different way, and in order to illustrate a very common arrangement the sketches in Fig. 211 are given. It will be seen from the section and plan that the end of the mast is closed by a circular plate fitted against, and connected with its outside plating. In the centre of this plate there is a square hole around which the angle-iron frame *a* is fitted, the vertical flange of the angle-iron thus forming the sides of a mortice in the heel. When in place the mast heel rests on a step-plate upon which is riveted a rectangular box-shaped frame of angle-iron *b*, and the tenon thus formed fits into the mortice in the heel of the mast. The plan and section of the step-plate given in Fig. 211 will illustrate these remarks. In many cases the preceding arrangement is modified, by working a circular rising ledge or ring of angle-iron upon the step-plate, and fitting it around the heel. An illustration of this mode of fitting is given in section and plan in Fig. 212. In the sketch the rising ledge is marked *c*, and the rectangular angle-iron tenon on the step-plate is marked *b*. The remainder of the heel-fittings are similar to those described above. In some ships instead of fitting the circular rising ledge close to the mast, there is a small space left between them into which wooden wedges are driven to keep the heel steady. In other ships the angle-iron tenon on the step-plate is dispensed with, and the mast is kept from rotating by being bolted to the vertical flange of the rising ledge. A man hole is usually cut at a few feet from the lower end of an iron mast in order to give access to the interior and to admit of ventilation; other openings are also made at various heights for the latter purpose.

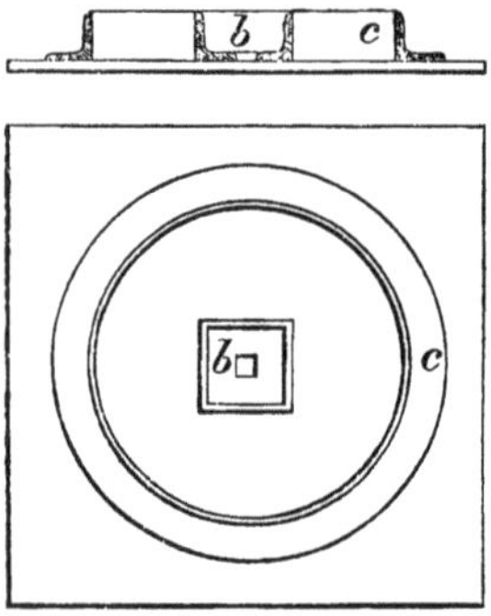

Fig. 212.

Before concluding these remarks on the heels of iron masts it may be well to call attention to a proposal made by Mr. Lamport in the Paper previously referred to. That gentleman considers that the usual plan of wedging should be abandoned, and that while the lower-mast, topmast, and topgallantmast should all be rigid in themselves they should yield with an articulated flexure to the elastic spring of the shrouds and stays. In order to combine rigidity with the power of yielding to the play of the rigging he proposes that each mast should oscillate from the heel, and to effect this he would apply "a cast-iron foot terminating in a ball a "little flattened in the fore and aft direction to prevent the mast "turning, and widening above to give a flat, even, but moveable "support to the plates of the hollow mast." He meets the objection which might be raised to the use of a ball-ended mast, by the statement that when the direction of the strain varies, as in a swaying mast, and the mast oscillates freely from the heel, the arrangement he proposes will have the same effect as a flat end has in the case of a fixed pillar and a strain of which the direction is constant, and will keep the strain in a line with the metal.

It was remarked at the commencement of this chapter that various plans had been proposed and adopted for letting iron masts go overboard (or cutting them away) if the safety of the ship required the loss of a mast. Messrs. Finch and Heath have proposed a *parting-joint* for iron or steel masts which is illustrated by Fig. 213.* The upper and lower parts of the mast are each strengthened at their junction by a flanged collar, the horizontal flanges of which are accurately fitted to each other. The ring marked R is grooved internally, and when in position (as is the case with one-half only in the sketch) this groove clasps the horizontal flanges of the collar. When both halves of the ring R are in position they are held together by the screw bolt B and the parting-joint is at least as strong as any other part of the mast. When the bolt is unscrewed the ring R falls off, and the mast goes overboard, while the deck remains uninjured. In the 'Aërolite' Mr. Lamport applied a cutting-away arrangement which he thus describes:—"The masts were in two lengths. "The lower piece reached about 2 feet above the deck and had "strong angle-iron riveted round, in the horizontal rim of which "oblong T-shaped slots were cut every 3 inches of the circum-"ference. The upper length had a similar angle-iron affixed, "holding T-ended bolts corresponding to and fitting into the

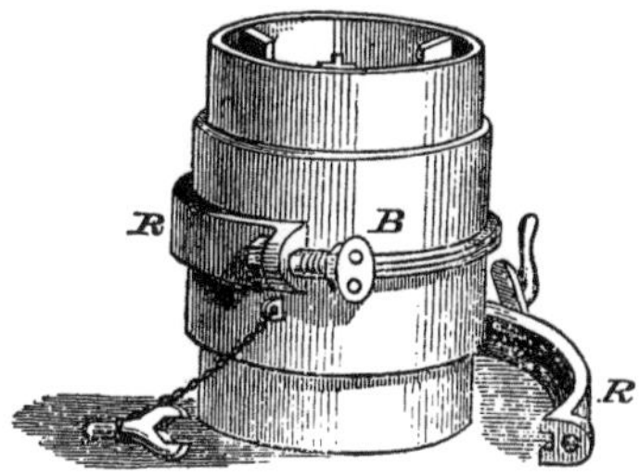

Fig. 213.

* This sketch is taken from 'Shipbuilding Theoretical and Practical,' edited by Dr. Rankine.

"slots. Below the angle-iron of the lower portion of the mast "was a loose flat ring with T-shaped slots through which the "bolts also passed. When therefore the ring was moved each "bolt was simultaneously locked. The ring was at once held "firmly in its place and turned when required by a horizontal "worm working into three or four teeth cut into the rim. A "few turns locked or unlocked the T-ended bolts at will, and "kept the mast firm, or severed it, as required. The ring should "be of brass to prevent the bolts and screw rusting from disuse." Mr. Lamport adds that a simpler arrangement, and one less liable to get out of order, would be to have broad angle-iron flanges on the two portions of the mast, with long pins secured to the upper flange and passing through holes in the lower, the pins being kept in place by a few lockings driven through the lower ends, or by a simple lashing. Mr. Roberts of Manchester proposed and patented a novel method of securing iron masts, in which the lower or housed part of the mast is built separately and is riveted to the upper part of a transverse bulkhead that passes up through the centre of the mast. The upper part of this fixed heel projects a little above the upper deck, and the lower end of the upper portion of the mast steps over, and is bolted to it with a double row of screw bolts. The inventor considers that by this arrangement the pressure of the mast will be distributed over the sides and bottom of the vessel, instead of being localized as he supposes it to be with the ordinary mast step. It is also thought that when the screw bolts connecting the lower end of the mast with the fixed heel are withdrawn the mast will go overboard, thus rendering the fitting of a parting-joint unnecessary.

Iron and steel have also been applied to the construction of topmasts, topgallantmasts, and yards, but in these spars the advantages resulting from the change from wood are not so great as in the case of lower masts, and consequently the employment has not been as general. The details of the construction of these various spars do not differ greatly in principle from those previously described for lower masts. The plating is usually flush-jointed and angle-iron or other stiffeners are generally fitted in the larger spars.

The following account of the weights of the masts and fittings of the sloop of war from which Figs. 208 and 209 are taken may be of interest. The foremast and the mainmast are 22 inches in diameter, and the extreme lengths are respectively 75 feet 6 inches and 80 feet 10 inches; the mizenmast is 15 inches in diameter and 62 feet 1 inch in length.

AN ACCOUNT SHOWING THE WEIGHTS OF MASTS AND FITTINGS FOR A BRITISH SLOOP OF WAR OF 1100 TONS BARQUE-RIGGED.

Description.	Foremast.				Mainmast.				Mizenmast.			
	tons.	cwts.	qrs.	lbs.	tons.	cwts.	qrs.	lbs.	tons.	cwts.	qrs.	lbs.
Shell of masts	5	11	0	1	5	19	3	25	2	9	0	16
Caps	0	6	1	6	0	6	1	6	0	2	2	11
Knees and angle-irons	0	5	0	12	0	5	0	12	0	3	0	3
Head hoops	0	2	2	18	0	1	2	6	0	1	0	14
Angle-iron necklace-hoop	0	0	2	4	0	0	2	4		..		
Awning hoops	0	0	2	20	0	0	2	21	0	0	2	10
Trysail hoops	0	0	3	25	0	0	3	24		..		
Belly-stay hoop	0	1	0	14		..				..		
Boat-davit hoop		..				..			0	0	2	15
Topmast fairlead hoop		..				..			0	0	2	27
Main topmast brace		..				..			0	0	2	12
Spider hoop		..				..			0	0	2	11
Brace-eyes	0	0	2	16	0	0	2	16		..		
Eyes for topping bowsprit	0	0	1	26		..				..		
Stay and sling cleats	0	0	1	14	0	0	1	14	0	0	0	18
Boom-topping lift eyes		..				..			0	0	0	22
Ventilators	0	0	3	16	0	0	3	16	0	0	2	9
Truss cleats	0	0	1	24	0	0	1	24		..		
Stepping-plates	0	1	0	9	0	1	0	9	0	0	2	10
Paunches and battens	For the three masts								0	3	2	15
Trestle and crosstrees	0	10	3	8	0	10	3	8	0	3	1	0
Screws for paunches, battens, and trestle, and crosstrees	For the three masts								0	1	0	8
Total weights .. .. tons	7	3	0	17	7	9	1	17	3	8	2	5

CHAPTER XV.

MISCELLANEOUS DETAILS.

THIS chapter is intended to supply information with respect to the arrangements of some of the minor parts of the structure, and a few of the more important fittings, of an iron ship. It has been considered desirable to reserve the notice of these details up to this point, rather than to interfere with the description of the principal features of iron ship construction which has been given in the preceding chapters, by going into the particulars of such subjects as the framing of hatchways or mast-holes, the arrangements of paddle beams, the construction of mast-steps, thrust-bearers, &c. No particular order has been attempted in the arrangement of the following descriptions, the endeavour being, as far as possible, to supply practical information concerning the various details.

FRAMING OF MAST HOLES.

The deck framing of an iron ship in wake of a mast hole usually consists of two fore and aft carlings, placed similarly to

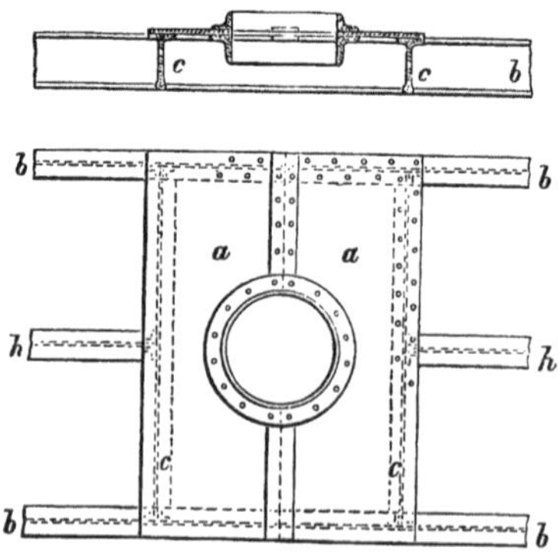

Fig. 214.

those marked *c, c* in Fig. 214, and secured at the ends to the beams *b b;* and when, as in the case illustrated by the sketch, the length of these carlings is considerable, half beams *h, h* are fitted. Fig. 214 is taken from the 'Queen,' a vessel of which mention has repeatedly been made, and shows the framing of a mast hole in plan and athwartship elevation. Upon the beams and carlings, plates, marked *a, a,* $\frac{3}{4}$-inch thick are worked, and are connected by a single riveted strip at the middle line. The mast hole is cut half out of each plate, and a circular ring or rising-ledge of angle-iron is worked around it both above and below the plating, thus forming a short tube, which gives a bearing for the wedges. The partner plates *a* are always fitted in iron ships, and are required by both Lloyd's and the Liverpool Rules, but in some cases wooden mast partners are fitted both above and below the plating, in order to give a greater depth of bearing surface for the wedges. In some ships the partner plates have been cut away in order to allow the corner chocks of the wood partners to pass down through in one length, but it is considered preferable not to cut the plates, and to fit the corner chocks in two parts. Another illustration of

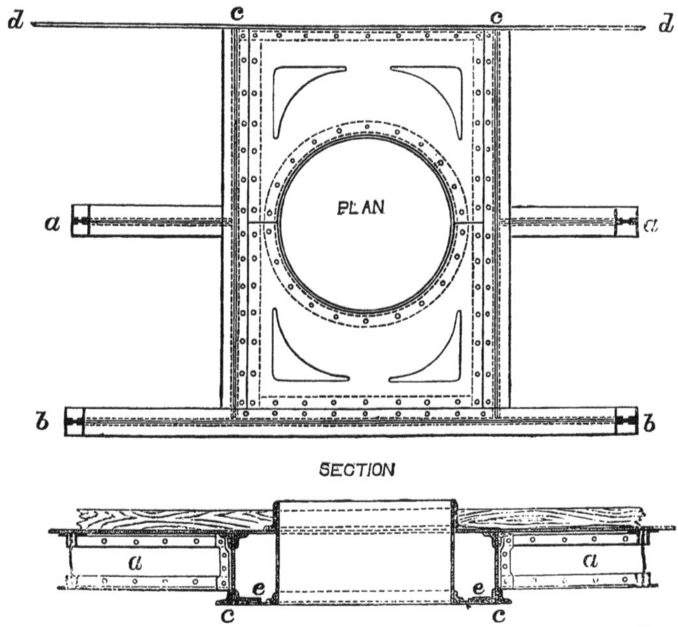

Fig. 215.

a mast hole is given in Fig. 215, which is taken from the iron-clad frigate 'Hercules,' and will serve to show the provision made for wedging large iron masts. The deck framing is of the usual character, except that the bulkhead marked *d d* in the plan occupies the place usually filled by a beam similar to *b b*. The carlings are marked *c*, *c*, and the half beams *a*, *a*; the section is taken through *a*, *a*. The mast hole is cut in the deck plating, and the lining or casing is formed of a wrought-iron tube $\frac{3}{4}$ inch thick, extending from the under side of the beams up to a sufficient height above the deck plating to be riveted to the vertical flanges of the angle-iron rising-ledge, worked around the mast hole. The plan given in the sketch is taken on the under side of the beams, and from it and the section, it will be seen that the lower end of the mast-hole casing is connected with a lightened plate, worked flush with the underside of the beams and fitted into the space enclosed by the deck framing and the bulkhead. This plate is connected with the horizontal flanges of the beam angle-irons by means of the strips marked *e*, *e* in the section; all the riveting in this plate has to be tapped. In some cases where very great rigidity is required the space between the tube and the mast framing is filled in with wood chocks. This is done in cases where the mast cannot be run down into the hold and is stepped on the lower deck; the upper deck wedging in such cases is obviously very important, and the mast hole needs to be made as rigid as possible. It may be observed in conclusion, that in both the arrangements given in Figs. 214 and 215, the deck plating and planking can be efficiently caulked, which is a matter requiring considerable attention in the arrangement of any deck fittings.

FRAMING OF HATCHWAYS.

The usual practice of iron shipbuilders is to fit wood coamings and headledges to all except the principal hatches, such as those over engines, boilers, and cargo holds. Iron carlings are generally fitted below the coamings, but as the upper flanges of the beams and carlings are too narrow to receive the bolts securing the headledges and coamings, and those which form the fastenings in the butts and edges of the deck planking, it is usual to work a broad plate and angle-iron in the manner shown in section in Fig. 216. The fastenings in the planks are brought clear of the beams and carlings, and the angle-iron *a* resists the pressure brought upon the coamings and headledges by caulking the deck planking, thus supplying the place of the dowels usually fitted when the framing is carried on wood carlings and beams. In the 'Serapis,' one of the new Indian troop-ships, instead of working a plate and angle-iron, a broad flanged angle-iron (8 by $3\frac{1}{2}$ by $\frac{1}{2}$ inches) is fitted around the hatchways, as shown in section in Fig. 217. By this arrangement the

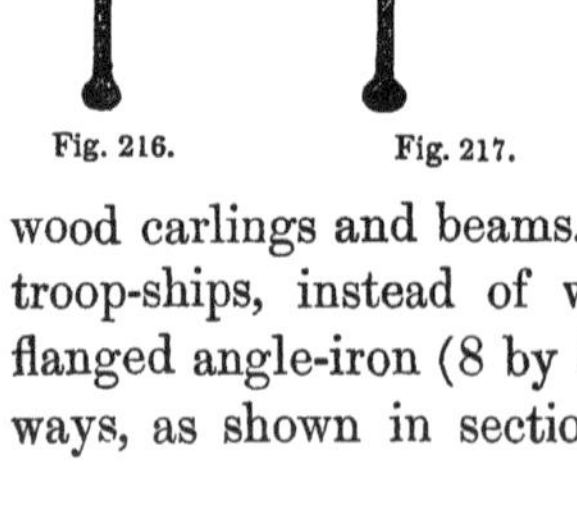

Fig. 216. Fig. 217.

butts of the planking are fastened, and the coamings and headledges bolted and supported as in the preceding case. When iron decks are worked upon the beams there is, of course, no necessity for fitting the plate and angle-iron, or the simple angle-iron arrangement just described. The section in Fig. 218 will illustrate the usual plan, and it will be seen that an angle-iron is fitted on the inside of the framing in order to resist the caulking.

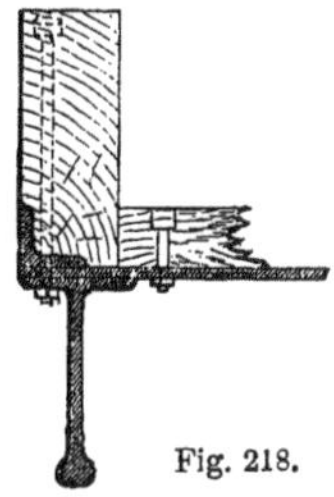

Fig. 218.

Iron coamings are generally fitted to the principal hatchways, and the sketch in Fig. 219 will illustrate an arrangement of this kind. It is taken from the boiler-hatch of the 'Hercules' on the main deck, and represents an athwartship elevation of the fore side of the

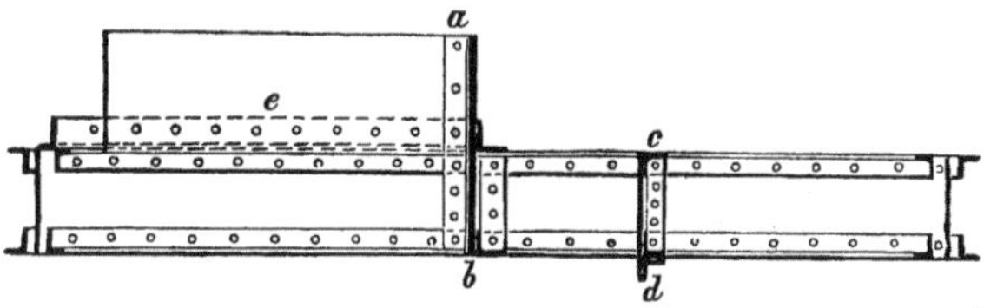

Fig. 219.

hatchway. The deep longitudinal plate *a b* is made to serve the purpose of the carling and coaming usually fitted, and the headledge *e* is also of plate, the two being connected at the angle of the hatchway by a vertical angle-iron as shown. In most instances half-round iron is riveted to the upper edges of the plate coamings and headledges. The inner ends of the half-beams forming the deck framing in wake of the hatchway are connected with the plate coaming *a b*, and a frame of angle-iron is worked around the hatchway upon the beams and half-beams, in order to connect the coamings and headledges with the beam angle-irons. The vertical stringer marked *c d* in the sketch, is fitted in order to secure the upper part of the coal bunker bulkhead. It is formed of intercostal plates fitted between the beams and half-beams, and secured to them by staple angle-irons, the lower edges of the plates being carried down sufficiently below the beams to be riveted to short angle-iron straps, which connect the different lengths together. It will be obvious that this vertical stringer also serves to keep up the strength of the deck in wake of the hatchway.

On the upper deck of an iron ship it is often considered desirable to arrange the framing of the principal hatchways in such a manner as to allow the openings in the deck to be partially closed after the engines, boilers, or cargoes, have been put in. In such cases the permanent deck framing around the hatchway is usually made up of a long carling on each side extending throughout the length of the hatch, the carlings being in most instances of the same form and dimensions as the deck beams. The inner ends of the half-beams are connected with these carlings by single or double angle-irons, usually the latter. What may be termed the portable part of the deck framing in wake of the hatchway, consists in most ships of plate coamings, similar to *a b* in Fig. 219, which are placed at a few feet from the fixed carlings, and of light transverse carlings placed so as to form continuations of the half-beams, and connected at the ends with the fixed carlings and the plate coaming. By this means the breadth of the opening in the deck is much diminished, and in some cases a similar plan is followed in order to reduce the length also by means of plate headledges and light fore and aft carlings. This portable framing is generally fastened with nut and screw bolts at such of the hatchways as frequently require to be opened.

BITTS AROUND MASTS.

The sketches in Fig. 220 illustrate a common method of securing the heels of bitts to iron beams. It will be seen from the side

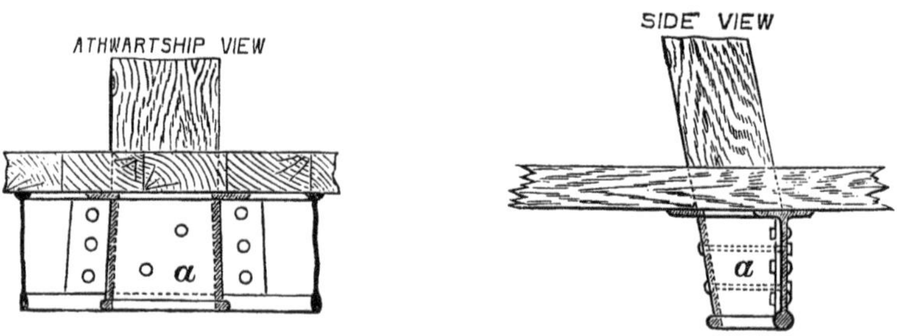

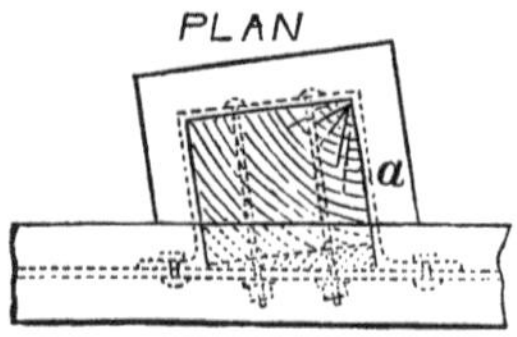

Fig. 220.

view of the bitt that its heel is faced in over the beam flange, and one edge is fitted against the web. In order to keep the bitt in its proper position, a triangular-shaped chock is fitted between the heel and the beam, as shown in the plan. The flanged plate *a*, which embraces the heel of the bitt, and is riveted to the beams, is slightly dovetailed as shown in the athwartship view, and this dovetail, together with the facing on of the heel, prevents the bitt from being drawn upward. The heel is further fastened by two through screw-bolts, as shown in the sketches. It will also be noticed that the horizontal flanges of *a* serve to receive the fastenings of the butts and edges of the deck planking and allow them to be efficiently caulked. In some ships, instead of having a flanged plate similar to *a*, the heel of the bitt is secured by wooden chocks fitted around it, and attached to the beams by short angle-irons.

RIDING BITTS.

Messrs. Brown and Harfield's patent riding bitt is that most commonly fitted in iron ships. It consists of a hollow cylindrical casting with a broad flanged base, and is secured to the deck by through bolts, which also pass through wood chocks fitted between the beams in wake of the bitt. It is usual to fit a plate into the under side of the chocks in order to prevent the nuts on the ends of the through bolts from working up into the wood when they are hove up, and to enable all the fastenings of the bitt to assist one another in resisting any strains. This very simple arrangement is found to answer all the requirements of the ships of the mercantile marine, and of some ships of war; but in the larger vessels of the Royal Navy it has been considered necessary to make special arrangements in the construction and support of the riding bitts. The sketches in Fig. 221 give the details of a riding bitt as fitted in the 'Bellerophon,' and several other iron-clads. The bitt is formed of a cylinder of wrought-iron made up of two plates, each of which is bent to a semicircular section. The lower part of the bitt rests upon a doubling plate worked upon the deck plating, and is secured by a ring of angle-iron. In addition to the strong connection thus made between the heel of the bitt and the deck,

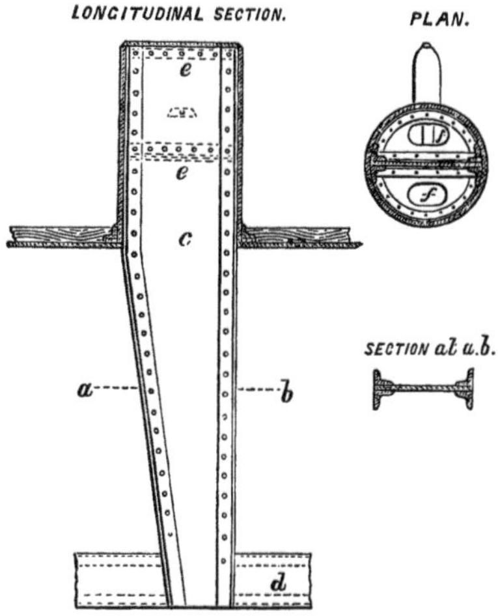

Fig. 221.

the bitt is stiffened by a vertical web *c*, formed of a plate with double angle-irons on the edges, as shown in the section at *a b*. This web extends from the top of the bitt down to the lower deck and the lower end is riveted to the carling *d*. The double angle-irons on the edges of the web plate are made to serve as edge strips to the two plates forming the bitt. The horizontal frames marked *e*, *e* are formed of semicircular plates fitted on each side of the vertical web, and attached to the plating by angle-irons worked as shown in the plan. The holes cut in these plates and marked *f*, *f* are made in order to admit of the bitt being used as a ventilator to the lower deck. It will be obvious that the frames *e*, *e*, and the angle-iron securing the heel of the bitt to the deck, form most efficient stiffeners, which resist any change of sectional form in the bitt, and the web *c* gives great strength to resist the tendency of the bitt to move forward at the head which is caused by the strain on the cables when the ship is riding at anchor.

PADDLE AND SPRING BEAMS.

Paddle-boxes are usually built upon a framing, of which the paddle-beams form the athwartship and the spring-beams the longitudinal boundaries. The paddle-beams are generally continued across the vessel, and the spring-beams lie parallel to the ship's side and are supported at their ends by the paddle-beams. In small paddle-steamers, however, the paddle-beams are frequently ended at a few feet inside the vessel, as it would be inconvenient to have a complete transverse tie. In iron ships the paddle-beams were at first of wood, and as an instance of this arrangement, we may mention the 'Dover,' in which ship the wood paddle-beams were strengthened at the outer ends by iron plates bolted to the sides. Now, however, paddle-beams are always of iron, and are usually of an I-shaped or box section, the latter being most common for the largest ships. The dimensions of paddle-beams are determined in great measure by the mode in which the paddle-shaft bearings are arranged. Common radial paddle-wheels usually have two shaft bearings, one of which is supported by the entablature of the engine-framing, and the other by the spring-beam. Feathering paddle-wheel shafts are sometimes carried on brackets secured to the ship's side, and in such cases the paddle and spring beams have only to carry the weight of the paddle-box and feathering gear, and can be made of about half the dimensions of the beams required for a radial paddle-wheel. Spring-beams are very often of wood, and in large ships are trussed considerably in order to give them the requisite strength. In large ships the spring-beams are sometimes of iron, and are either of I-shaped or box section, their depth in most cases being greatest at the middle and being gradually diminished toward the ends where it is equal to the depth of the paddle-beams.

The sketch in Fig. 222 will serve to illustrate a mode of securing the paddle-beams of a small tug-vessel. The beam is formed of a vertical plate with double angle-irons on the edges, the depth being greatest at the ship's side, and gradually reduced towards the ends. The length outboard is 8 feet 6 inches, and that inboard about 3 feet. The transverse tie is thus incomplete, but, as before remarked, it would be very in-

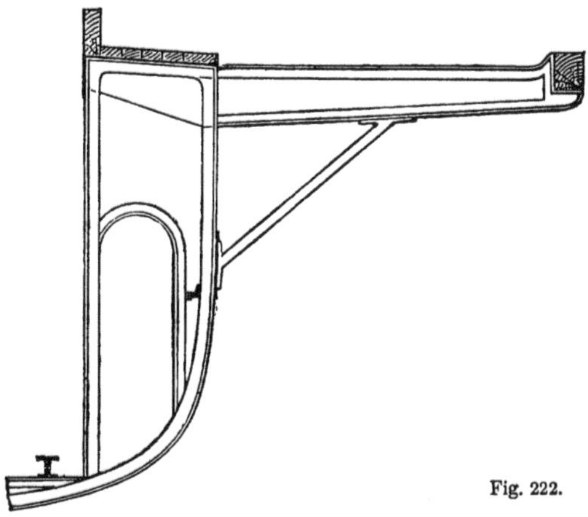

Fig. 222.

convenient to have the paddle-beam continued across the ship when the depth in hold is so very small. The inner end of the paddle-beam is secured to a partial bulkhead fitted in the coal-bunker, and the deck is plated over above it. The outer part is supported by a diagonal iron stay 3 inches in diameter (fitted as shown in the sketch), which also serves to hold down the paddle-beam in case the framing of the paddle-box should be struck by a rising sea. The spring-beam is of wood, and its ends rest upon shoulders formed in the outer ends of the paddle-beams. The two last-mentioned plans for supporting paddle-beams, and receiving the ends of spring-beams, are very commonly employed.

The second illustration of a paddle-beam, given in Fig. 223, is taken from a steamer of 1500 tons. In this case there is a complete athwartship tie at each paddle-beam; but in order to economise material, the plate which forms the beam-web inboard only extends 3 feet beyond the side, and the remaining 8 feet of the length of the arm is made up of a separate plate secured to

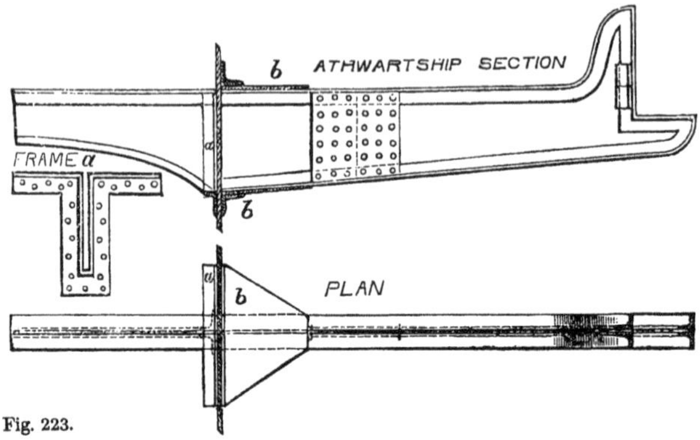

Fig. 223.

the main plate by double-butt straps, treble-chain riveted as shown in the athwartship section. The beam-arm is of I-shaped section, the web-plate being $\frac{3}{4}$-inch thick, and the double angle-irons on the edges 4 by 4 by $\frac{5}{8}$ inches. Inboard the beam is of uniform depth except for a short length close to each side, where it is slightly increased, and the outboard portion is tapered a little in depth toward the end. Where the beam passes through the side, the plating is strengthened by a frame of angle-iron marked *a*, of which a front view is given. The horizontal knee-plates, *b*, *b*, are fitted in order to increase the strength of the connection between the side and the beam, and to give lateral stiffness to the beam at this part. In this ship also the spring-beam is of wood, and the shoulder for the reception of its end is formed as in the preceding case. Before passing on to consider a third mode of fitting paddle-beams, it may be added that where box-beams are employed and are continued out through the ship's side, as is generally done in large vessels, the plating is stiffened by an angle-iron frame, corresponding to that marked *a* in Fig. 223, which is fitted around and riveted to the beam-plates.

The last illustration of a paddle-beam which will be given is taken from a larger vessel, and differs from the preceding cases both in its construction and connections. In Fig. 224 the details of the arrangement are given, and it will be obvious from a study of the sketches that by this method of fitting a very strong transverse tie is secured, and the ship's framing is specially strengthened to support the strains brought upon it. The section of the paddle-beam is marked *b* in the sketch, and it will be observed that the ordinary I-shaped section, formed of a plate with double angle-irons on both edges, is modified by adding a horizontal plate to both the upper and lower flanges. At the middle line of the ship the beam is 15 inches deep, the web-plate is $\frac{1}{2}$ inch thick, the double angle-irons on both edges are 4 by 4 by $\frac{1}{2}$ inches, the upper flange-plate is 12 by $\frac{3}{8}$ inches, and the lower $8\frac{1}{2}$ by $\frac{3}{8}$ inches.

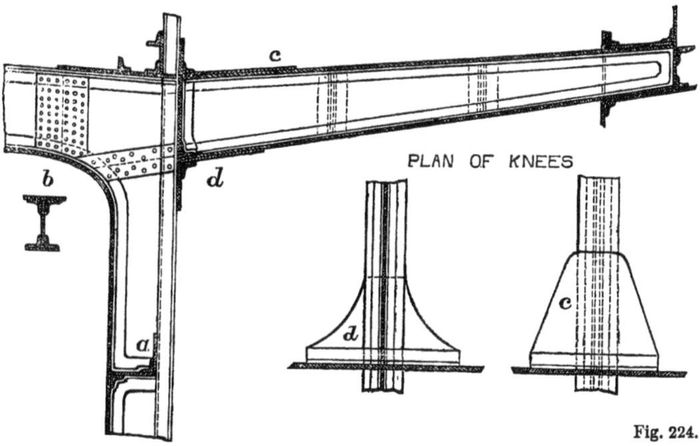

Fig. 224.

The web of the beam-arm is formed of a separate plate of which the thickness is $\frac{5}{8}$ inch full, the double angle-irons on the edges are of the same dimensions as those inboard, but the flange-plates are increased in thickness to $\frac{1}{2}$ inch. It will be remarked from the athwartship view that the butt of the web-plate in the beam-arm with the main web-plate is placed 3 feet inboard, and is secured by means of double butt-straps treble chain-riveted. A very important feature of the arrangement which requires notice is the addition of a plate-frame 20 inches deep to the ordinary angle-iron frame upon which the paddle-beam comes, thus converting it into a partial bulkhead, to the upper part of which the lower edge of the beam is connected by double-riveted strips. It will be noticed that the double angle-irons on the upper edge of the inboard portion of the beam are continued out upon the beam-arm, and that the double angle-irons and flange-plate on the lower edge of the beam are turned down on the inside of the 20-inch frame. The strong connections thus made are strengthened by the addition of the horizontal knee-plates *c* and *d* fitted above and below the beam-arm at the side. The details of these knees are shown in the athwartship view and the plans. It need only be added with respect to these connections that a very efficient arrangement of upper-deck stringer and clamp plates is made, and that the stringer marked *a* in the sketch forms a continuation of the lower-deck stringer through the engine-room. The spring-beam in this ship is formed of iron, and is of a box section as shown in the athwartship view. The side plates are $\frac{1}{2}$ inch thick, the top $\frac{3}{8}$ inch, and the bottom $\frac{7}{16}$ inch, and it will be remarked that the angle-irons connecting the various plates are all placed outside the box, and can consequently be much more conveniently riveted than if they were placed inside. The two angle-irons on the outer end of the beam-arm serve to receive the fastenings of the wood rubbing-piece worked to protect the outside of the paddle-box.

Before and abaft the paddle-boxes light platforms or wings are usually fitted, of which the plan is nearly a triangle, and are supported by light brackets formed of plate and angle-iron and placed at intervals of about 3 feet. The outer edges of the wings are fitted with rubbing-pieces or fenders which form continuations of the spring-beam and are scarphed to it, their foremost and aftermost ends being secured to the ship's side. Diagonal iron stays are often fitted underneath the wings, in a manner similar to that shown in Fig. 222, and act as ties or struts according to circumstances. In small vessels the only framing of the wings is often formed by the paddle-beams, and the fenders on the outer edges.

WATERTIGHT SCUTTLES.

When watertight flats are fitted below the lower deck in order to divide the extremities of the ship into separate compartments it is necessary to provide means of access to the spaces thus enclosed. For this purpose watertight scuttles are fitted in some ships, and in others Mr. Lungley's plan is adopted, and watertight trunks are fitted around the openings and extended up to the main deck. It will be remembered that Mr. Lungley's plan has been carried out in the 'Bellerophon' and other iron-clad frigates; but in the earlier iron-clads watertight scuttles are fitted, and an illustration of their arrangements is given in Fig. 225. The sketches are taken from the 'Defence,' and consist of a plan and two sections. The framing of the scuttle is formed of $3\frac{1}{2}$ by $2\frac{1}{2}$ by $\frac{3}{8}$ inches angle-iron riveted to the plating of the flat, and having a 2 by 2 by $\frac{5}{16}$ inches angle-iron on the upper edge. The flap-cover to the scuttle is hinged to this frame, an additional plate and angle-iron being worked upon the upper edge of the 3-inch angle-iron in wake of the hinges in order to receive the fastenings. When closed, the cover rests upon brass bearings which are shown in the sections at *a b* and *c d*, and is pressed upon the bearings by the catches or buttons *f*,*f*, which also prevent the cover from being lifted in case the compartment below the flat is filled.

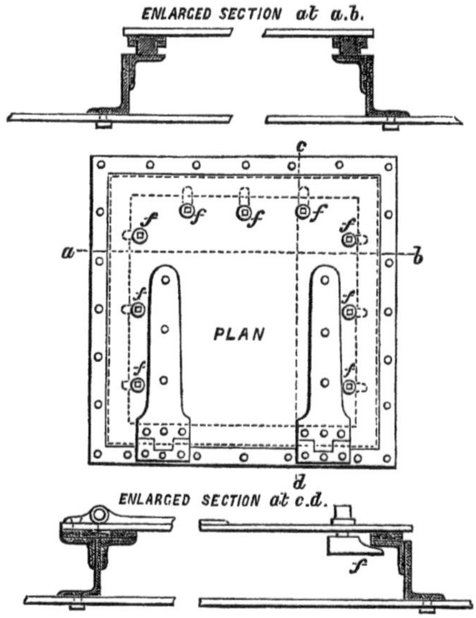

Fig. 225.

In some ships the framing of the scuttle is formed of a solid forging with a rabbet in the edge into which the cover fits, and is prevented from rising under pressure from beneath by numerous catches, which are bolted to the frame and are turned in over the cover when it has been closed.

In parts of an iron ship where it is not often necessary to obtain admission to spaces below watertight flats (as for instance the lower parts of the wing passages of iron-clads), it is usual to have man-holes instead of scuttles, and to fit simple plate-covers. These covers are of rather larger dimensions than the man-holes, and are secured directly upon the plating of the flat by screw-bolts, a thick coating of red lead being interposed between the plating and the covers before the nuts are hove down in order to ensure watertightness.

MAST STEPS.

In small iron ships the mast steps are sometimes formed by fitting large wood chocks upon the floors, and bolting them to the reversed angle-irons. The step-plate upon which the mast-heel rests is bolted to the upper surface of the bed thus formed, the arrangements of the plate being, in most cases, similar to those described in Chapter XIV., and illustrated by Figs. 211

and 212, pp. 267, 268. The more common practice in vessels of the mercantile marine is to heel the mast upon a plate worked directly upon the floors, the length of the plate being such as to allow it to be riveted to three or more floors. The middle-line keelson serves to distribute the vertical thrust caused by the weight of the mast, spars, sails, and rigging, and by the tension of the shrouds, and it also resists the tendency of the floors to fold down under longitudinal strains, both of which services are very important. A very similar arrangement has been adopted in some of the mast-steps of the new Indian troop-ships, in which the step-plate has been worked directly upon the inner-skin plating, but as the frames are 3 feet 6 inches apart, it has been considered necessary to work transverse brackets formed of plate and angle-iron midway between the frames over which the step-plate extends, in order to support it efficiently.

In order to illustrate more fully the arrangements of mast-steps in a large iron ship, we have given the sketches in Figs. 226, 227, and 228, which are respectively taken from the fore, main, and mizen steps of the 'Hercules.' The step of the foremast is built upon the inner bottom, the framing consisting of three

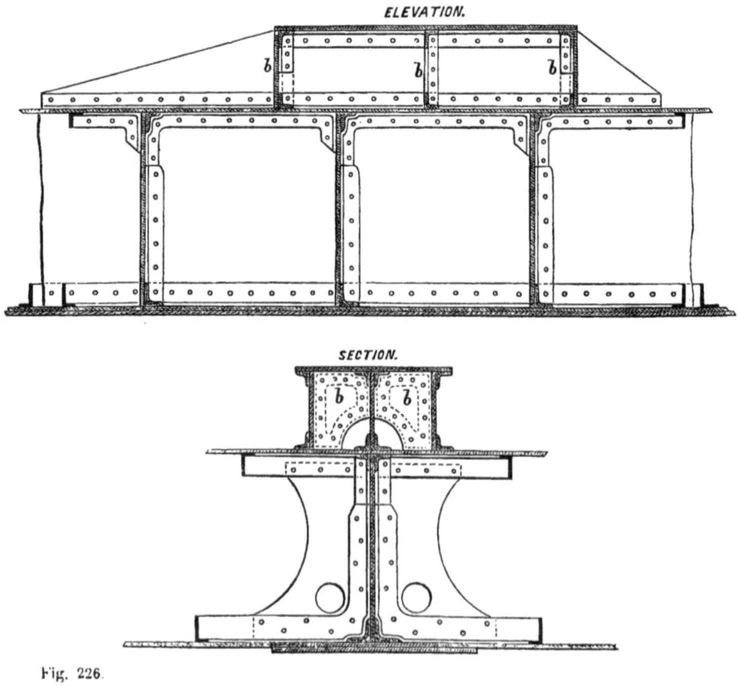

Fig. 226.

longitudinal bearers placed as shown in the section in Fig. 226, and stiffened by the transverse frames *b*, *b*, *b*. The longitudinal bearer

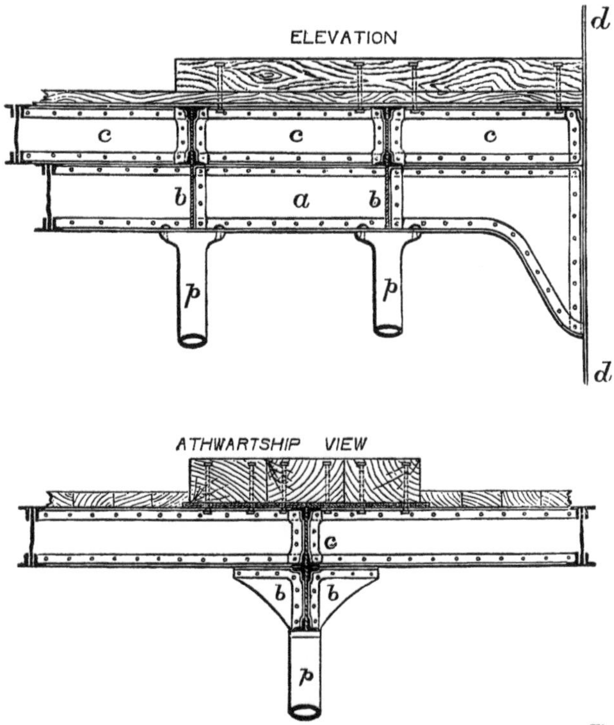

Fig. 227.

at the middle line is formed of $\frac{3}{4}$-inch plating with double angle-irons on both edges, and the other two bearers are $\frac{11}{16}$ inch thick with single angle-irons on the edges. It will be remarked in the elevation that the longitudinal bearers are continued for some distance before and abaft the horizontal platform on which the heel of the mast rests, and by this means the vertical thrust of the mast is distributed, and much more than the requisite strength is obtained to resist the small horizontal thrust due to the rake of the mast.

The mainmast step is on the lower deck, the reason for not continuing the mast down into the hold being that it is considered desirable that the passage between the engine and boiler rooms should be kept clear. The details of the step are given in elevation and athwartship view in Fig. 227. In order to strengthen the deck-framing in wake of the step, middle-line carlings, *c*, *c*, are fitted between the beams, and a deep longitudinal girder *a*, formed of $\frac{3}{4}$-inch plate with double angle-irons on both edges, is worked below the beams, and extends from the transverse bulkhead *d d* to

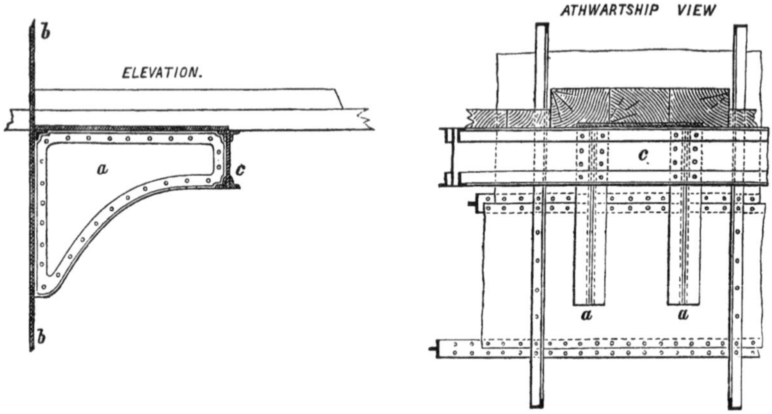

Fig. 228.

the bulkhead next abaft it. The knee-ends of the girder *a* are connected with the bulkheads by double vertical angle-irons, the double angle-irons on the upper edge are riveted to the angle-irons on the lower edges of the carlings *c*, *c* and of the beams, and at each beam the girder is stiffened by brackets marked *b*, *b* in order to enable it to transmit the load to the pillars *p*, *p*, and to prevent it buckling. The pillars are hollow cylinders of wrought iron $\frac{1}{2}$ inch thick and 7 inches external diameter, the heads and heels being welded in solid and secured to the girder *a* and the middle-line keelson respectively. A strip of plating, 5 feet wide and $\frac{3}{4}$ inch thick, is worked upon the beams and carlings and runs throughout the compartment. This completes what may be termed the framing of the mast-step, and the only other part requiring notice is the manner in which the teak chocks are fitted and bolted that form the bed to which the step-plate is secured. These chocks are 12 inches thick in the neighbourhood of the mast-step, and are screw-bolted to the deck-plating. It will be obvious that by these arrangements the vertical load is well sustained and distributed, and the small longitudinal thrust efficiently provided against.

The mizenmast step is also on the lower deck and is placed just before the transverse bulkhead shown by *b b* in the elevation in Fig. 228. The dimensions of this mast, and the vertical and longitudinal thrusts to be provided for, being so much less than for the mainmast, a very much simpler arrangement is made.

In order to support the deck in wake of the step two fore and aft bracket-knees *a, a* are worked between the bulkhead and the beam *c*. Each of the brackets is formed of $\frac{3}{4}$-inch plate with double angle-irons on the edges, and is placed at about a foot from the middle line. Upon the brackets a $\frac{5}{8}$-inch plate is worked, and its fore end is secured to the beam *c*. This plate serves to support the teak chocks forming the bed on which the mast-heel rests, and receives their fastenings.

THRUST-BEARERS.

In a screw steam-ship it is necessary to make some arrangement by means of which the thrust of the propeller-shaft shall be transmitted to the ship, and the injurious effects prevented which would result from the direct action of the thrust upon the machinery. For this purpose thrust-bearers are fitted, and the example we have chosen to illustrate the details of the construction of such a bearer is taken from the 'Agincourt,' and shown in elevation, plan, and transverse section in Fig. 229. The bulkhead *a a* forms the boundary of the engine-room, and the shaft-passage bulkheads are marked *b b*. The thrust-bearer is placed just abaft the bulkhead *a a*, and is built upon the inner bottom-plating, which extends out as far as the lowest longitudinal frame on each side, as shown in the section at *a a*. The shaft-passage bulkheads are brought down upon and connected with the inner skin, and a crown of watertight plating is worked upon their upper ends, thus converting the passage into a compartment, as has been previously explained. The flat of the shaft-passage is carried by the middle-line keelson *c*, and its edges are connected with the side bulkheads. In wake of the thrust-bearer additional support is given to the flat by means of the transverse frames *e, e, e, e* formed of lightened plates with double angle-irons on the edges, as shown in the elevation and section. Upon the plating of the flat three longitudinal bearers *d, d, d* are worked, and upon their upper edges a horizontal plate $1\frac{1}{4}$ inch thick is secured, to which the bearing is afterwards bolted. The bearers are about 15 feet long, and being secured to the flat by double angle-irons, give ample strength to resist the strains consequent on the transmission of the thrust of the propeller. It should be remarked that this very efficient bearer only requires for its construction the addition of the transverse frames *e, e, e*, the longitudinal bearers *d, d, d*, and the hori-

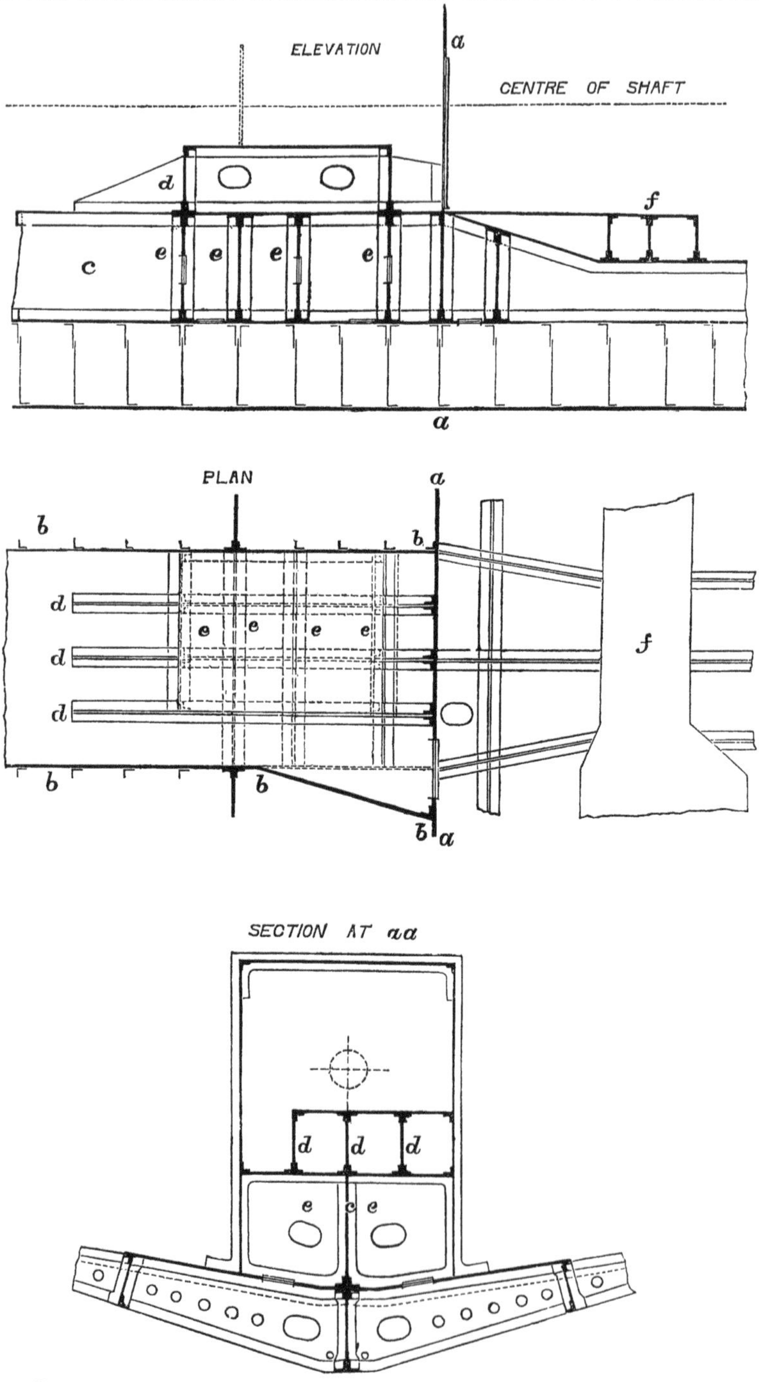
ELEVATION
a
CENTRE OF SHAFT
d
f
c
e
e
e
e
a
PLAN
a
b
b
d
d
d
e
e
e
e
f
b
b
b
a
SECTION AT aa
d
d
d
e
c
e

Fig. 229.

zontal plate to which the bearing is bolted, together with their connecting angle-irons; as the keelson *c* and the flat are continuous throughout the shaft-passage. It may be added that the increased breadth of the fore end of the shaft-passage, shown in the plan, is given in order to make room for the watertight door by which admission is obtained to the passage from the engine-room; and that the bearer *f* is the aftermost one in the engine-room. It will be obvious that the bearers similar to *f* are useful not only in supporting the engines and boilers, but in giving great transverse strength to the ship at a part where the transverse tie of the decks is seriously reduced. This feature of construction is of especial importance in small ships in which the lower deck is entirely discontinued in wake of engines and boilers.

The thrust bearers of small screw-steamers are, of course, of a very much simpler and lighter character, but in most cases the framing consists of longitudinal bearers formed of plates and angle-irons worked upon the floors, and stiffened by transverse plate frames, which are usually stationed at the floors and connected with the reversed angle-irons.

CHAIN PLATES.

In order to illustrate some of the modes of securing the chain plates of iron ships we have given the three following sketches. The first is taken from the 'Agincourt,' and is shown in Fig. 230. This ship has no channels, the shrouds being led down to the side inboard, and pin-racks being fitted below the lower dead-eyes. The chain plates are fitted upon and riveted to the skin plating behind armour, the rivets connecting the plates also passing through the frame angle-irons. A very strong attachment is thus secured, but the chain plates have to be put in place and fastened before the teak backing is worked, and it is impossible to examine their condition or to remove them at any time without incurring a large amount of work. It should be added, however, that being protected as they are for nearly the whole length the

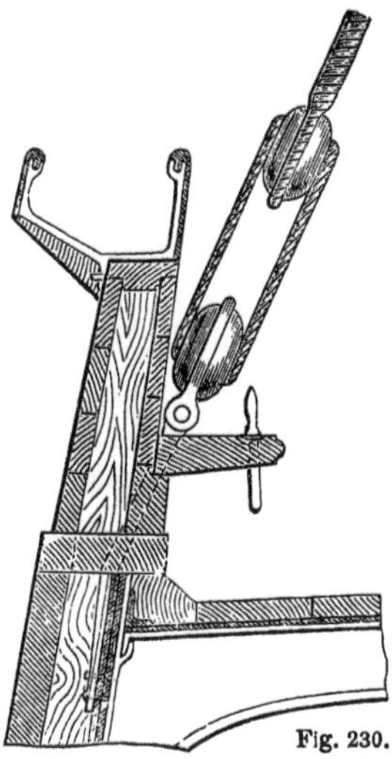

Fig. 230.

chain plates are much less liable to rust than they would be if exposed to the weather as is commonly the case.

The second sketch, Fig. 231, shows the manner in which the chain plates of the 'Bellerophon' are secured. There is a very

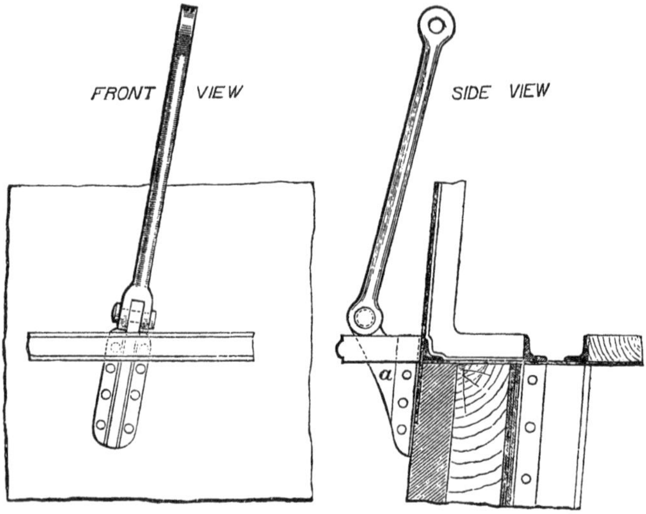

Fig. 231.

narrow channel fitted in this case, and the lower end of each chain plate is shackled to the upper part of brackets similar to *a*. The brackets are formed of 1-inch plates, and are connected with the armour plating by double angle-irons, tap riveted to the armour, and through riveted to the bracket. The main pieces of the channel are worked in short lengths between the brackets, and the face piece or channel rail is worked in one length. The chain plates are of such a length as to allow the dead-eyes to be housed in the hammock berthing. This mode of securing chain plates is the one now followed for the iron-clads of the Royal Navy, and is also applicable to unarmoured iron ships, as the brackets might be through riveted to the skin plating instead of being tap riveted to the armour. Of late wooden dead-eyes have been dispensed with, and fair-leads or dead-eyes of malleable cast-iron have been employed.

A third mode of securing chain plates is illustrated by Fig. 232. The sketch is taken from the wood-built iron-clad 'Lord Warden,' but it will be evident that the plan is suited to any iron-clad ship whether of wood or iron. In this case there is a channel fixed at the height of the covering board. The chain plate is forged to such a form as to serve the purposes for which chain and preventer chain plates are usually fitted, as well as to support the channel instead of having T-plates, as is customary in a wood ship. The most remarkable feature of the plan consists, however, in making the armour bolts serve as fastenings for the chain plates. The head of the bolt is formed, as shown by *b* in the sketch, so that when the chain plate has been shipped over the projecting points, it may be secured by metal nuts hove up on the thread cut in the bolt points. There are two bolts in each chain plate, as shown by the side view, and by the front view of the lower part of the plate, marked *a*. It will be obvious that great care is needed in driving the bolts, and it has been found desirable to leave the projecting part of the head unfinished until the bolt has been put in, when the finishing of the point and the cutting of the screw thread can be readily accomplished.

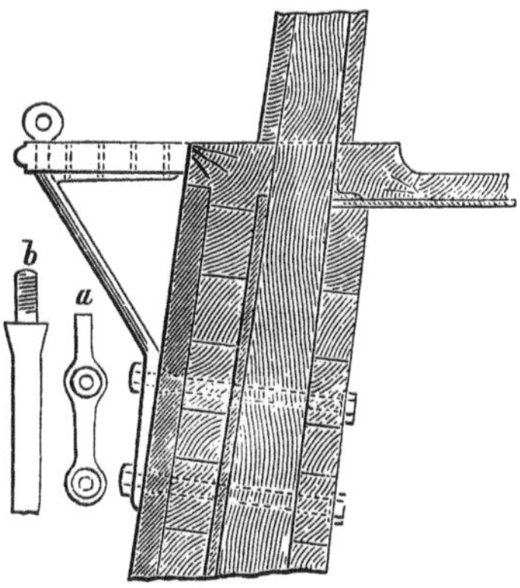

Fig. 232.

BOLLARD HEADS, &c.

In iron ships bollard heads and towing bollards are frequently of cast-iron, and are bolted to the gunwale. In cases where the frames are run up above the upper deck-stringer the bollard heads are sometimes formed by bolting wood chocks to the sides of the frames. In some small vessels the bollard heads have been formed of wrought-iron plates and angle-irons, and an instance of such an arrangement is given in Fig. 233. The section at *a b* shows that the bollard is formed of two bent plates ½ inch thick, connected by single riveted edge strips. The dimensions of the bollard are 9 by 6 inches. The lower edge of the plate forming the outer part overlaps and is riveted to the upper edge of the sheer strake, and the inside plate of the bollard

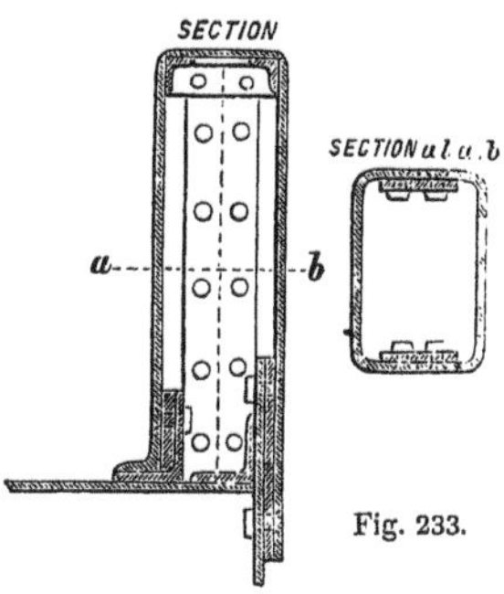

Fig. 233.

is formed with a flanged foot which rests upon the stringer angle-iron, and is strongly riveted to it. At the upper edge of the bollard there is an angle-iron frame upon which the plate cover is fitted. These arrangements are clearly shown in the sketches.

CATHEADS.

The catheads of iron ships are usually solid forgings bolted to the ship's side and to the gunwale. It will be evident that the dimensions and weight of the forging required for the cathead of a very large ship must be considerable, and the time and expense required for its manufacture must be great. In order to reduce both the weight and the cost of the catheads of the iron-clad frigates and some other ships of the Navy, box catheads have been introduced instead of solid forgings. The construction of a cathead of this description is fully shown in the sketches given in Fig. 234. The section through *c d* shows the general arrangement of the box section, the plates forming the sides being $\frac{1}{2}$ inch, and the connecting angle-irons 3 by 3 by $\frac{1}{2}$ inches. The form of the cathead and its connections with the side will be fully understood from the side view and plan. The angle-irons connecting the sides of the cathead with the outside plating are $3\frac{1}{2}$ by $3\frac{1}{2}$ by $\frac{5}{8}$ inches, and an angle-iron of the same dimensions secures the top plate of the cathead to the side. It will be observed from the plan and the section through *e f* that the top plate is considerably increased in breadth as it approaches the ship's side, the object of this arrangement being to distribute the fastenings of the angle-iron connecting the top with the side, and thus to make a very strong connection, by means of which the tendency of the upper part of the cathead to move outward when the anchor is catted is effectually resisted. The space for the sheave is enclosed by two $\frac{7}{8}$ inch plates fitted into the cathead, and connected with the side plates and top and bottom by forged angle-irons. The block is of iron, and is carried by two spindles which rest in metal bearings bolted to the $\frac{7}{8}$ inch plates. These arrangements are illustrated by the section through *a b*. The object of thus allowing the block to swivel, is to give a fairer lead to the cat pendant when it is led forward to the hawse-hole, and to prevent the edge of the block from being chafed. This arrangement is specially needed in ships with ram-bows in which the distance from the catheads to the hawse-holes is considerable. Having thus briefly

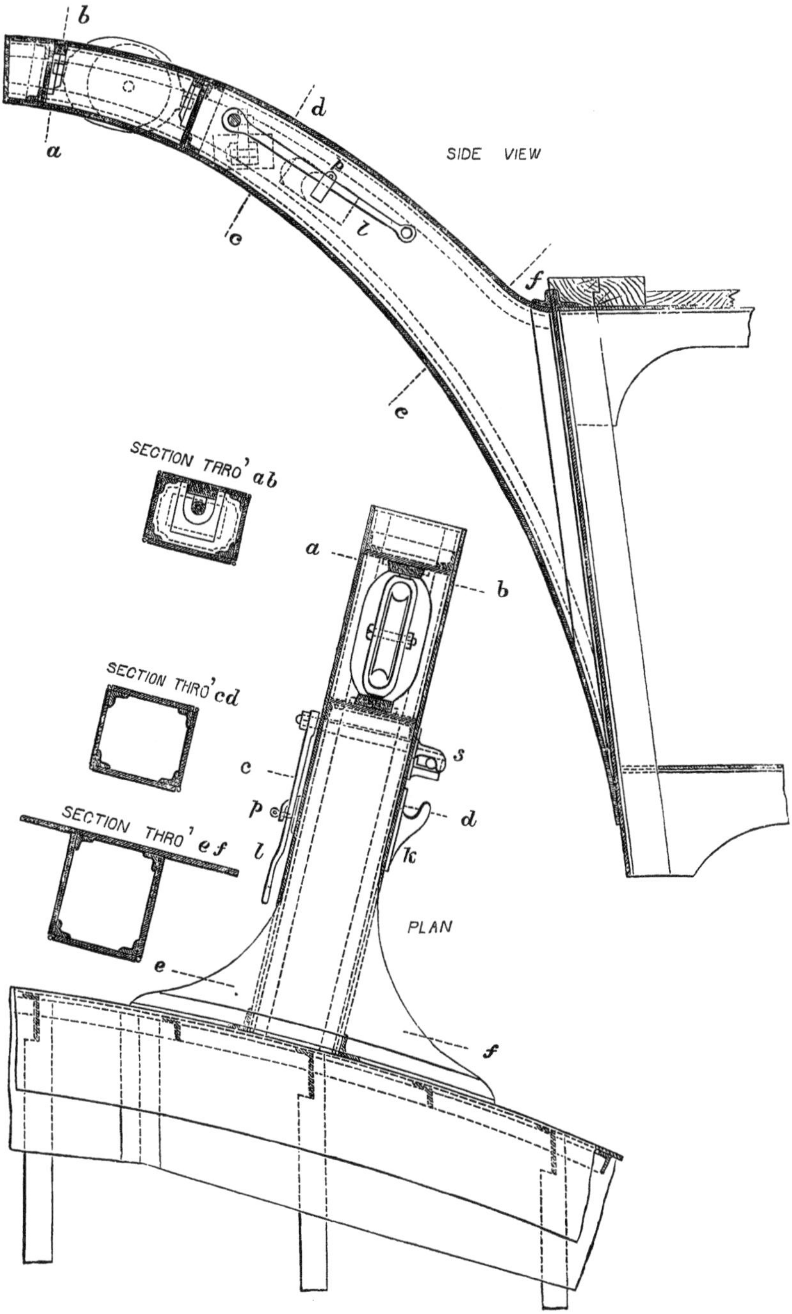
b
a
d
SIDE VIEW
c
l
f
e
SECTION THRO' ab
a
b
SECTION THRO' cd
c
s
p
d
SECTION THRO' ef
l
k
PLAN
e
f

Fig. 234.

described the construction and connections of the cathead, it may be of interest to notice a few of the fittings. When the anchor is suspended from the cathead, one end of the chain on which it hangs is secured to the slip stopper *s*, and after being passed through the anchor-ring the chain is led over the bracket *k*, and its inner end is secured to the bollard head. The slip-stopper *s* is worked by means of a lever *l*, on the opposite side of the cathead, and when the anchor is to be let go the lever is raised by means of a lanyard attached to its inner end, and the toggle bolt of the stopper being released, the end of the chain to which the anchor hangs is freed. In order to prevent the lever from being drawn upwards by the catching of a rope or any other accident, the pin *p* is put in above it when the anchor is catted.

In conclusion it may be remarked, that in addition to the great increase of strength obtained by this mode of constructing large catheads, there is a considerable saving both in weight and cost as compared with a forged cathead. Thus the actual saving in weight on a box cathead of the dimensions of that shown in Fig. 234, amounts to one-fourth the weight of a forged cathead for the same class of ship, and the saving in cost amounts to nearly one-third.

DECK HOUSES.

We shall conclude this chapter with an account of the special arrangements made for supporting the spar and awning decks, and giving transverse strength to the Pacific Steam Navigation Company's ships, 'Pacifico,' 'Limenia,' and 'Santiago,' built by Messrs. Randolph and Elder. These ships are 260 feet in length between the perpendiculars, and have a complete spar deck throughout their length, and an awning deck above this extending from the stern forward to about 60 feet from the bow. As far back as the fore end of the awning deck, and for about 45 feet before the stern, the side plating is carried up to the height of the spar deck, and for 63 feet of the amidship length, in wake of the paddle-boxes, the plating is continued up to the height of the awning deck. At the intermediate parts of the ship the framing is of the character shown in the section in Fig. 235, the frame angle-irons and the plating being ended at about 4 feet above the upper deck, while the spar deck beams are carried by light iron stanchions riveted to the frames and supported by three tiers of pillars, and the awning deck is carried by three tiers of pillars. The diagonal stays marked *s*, *s* are fitted in order to give rigidity to the structure, and efficiently connect the various decks and the hull. There are

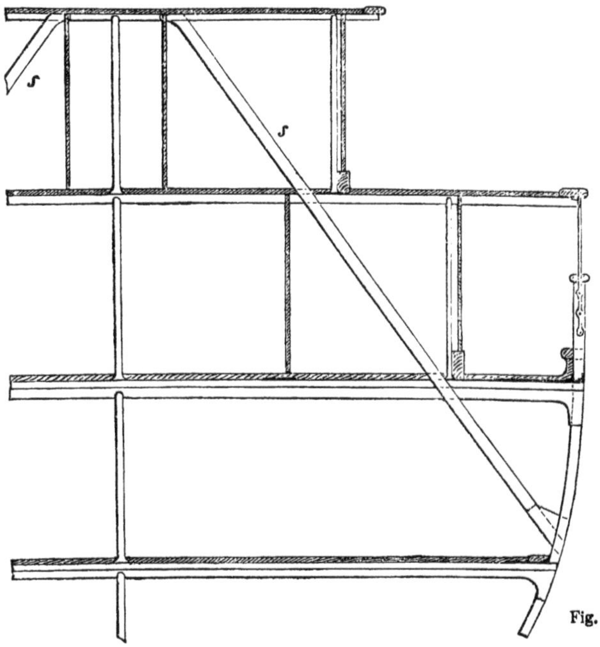

Fig. 235.

five of these stays in wake of the awning deck, two being placed before the paddle-boxes, and three abaft them. They are formed of two bars of wrought-iron 6 × 1 inches, enclosing the deck beams, and being secured by through bolts to both beams and frames. The details of the stays *s*, *s* are given in Fig. 236, next page. It will be seen that the upper ends of the plates forming the stay *s* are brought on opposite sides of the vertical flanges of the T-iron awning deck beams *b*, and are secured by through bolts. At the spar and upper decks the plates have to be kept sufficiently far apart to clear the beam flanges, and consequently the forgings *a*, *a* have to be introduced between the beam web *b* and the stays *s*, *s* in order to receive the screw bolts. Below the upper deck the plates are brought close together as shown at *d*, and the lower end of the stay is connected with the outside plating by means of the bracket plate *c* and the angle-iron *e*. These arrangements afford a most instructive example of the manner in which lightness and strength may be combined in the construction of vessels designed for special services.

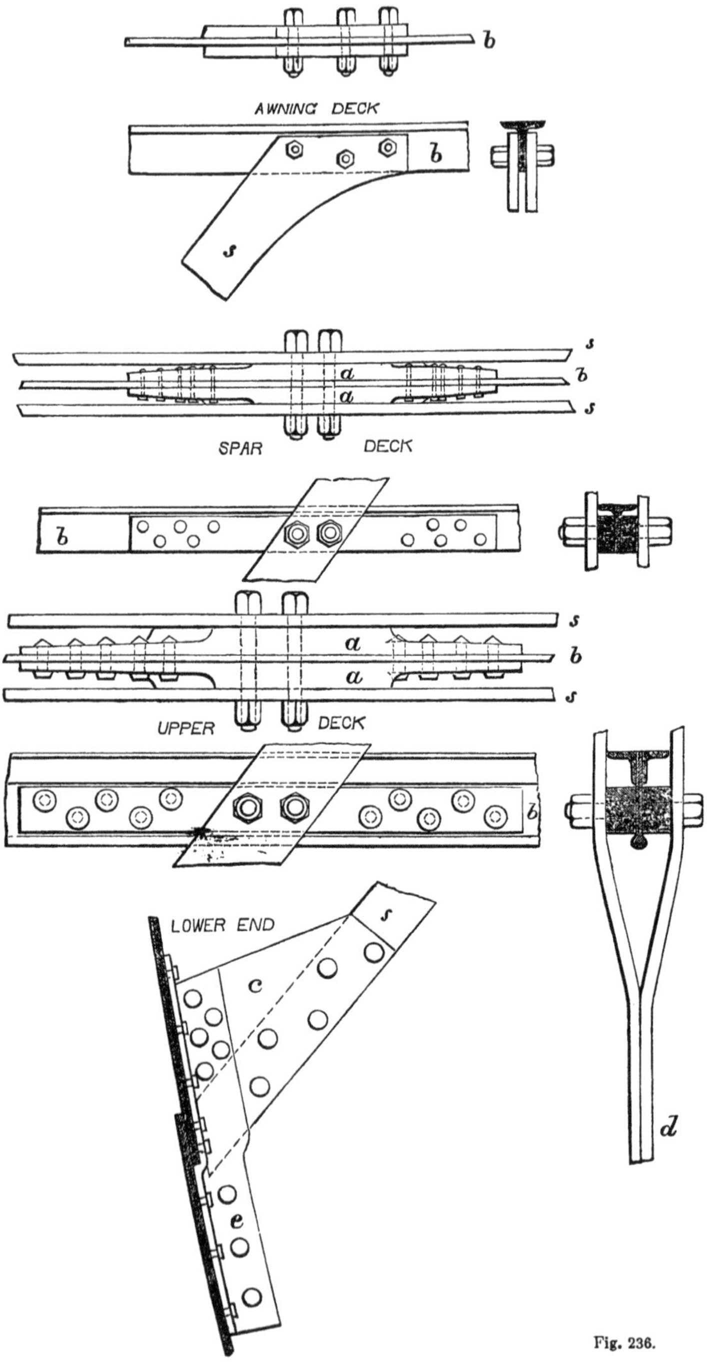
b
AWNING DECK
b
s
s
a
b
a
s
SPAR
DECK
b
s
a
b
a
s
UPPER
DECK
b
LOWER END
s
c
e
d

Fig. 236.

CHAPTER XVI.

STEEL PLATES FOR SHIPBUILDING.

NEARLY all the general considerations which have been set forth in the preceding chapters respecting combinations of materials, and modes of operation, are as applicable to steel as to iron. But there are essential differences in the two materials of great importance to the shipbuilder, and it is proposed to consider them at some length in this chapter.

Steel may be defined chemically as iron combined with a small but definite proportion of carbon. It is distinguished on the one side from pure iron, by the presence of carbon; and on the other side from cast iron by the smallness of its proportion of carbon. Steel may be said, in another aspect of it, to be iron in such a condition, that when heated to redness, and suddenly cooled, it becomes very hard but may be again softened by heat. Taking advantage of this peculiarity the shipbuilder, like others, has long employed it for tools for cutting iron and softer metals. It is but recently, however, that it has come to occupy the important position in which we have now to consider it, which may be said to be that of wrought iron possessing extraordinary ductility and strength.

Steel is now regarded by the shipbuilder as a material which may with care be made to possess greater ductility, both hot and cold, than the best wrought iron, in combination with a tensile strength 50 per cent. greater than that of iron. It has on this account come to be largely used by shipbuilders instead of iron for plates and angles, and to a great extent for rivets also. There is but little room for doubt that it is destined to a far more extended use than it now has for these purposes; and it is possible that it may ultimately displace iron for such uses. But, in the present state of the manufacture, steel ship-plates possess some very dangerous peculiarities. There is ample experience to prove that ships built of steel may be weaker, both structurally and locally, than ships built of iron of the same scantlings, and with precisely similar arrangements of framing and fastening. It may be said indeed, with truth, that if steel supplied by first class makers is treated in the same manner as iron in working it into a ship, it will require to be of the same thickness as the best iron in order to obtain the same strength, and that as the practice has been to reduce the thickness in nearly inverse proportion to the tensile strength of the perfect plate, steel ships so built are by so much inferior in this respect to ships built of iron of unreduced thickness.

Several kinds of steel have been used in shipbuilding in the form of plates and angles, but there are only two which have had any extensive use, viz.: Puddled and Bessemer steels. These two materials differ widely from each other, not only in the mode of manufacture, but in their qualities. Puddled steel plates and bars, like iron, are made from a pile of small pieces, welded together under the hammer, and between the rolls, and are subject to those well known and troublesome defects produced in these processes, to a greater extent even than iron. Each Bessemer steel plate or bar is, on the contrary, made from a single ingot, and is therefore free from these defects. Large plates can thus be made by the latter process with the same precision, and almost with the same ease as small ones, but in puddled steel this is not so. It also appears to be more difficult to obtain uniformity of temper in a batch of puddled steel plates than in Bessemer steel, probably because of the extreme care required in selecting the puddled bars of which each pile is made, as these bars differ greatly from each other in temper, and the selection is made from observation of the nature of the fracture when they are broken. In the manufacture of Bessemer steel, the selection which requires to be made is dependent on precise chemical analysis, which is part of the daily operation of the mills, and is altogether independent of the workmen. Puddled steel is not necessarily inferior in strength to Bessemer steel, but that made in England is generally of lower tensile strength, as well as less uniform in strength. The following are examples of batches of puddled and Bessemer steel sent in by different makers to stand the Admiralty forge tests, and to have a tensile strength of 33 tons per square inch lengthways or with the grain, and 30 tons across the length or grain.

PUDDLED STEEL, ¼ inch thick.		BESSEMER STEEL, ¼ inch thick.	
Tons per sq. inch of original section.	Elongation in 12 inches.	Tons per sq. inch of original section.	Elongation in 12 inches.
	inches.		inches.
*L 32·485	2		
L 30·203	2		
A 30·300	3/4	43·319	1 1/4
A 29·409	3/4	43·617	1 1/4
L 26·593	1 3/8	45·572	9/16
L 27·377	1 3/8	49·218	5/8
A 27·306	3/4	42·897	1 1/4
A 27·189	11/16	41·851	1 3/16
L 34·466	1 1/8	45·277	1 3/8
L 32·743	1 3/16	42·5	1 1/16
A 27·212	9/16		
A 23·893	7/16	½ in. thick, by the same maker as above.	
L 26·247	1 3/8		
L 26·2	1 7/16	34·985	2
A 25·678	3/8	35·841	1 7/8
A 24·321	3/8	36·386	1 3/4
L 35·664	1 9/16	35·862	1 9/16
L 32·867	1 3/8	38·297	1 11/16
L 29·699	1 7/8	38·122	1 3/8
L 27·656	1 13/16	37·89	1 5/8
A 26·89	11/16	38·151	1 13/16
A 28·093	13/16		
		½ in. thick, by another maker.	
L 36·269	1 1/4	32·649	2 3/8
L 35·576	1	33·552	2
A 24·038	3/8	32·345	2 1/2
A 25·314	3/8	31·841	2 1/8
L 33·26	1 3/16	34·552	1 15/16
L 36·754	1 1/8	34·765	1 5/8
A 31·958	1/4	34·217	2
A 33·006	7/16	34·303	1 3/4

From these figures it will be seen that in the puddled steel there resulted a variation in the strength, lengthwise, between 26·2 and 36·754, in the attempt to secure 33 tons; and in the strength crosswise, which should have been 30 tons, there was a variation between 23·893 and 33·006. The great want of elongation in some of the crosswise specimens is undoubtedly due to the defective welding of the pieces composing the pile. Subsequent plates of the same thickness, made by the same makers, but by a different system of piling, gave the following results as tested by the manufacturer.

* In this and the following tables the letters L and A stand for "lengthwise" and "across" respectively.

TESTS OF ¼-INCH PUDDLED STEEL.

Breaking strain per sq. inch.	Stretch in 4 inches.	Breaking strain per sq. inch.		Stretch in 4 inches.
tons.	16ths of an inch.	tons.	cwt.	16ths of an inch.
L 37·75	9	L 35	0	11
A 35·00	4	A 32	16	8
L 38·30	12	L 32	16	11
L 35·00	12	A 29	10	5
A 33·90	7	L 35	0	9
L 36·65	14	A 30	12	5
A 32·25	5	L 33	18	10
L 32·80	5	A 29	10	10
A 32·25	7	L 32	16	11
L 36·65	10	A 30	12	4
A 33·9	6			

In the case of the tests of the Bessemer steel the decrease in the elongation as the tensile strength increases is very noticeable. The variation in the tensile strength of the ½-inch and ¼-inch plates supplied by one maker is between 34·985 tons and 49·218 tons, and the corresponding elongations are 2 inches, and ⅝ of an inch. These examples of Bessemer steel are not so good or so uniform as might have been given. The uniformity in strength in some cases, where the manufacturers make Bessemer steel plates an important part of their business, is admirable.

Every one acquainted with these steels, or steel irons as they are sometimes called, knows that their ductility when hot is very remarkable. The angles to which ordinary or second-class iron ship plates are expected to bend, when hot, are 60° and 90° breadthwise and lengthwise respectively. In first-class iron plates the standard is raised to 90° and 125°; but these steels will stand, if of good quality, not only the 110° and 140° prescribed by the Admiralty tests, but will, up to ¾ of an inch in thickness, bend double twice without fracture. The superior ductility of this material when cold, as compared with iron, is also very marked, but not so considerable as when hot. Bessemer steel plates may be said to vary in tensile strength from about 30 to 50 tons per square inch, but they are most useful for shipbuilders' purposes when they are between 32 and 38 tons, say 35 tons per inch, or about 50 per cent. stronger than first-class iron. At this temper the plates are more ductile when cold than first-class iron, as indicated by the following comparative table of tests taken from the Admiralty Code.

COMPARATIVE ANGLES of BENDING of "BEST BEST" IRON and BESSEMER STEEL PLATES when cold, without fracture.

With the grain.			Across the grain.		
Thickness.	Best Best Iron.	Bessemer Steel.	Thickness.	Best Best Iron.	Bessemer Steel.
	Degrees.	Degrees.		Degrees.	Degrees.
1 inch and 15–16ths inch	15	30	1 inch and 15–16ths inch	5	20
$\frac{7}{8}$,, 13 ,,	20	40	$\frac{7}{8}$,, 13 ,,	5	25
$\frac{3}{4}$,, 11 ,,	25	50	$\frac{3}{4}$,, 11 ,,	10	30
$\frac{5}{8}$,, 9 ,,	35	60	$\frac{5}{8}$,, 9 ,,	15	35
$\frac{1}{2}$,, ...	35	70	$\frac{1}{2}$,, ...	15	40
$\frac{7}{16}$,, ...	50	75	$\frac{7}{16}$,, ...	20	50
$\frac{3}{8}$,, ...	50	80	$\frac{3}{8}$,, ...	20	60
$\frac{5}{16}$,, ...	70	85	$\frac{5}{16}$,, ...	30	65
$\frac{1}{4}$,, ...	70	90	$\frac{1}{4}$,, ...	30	70
$\frac{3}{16}$ and under	90	90	$\frac{3}{16}$ and under	40	70

If the ductility here indicated could be secured as uniformly in a batch of steel plates as the average qualities of "best best" iron are secured in a batch of plates of this material, the higher price of steel would be but little bar to its use; but at present one is obliged to say that it cannot be, for the manufacturers with whom we are acquainted—and they are the best in England—have so far as our experience goes, failed to secure this uniformity. The following is the result of tests with samples from deliveries of Bessemer plates for which a high price was given.

BESSEMER STEEL PLATES, manufactured by two Firms to comply with the Admiralty conditions. Angles of bending, *cold.*

Thickness of plate.	With grain.	Across grain.	Remarks.	Thickness of plate.	With grain.	Across grain.	Remarks.
	Degrees.	Degrees.			Degrees.	Degrees.	
$\frac{1}{2}$-inch ..	70	40	No fracture.	$\frac{3}{8}$-inch ..	25	45	Fracture.
,, ..	70	20	Fracture.	,, ..	47	60	,,
,, ..	8	15	Fracture at 8°.	,, ..	88	62	Slight fracture.
,, ..	18	40	Fracture.	,, ..	87	65	,,
,, ..	20	10	,,	,, ..	85	72	,,
,, ..	35	25	,,	,, ..	85	73	,,
,, ..	34	17	,,				

Puddled steel is usually found to be more ductile than Bessemer steel, and is on that account often strongly recommended instead of Bessemer steel, but disappointments are sometimes experienced in this respect, as well as in its tensile strength.

The course recently taken by Lloyd's Committee with reference to the use of steel is a very important one. It will doubtless be productive of much good on the whole, but it will need extreme care on the part of the Surveyors to avert in many cases disastrous consequences. It has been resolved as follows :—

" That ships built of steel of approved quality, under special " survey, be classed in the Register Book with the notation 'Ex- " perimental' against their characters.

" In all cases, however, the specifications for the ships must be " submitted to the Committee for approval.

" That a reduction be allowed in the thickness of the plates, " frames, &c., of ships built of steel, not exceeding one-fourth from " that prescribed in Table G for iron ships.

" (In no case, however, are the rivets to be made of steel, nor " will any reduction be allowed in the sizes of rivets from those " prescribed in Table G for ships of the same tonnage, built of iron.)

" In other respects the rules for the construction of iron ships " will apply equally to ships built of steel."

The reduction thus made in scantlings will bring the cost of a steel ship down to about that of a similar ship built of good iron, and there will probably be a saving in weight of about 100 tons for 1000 tons of builders' measurement. It is understood that the steel which has commended itself to the surveyors of Lloyd's is Bessemer steel, but as the price of puddled steel is about 1*l.* per ton less than that of Bessemer, the resolution will be likely to give a great impulse to the manufacture of the cheaper material. Its ductility, and its great resemblance to iron in its characteristics, the absence also of certain peculiar and disagreeable defects to which Bessemer steel is liable, and what is perhaps still more important, the fact that all the good iron-makers can produce this material—all these considerations will probably give to it, for the present at least, a run of popular favour. Hereafter, when the peculiarities of the Bessemer steel are better understood by makers and shipbuilders, and when the price is reduced by the termination of the present royalty, it will in all probability be used almost exclusively by shipbuilders and engineers. So far as angles are concerned, Bessemer steel is now admitted to be superior to every other material. The amount of working which the ingot receives in the course of manufacture into a bar appears to perfect it, and the material has proved itself to be, for angle-bars, unequalled. In order to explain the reservation with which the good qualities of Bessemer plates have been set forth, it is necessary to give the results of experience in its use.

In the year 1864 we had a series of experiments on steel made in Chatham Dockyard. The material was of Bessemer manufacture, the conditions being that it should stand 33 tons per square inch lengthwise and 30 tons crosswise. The results are recorded in the following table.

TESTS of ½-Inch BESSEMER STEEL.

No. of Specimen.	Breadth of Specimen.	Breaking strain per square inch in tons. Original area.	Fractured area.	Elongation* in inches.	Remarks.		
	inches.						
1	1	42·55	46·37	1·18	Lengthwise		
	1	43·5	55·49	1·94	Across		
2	1	43·25	59·4	2·00	Lengthwise		
	1	43·75	55·81	1·94	Across		
3	2	41·75	50·88	1·75	Lengthwise		
	2	40·25	43·85	1·69	Across		
4	2	40·25	48·02	1·69	Lengthwise		
	2	40·50	52·89	2·12	Across		
5	3	40·16	37·28	0·56	Lengthwise	Rivet-holes punched 1¼ ins. from edge.	Broke through narrowed part, but not at narrowest place.
	3	38·60	15·73	0·44	Across ..		Through rivet-holes in the head.
6	3	40·58	16·51	0·56	Lengthwise		Ditto.
	3	36·50	35·05	0·37	Across ..		Through narrowed part, but not at narrowest place.
7	4	31·0	16·0	0·25	Lengthwise		Through head.
	4	31·5	16·0	0·19	Across ..		,,
8	4	32·75	17·64	0·19	Lengthwise		,,
	4	28·0	15·18	0·12	Across ..		,,
9	5	29·2	14·97	0·12	Lengthwise		,,
	5	29·8	16·55	0·12	Across ..		,,
10	5	28·0	15·34	0·12	Lengthwise		,,
	5	30·6	16·76	0·19	Across ..		,,
11	6	28·0	17·01	0·19	Lengthwise		,,
	6	27·5	16·5	0·19	Across ..		,,
12	6	27·5	16·5	0·12	Lengthwise		,,
	6	25·3	14·82	0·12	Across ..		,,
13	5	38·1	16·38	0·56	Lengthwise	Punched 2¼ ins. from edge.	,,
	5	34·8	14·96	0·50	Across ..		,,
14	5	38·15	16·41	0·50	Lengthwise		,,
	5	36·3	15·95	0·44	Across ..		,,
15	5	35·2	16·0	0·50	Lengthwise	Drilled 2¼ ins. from edge.	,,
	5	42·8	47·32	1·62	Across ..		Through narrowest part.
16	5	42·7	47·55	1·50	Lengthwise		,,
17	5	41·05	43·59	1·37	Lengthwise		,,

* The elongation in this case and in cases following, where not otherwise specified, was taken in a length of 24 inches, that being the distance between the bolts by which the pieces were pulled. This was done because it was impossible to tell where, within these points, the pieces would break. Nearly the whole of the elongation

The thirty-two pieces tested were cut from four ½-inch plates, viz. Nos. 1, 2, 3, 4, 9, 10, 11, and 12 in the table lengthwise, and the same numbers crosswise, from one plate; 5, 6, 7, and 8 lengthwise and crosswise from another; 13, 14, and 15 lengthwise and crosswise from a third; and 16 and 17 lengthwise from a fourth. The lengthwise and crosswise tests are bracketed together in all the cases in which they were made upon similar pieces cut from the same plate. The pieces numbered 1, 2, 3, and 4 were parallel from end to end, and were gripped by the machines; the others had a head formed on them with strengthening pieces or clamp-plates riveted on each side with four ¾-inch rivets placed symmetrically around the hole formed to receive the centre bolt, to which the chain was attached. This hole was in each case centred with great care.

The results were very remarkable. The material was shown to have one-third more strength than was expected, when it fractured fairly; but it was also shown to have an erratic mode of fracturing, which caused a variation in the breaking strain per square inch of original area between 43¾ tons and 25⅓ tons. Or, regarding the fractured area, there is a variation between, say, 15 and 60 tons.

There appears to be nothing in the circumstances to account for these eccentricities. When the holes in the head, which determined the fracture in so many cases, were punched 1¼ inches from the edges (*i. e.* nearly two diameters), the plates would not always break there, but fractured through a portion of the unpierced plate, which was not the least in section. And when the holes were drilled, one of the specimens broke through the drilled holes at 16 tons to the inch, while its companion piece, similar to it in every respect, except that it was cut crosswise of the plate instead of lengthwise, stood more than 47 tons to the inch.

The tests following, with steel manufactured by another firm, were made under precisely the same circumstances as those above recorded. The first four pieces, being only 2 inches wide, were gripped; and the others had a clamped head with rivet-holes punched 1¼ inches from the head.

There is certainly more uniformity in these cases, inasmuch as all the riveted pieces broke at a low strain, the variation lying between 13·16 and 15·3 tons per square inch.

must, however, have taken place in the reduced parallel part, which was in every case 9 inches long.

FURTHER TESTS of ½-Inch BESSEMER STEEL PLATES.

No. of Specimen.	Breadth in inches.	Breaking strain per square inch in tons.		Elongation in inches.	Remarks.	
		Original area.	Fractured area.			
1	2	35·0	59·73	2·5	Lengthwise.	
	2	36·0	58·21	2·25	Across.	
2	2	35·25	57·0	2·87	Lengthwise.	
	2	36·25	57·12	2·75	Across.	
3	4	33·5	15·09	0·75	Lengthwise ..	All broke through rivet-holes in the head.
	4	32·5	13·16	0·812	Across	
4	4	33·5	15·098	1·0	Lengthwise ..	
	4	34·12	14·18	1·25	Across	
5	5	32·2	15·3	0·5	Lengthwise ..	
	5	30·2	15·1	0·5	Across	
6	5	30·4	14·65	0·5	Lengthwise ..	
	5	31·9	14·6	0·5	Across	

These experiments might by many persons have been held sufficient to prove that Bessemer steel plates were altogether unsuited to the use of the shipbuilder. But as there was some indication that the plates suffered much less from drilling than from punching, further experiments were made in this direction before any decision was arrived at, as the material was held to be worthy of the most careful and patient trial. Four pieces of a ½-inch steel plate were therefore next cut out lengthwise, and formed with heads, the first two being unclamped, and the latter two clamped. The former were 3 inches wide in the narrowed part and had two ¼-inch holes drilled in the centre of the length of this part: the latter were 6 inches wide and had two ½-inch holes similarly drilled. The former two broke through the drilled holes at 42·91 and 40·83 tons respectively per square inch of fractured area. The others broke, not through the ½-inch holes which had been drilled through the smallest section, but through the connecting holes at the head, at 17·11 and 17·94 tons respectively per square inch of fractured area. Two other plates of ½-inch steel were then taken, supplied by the same maker as those first above referred to, and four pieces were cut from one plate, and two from another, all of them lengthwise. They were all clamped at the ends, and the holes fastening the clamps were *drilled* at 1¼ inches from the edge. In these experiments, notwithstanding the drilling of the holes, the first piece broke, like those in the first experiments, at 14·15 tons per square inch of fractured area, and with an elongation of 0·375 of an inch. The next three pieces broke fairly in the reduced part at the following strains respectively, viz., per square inch of original area 41·375, 29·2, and 39·8 tons, with an elongation of 1·37, 2·25, and 2·12 inches; the breaking strain per square inch of fractured area being 41·54, 55·42, and 56·27 tons. The other two pieces, which were 6 inches wide, had two ½-inch holes drilled in the middle of them. The plates broke through these drilled holes at 39·2 and 37·8 tons per square inch of original area, exclusive of holes, and 41·26 and 39·52 tons per square inch of fractured area, the elongation being 0·37 and 0·5 inch respectively.

The next plates tried were by the other maker. Two samples were taken precisely similar in every respect to the last two, *i. e.* they were ½-inch plates cut lengthwise, clamped, with punched holes for the rivets, and they had two ½-inch holes drilled in the centre of the straight reduced part. They broke through the drilled holes at 32·55 and 33·90 tons per square inch of original area, exclusive of holes, and 34·92 and 35·91 tons of fractured area, the elongation being 0·5 inch in both cases.

Some further experiments were then made with pieces 4 inches and 6 inches wide, and ½ inch thick, clamped at the ends, but having a third hole put through the clamps in a line with, and midway between the two holes through which the fracture always occurred, the object being to reduce the strength still further, to ascertain what this reduction would be, and the disadvantage of punching as compared with drilling under these circumstances. The results were as follows:—

TESTS of ½-inch BESSEMER STEEL PLATES, 4 inches broad.

	Breaking strain per square inch in tons.		Elongation in inches.	Remarks.
	Original area.	Fractured area.		
By one manufacturer.	34·19	13·34	0·31	Broke through two of the holes in the head. Holes punched.
	34·87	16·66	0·31	Ditto. ditto.
	40·25	49·69	1·18	Broken in centre of reduced part. Holes drilled.
	40·44	53·8	1·62	Ditto. ditto.
By another manufacturer.	32·31	14·98	0·25	Broke through two of the holes in the head. Holes punched.
	30·03	14·13	1·375	Broke through the three holes in the head. Holes punched.
	35·43	47·88	1·875	Broke in centre of reduced part. Holes drilled.
	35·43	44·21	2·125	Ditto. ditto.

In the six cases following two holes were put through the centre of the reduced part. The first two (½-inch) punched; the second two (½-inch) drilled; and the third two punched to ½ an inch and reamed to ⅝-inch, ¾ inch from the edge.

TESTS of ½-inch BESSEMER STEEL PLATES, 6 inches broad.

Breaking strain per square inch in tons.		Elongation in inches.	Remarks.
Original area.	Fractured area.		
22·30	22·30	0·016	Broke through the two holes in reduced part.
20·0	20·0	0·125	Ditto. ditto.
32·45	33·8	0·5	Ditto. ditto.
32·35	33·4	0·437	Ditto. ditto.
29·13	26·95	0·5	Ditto. ditto.
30·45	24·54	0·625	Broke through one of the holes, and also at the same moment through the solid reduced part.

Further experiments were made at about the same time at Chatham, and some others at Pembroke. The only ones among them which appear to possess additional interest were made at Chatham, to ascertain the comparative effect of a falling weight upon a Bessemer steel, and an iron plate. A deep angle-iron was bent into an ellipse, about 6 feet long, and 4 feet 6 inches wide. A piece of Bessemer plate 15 inches wide and ½ an inch thick was placed on this, in the direction of the longest diameter, and was riveted to the flange at each extremity with five rivets, there being thus nearly six feet of the plate left unsupported between. This piece had been butted in the middle, and strapped with double-straps, ¼ of an inch thick, treble-riveted, with five rows of iron rivets, the alternate ones nearest the butt being omitted. All the holes for these rivets were drilled. A weight was then allowed to fall on the centre of the plate, *i.e.*, on the butt strap, from a height of 32 feet.

A 32-pound shot was first tried, this slightly loosened one rivet. A 68-pounder followed; this slightly loosened another rivet; then an 84-pounder was dropped, and it was found that the total damage was seven rivets slightly loosened, one point of a rivet off, and the plate bent somewhat below a level. An elongated ball having a round bottom, and weighing 16 cwts., was then dropped. This broke out a piece of the solid steel 21 inches long, short of the rivet holes at both ends. The rivets in the frame were in fact perfect, and there were no fractures at the butt-strap. The shape of the frame remained unaltered.

The rivets were then cut out and an iron plate riveted on. This was similar in every way to the steel plate, except that the rivet holes had been punched. The 32-pounder appeared to have no effect. The 68-pounder slightly loosened one rivet. The 84-pounder bent the plate down about 2 inches, but there was no further injury to the rivets. The 16-cwt. ball cracked the butt-strap in wake of several of the rivets, and the frame in front of several of the connecting rivets. The plate did not break, but it drew the frame out of shape, shortening one diameter, and lengthening the other by about nine inches.

The issue of these various experiments was that the use of Bessemer steel was chiefly limited to those portions of the ship in which the strain on the plates was in the direction of their length, *i.e.* in which dangerous cross-strains were not to be expected, such, for example, as upper deck plating, plating of inner bottom, and longitudinal frames. The rule was also made imperative that the rivet-holes should be drilled.

But with these precautions there was still danger, as was made apparent by the sudden snapping off of a ½-inch Bessemer plate which formed one of the upper deck plates in the 'Hercules,' without warning, and without any apparent cause. This plate was delivered in a breadth of 4 feet, and it was so placed as to form one length of the outer or side stringer plate to the upper deck. It lapped half over the centre battery, and had therefore to be narrowed to about 2 feet for half its length. A joggle or abutment was formed in doing this, which joggle was punched out by the workmen, and an angle was formed in the corner. When the plate had been fitted, it was laid down on the beams, and riveted to them. After having been so fastened for several days, during which it was exposed without cover to the weather, it snapped suddenly one morning (succeeding a cold night) across the breadth of 2 feet, the fracture commencing at the angle. Tests made on several pieces of this broken plate gave the following curious results.

TESTS of Fractured ½-inch BESSEMER STEEL PLATE.

Size of sample.	Breaking strain per square inch of original area.	Elongation in 6 inches.	Direction.
inches. inch.	tons.	inches.	
2·12 × ·485	35·86	1¼	Lengthwise.
2·11 × ·485	35·67	15/16	,,
2·14 × ·495	24·07	⅛	Across.
2·12 × ·5	35·73	7/16	,,

FORGE TESTS.

HOT TESTS ALL GOOD.

COLD TESTS.

1st piece ..	With the grain ..	Proof 70° ..	.. Fractured badly at 24°.
2nd ditto ..	,, ..	,, ..	.. Bent to 70°.
1st ditto ..	Across the grain ..	Proof 40° ..	.. Fractured badly at 24°.
2nd ditto ..	,, ..	,, ..	.. Bent to 40°.

Other cases of fracture have occurred, where an angle has been the evident origin, and the necessity for rounding the angle has become apparent.

It was conjectured with regard to this plate that it had not been annealed after manufacture, and that while cooling from the rolls a current of cold air had perhaps been allowed to pass over it, or it might have been laid on cold iron, or in a damp place to cool. But whatever was the cause it was evident that some of the best English Bessemer steel makers were not then to be trusted in the performance of this most important duty—the careful annealing of the plates after manufacture; and it was thought that the only absolute security was to be found in annealing the plates after receipt.* While this was under consideration it was suggested by Mr. Sharp, of the Bolton Iron and Steel Works, that if annealing were adopted in the dockyards it would be found to restore a large

* "Steel is annealed in a variety of ways. Some artists anneal steel by heating it "to redness in the open or hollow fire, and then burying it in lime; others heat it, "and bury it in sand; others heat it and bury it in cast-iron borings; others heat "it and bury it in dry sawdust, and some anneal it by surrounding it on all sides "in an iron box, with carbon, and then heat the whole to redness. This latter "process is undoubtedly the most effectual method of annealing steel; that is, "providing the steel is not heated to excess. When this method of annealing steel "is adopted, a layer of wood charcoal, coarsely powdered, is placed at the bottom of "an iron box, and then a layer of steel, upon this another layer of charcoal, and "upon that again another layer of steel, and so on until the box is nearly full, "finishing with a layer of charcoal. The lid of the box must then be put on, "and the box luted with clay or loam in order to exclude the air. The whole "may then be placed in a furnace or hollow fire and gradually heated to redness." —Ede 'On the Management of Steel,' chapter v. *See* also p. 318.

portion of the strength lost by punching. Experiments were made to ascertain to what extent this was the case.

Two ½-inch Bessemer steel plates were taken, manufactured by the Bolton Company, and they were punched throughout to receive the regular fastening of deck plates. One of these plates was then cut into two, and while one half was laid on one side, the entire plate and the other half plate were annealed. The entire plate was then tried at its work to ascertain whether the holes had been displaced in any way by the process. It was ascertained that no displacement had occurred, and that no inconvenience would arise in this respect from annealing after punching.

Pieces were then cut from the annealed and the unannealed halves of the separated plate. All these pieces were similar in size and shape, and they each had two ⅝ holes similarly situated in them. The pieces were 3·6 inches wide, and there was a little more than a diameter outside each hole. The results were as follows:—

TESTS of BESSEMER STEEL, Annealed and Unannealed after Punching.

Size of Sample exclusive of holes.	Breaking strain per square inch in tons.		Direction of grain.
	Unannealed.	Annealed.	
inches. inch.			
2·28 × ·475	19·39	..	Across.
2·28 × ·475	..	33·01	,,
2·33 × ·47	26·369	..	,,
2·23 × ·475	..	29·98	,,
2·18 × ·47	22·46	..	Lengthwise.
2·23 × ·48	..	34·929	,,
2·23 × ·47	22·781	..	,,
2·23 × ·48	..	35·397	,,
2·23 × ·475	18·885	..	Across.
2·23 × ·48	..	30·724	,,
2·28 × ·475	18·005	..	,,
2·28 × ·475	..	31·278	,,
2·23 × ·475	16·997	..	Lengthwise.
2·23 × ·475	..	33·522	,,
2·28 × ·475	23·891	..	,,
2·23 × ·475	..	33·876	,,
Mean	21·09	32·84	
Percentage		Gain 56	

It was found also, from the cold forge tests, that the ductility had been increased considerably.

Mr. Sharp, in a paper read at the Institution of Naval Architects, in April 1868, gave the results of some further experiments made by him as to the relative damage done by drilling and punching, and the recovery of strength due to annealing in Bessemer steel.

He first ascertained, from six experiments made on $\frac{3}{16}$-inch Bessemer plates, three of which were punched, and three drilled for $\frac{5}{8}$-inch holes, that the drilled plates broke at 35·22, 37·27 and 36·40 tons per square inch of original section, exclusive of holes; and that the punched plates broke at 26·69, 23·735 and 22·57 tons per square inch.

A number of steel plates were then prepared with riveted joints of various construction, as described in the table given below. The pieces were all cut from one plate, $\frac{5}{16}$ of an inch thick, lengthwise of the plate. One half were drilled, as shewn below, and the other half punched and annealed. They were then riveted up by a Garforth Machine, and reduced in the middle on a shaping machine, to an uniform width of $4\frac{1}{8}$ inches, the length of the straight reduced part being 6 inches. The rivets were $\frac{9}{16}$-inch in diameter, and were placed $1\frac{3}{8}$ inches from centre to centre. The plates were first riveted with "best best" double worked rivet iron, but the rivets gave way in every case but one. In this case there was double-riveting with double butt straps, the holes had been punched and the piece subsequently annealed. The plate broke through the upper row of holes, under a tensile strain of 39·12 tons per square inch of original area, exclusive of holes.

The iron rivets having thus proved to be too weak, the plates were again riveted up with mild steel rivets, and the results were as follows:—

TESTS of BESSEMER STEEL, Annealed after Punching, and Unannealed after Drilling.

Area of original Section, exclusive of holes.	Breaking Strain per square inch of Section.	Remarks.
Sq. inches.	Tons.	
·78125	36·22	Not perforated; elongated $1\frac{3}{16}$ ins. in 6 inches.
·7617	24·928	Drilled holes. Lap joint; three rivets; rivets sheared.
·754	26·254	Punched and annealed. Same as above.
·7617	42·33	Drilled. Lap joint; double riveted; 6 rivets; rivets sheared.
·754	37·0	Punched and annealed. Same as above.
·7617	23·68	Drilled. Single butt strap; single riveted; three rivets each side of butt; rivets sheared.
·754	24·53	Punched and annealed. Same as above.
·7617	39·25	Drilled. Single butt strap; double riveted; seven rivets each side of butt; plate and rivets gave way together.
·754	43·63	Punched and annealed. Same as above, except that only plate gave way.
·7617	36·62	Drilled. Double butt strap; single riveted; three rivets each side butt; rivets sheared.
·754	40·98	Punched and annealed. Same as above, but plate broke.
·7617	42·93	Drilled. Double butt strap; double riveted; seven rivets each side of butt: plate broke.
·754	39·11	Punched and annealed. Same as above.

Taking all the plates that gave way through the holes, we find that the drilled unannealed plates gave an average of 41·075 tons per square inch; and the punched annealed plates 41·24 tons per square inch.

It may now be interesting to state the result of tests similar to the foregoing applied to puddled steel (unannealed), the average tensile strength of which was 31½ tons lengthwise, and 27½ tons crosswise.

TESTS of PUDDLED STEEL PLATES, Comparative Effects of Punching and Drilling.

Size of Test-piece with holes deducted.	Breaking Strain per square inch of this Section.	Remarks.	
ins. in.	Tons.		
L 2·06 × ·27	25·17	One ⅝ in. hole.	Drilled.
L 2·07 × ·26	23·69	,,	Punched.
A 2·06 × ·27	23·822	,,	Drilled.
A 2·10 × ·27	19·217	,,	Punched.
L 2·06 × ·26	28·939	,,	Drilled.
L 2·03 × ·26	25·104	,,	Punched.
A 2·08 × ·26	27·644	,,	Drilled.
A 2·06 × ·26	22·404	,,	Punched.

In each of the above cases the test-piece was 2·7 inches wide, so that the ⅝-inch hole removed nearly one-fourth of the material.

In the cases following the test-piece was 4·24 inches wide and two $\frac{9}{16}$-inch holes were put through in a line with each other across the middle of the piece. Thus a little more than one-fourth of the material was removed.

FURTHER TESTS of PUDDLED STEEL PLATES, Comparative Effects of Punching and Drilling.

Size of Test-piece with holes deducted.	Breaking Strain per sq. inch of this Section.	Remarks.	
ins. in.	Tons.		
L 3·12 × ·245	31·07	Two $\frac{9}{16}$ in. holes.	Drilled.
L 3·12 × ·255	26·08	,,	Punched.
A 3·12 × ·255	28·751	,,	Drilled.
A 3·12 × ·26	20·95	,,	Punched.
L 3·12 × ·245	29·925	,,	Drilled.
L 3·12 × ·25	23·076	,,	Punched.
A 3·12 × ·255	31·894	,,	Drilled.
A 3·12 × ·25	25·474	,,	Punched.

The following tests were made to ascertain the benefit of annealing after punching in puddled steel plates ¼ of an inch thick, of which the tensile strength lengthwise was 34 tons, and crosswise 30¾ tons. The test-pieces were 4·06 inches wide and had two ½-inch holes punched in them placed at about the same distance from each other as from the edges. Four of the pieces were annealed after they were punched, and four were left unannealed.

TESTS of PUDDLED STEEL PLATES, Annealed and Unannealed after Punching.

Size of Test-piece with holes deducted.	Breaking Strain per square inch of this Section in tons.	
	Unannealed.	Annealed.
ins. in.		
L 3·0425 × ·27	26·644	..
L ,, × ·26	..	27·496
L ,, × ·28	25·705	..
L ,, × ·27	..	24·36
A ,, × ·27	22·99	..
A ,, × ·275	..	21·531
A ,, × ·27	24·208	..
A ,, × ·265	..	23·107

The cold forge tests were found to be improved by annealing. These experiments show that this mild material loses less of its tensile strength by punching than Bessemer steel does; and that it is not benefited in this respect by subsequent annealing. No experiments have been made with stronger puddled steel.

Some tests exactly similar to those last described were made with some $\frac{5}{16}$-inch and ⅜-inch crucible cast steel, the average tensile strength of which was lengthwise 26·63 tons, and crosswise 26·21 tons per square inch, the actual strengths being as follows:—

TESTS of ⅜-inch CRUCIBLE CAST STEEL PLATES Unpunched.

Size of Test-piece.	Breaking Strain per sq. inch of this Section.	Elongation in 6 inches.
ins. in.	tons.	ins.
L 3·125 × ·37	25·843	$1\frac{13}{16}$
L ,, × ·37	26·167	$1\frac{7}{8}$
L ,, × ·37	25·735	$1\frac{13}{16}$
L ,, × ·35	27·104	$1\frac{13}{16}$
A ,, × ·37	27·249	$1\frac{7}{8}$
A ,, × ·37	26·492	$1\frac{9}{16}$
A ,, × ·36	26·222	$1\frac{7}{16}$
A ,, × ·36	26·555	$1\frac{3}{4}$

These tests were made it will be seen with the $\frac{3}{8}$ inch plate; those following were made with the $\frac{5}{16}$ inch. Both sets of plates were made at the same time, and were of the same temper.

The tests with the $\frac{5}{16}$ inch plate as to the loss by punching, and the benefit of reheating after punching were as follows:—

TESTS of CRUCIBLE CAST STEEL PLATES, Annealed and Unannealed after Punching.

Size of Test-piece with holes deducted.	Breaking Strain per square inch of this Section.	
	Unannealed.	Annealed.
ins. in.	Tons.	Tons.
A 3·03 × ·32	26·315	..
A ,, × ·32	..	28·766
A ,, × ·32	24·122	..
A ,, × ·32	..	28·508
L ,, × ·31	25·159	..
L ,, × ·315	..	29·612
L ,, × ·31	24·361	..
L ,, × ·315	..	28·432

Here again the material is mild, the loss by punching is remarkably little, and the advantage of annealing is not very marked. It is, however, sufficient to command attention. The loss per cent. in punching is lengthwise 7, and crosswise $3\frac{3}{4}$. The gain per cent. of annealed over unannealed is 14 lengthwise, and 12 crosswise.

The cold forge tests were extremely good unannealed, and somewhat better annealed. This crucible steel has been used for boilers, but it is too costly for use in shipwork. It costs nearly twice as much as Bessemer steel.

The foregoing results of experience with steel in H. M. Dockyards, show what just ground there is for extreme caution in its use in the form of plates.* The injury sustained by Bessemer

* In a paper 'On the Employment of Steel in Shipbuilding and Marine Engineering,' published in the 'Transactions of the Institution of Engineers in Scotland for 1866,' Mr. Barber, surveyor to the Board of Trade, states that in his visits to various yards he has seen steel plates buckle and fly under the hammer, and crack during the process of riveting up; and that much greater care is required on the part of the workmen in working, countersinking and riveting steel plates, frames and beams, than in working those of iron. He adds that "having regard to the "facts which have come to my own knowledge, I should hesitate to recommend, "except under carefully considered regulations, its adoption for the *entire* con-"struction of *sea-going* ships;" but he considers that steel may be advantageously employed for sheer-strakes, deck-stringer, and tie-plates, and in the construction of masts and yards, for sea-going ships, and in the entire construction of river steam-boats with light draught of water.

steel in punching is shown to be very great, and the necessity for drilling, or for annealing after punching is evident.

There has not yet been sufficient experience in annealing punched Bessemer plates to bring out the special difficulties which may practically attend it. It may, however, be as well to remark, that when the plates are obliged to be removed from the furnace for cooling, and are covered with sand or ashes, there is said to be some risk of a kind of cementation or chemical combination between the constituents of the sand or ashes and the heated plate, which will induce brittleness in the plate. There may possibly be no good foundation for this suspicion in the case of such a low heat as that which would be employed, and should there be, the danger might probably be avoided by using a top plate to prevent actual contact between the plates to be used and the covering material.

Mr. Rochussen, of the Höerde Iron and Steel Works, Prussia, recommends a melted lead bath as superior to an ordinary reheating furnace.* His remarks on the subject, forming part of a Paper read by him before the Institution of Naval Architects, April, 1868, are as follows :—

" We often hear, and probably with justice, that steel is not " reliable, that it is not homogeneous, and people who have spent " a life in successfully treating iron, point with scorn at a steel " plate which has split or snapped, under circumstances where " iron would not have sustained any injury. Thus steel yards " have snapped in the truss, topmasts split in the fid-hole, plates " cracked on sharp curves, and saving the possibility of bad " material, inherent to all human production, the quality of the " steel, may for all that, have originally been unimpeachable. " Steel, as many a young beginner in life, had to be saved from " its friends; the belief in its breaking strain was at first un- " fortunately based upon the knowledge of tool steel, and it was " not uncommon to specify in construction, steel equal to 42 to " 45 tons per square inch. That metal, supplied by ambitious or " sanguine makers, did not work well, or committed suicide after " working, and the effect of such failures has taken some time to " work off. The fault did not always arise from want of homo- " geneity, because of all the varieties of iron manufacture, that of

* Lead baths have been used for many years for annealing and tempering wire.—E. J. R.

"making steel ensures more than any other an even distribution "of component particles. Having split upon the rock of hardness, "and knowing what steel could not do, it seems to be agreed by "makers and ship constructors, that we are within bounds of "safety by exacting a tensile strength of 32 tons per square inch. "This mild steel is adapted to every purpose intended by the "builder, viz., bending to curve and punching, but with greater "care than would be observed with iron, inasmuch that we have "not the free command of heat which iron allows with impunity. "And it is just this important point of heat which has to decide "the part which steel can play in shipbuilding, and the careless "application of which has been the primary cause of such mishaps "as may have occurred. Heating for the purpose of bending "through rolls where the material does not receive any elongating "pressure, under all circumstances entails a loss of tensile strength "both in iron and steel. But with steel we run two risks, either "the steel may become overheated, and, slowly cooling after "bending, remain very soft and weak; or, on the passage of the "heated plate from the furnace through the rolls, a keen draught "of wind, a shower of rain, or dragging over wet ground, may "chill the metal, and, while making it wholly, or in part, hard, "render it unfit for a construction requiring elasticity. Build-"ing yards are, as a rule, constructed for the requirements of "wooden ships, viz., plenty of room for bulky timber. There is "free access to wind and weather, and almost every operation "of iron and steel is carried on in the open air. The bending "rolls are seldom roofed over; the reheating furnaces are fixed "from the back or sides; thus when opening the doors there is "a rush of cold air upon the hot plate, which, then, getting a "chill, cracks when bending through the rolls, or when ham-"mered to a curve. Above all things, it is therefore necessary "either so to control the heat given in the yard within a limit which "*cannot* be exceeded, or to dispense with heat altogether. Pre-"vious annealing has been suggested as the sovereign cure for over-"heating hard steel, or as affording indemnity against the danger "of cracking when steel is worked too cold in the building yard. "But looking at this question from an economical point of view, "it must be settled whether the shipbuilder is expected to erect "furnaces sufficient for the annealing of a large number of plates, "and devote attention to the careful issue of an operation on a

" metal to which he is a stranger, or is the annealing to be done " by the steel manufacturer? Experience teaches that the anneal- " ing of cast steel gun barrels, cannon, &c., in furnaces of from " 10 to 18 ft. long, with charges of from $2\frac{1}{2}$ to 7 tons, takes five " to seven days. Assuming that ship plates from $\frac{1}{4}$ to $\frac{9}{16}$ in. thick, " in sizes up to 30 ft long and 6 ft. wide, will require only half- " day firing, and a day and a half cooling down, the annealing of " 300 tons of plates per week would entail such an extension " of plant and labour as materially to affect the price of the steel, " independent of which the annealed plate always has a rougher " surface, unsightly to the eye, and decidedly to be avoided in " a ship's skin. Annealing of large masses therefore being im- " practicable, the builder must have furnaces which cannot be " overheated, or he must work the steel cold. In order to settle " both points I have conducted a number of experiments with 127 " plates of all thicknesses of the quality usually supplied to ship- " builders. Collecting opinions on the Clyde and in England, I " found that in the different yards, the same steel in one yard, " heated simply to a temperature touchable by hand, in the other " yard to a cherry red, was reported to yield the same results, and " that therefore a low temperature, say of molten lead, would be " sufficient for all purposes. Operating with plates, equal to a " breaking strain of 38 tons, and an elasticity modulus of 21 tons, " the dipping into a bath of molten lead made no difference in the " strength of the steel, while it worked well in punching or bend- " ing, and I therefore conclude that while this heat could not " possibly hurt a good material, it may serve to let down the " temper of hard steel, while the expense of a lead bath, involving " scarcely any consumption of that metal, would be only a trifling " increase to the plant of a building yard, the more so since the " heat communicated by molten lead is instantaneous, though " limited to an unvarying temperature, while a coal furnace is " less certain, and the heating of one plate takes from eight to " ten minutes."

The toughening effect produced on a mass of steel when it is heated and then plunged into a bath of oil, has also been made the subject of experimental enquiry. When the steel is brought to a bright red heat and suddenly plunged into cold water, the effect produced, as is well known, is to harden the material and to render it brittle; but the effect of an immersion in a bath of oil is to increase its tenacity as well as to harden it. Mr. Anderson, of the Royal Arsenal at Woolwich, has introduced this mode of toughening the large masses of steel used for gun tubes, &c. The process is conducted as follows:—The mass of steel is placed in a furnace heated with wood, special precautions being taken to prevent the direct access of cold air to the lower part of the block or tube, and to secure uniformity of temperature throughout the mass. When the steel arrives at a bright red heat and has acquired a uniform temperature, it is drawn out of the furnace and immersed in a large tank containing oil, the tank being surrounded by a water space. By means of the water in this space the temperature of the oil is kept from rising greatly when the steel is immersed, and a constant supply of cold water and withdrawal of the heated water are kept up, in order to render the cooling as uniform as possible. The operation of cooling usually lasts about 12 hours. Full particulars of the operation will be found in Mr. Ede's interesting and valuable little book before referred to,* from which the preceding description has been drawn. He observes, that "this operation has in many respects the character of annealing yet it is something more; for it is quite "certain that a different change of the particles takes place, as it "leaves the steel in an intermediate state between hard and soft; "and when mild cast steel is required in this particular state, it "can only be accomplished by a slow process of cooling in oil, or "some other liquid of the same conducting quality, and which "requires as high a temperature to convert it into vapour." It is probable that the last-named characteristic of oil,—the very high temperature at which it becomes vaporized—is that which constitutes the principal difference in the effect produced by it on the steel from that produced by water; for the water is so rapidly converted into steam when the heated steel is immersed in it, and the heat is thus abstracted so suddenly from the steel, as to render its cooling very rapid, and consequently to reduce the toughness. Mr. Ede remarks on this point:—"It must not be imagined that "the oil penetrates into the pores of the steel and causes it to be "more tough; because if it were possible for the oil to enter the "pores, it would then lessen the strength of the attraction of "cohesion between the particles, and the tenacity of the steel

* London, Tweedie, Strand.

"would be in a measure destroyed. The effect is not in the "least owing to the penetrating quality of the oil; but the effect "is owing to its imperfectly conducting quality, which causes the "steel to part with its heat so slowly, and the elevated tem-"perature it demands to be converted into the vaporous state. "A covering of coal is also formed round the steel by the burned "oil, which greatly retards the transmission of heat. This slow "rate of cooling is necessary to favour a uniform degree of con-"traction, and give the steel a much longer time for the re-"arrangement of its particles, and to make the strain more uniform "throughout the body of the steel." The writer also states that the effect produced on steel by this operation, is to increase the tensile and compressive strengths, to render it harder and more elastic, to enable it to stand a much heavier blow from a hammer, and to make it less liable to be worn or indented than when received from the tilt or annealed in the usual way. Mild cast steel can be turned, bored, planed, slotted, chipped, or filed with well tempered tools, after it has thus been treated.

In order to give some more definite idea of the effect produced, we subjoin a table of the results of an experiment made at Woolwich, with which we have been favoured by Mr. Anderson, of Woolwich Arsenal, and which may be taken as a fair average of hundreds of similar testings. The length of the part operated on was 2 inches:—

TESTS of STEEL, Toughened in Oil.

Treatment of specimen.	Weight applied in tons per sq. inch.	Elongation per specimen.		Elasticity.	Permanent elongation per inch.
		Visible.	Permanent.		
		inch.	inch.	inch.	inch.
Tested in the soft state (as received)	13·98	·004	·001	·003	·0005
	15·46	·005	·003	·002	·0015
	34·79*	..	·372	..	·186
Tempered in oil at a low heat and then tested	15·01	·003	..	·003	..
	20·01	·0035	..	·0035	..
	25·01	·00425	..	·00425	..
	32·52	..	·009	..	·0045
	51·17*	..	·28	..	·14
Tempered in oil at a high heat and then tested	15·01	·0025	..	·0025	..
	19·78	·003	..	·003	..
	25·01	·0035	..	·0035	..
	32·06	·005	·0005	·0045	·00025
	32·29	·006	·0015	·0045	·00075
	33·66	·0085	·0035	·005	·00175
	53·78*	..	·225	..	·1125

* Breaking weight.

Mr. Kirkaldy has also made experiments on the effect produced on steel by cooling in oil, and has given the results in his very valuable work. He concludes from these experiments that the strength is greatly increased; that the higher the steel is heated (without injuring it by "burning") the greater is the increase of strength; and that a highly converted or hard steel receives a greater increase in strength and hardness from the operation than a soft steel does. He also considers that the material is toughened as well as hardened. Perhaps the most interesting portion of these experiments, from a shipbuilder's point of view, is that connected with steel plates. On this part of the subject Mr. Kirkaldy remarks that steel plates which have been hardened in oil and then riveted together, are fully equal in strength to an unpierced plate not treated in this manner; the loss of strength due to the riveting, being more than counterbalanced by the increased strength due to the hardening. Mr. Kirkaldy states that these are to be considered as mere preliminary experiments, and from their limited number it would, perhaps, be hardly proper to receive the above stated deduction as conclusive; although the interesting nature of the statement is such as to well merit further experiments in order to test its general applicability.

There is a point in connection with punching steel plates, which is deserving of attention. Mr. Sharp, the gentleman previously referred to, in the Paper read by him at the Institution of Naval Architects, says:—"It was suggested to me that steel plates might "be punched with holes sufficiently taper to do away with the "necessity of countersinking the holes with a drill, and also that "this method of punching would injure the plates much less than "that generally adopted. In order, therefore, to test it, a steel "plate $\frac{1}{2}$ in. thick was taken and cut in two. One piece was "punched across the middle with holes $\frac{11}{16}$ in. diameter, the punch "and die used being of the usual proportion, the clearance being a "bare $\frac{1}{16}$ of an inch; the other piece was punched with the same "punch, but the die had a clearance of $\frac{3}{16}$ of an inch, being $\frac{7}{8}$ in. "diameter, the holes formed being taper. The plates were then "cut on a planing machine into strips, and, when tested, gave the "following results:—

TESTS of STEEL PLATES, with Punched Holes of different Tapers.

No. of Plate.	Diameter of Punch.	Diameter of Die.	Net width across solid.	Net section in sq. inch.	Load on Plate.	Load per sq. in. net section.	Average load per sq. inch.
	in.	in.	in.		tons cwt.		
1	$\frac{11}{16}$	$\frac{7}{8}$	1·75	0·8476	27 6	32·208	32·527
2	$\frac{11}{16}$	$\frac{7}{8}$	1·6562	0·8022	26 7	32·847	
3	$\frac{11}{16}$	$\frac{3}{4}$ bare.	1·8593	0·9007	23 7	25·924	26·004
4	$\frac{11}{16}$	$\frac{3}{4}$ bare.	1·8125	0·8779	22 18	26·0849	

	in.
Thickness of plates	$\frac{31}{64}$
Pitch of holes	$1\frac{5}{8}$

Plates not annealed after being punched,

"from which it will be seen that the ultimate tensile strength "of the strips with ordinary holes averaged 26 tons per square "inch of net sectional area, whereas that of the strips with the "more taper holes was 32·527 tons per square inch, showing a "difference of 25 per cent. in favour of the taper punching.

"The fractures of the strips with taper holes showed much "tougher and more fibrous than the others, and it was observed "that it took much less power to punch them."

Experiments made at Chatham for the purpose of ascertaining to what extent the fact referred to above could be usefully applied in punching steel and iron showed that while there was no advantage in the increased clearance in iron, but rather a disadvantage, there was an appreciable gain in steel. The gain in $\frac{1}{2}$-inch Bessemer steel plates when the clearance was increased from $\frac{1}{16}$ to $\frac{3}{16}$ of an inch is shown in the following table:—

FURTHER TESTS of STEEL PLATES, with Punched Holes of different Tapers.

Thickness of Plate.	Sectional area after punching.	Breaking strain per square inch of this area.	Remarks.	
inch.	sq. inch.	tons.		
·50	1·08	22·222	Two $\frac{5}{8}$-inch holes.	Die $\frac{1}{16}$-inch larger than punch.
·49	1·024	26·123	,,	,,
·495	1·029	26·239	,,	,,
·495	1·029	24·538	,,	,,
·50	·99	27·777	,,	Die $\frac{3}{16}$-inch larger than punch.
·49	·98	27·295	,,	,,
·495	1·039	24·891	,,	,,
·495	1·039	29·461	,,	,,

These experiments appear to show that much of the injury which is done in punching Bessemer steel is due to the strain at the under side of the hole. Indications of this may be found on close examination in minute cracks round the hole. A little increase in clearance removes these, and gives the good result above indicated. A similar result is obtained by riming or enlarging the hole after punching, as shown by some of the earliest experiments recorded in this chapter.

This curious fact is analogous to what may be observed in the cold bending test of both iron and steel, but particularly the latter, viz., that if the piece has been sheared, and the ragged edge is left on, the piece will break sooner when the ragged edge is upward in bending than when it is downward. This disadvantage may be removed by merely filing off the edge.

Mr. Krupp says with regard to the treatment of cold cast-steel boiler-plates:—" In working the plates cold, all sharp turns, corners, "and edges must be avoided or removed. The surfaces of cuts "and rivet-holes must, before bending and riveting, be worked and "rounded off as neatly as possible, so that no rough and serrated places "remain after cutting and punching." He also recommends as a general rule that the plates should be thoroughly and equably annealed at a *dark-red* heat after every large operation, and that they should certainly have such annealing at the conclusion of all operations.

The directions given by him as to bending hot are as follows. "The plates should be heated, preparatory to bending, to a heat "*not exceeding* a bright cherry-red. Also the greatest possible "portion of the surface should be heated, and not merely the edge, "and even, where practicable, the whole plate should be equally "heated. By this means the strains which arise from local heating "and cooling, and which are much greater in cast-steel plates, on "account of their higher absolute and reflex density, than in iron "are, by the general heating of the plate, more equably distri- "buted. The thickest and toughest plates can be broken by local "heating, bending, and cooling.

"Bends which cannot be completed in one, or at most in two "immediately consecutive heatings, must be made gradually and "equably over the whole extent to be operated on."

In bending, for example, to an angle of 90°, the whole plate should first be bent through about one-third of the angle, then through another third, and finally to the complete angle.

"After the whole of these operations, the plate is to be equably "annealed at a *dark-red* heat, which will thus equalize the strains "caused by the previous working."

We have in a previous part of this chapter referred to the reduction in the scantlings of steel-built ships now permitted by Lloyd's Committee, which reduction must not exceed one-fourth the thickness prescribed for iron ships. In the steel ships previously built the reduction made in the thickness of the plating, &c., has been, in most cases, about one-third the thickness used in iron ships, or, in other words, the thickness of iron and steel have been nearly in the inverse proportions of their tensile strengths. In some ships the reduction made has amounted to about one-half the thickness of iron for a ship of the same dimensions, but this has been found to exceed the limits of safety.

In considering the amount of reduction which may be made in a steel-built ship, it is not sufficient to take account simply of the tensile strength of the material. As Mr. Scott Russell has ably pointed out, thin plating is more liable to failure from a compressive than from a tensile strain; and experience proves that the skin plating and the bulkheads of a steel ship are, when thin, very apt to be buckled or crumpled up under compression. The necessity for supplying proper local strength to the various parts of the structure should also regulate the proportioning of the thicknesses. For instance, in ships which often have to take the ground, it has been found that a thickness of plating amply sufficient to supply the required longitudinal strength, is altogether inadequate to supply the necessary resistance to the penetration of the bottom by stones or other hard substances; and in other vessels accidents have occurred through the side plating being broken in by striking a pier-head, or some floating body. The countersinking of the rivet-holes in thin steel plates is difficult, and the rivets have but comparatively little hold. It is also found that with the ordinary frame-space it is difficult to get the bottom plating fair when thin steel plates are used, the plates springing inwards between the frames, and the lines of the framing being distinctly shown by projections on the plating. With very thin plates there is the additional objection of the proportionally great reduction in thickness caused by corrosion on the inside and outside of the bottom. Attempts have been made to obviate this by galvanising the plates of some vessels with a very small draught of water, but the results of these experiments are not known. These are some of the considerations which, with those given previously depending on the quality of the material, prevent the reduction in the thickness of the steel used in shipbuilding from being carried to a greater extent than it is at present.

The following outline specification will give an idea of the scantlings which have been considered sufficient for vessels built in Liverpool during the late American war, designed for running the blockade. The vessel referred to is a paddle-steamer of 30 feet breadth of beam, and 1070 tons burthen, O. M. :—

	Material.	Scantlings.
		inches.
Keel-plates	Cast steel ..	$\frac{1}{2}$.
Centre keelson-plates, vertical ..	,,	$\frac{3}{8}$, tapering to $\frac{5}{16}$.
,, horizontal ..	,,	$\frac{3}{8}$.
Keel-angles	Puddled steel ..	$3 \times 3 \times \frac{3}{8}$ bare.
Frame-angles	,,	$4 \times 2\frac{1}{2} \times \frac{5}{16}$ bare, and $4 \times 3 \times \frac{5}{16}$.
Angles for bilge-keelsons, gunwales, fore and aft bulkheads, engine and boiler bearers and beams ..	,,	$3 \times 3 \times \frac{5}{16}$.
Angles for reversed frames	,,	$3 \times 2\frac{1}{2} \times \frac{5}{16}$ and $2\frac{1}{2} \times 2\frac{1}{2} \times \frac{1}{4}$.
Bulkhead stiffening bars	,,	$2\frac{1}{2} \times 2\frac{1}{2} \times \frac{1}{4}$.
Beams, bulb	,,	$6 \times \frac{5}{16}$.
,, angles for	Iron	$2\frac{1}{4} \times 2\frac{1}{4} \times \frac{3}{16}$.
Bulkheads, transverse	Bessemer steel	$\frac{8}{32}$ to $\frac{4}{32}$.
,, fore and aft	Puddled steel ..	$\frac{2}{16}$.
,, top and bottom plates	,,	$\frac{5}{16}$.
Floors, after body	Bessemer steel	$\frac{1}{4}$.
,, fore body	,,	$\frac{5}{16}$.

OUTSIDE PLATING OF BESSEMER STEEL.

	Scantlings.		
	Amidships.		Fore and aft.
	inch.		inch.
Row No. 1	$\frac{12}{32}$		$\frac{10}{32}$
,, 2	$\frac{11}{32}$		$\frac{10}{32}$, $\frac{9}{32}$, $\frac{8}{32}$
,, 3	$\frac{11}{32}$		,,
,, 4	$\frac{10}{32}$		$\frac{9}{32}$, $\frac{8}{32}$, $\frac{7}{32}$
,, 5	,,		,,
,, 6	$\frac{10}{32}$	to	$\frac{7}{32}$
,, 7	,,	,,	,,
,, 8	$\frac{9}{32}$	,,	,,
,, 9	,,	,,	,,
,, 10	,,	,,	,,
,, 11, or sheer-strake	$\frac{12}{32}$	,,	$\frac{9}{32}$
Butt-straps	$\frac{11}{32}$	,,	$\frac{8}{32}$
Liners	$\frac{11}{32}$	,,	$\frac{7}{32}$

	Amidships.	Fore and aft.
	inch.	inch.
Bulwark plating	$\frac{3}{8}$	$\frac{3}{16}$
Inside of paddle-boxes ..	$\frac{1}{4}$	$\frac{3}{16}$

	Material.	Scantlings.
		inch.
Beams : engine and boiler	Bessemer steel ..	$\frac{3}{8}$ to $\frac{1}{4}$
,, paddle	,,	$\frac{3}{8}$
Stiffening plates for 6 frames amidships ..	,,	$\frac{5}{16}$

On comparison with the scantlings required by Lloyd's Rules for a ship of the same dimensions, it will be found that the reductions of thickness made in this vessel amount to about one-half for outside and bulkhead plating, floors, frame and reversed angle-irons, &c.

CHAPTER XVII.

RIVETS AND RIVET-WORK.

In a structure such as an iron ship, where the connection of the various parts is almost entirely made by riveting, it is most important that the dimensions and positions of the rivets should be governed by a knowledge of the true value of rivets and riveted work. Numerous experiments have been made with the object of ascertaining the best dispositions and proportions of rivets in the joints of wrought-iron plates, among which those recorded by Mr. Fairbairn, Mr. Clark, and Mr. Doyne are the most valuable. Up to the present time, however, our information on this subject is far from being perfect or satisfactory, as no complete set of experiments has been made which would fairly represent the practice of different builders, or enable a comparison to be made between the various modes of riveting; the size, pitch, and arrangement of rivets; and the breadths of laps in joints. We are consequently compelled in arranging the fastenings of an iron ship to avail ourselves to the utmost of the few facts which have been proved by experiment, in order to make all the details of the structure conduce, as much as possible, to uniformity of strength.

In Chapter X., when describing the various arrangements of outside plating and its fastenings, we briefly noticed some of the features of punching and riveting which are common to all riveted work, and referred to those which are specially connected with the shell of a ship. In this chapter we propose to treat of riveted work generally, apart from any particular portion of the structure, and shall not scruple to repeat where necessary some of the statements previously made.

Rivets are generally made from the best quality of iron which is procurable, such as Lowmoor, Bowling, and other irons of the highest class. The mean tensile strength of rivet-iron is given by Mr. Clark as 24 tons per square inch of sectional area, and by Mr. Fairbairn's table in his treatise on Iron as 26·33 tons, while from the summary of the results of Mr. Kirkaldy's experiments it is found to be 25·98 tons.

Rivets were formerly made by hand, the bar-iron being cut into lengths while cold, and then heated in a furnace. After being heated the pieces were dropped into holes drilled in a cast-iron block, the depth of the holes being so much less than the length of the pieces as would allow the latter to project sufficiently above the surface of the block to have the heads formed. This operation was effected by beating down the projecting parts of the pieces with a few blows from a heavy hammer. Rivets are now made by machinery, several machines having been patented by different makers. The simplest of these is the Oliver machine, which is represented in the annexed engraving (Fig. 237). A is a spring-beam made of hickory or lancewood, B a hammer carrying a suitable die, C a foot-board which acts against the beam, and lowers the hammer on to the rivet, and D a lever for driving the rivet out of the lower die when finished. These machines are exclusively used for making rivets in some shipbuilding establishments, that of Messrs. Laird for example. Each machine is capable of making from $3\frac{1}{2}$ to 4 cwt. of $\frac{3}{4}$-inch rivets $2\frac{1}{2}$ inches long per day. Steam rivet-making machines of a larger and more elaborate nature are now, however, in use. In employing these,

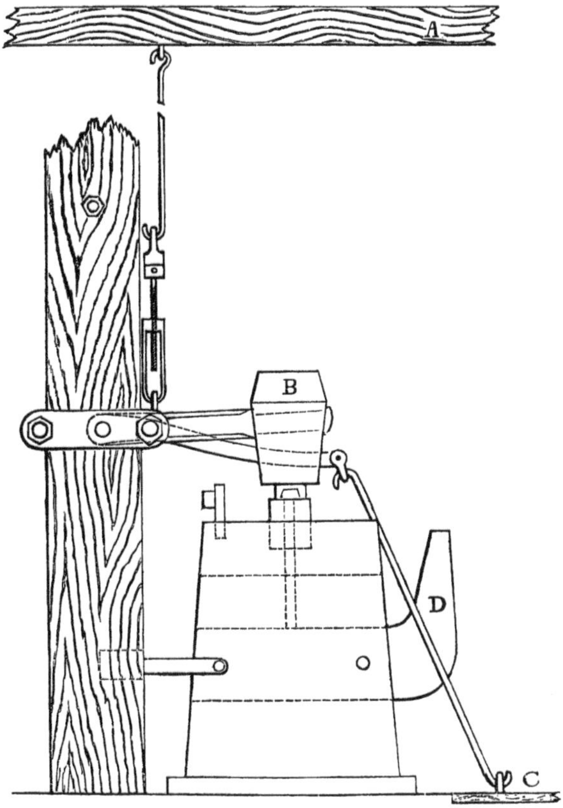

Fig. 237.

the bars of iron from which the rivets are to be made are generally heated in furnaces placed near the machine, and then cut into the required lengths by shears attached to it. These lengths are then placed, either by hand or by mechanism, into dies fixed either in a horizontal disc, or in the rim of a vertical wheel, and are thus brought under another die which descends upon the projecting portion of the length and flattens it out into a head. The motion given to the horizontal disc, or the vertical wheel in which are fixed the dies by which the rivet-shank is held, is of such a character as to bring the fixed dies consecutively under the die which forms the head, and there allow them to remain at rest while that operation is being performed. The desired form of head regulates, of course, the form of the die.*

The common form of rivet-head employed for shipbuilding is that shown in Figs. 238 to 240, and is known as a "pan" head; but hemispherical or "snap" heads are also used in some cases, especially for machine-riveting. In the outside plating, deck-stringers, &c., where the holes are punched, it is very desirable that the rivet should be formed with a conical enlargement under the head (corresponding in amount with the dies used with the

* The following is a brief description of a rivet-making machine in use at the Works of Palmer and Co. (Limited), Jarrow-on-Tyne. It was made by the patentees, Messrs. Brown, of Hylton, near Sunderland, who are large manufacturers of rivets, the cost being 350*l.* The rods are heated in a small furnace placed near the machine, and are drawn out by hand and fed into the shears attached to the machine, by which they are cut off to the required lengths, correctness being ensured by means of a stop which can be shifted into any position required. As soon as a length is sheared off the rod, it drops into a trough along which it is led to a vertical wheel. In the rim of the wheel there are 16 circular recesses or dies, each of which is brought in turn, by the revolution of the wheel, under a die moving vertically. The short lengths of rivet-iron having been guided into these recesses in the rim of the wheel, are driven down into them by a workman when they are not properly lodged, after which they are brought under the vertical die and have the heads formed. Both the vertical die, and the dies fixed on the wheel can be changed when it is desired to make a different form of rivet from that previously manufactured. It should be added that a stationary ring of iron within the rim of the wheel immediately beneath the vertical die, prevents the short lengths of iron from being forced in by the pressure which flattens out the end of the bar into the head; and that after the head has been formed, the finished rivet is driven out of the recess in the wheel by means of a punch working radially from the centre of the wheel. The wheel is moved through a quarter of a revolution between the time of the descent of the vertical die and the action of the radial punch. The rivets fall into a trough when they are completed, and a small stream of water is made to play upon them in order to cool them. Two men and a boy work the machine; the boy puts the cold rods into the furnace, one of the men feeds the heated rods into the shears, and the other works the lever which regulates the machine and drives the short lengths down into the dies. When in full work, the machine will make 50 rivets per minute.

punch), as shown in Fig. 142, p. 196, in order that the countersink formed in punching may be completely filled. This form is arrived at by a corresponding countersink being made in the dies in which the rivet-shank is held, and the pressure of the die which forms the head also serves to form the cone under the head. In beam-work, and the riveting of frames and reversed angle-irons, &c., the rivet-shank is generally of uniform diameter throughout. The length allowed for forming the rivet-head varies from two to two and a half diameters, the thickness of the head is usually from two-thirds to seven-eighths the diameter, and the diameter of the head ordinarily approximates to one-and-a-half times the diameter of the shank. When a conical form is given under the rivet-head an additional length of about $\frac{1}{4}$ inch is usually allowed in making the rivet.

It has been proposed recently to do away with the pan-head, and to have the rivet-head formed by a continuation of the cone which generally exists under the ordinary rivet-head. It will be obvious that the proposed kind of rivet, if formed with a suitable taper, would hold the plates together as effectively as that which is now employed; but that in the contraction consequent on cooling the rivet-head would require to be hammered into the hole to ensure its being filled, and that the laying up of the rivet-head, on which some builders so much insist, and which is required by the Liverpool Rules, would be rendered impossible with the proposed form of head.

Having thus briefly described the process of manufacturing rivets, we pass to the consideration of the various modes of forming the rivet-point, or, in technical language, of "knocking-down" the rivet. The first mode, which is known as "countersunk" riveting, is illustrated in Fig. 238, and is very largely employed in iron shipbuilding for outside and deck plating, and other work where a flush surface is required. A second form of point known as the "snap" point is shown in Fig. 239, and this kind of riveting is very commonly employed in the work in the interior of a ship, such as the riveting of the reversed frames and floor-plates to the frames, the fastenings of keelsons, stringers, bulkheads, beam-knees, &c., and the construction of made beams. The snap-point is sometimes formed on snap-headed rivets, and nearly always so in

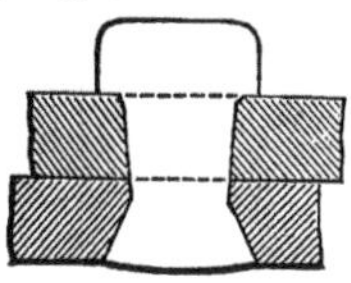

Fig. 238.

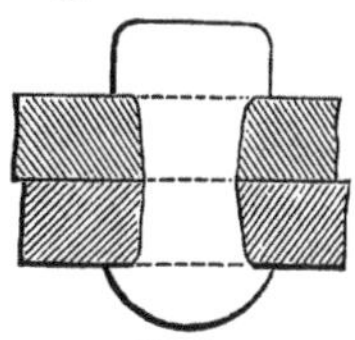

Fig. 239.

machine-riveting. The more common case is, however, that shown in the sketch where the rivet has a pan-head and a snap-point.

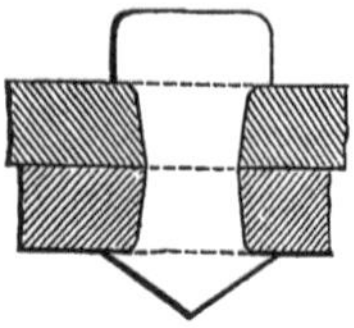

Fig. 240.

A third form of rivet-point is shown in Fig. 240, and is known as the conical or hammered point. This description of riveting is often used instead of snap-riveting for the work in the interior of a ship, especially where the diameter of the rivet exceeds $\frac{3}{4}$ inch, as snap-rivets of larger diameter would require very heavy hammers to be used in order to knock down the points quickly. Conical-point rivets are thought by many builders to be especially adapted for bulkhead and other watertight work, as they are supposed to draw the joints closer than they would be drawn by snap-pointed rivets. An objection has been made to the use of conical points on the ground that the great amount of hammering they have to undergo tends to injure the iron and make it crystalline; but their general use, especially in boiler-work, removes much of the force of this objection.

In some parts of the interior of an iron ship where the space in which the work has to be performed is very confined, a mode of riveting is adopted which differs from the preceding. It is illustrated in Fig. 241, and the rivet-point, which is conical, is that shown on the under side of the sketch. It will be remarked that the hole is much more countersunk under the head of the rivet than is usual with conical-pointed rivets, although not as much as is common at countersunk points. The rivet-head is nearly conical, and the shank is conically shaped under the head in order to fill the countersink. The object of the arrangement seems to be to make up for the reduced size of the rivet-head by increasing the cone under the head, and thus preventing any reduction in the holding power of the rivet.

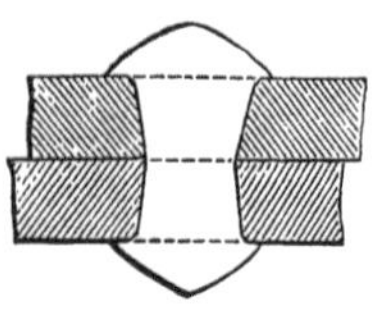

Fig. 241.

For countersunk points it is usual to have the length of the rivet about one diameter greater than the length of the hole if the rivet passes through two thicknesses, but when it passes through three thicknesses an additional $\frac{1}{8}$ inch is allowed. The allowance usually made for forming a snap-point is a length of one and a quarter times the diameter of the rivet, or, as other builders prefer to put it, about $\frac{1}{4}$ inch more than the allowance for a countersunk point. In knocking down a snap-point the workmen roughly beat the point into shape with their hand-hammers, and the formation of the point is completed by means of a cup-shaped die, held by one riveter and struck with a heavy hammer by the other. The depth of a snap-point is usually equal to the thickness of the rivet-head, and the shoulder around the hole is about $\frac{3}{16}$ inch or $\frac{1}{4}$ inch. Conical or hammered points are made entirely by the workmen with their hand-hammers. The allowance of length made for forming a conical point is in many yards the same as for a snap-point, but some builders allow one-fourth the diameter more for a conical than for a snap-point. On the Clyde it is customary to allow from $1\frac{1}{8}$ to $1\frac{3}{8}$ inch for all sizes of rivets for both snap and conical points.

In order to illustrate more fully the foregoing statements, we have given the following particulars of the riveted work of the 'Northumberland,' built at the Millwall Iron Works. Rivets with countersunk points were employed for the bottom plating up to the armour-shelf, the deck-stringers and tie-plates, the inner bottom, the bulkheads of the wing and shaft passages, shell-rooms and magazines, and the middle-line bulkhead aft. Snap-pointed rivets were used in the made beams, the transverse plate-frames, the transverse watertight bulkheads, the plating behind armour, the flat keelson-plate, and a large portion of the web-plates of the inner bottom. Conical-pointed rivets were employed in the middle-line bulkhead forward, a small portion of the web-plates of the inner bottom, the butts and angle irons of the vertical keel and keelson plates, the reversed or continuous frames, and the beam-knees. The following table shows the lengths and sizes of the rivets used at the Millwall Iron Works for riveting together two plates of the various thicknesses therein named. The thicknesses, diameters, and lengths are all given in sixteenths of an inch.

Thickness of Plates.	Rivets.		
	Diameter.	Length with countersunk point.	Length with snap point.
4	8	16	20
5	10	18	24
6	10	20	26
7	10	22	28
8	12	24	32
9	12	27	35
10	14	31	38
11	14	35	42
12	14	38	44
13	14	40	46
14	16	42	48
15	16	44	50
16	16	46	52

The proportion of the diameter of the rivet to the thickness of the plates it passes through, is a subject requiring some attention. Both Lloyd's and the Liverpool Rules give tables of the diameters of rivets required for various thicknesses of plating up to 1 inch, and in order to contrast their regulations with each other, and with the practice of H.M. Dockyards we have compiled the following table, in which the thickness of the plates and the diameters of the rivets are given in sixteenths of an inch.

TABLE of DIAMETERS of RIVETS for Plates of different thicknesses.

Thickness of Plates.	Diameter of Rivets.		
	Lloyd's Rules.	Liverpool Rules.	H. M. Dockyards.
5	10	8	8
6	10	10	10
7	10	12	12
8	12	13	12
9	12	13	14
10	12	14	14
11	14	14	14
12	14	15	16
13	14	16	16
14	16	18	18
15	16	19	18
16	16	20	18

Lloyd's and the Liverpool Rules may be regarded as representing the practice of private shipbuilders, and it will be seen that the most marked differences between the diameters required by the two sets of Rules are found in those given for the thickest plates, where the Liverpool Rules require much larger rivets than Lloyd's. The practice of H. M. Dockyards agrees more nearly with the Liverpool than with Lloyd's Rules, and in this case also the thickest plates have larger rivets than Lloyd's require, although not as large as those which are required by the Liverpool Rules. It need hardly be added that heavier plates than those provided for in the preceding table would not be employed in the construction of most merchant ships. For plates of more than 1 inch in thickness used in H. M. Ships, for flat keels, &c., it is usual to have the diameter of the rivet about $\frac{1}{8}$-inch or $\frac{3}{16}$-inch greater than the thickness of the plates. In cases where two plates, or a plate and angle-iron, of different thicknesses are riveted together, it is usual to estimate the diameter of the rivet from the greater thickness.

Mr. Fairbairn gives a table* in his work on 'Iron Shipbuilding' founded on his experience in iron construction, in which he states that for plates from $\frac{3}{16}$ to $\frac{3}{8}$ inch the diameter of the rivet should be twice the thickness of the plates, and that for plates from $\frac{1}{2}$ inch to $\frac{3}{4}$ inch the diameter should be once and a half times the thickness.

All the preceding regulations and information apply to the riveting of two thicknesses only. If there are more than two thicknesses the diameter of the rivet is increased, the usual increase being $\frac{1}{8}$ inch. In many cases the increase in diameter is governed by the breadth of iron in the lap, or by the breadth of an angle-iron flange; but where these considerations do not come in, a still greater increase is often made with advantage.

Having thus considered the proportions usually adopted for the diameters of rivets, it may not be amiss to notice some considerations partly of a practical and partly of a theoretical character, which, in some measure, fix the maximum and minimum diameters that may be employed. In iron shipbuilding nearly all the rivet holes are punched, and it is found, that, as a rule, it is not practicable to punch a plate with a punch (of ordinary temper) of less diameter than the thickness of the plate, nor even with a diameter equal to the thickness. This practical consideration therefore fixes approximately the minimum diameter of the rivet, although it is obvious that by drilling holes instead of punching them, this objection might be removed, and smaller rivets might be used if it were considered desirable. This, however, is not the case, and in practice the diameter of the rivet is hardly ever less than eleven-tenths of the thickness of the plate. In determining the maximum diameter of the rivet for a certain thickness of plate, we must avail ourselves of two experimental facts to which we shall refer more particularly hereafter. The first fact is that the shearing strengths of rivets are proportional to the sectional areas; and the second that the single shearing strength of a $\frac{3}{4}$-inch rivet made of Lowmoor or Bowling iron is, as nearly as possible, 10 tons. It consequently follows that for a rivet of d inches diameter, the shearing strength would be determined by the proportion:—

$$\text{Shearing strength} : 10 \text{ tons} :: d^2 : (\tfrac{3}{4})^2;$$

whence we have

$$\text{Shearing strength (in tons)} = \frac{160}{9} d^2.$$

* Also published in his 'Useful Information for Engineers.'

Now it is desirable to proportion the diameter d of the rivet to the thickness t of the plate, in such a manner that the rivet shall shear before the piece of iron between the rivet hole and the plate edge, known as the "bearing surface," is torn out. In iron ship-building it is customary to have the rivet-hole at least its own diameter clear of the plate edge, and consequently in tearing out the bearing surface the sectional area of the fracture would be approximately equal to $2\,d\,t$. Supposing the strength of the iron to be 18 tons per sq. inch of sectional area (which is a very fair value for the strength of the best iron plate when punched), we obtain

$$\text{Strain required to shear out the bearing surface (in tons)} = 18 \times 2\,d\,t = 36\,d\,t.$$

In order therefore that the rivet shall shear before the bearing surface is torn out, we must have the shearing strength of the rivet less than the shearing strength of the bearing surface or, substituting the values found above, we must have

$$\frac{160}{9}\,d^2 \text{ less than } 36\,d\,t,$$

$$\text{Or, } d \text{ less than } 2\tfrac{1}{40}\,t.$$

It thus appears that the maximum diameter of the rivet, for iron of this quality, should not much exceed twice the thickness of the plate. It will be evident from the preceding investigation that a variation of the strengths of the iron in the plate and rivet would give a different result. This does not, however, in the least affect the principle of the investigation, and it is proper to add that the values used in the calculation are such as are known to be very fair ones for iron of good quality. It may be of interest to state that the conclusions thus independently arrived at were confirmed by a reference to Mr. Fairbairn's experiments recorded in the first series of 'Useful Information for Engineers.' In the experiments numbered 22, 23, and 25 respectively, the thickness of the plates was ·22 inch, the diameter of the rivets was ½ inch, and the breadth of the lap three diameters. The diameter of the rivets was thus rather more than two and a quarter times the thickness of plates, and in all three instances fracture took place by the shearing out of the bearing surfaces. In the other experiments this result was avoided by increasing the breadth of the lap, but it is obvious that a decrease in the diameter of the rivet would have been equally effective. It will be remembered also that Mr. Fairbairn in his table fixes the maximum diameter of the rivet at twice the thickness of the plates.

M. Dupuy de Lôme gives some calculations of the theoretical proportions for riveting in his Report on iron shipbuilding previously referred to, and M. de Freminville has republished them in his recent work on practical shipbuilding. In these calculations a maximum value is obtained for the diameter of the rivet, from the consideration that its tensile strength should at least equal the pressure required to punch out the hole, into which the rivet point is beaten. The final result arrived at is, that the diameter may be made four times the thickness; but as shown by the preceding investigation, and by the experiments above referred to, this would make the shearing strength of the rivet far greater than that of the bearing surface, and the plate edge would be torn.

The question of the proper pitch of rivets, *i.e.* their distance apart from centre to centre, requires some consideration. Originally in boiler-making $\frac{3}{4}$-inch rivets were generally used, and the pitch was 2 inches, thus giving 18 rivets to a yard; hence "a yard of rivets" usually means 18 rivets. Mr. Fairbairn's table, previously referred to, gives the following pitches of rivets as the best that can be adopted in the joints of steamtight and watertight work. For $\frac{3}{16}$-inch and $\frac{1}{4}$-inch plates the pitch should be three diameters of the rivet, or six thicknesses of the plates; for $\frac{5}{16}$-inch and $\frac{6}{16}$-inch plates, two and a half diameters, or five thicknesses; and for $\frac{8}{16}$-inch, $\frac{10}{16}$-inch, and $\frac{12}{16}$-inch plates, two and two-third diameters, or four thicknesses. We have previously given Lloyd's and the Liverpool Rules for the pitch of rivets, but for convenience we may repeat them here. The pitch required by Lloyd's is not less than four diameters nor more than five, except in the riveting of the frame angle-irons to the reversed frames and bottom plating where the pitch required is nine diameters. The Liverpool Rules give eight diameters as the pitch of the rivets in the framing, and four diameters as the pitch for other work, in seams and butts, except in the butts where treble riveting is required, when the rivets in the rows farthest from the butts may have a pitch of eight diameters. It thus appears that the Liverpool Rules, requiring a smaller pitch and a larger rivet than Lloyd's, would considerably increase the amount of the fastenings in a ship. Mr. Price, the Chief Surveyor at Sunderland for the Liverpool Underwriters, states* that in a ship of 1149 tons there would be 12 per cent.

* In a paper in the 'Transactions of the Institution of Engineers in Scotland for 1866.'

more rivets if built according to the Liverpool Rules than if built according to Lloyd's, in addition to the greater diameter of the rivets used. Both the Rules give a considerably greater pitch than that given by Mr. Fairbairn; and it appears probable that the difference is accounted for by his table having been principally based on boiler work. It may be added, that since the stiffness of plates increases very rapidly with an increase of the thickness, it is probable that the pitch of the rivets might be made greater in proportion to their diameter as the plates become thicker. Mr. Fairbairn's table, on the contrary, gives a smaller pitch for the thicker plates.

The general practice of shipbuilders is in accordance with the regulations of Lloyd's and the Liverpool Rules, but it has been thought desirable to test experimentally what pitch of rivets would secure watertightness in a joint. Mr. Samuda made experiments having this object, on a box about 33 inches square, of which the top and bottom were formed of $\frac{3}{4}$-inch plates, and the sides and ends of an angle-iron 6 by 4 by $\frac{5}{8}$ inches having a reversed angle-iron 4 by 4 by $\frac{5}{8}$ inches riveted to the upper edge. The rivets used were $\frac{7}{8}$-inch, their pitch in the bottom plate being eight diameters, and in the top plate four diameters. All the joints of the plates and angle-irons were carefully caulked, and the box having been filled with water, a pressure of about 10 lbs. to the sq. inch was obtained by means of a pipe leading to a reservoir about 23 feet above the box. It was then found that the box was perfectly watertight, and no indications of a leak were observed where the pitch of the rivets was eight diameters.

Other experiments have since been made in H.M. Dockyard at Chatham on a tank specially constructed for the purpose. The length of the tank was about 8 feet, its breadth 5 feet, and its depth 3 feet 8 inches. The ends were formed of $\frac{5}{8}$-inch plates lap-jointed as is usual in bulkheads, with a frame of angle-iron 4 by $3\frac{1}{2}$ by $\frac{5}{8}$ inches connecting the plates with the sides, top, and bottom of the tank. The rivets used in the ends were $\frac{7}{8}$-inch, the pitch in one end being from four to five and a half diameters, and the pitch in the other end from six to eight diameters. The riveting of the 4-inch angle-iron frame was performed with 1-inch rivets, having a pitch of eight diameters. The plating in the top, bottom, and back was $\frac{7}{8}$-inch thick, and the rivets were 1-inch. In both the top and bottom there were three strakes of plating lap-jointed, and at the centre of the length of the middle strake there was a single riveted butt. The back was formed of one plate, and fitted with a watertight man-hole. One seam of the top and bottom plating was single-riveted, and one seam double riveted, the pitch of the rivets being varied from four to eight diameters. The $\frac{7}{8}$-inch rivets used had snap-points, and the 1-inch had countersunk points. The front of the tank was arranged and riveted similarly to the top and bottom, but the plates forming it were put together out of place, and when the riveting had been completed the front was secured, by means of 1-inch nut and screw bolts having a pitch of four diameters, to the flange of an angle-iron frame worked on the edges of the tank. The experiments conducted with this tank were made under pressures varying from 5 lbs. up to 20 lbs. to the sq. inch. The results were as follows:—

I. With 5 lbs. pressure, a very slight leak was observed in the riveting of the butt of the front where the pitch was about four and a half diameters; and another slight leak was observed at the end where the pitch was eight diameters.

II. With 10 lbs. pressure, these two leaks slightly increased, and the butt of the bottom plating commenced to leak, the pitch of the rivets at the new leak being about five and a quarter diameters.

III. With 15 lbs. pressure, the leaks above described increased, but no additional leaks were observed.

IV. With 20 lbs. pressure, the leaks were all increased, and when the pressure had been kept up for an hour and a half, the end in which the pitch of the rivets was from six to eight diameters became very leaky.

No leak was observed in the end where the pitch did not exceed five and a half diameters, and hence it is fair to conclude that the leak in the butt of the front plating, which began under a pressure of 5 lbs., was due to defective workmanship. On the whole, we may infer from these experiments, that within the limits of these thicknesses of plating, and under a continued pressure of from 10 to 20 lbs. per sq. inch, the pitch of from four to five and a half diameters is required for watertight work; and that the pitch of from six to eight diameters is too great for a watertight joint under a pressure of 5 lbs. to the sq. inch. These experiments clash in their results with Mr. Samuda's, but having been conducted on a larger scale, and in a manner more corresponding to the actual construction of a ship, may, we venture to think, be regarded as more conclusive. It appears therefore that the general practice of iron shipbuilders in adopting a pitch of from four to five diameters, is probably not far removed from the best arrangement, and it certainly has the merit of being safe.

We now come to the consideration of the practical part of the operations of punching and riveting. When the work has been fitted and the fastenings marked, the holes for the rivets are usually punched, care being taken to punch from the faying-side, the operation of punching being conducted in the manner previously described in Chapter X. for outside plating. It is unnecessary to repeat the description there given, but it may be added that in cases in which the holes cannot be conveniently punched at the machine, a portable screw press, known as a "bear," is employed to punch them, or the holes are drilled, after the work is in place. According to Mr. Clark the pressure required to punch a hole 1 inch in diameter in a $\frac{3}{4}$-inch plate amounts to 46 tons.* As the surface to be sheared in punching a hole very nearly equals the product of the circumference of the hole by the thickness of the plate, it seems probable that the pressure required to punch a hole d inches in diameter in a plate t inches thick would be found from the proportion :—

$$\text{Pressure required} : 46 \text{ tons} :: d'' \times t'' : \tfrac{3}{4}'' \times 1'',$$

$$\text{Whence we have, as the required pressure, } \frac{184}{3}\, d\, t \text{ tons.}$$

This investigation is of no practical importance as there is always a considerable surplus of power in the punching machines employed, and in plates exceeding $1\frac{1}{8}$-inch in thickness it is a very common practice to drill the holes. Drilled holes are almost

* In the 'Practical Mechanics' Journal for 1853' there is given an elaborate table of the results of experiments made by Mr. Jones at the Great Western Steam-ship Works, Bristol, on the pressures required to punch holes of various sizes in plates of various thicknesses. Mr. Jones gives about 60 tons as the pressure required to punch a hole 1 inch in diameter in a $\frac{3}{4}$-inch plate, this result differing considerably from that obtained by Mr. Clark. The experiments made were very numerous, the thicknesses of the plates and the diameters of the holes punched varying from $\frac{1}{8}$ inch to 1 inch. The tabulated record of the results will well repay a careful study, but need not be given here.

always adopted in thick keels and garboards where the rivets used are of large diameter. It will be remembered that we have previously stated the arguments adduced by the advocates of the substitution of drilled holes for punched holes, and have noticed the more extended use of drilling machines; but as punching is now, and seems likely to remain, the common and almost universal practice of iron shipbuilders, it may be well to add a few remarks on the subject here.

It has been established by direct experiment that when a row of holes has been punched in a plate, the tensile strength of the iron left between the holes is considerably reduced, whereas with drilled holes the strength remains almost unaltered.* This is still more strikingly the case when holes are punched in steel plates as shewn in the preceding chapter. In iron plates the reduction in strength made by punching is considerable. For example, with the pitch of rivets usually adopted for watertight work the tensile strength of the iron taken along a section through a row of rivet holes has been found to be from 16 to 18 tons per sq. inch of sectional area, whereas for the unpunched plate it has been upwards of 22 tons per sq. inch. It will be obvious that the large factors of safety employed in proportioning the scantlings of the various parts of ships, and other wrought iron structures, render these experimental facts comparatively unimportant in relation to the consideration of the ultimate or breaking strengths. But with regard to the arrangement and proportions of butt fastenings, where the principal object aimed at is the preservation of continuity of strength, it becomes absolutely necessary to take into account the reduction of strength caused by punching, in order to arrive at a

* Mr. W. H. Maynard, who has made experiments upon punched and drilled bars cut from the same plate, parallel to each other, states that the holes were about 1 inch in diameter, and drilled in two of the bars, the other two being punched so as to leave exactly the same section of metal in the plate as in the case of those drilled, viz., about 1·5 square inches at the part reduced; and that the results were as follows showing a mean of 19 per cent. in favour of the drilled plates:—

	Drilled bar broke with a tensile strain in tons.	Punched bar broke with a tensile strain in tons.	Difference in tons.	Difference per cent. in favour of drilled iron.
1st,	30½	26	4½	17
2nd,	31½	26	5½	21
Mean,	31	26	5	19

correct result. It may be added, that although this is not commonly done in arranging butt fastenings, in the investigations of riveted work given further on in this chapter we shall take this reduction of strength into account, and assuming the tensile strength of the plate iron lengthwise to be 22 tons per sq. inch before being punched, shall, in most cases, take 18 tons as the strength of the iron left between the holes, after the punching has been performed.

When the punching or drilling has been completed, the various parts of the work are fixed in position and temporarily secured, either by nut and screw bolts, or, as is far more commonly the case, by means of pins and cotters, or forelocks. When these operations have been completed and the work "closed," the riveting is commenced. Each "set" of riveters consists of two riveters, a "holder-up," and one or two boys. The rivets are heated in a portable hearth placed as near as is convenient to the work. The blast required is nearly always supplied by hand-bellows attached to the hearth, which are so constructed as to keep up a continuous blast, and are worked by one of the boys. In the construction of one or two large ships the rivet-hearths have been put in communication with a blowing-engine and the hand-bellows dispensed with; this can, of course, only be done under special circumstances, and in the cases referred to the vessels were built close alongside of the machine-sheds, so that the communication with the blowing-engine was readily effected. The rivets are, in some instances, put into holes in a plate placed in the fire, by means of which the heads are in some measure protected from the great heat, while the points are brought to the required temperature; but, in other cases, the rivets are simply put into the fire, the plate being dispensed with. The heads should be made moderately hot in order to be made to bear fair on their work. The shank is brought nearly to a welding heat, and then the rivet is taken or thrown by one of the boys to the holder-up, who places it in the hole, and, after having driven the head well up by a few heavy blows, holds upon it with a large hammer or a tool called a "dolly." The riveters immediately commence their work by striking a few blows around the rivet in order to bring the plates into close contact, and then beat down the point. If the hole is countersunk, any superfluous iron which may remain after the countersink is filled is cut off, and the riveters then strike again. If the rivet has a snap-point, the riveters beat it down roughly to shape, and then finish it by means of a cup-shaped die, as previously explained. If the point is conical, the riveters complete the operation with their hammers. When the point has been roughly finished, the riveters hold their hammers on the point, one behind another, while the holder-up strikes a few blows on the head. In some yards all rivet heads are "laid-up," *i. e.* are beaten all round the edge in order to make them fit closely to their work; but in other yards only the rivets in watertight work, or in some cases but a portion of them, are thus treated. After this has been done the dolly or hammer is replaced, and a few more blows are struck on the plates and the rivet point by the riveters. It is generally found that the adjacent rivet last put in requires dressing up after the new rivet is knocked down. The hammers used by the riveters vary from 2 to 7 lbs. in weight, according to the character of the work, and the size of the rivets. The holding-up hammers weigh from 10 to 40 lbs. The dolly weighs from 10 to 30 lbs., and consists of a short bar of iron which the workman holds in his hand. It is especially suited for light work, and can only be used in places where the holder-up can gain access to the rivet head; but for the heaviest work, and in places difficult of access, the holding-up hammer is employed.

The preceding description assumes that the holes in the two or more thicknesses through which a rivet passes are coincident, but in practice it is often found that holes are partially "blind," as described in Chapter X., and illustrated by Fig. 143, p. 197. The manner in which the workman usually gets over this difficulty by the use of a steel drift punch,* has been previously explained, and its evil effects have been noticed. It will be sufficient, therefore, to add, that in cases where the drift punch will not make the hole good, it must be "rimed" out with a rimer, and if this will not suffice a "rose-drill" must be employed.

Hand riveting is necessarily adopted for the greater part of the work of an iron ship, and in the great majority of yards it is exclusively employed. Machine riveting is, however, employed in some yards for such parts of the work as can be conveniently brought to the machines, as for instance, frames and reversed frames when put together before being hoisted, beams, &c. It becomes necessary, therefore, to give a brief sketch of the process

* A serrated drift punch, suggested by Mr. Willing, is being tried in H. M.s dockyards.

of machine riveting. Riveting machines were, we believe, introduced by Mr. Fairbairn. Mr. Clark gives a description of those used in the construction of the Conway tubular bridge, where they were found particularly useful. In these machines the piston was 48 inches in diameter, but had only a 9-inch stroke, working horizontally. The end of the piston terminated in a cup-shaped die, and was pressed against a corresponding die fixed on the head of a cast-iron pillar which sprang from the base of the machine. The steam was used at a pressure of 40 lbs. per sq. inch, and exerted a pressure of 32 tons upon the rivet. Other machines—including several hydraulic machines—have been patented and brought into use, but we need only add with respect to their construction that the principle of having a moveable die on the end of the piston rod, and pressing it against a fixed die, is common to all, or nearly all, of them. The work to be riveted is temporarily secured by means of cotters and pins, and is brought into the position required for riveting by means of cranes. The rivets are placed in the holes by hand; when this has been done and the rivet head placed in the fixed die, the steam is let into the cylinder, and the die on the piston is pressed forward upon the red hot rivet point and squeezes it into form. As stated previously, the rivets used for machine riveting generally have snap-heads, and the points are also snap-shaped.

Considerable discussion has taken place as to the superiority of the riveted work done by machines over that done by hand. Mr. Stephenson had specimens of both kinds of work planed down through the centre of the rivets, and it was found that the quality of the work was very nearly the same in both. There can be little doubt that, with proper care, rivets closed by the machines must fill the holes and make sound work, whereas in hand riveting much depends on the skill and strength of the workman. Experiments were made in the construction of the Conway bridge which showed that holes made purposely untrue were completely filled by rivets closed by the machine, and the fact that the work can be performed so much more rapidly than it can be by hand, gives an additional advantage to machine work, as the rivet cools after the point is formed, the joints are drawn closer by the contraction of the rivet, and the friction of the surfaces is greatly increased. Mr. Fairbairn states that 12 rivets can be put in per minute, and that the machine saves time and labour in the proportion of 12 to 1. Mr. Grantham gives a description of Messrs. Garforth's riveting machine, and says that the makers state that 6 rivets can be put in per minute, while 20 rivets per hour is the work of a set of riveters. One advantage of hand riveting, pointed out by Mr. Clark, consists in the close contact into which the plates are brought by occasional blows struck on the plates, and the consequent prevention of the oozing away of the rivet in thin laminæ between the plates, as sometimes happens with machine riveting unless the plates are struck around the rivet hole before the rivet point is formed. On the whole, machine riveting seems to make better and cheaper work than hand riveting; but as before remarked, it can only be used for such work as can be brought to the machine, and nearly all the riveting is still done by hand.*

Before leaving the subject of riveting machines, it may be well to state that a compressed air machine is now coming into use for riveting purposes. The air engine is mounted upon wheels for convenience in moving it about, motion being given to a small piston by means of gearing driven by a belt from a steam-engine shaft. The air which is in this way compressed is led off by elastic tubes to hand machines, in each of which a small piston is driven at a very high velocity. This piston strikes upon a die at the end of the hand machine, and by its intervention beats up the rivet point. The die is, of course, moveable, and fitted to suit any required size of rivet. We have timed a machine of this description at Messrs. Forrester's Works, Liverpool, and have satisfied ourselves that with two rivet fires, or a small heating furnace, and the rivets quickly served, four rivets a minute can be knocked down in straight-

* Mr. Robert Harvey, of Cook's Engine Works, Glasgow, in a letter to the 'Engineer' of September 17, 1858, reasons as follows in favour of steam-riveting:— "It is admitted on all hands that hammering iron when nearly cold has a tendency "to destroy, more or less, its fibrous character; thus, a crystalline character must be, "to a certain extent, assumed in all rivets worked by hand in the usual manner, and "hence, besides aiming at economy by the use of steam for riveting, it is very de-"sirable that the rivets should be finished in as short a time as possible, and without "that succession of blows so destructive to the texture of the iron, and without which "hand-riveting cannot be effected; and, further, the conical form of the head generally "given to hand-work does not present the same strength as the semicircular head "which the steam-riveting die gives such a facility for producing, and these being "finished *red-hot*, the contraction which takes place brings the plates more closely "together. The strong form of the head makes this contraction peculiarly efficient, "so that the 'nip' between the heads goes largely to make up for the weakening "effected by the piece bitten out by the punch. I have frequently planed up to the "centre of the rivets portions of plates so riveted together, and in most cases have "only been able to detect the rivet from the plate by moistening the surface with an "acid, which at the same time revealed the beautiful compressed rose-like form of the "head."

forward work. One man is required to work the portable part of the machine, which weighs 23 lbs., and another to hold up behind.

The following information with respect to the number of rivets which are put in for a day's work may prove of interest. Mr. Grantham states that a set of riveters can put in 100 rivets in a day's work of ten hours, and will increase the number to 140 if employed on piecework. From information obtained from various sources we conclude that from 130 to 150 rivets may now be considered as a fair average day's work of a set of riveters, when allowance has been made for the various sizes of the rivets, and the different positions of the work. It also appears that on piecework the men can earn a day and a half's pay per day. In order to supply more definite information on this matter, we have given the following tables, of which the first represents the practice of one of the largest private firms, and the second that of H. M.'s dockyards.

Position of Rivets.	Diameter of Rivets and description of Points.	Number of Rivets for a day's work.
In outside plating	1-inch, countersunk.	85 to 90
In bulkheads, &c.	$\frac{3}{4}$-inch, snap.	180
In made beams, &c. ..	$\frac{5}{8}$-inch, snap.	200
In beam ends	1-inch to $\frac{13}{16}$-inch, hammered.	50
In deck plating, &c. ..	$\frac{7}{8}$-inch, countersunk.	140

The average week's wages received in private yards, when on day work, are 28 shillings for each riveter, from 18 to 24 shillings for the holder-up, and 6 shillings for each boy. The prices per hundred allowed vary in different yards, and, of course, are always governed by the sizes and positions of the rivets. At the present time the riveters on the Clyde, who are nearly always on taskwork, take the work at about the following prices:—$\frac{3}{4}$-inch rivets in frame and reversed angle-irons at from 5 to 6 shillings per hundred; $\frac{5}{8}$-inch rivets in beam work at from 3 to 4 shillings per hundred; rivets in outside plating (averaging $\frac{3}{4}$-inch rivets) at from 7 to 8 shillings per hundred. The average price per hundred may be taken at from 5*s.* 6*d.* to 6*s.* 6*d.*, and it must be stated that the Clyde prices are now very low,* and cannot be taken as the fair average of those given in private yards. In the Royal dockyards the rivets are allowed for per dozen instead of per hundred, and a

* This was written in April, 1868.

set of riveters consists of two shipwrights at 27 shillings per week for each, one holder-up at 18 shillings, and two boys, each of whom receives 7 shillings per week. In the following table a detailed statement is given of the prices at present allowed per dozen rivets, number of rivets for a day's work, &c.

Description of Work.	Diameter of Rivets.	Number of Rivets for a day's work.	Rate per dozen.
	in.		s. d.
Riveting of bottom plating, &c..	$1\frac{1}{8}$	48	3 8
	1	66	2 8
	$\frac{7}{8}$	80	2 2
	$\frac{3}{4}$	116	1 6
	$\frac{5}{8}$	150	1 2
Riveting of bulkheads, topsides, stringers, &c., ...	1	67	2 7
	$\frac{7}{8}$	87	2 0
	$\frac{3}{4}$	122	1 5
	$\frac{5}{8}$	160	1 1
	$\frac{1}{2}$	188	0 11
Riveting of beams, carlings, frames, etc., out of place	1	84	2 1
	$\frac{7}{8}$	100	1 9
	$\frac{3}{4}$	147	1 2
	$\frac{5}{8}$	190	0 11
	$\frac{1}{2}$	230	0 9

It will be seen that the number of rivets in a day's work is just now less in H. M.'s dockyards than in private yards, but it should be remembered that this is in part explained by the fact that the average length of the day's work in the Royal yards is considerably less than 10 hours, which is the recognised length of the day in private yards, that the supervision is much more stringent, and that the prices never rise.

One of the most important features of riveted work is the shearing strength of rivets under various circumstances. This matter has been made the subject of experiment by Mr. Fairbairn, Mr. Clark, and others, and the conclusion arrived at by all experimenters has been, that the shearing strength is directly proportioned to the area of the section sheared through. In his work on the 'Britannia and Conway Bridges,' Mr. Clark gives the full details of the experiments made on the shearing of rivets, and rivet-iron, and points out the fact that while a rivet passing through the lap of two plates has to be sheared once only, a rivet passing through three thicknesses—united by a "chain-joint," as shown in section in Fig. 243, page 353,—would require to be sheared twice, and, generally, if n be the number of plates combined by a pin or rivet, and the pin that unites them fails, it must be sheared in n-1 places. These considerations are based upon the assumption that the pin or rivet is well fitted in all the holes, for if the hole is not filled and the pin can be inclined in the hole, the strain ceases to be uniformly distributed over the section of the pin, and the effective shearing strength becomes reduced. In shipbuilding, the principal application which can be made of these experimental results consists in the use of double butt-straps for keelsons, stringers, deck plating, and other work in the interior of a ship, by which means the shearing strength of the butt fastenings is very nearly doubled. The same consideration enters into the strapping of sheer and other strakes when worked in two thicknesses. In bridge construction, however, the pins or rivets pass through several thicknesses of plates, and then the importance of these experimental facts is fully recognised and acted upon.

Sir Charles Fox has made an important and interesting discovery with respect to the size of pins used for connecting the flat links of the chains of suspension bridges, which has relation to this question of shearing. He communicated the results of his experiments to the Royal Society in 1865, and as they are applicable, in some measure, to the arrangement of some of the details of iron-ship construction, we propose to give a brief account of them here, taken from a reprint from the Society's 'Proceedings.'

Chains for suspension bridges are now usually composed of several flat bars of equal thickness throughout, placed side by side, and having the ends swelled edgeways so as to form heads,

Fig. 242.

through holes in which pass the pins that couple the bars together. A sketch of one of these links is given in Fig. 242. The rule commonly employed in determining the size of the pins, before these experiments were made, was to make the cross section of the pin at least equal to the sectional area of the smallest portion of the link. This rule was based on the consideration that about the same force is required for shearing, as for breaking wrought-iron by extension. In the manufacture of the chains for the great suspension bridge over the Dnieper at Kieff, it was considered very desirable to determine experimentally whether or not they were well proportioned, and accordingly a proving machine was prepared for the purpose of testing them. The links were 12 feet long from centre to centre of the pin-holes, 10¼ inches wide by 1 inch thick in their body or smallest part, with a head at each end also 1 inch thick swelled out to 16½ inches in width, so as to allow for pins 4½ inches in diameter. The cross sectional area of the pins was 15·9 sq. inches, or rather more than 50 per cent. in excess of the cross-sectional area of the link at its smallest section, thus giving them upwards of one-third more section than would be required by the ordinary rule. The iron in the links was of a very high quality, its tensile strength being about 27 tons per sq. inch, so that the strain of 270 tons would have been required to break the link at its smallest section. When proved in the machine, however, it was found that with the 4½-inch pins a strain of about 180 tons only was required to break the link, and the head tore across in the widest part, in wake of the centre of the pin-hole. This unexpected result led to the belief that the size of the heads was insufficient, and a few experimental links were prepared with the heads 2 inches wider than before, but these were found to require no additional force to tear them asunder, and it became obvious that fracture arose from some other cause. It was observed, on attempting to adjust the piece broken off to the position it occupied before the fracture took place, that while the fractured surfaces came into contact at the outside of the head, they were a considerable distance apart at the edge of the pin-hole. This proved that the various portions of the head had been subject to very unequal strains during the application of the tensile force. Upon careful examination it was found that the pin-hole which originally was round had become pear-shaped, and that the iron in the bearing surface of the hole, which during the application of the load was in a state of compression, had become considerably thickened, while the other portion of the iron around the hole, being subject to tension, had been thinned down. Fracture commenced in the thinned part of the head, and when it had once commenced, the same strain would obviously suffice to rend through the head, if its application were continued, no matter what the width of the head might be. It thus became evident, that there was a certain area of the semi-cylindrical surface of the hole having a bearing on the pin, proportionate to the transverse section of the body of the link, quite essential to its having equal strength in all its parts; and that any departure from this proportion must bring about, either waste of iron in the body of the link, if the pin were of insufficient size, or waste of metal in the heads of the links and in the pins, if the latter were larger than necessary for obtaining the fixed proportion of areas. This is the statement of the principle put forward by Sir Charles Fox in the paper above referred to, and additional experiments fully proved the truth of the deductions made from the former experiments. The links used in these further experiments were of exactly the same dimensions as those previously broken, but the pins and pin-holes were increased in diameter to 6 inches, and consequently the bearing surfaces were increased from 7 to 9·4 sq. inches. By this alteration the force necessary to produce fracture was brought up to nearly 240 tons, and from subsequent experience it has become evident that had the pins been increased to $6\frac{1}{2}$ inches diameter, and the bearing surface to 10·2 sq. inches, the proper proportion would have been arrived at, while with the 6-inch pins about an inch of the body of the links was wasted. From these experiments it appears that in order to obtain the full efficiency of a link the area of the semi-cylindrical bearing surface must be a little more than equal to the transverse sectional area of the smallest part of the body. In order to make allowance for the fact that the iron in the head of a link is never quite as strong as that in the body, Sir Charles Fox states that he would make the diameter of the pins two-thirds the width of the body of the link. He also considers it desirable to have the sum of the widths of the iron on both sides of the hole 10 per cent. greater than the width of the body. As the pins are of much greater diameter than is required for the shearing strength, he states that they might, with advantage, be made hollow, and of steel.

At the commencement of this description it was observed, that the principle thus experimentally proved was applicable to some parts of iron ship construction. As a case in point we may refer to the balanced rudder of the 'Bellerophon,' illustrated in Fig. 87, p. 120. In this ship there is no sole piece forged on the lower end of the stern post, the flat keel plates being continued aft in place of it. In order to steady the rudder-heel a large pintle is fitted, which passes through a hole in the keel-plates, and is secured by a nut hove up underneath. In determining the diameter of this pintle it would, but for the principle stated above, have been considered sufficient to make the shearing strength of the pintle equal to the breaking strength of the keel-plates at the weakest section; but taking into consideration the necessity for increased bearing surface, as set forth by Sir Charles Fox, it was deemed imperative to add considerably to the diameter which would have been required merely to give the necessary shearing strength, and the present diameter was adopted. The bearing surface was also increased in this case by means of a collar riveted to the plates.

Reverting to the consideration of the shearing strength of rivets, we will first direct attention to Mr. Clark's experiments on the shearing strength of rivet-iron. These experiments were made with a lever working in a slot in a block of cast-iron. The $\frac{7}{8}$-inch bars which were to be sheared were introduced through a hole in the side of the block, and the lever was pressed down upon them by weights suspended at the end. This served to determine the force required for a single shear, and when that for a double shear was required the inner end of the bar was introduced into a hole in the cheek on the side of the slot opposite to that where the bar was entered. As the mean result of these experiments, Mr. Clark gives 24·15 tons per square inch of sectional area sheared as the single shearing force, and 22·1 tons as the double shearing force, for rivet-iron of which the mean tensile strength was 24 tons. These experiments are hardly applicable to cases in which the strengths of riveted work have to be considered, for in such cases the rivets are put in hot, and beaten down to fill the hole and form the point, consequently the strength of the iron is in some measure reduced, whereas in all the experiments with the lever machine the bar-iron was as strong when sheared as when it came from the manufacturer. We have also to consider the tension produced by the contraction of the rivet in cooling, which probably further reduces its shearing strength by subjecting the iron to a considerable stress before the shearing strain commences. It is also to be observed, on the other hand, that this tension produces great friction between the surfaces of the plates, and thus increases their resistance to separation. We shall refer more fully hereafter to this subject, but it is mentioned here to illustrate the statement that it would not be proper to regard the shearing strengths of bar-iron determined by Mr. Clark as the shearing strengths of rivets when in place. Mr. Clark himself records other experiments (made with the view of avoiding any anomaly arising from the use of the lever machine), which will illustrate the truth of this. He tested the actual shearing strengths of a $\frac{7}{8}$-inch rivet passing through two and three thicknesses of plates respectively, and found the single shearing force to be 20·4 tons and the double shearing force 22·3 tons per square inch of sectional area sheared. It will thus be seen that the shearing strengths obtained, including friction, were for a single shear less than, and for a double shear nearly the same as, the strengths obtained for bar-iron; and consequently the strength of the iron, apart from the friction, must have undergone great reduction. In another place Mr. Clark gives 16 tons as the single shearing strength of a 1-inch rivet, and states that for a double shear the force required would be nearly double, thus incidentally confirming the foregoing view.

Mr. Doyne has also made experiments on the shearing strength of rivets, and appears to have employed actual riveted work for the purpose of determining the strengths. He states that the average shearing strengths, drawn from experiments made with rivets of different sizes and under different circumstances, are 18·82 tons per square inch of sectional area sheared for a single shear, and 17·55 tons for a double shear; which are equivalent to 14·78 tons for a single shear of a 1-inch rivet, and 27·56 tons for its double shear. The tensile strength of the iron in the rivets experimented upon is not stated.

Mr. Maynard has also made experiments upon this subject, and has published the following as the results, observing that his experiments were made to test what difference in value, if any, there was between rivets in punched holes and similar rivets in drilled holes.

$\frac{5}{8}$-inch RIVETS in DRILLED HOLES.

Single shear = 26 tons per sq. in.	Double shear = 39·2 tons per sq. in.
Do = 26·4 do.	Second experiment failed.

$\frac{5}{8}$-inch RIVETS in PUNCHED HOLES.

Single shear = 27·2 tons per sq. in.	Double shear = 45·6 tons per sq. in.
Do. = 26·0 do.	Second experiment failed.

Mr. Maynard says: "I considered the above as conclusive that "rivets in drilled holes, subject to shearing strain, were about

"4 per cent. weaker than rivets in punched holes under similar "strain; and think that the sharp edges of the drilled plates have "a greater tendency to snip off the rivets than the rounded edges "of punched holes. The rivets appeared cut off cleaner by the "drilled plates than by the punched." *

Carefully conducted experiments on the shearing strength of rivets have also been made by the Author's desire in H. M. Dockyard at Chatham, with the sanction of the Board of Admiralty and the Controller of the Navy, and have been so arranged as to fairly represent the pitches of rivets, breadths of lap, &c., used in shipbuilding. The chances of error were reduced by making the experiments in duplicate, and by having one, two, and three rivets respectively in the laps. The results of these experiments agree with those of both the preceding sets in showing that the shearing strength of the rivets is proportional to the sectional area. The rivet-iron was either Lowmoor or Bowling, and the mean shearing strength of a $\frac{3}{4}$-inch rivet passing through two plates was 10 tons, and the mean of the double shearing strengths of a $\frac{3}{4}$-inch rivet passing through three plates was about 18 tons. This would be, for a 1-inch rivet, single shear $17\frac{7}{9}$ tons, double shear 32 tons. These are the values which will be employed in the investigations which are given hereafter. It will be remarked that in these experiments, as in those of Mr. Doyne and Mr. Maynard, friction is included in the shearing strengths given; the differences in the results obtained in the three cases are no doubt chiefly due to differences in the quality of the rivet-iron, that used by Mr. Maynard apparently having been of a very superior quality.

As previously stated, rivets are nearly always put in when they are at a high temperature, and Mr. Clark gives some valuable information with respect to their contraction in cooling, and the consequent closeness of the joints in riveted work. He states that the contraction is about $\frac{1}{10,000}$th of the length for a decrease of temperature of 15° Fahrenheit, and the strain thus induced is about

* It must be observed that these experiments are confined solely to the value of the rivet, and not of the plate, which latter, Mr. Maynard admits, is stronger when drilled than when punched (*see* foot-note on p. 339). In comparing the strength of punched and drilled work together, on the whole, Mr. Maynard considers, 1st, That drilled plates are stronger than punched by 19 per cent.; 2nd, That rivets are weaker in drilled holes than in punched by 4 per cent.; 3rd, That the difference is in favour of drilled work by 15 per cent.

one ton per square inch of sectional area. Thus, if a rivet were closed by a machine at a temperature of 900°, the strain resulting from its cooling would greatly exceed the tensile strength which the iron can endure without stretching, and there would consequently be a permanent elongation in the length of the rivet; while the resulting tension on the plates when the rivet had cooled would approximate more or less nearly to the tensile strength of the rivet-iron. When rivets are more than 6 or 8 inches in length the head is frequently drawn off by the cooling of the rivet, and Mr. Clark remarks that the explanation of this fact is not very obvious as the contraction is always in proportion to the length of the rivet, and there is consequently no reason why a long bar should be injured more than a short one by a proportionate extension. This was proved to be the case by experiments made with rivet-bars 8 feet long, which were inserted in some castings of great strength, and it was found that in all cases these rivets remained perfectly sound after cooling, having undergone a permanent extension proportional to the temperature. The explanation Mr. Clark gives, as being the most probable, of the breaking off of the heads of long rivets on their cooling, is that the head is somewhat damaged by hammering; and he also states that it was found that the various portions of the length of the bars experimented upon had stretched very irregularly in cooling, thus showing that, in all probability, the cooling had been unequal at different parts of the length. There can be no doubt that a long rivet passing through a number of thicknesses of plates will become less distressed in cooling than if it passed through fewer thicknesses, owing to the closing together of the plates. One other matter mentioned by Mr. Clark deserves especial attention, as it is of great importance in the putting in of the large rivets or bolts in stems and stern posts. In the construction of some of the beams of the tubular bridge, most of the 12-inch rivets which were used broke at the head in cooling, and it was found that in order to prevent this breaking, the centre part of the rivet had to be cooled previously to its being put in, the head and point alone remaining red-hot.

The close union of the plates due to the contraction of the rivets is not only beneficial in respect of making good joints, but is also useful in preventing oxidation at the joints, as the rust is entirely superficial; it also adds considerable strength by causing friction at the surfaces in contact. We have now to advert more fully to this friction induced by the cooling, and the consequent contraction of the rivets in a joint. Mr. Clark gives a very interesting account of some simple experiments made by him to determine its amount. Three $\frac{5}{8}$-inch plates were riveted together with a single $\frac{7}{8}$-inch rivet, the hole in the centre plate being oval, and very much larger than the rivet. Weights were then suspended from the centre plate until it slipped, and the motion commenced abruptly under a load of 5·59 tons. The experiment was repeated with the addition of a $\frac{1}{2}$-inch plate riveted on each side between the rivet-head and point and the $\frac{5}{8}$-inch plates, thus making the rivet-shank $2\frac{7}{8}$ inches long; 4·47 tons caused the plates to slide. A repetition of this experiment gave 7·94 tons as the weight required to cause the plates to slip, the difference being due to the faulty character of the rivet in the previous experiment. In the next experiment a $\frac{7}{8}$-inch rivet was put through two $\frac{5}{16}$-inch plates with large holes, a $\frac{5}{16}$-inch washer being placed on each side next the rivet-head and point respectively. This combination supported 4·73 tons before it gave way. Mr. Clark supposes that he is warranted in inferring from these experiments that the tubes in the Britannia bridge would not deflect more than they do at present, if all the holes were much too large for the rivets, so as not to be in contact anywhere except at the heads, because the strain of $4\frac{1}{2}$ tons, which was the smallest weight that occasioned sliding in the experiments, is a greater strain than any of the rivets in the tubes sustain. He also supposes that it is possible by judicious riveting to make the friction nearly counterbalance the weakening of the plate from the punching of the holes, and to bring the strength of a riveted joint up to the strength of the solid plates united. We are unable to concur in these views for reasons which will presently appear.

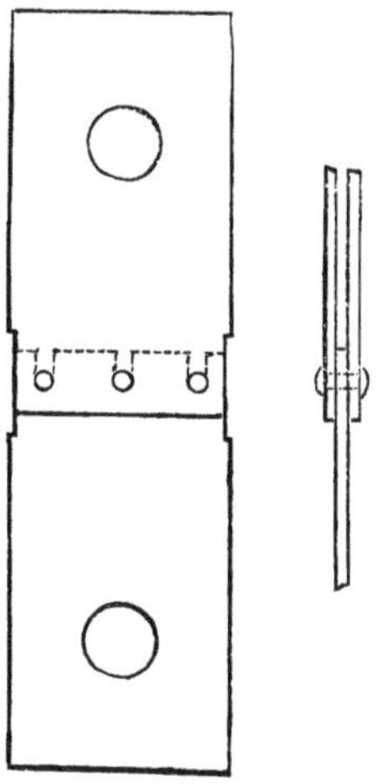

Fig. 243.

Experiments of a more detailed description, and more closely resembling the work found in practical shipbuilding, have since been made by the Author's directions in H. M. Dockyard at Pembroke. In this case three plates were united by what is known as a "chain-joint"—that is, the ends of the two outer plates overlapped the end of the middle plate—as shown in Fig. 243. The connection of the plates was made by three rivets passing through the lap, the rivet-holes in the outer plates being filled by the rivets, but the bearing surface of the holes in the middle plate being slotted out as shown in the sketch. It will thus be obvious that when a tensile strain was brought upon the middle plate, the amount of the friction could be measured by the force just able to produce a sliding motion. The breadth of the lap was three diameters, the rivets were a diameter clear of the edges of the plates, and their pitch was four diameters. There were two sets of experiments made with iron plates and rivets, and in each set two experiments were made with rivets having heads and points snap-shaped; two others with rivets having pan-heads and conical points; and the remaining two with rivets having countersunk heads and points. The experiments were made in duplicate in order to reduce the chance of error. The first set of experiments was made with ½-inch plates, 8¼ inches wide, the rivets being ¾-inch. The results were as follows:—

Description of rivet.	Friction per rivet.		
	1st Experiment.	2nd Experiment.	Mean.
	Tons.	Tons.	Tons.
Snap heads and points	5·14	4·21	4·67
Pan-heads and conical points	5·26	4·81	5·0
Countersunk heads and points	4·56	3·74	4·15
Mean of the three	..	..	4·61

The second set of experiments was made with plates 11 inches wide and ⅞-inch thick, the rivets used being 1-inch. The following results were obtained under the above-stated conditions of pitch of rivets, lap, &c. :—

Description of rivets.	Friction per rivet.		
	1st Experiment.	2nd Experiment.	Mean.
	Tons.	Tons.	Tons.
Snap heads and points	5·84	5·61	5·7
Pan-heads and conical points	6·87	7·24	7·0
Countersunk heads and points	4·56	4·09	4·3
Mean of the three	..	..	5·6

In addition to these experiments with iron plates and rivets, two other sets of experiments were made with steel plates and rivets of exactly the same dimensions as those used in the former experiments, the pitch of rivets, breadth of lap, &c., being in each case identical with those previously given. With $\frac{1}{2}$-inch plates and $\frac{3}{4}$-inch rivets the results obtained were as follows:—

Description of rivet.	Friction per rivet.		
	1st Experiment.	2nd Experiment.	Mean.
	Tons.	Tons.	Tons.
Snap heads and points	3·86	4·09	3·98
Pan-heads and conical points	4·79	4·79	4·79
Countersunk heads and points	3·63	3·43	3·53
Mean of the three	..	..	4·1

With $\frac{7}{8}$-inch plates and 1-inch rivets the following results were obtained:—

Description of rivet.	Friction per rivet.		
	1st Experiment.	2nd Experiment.	Mean.
	Tons.	Tons.	Tons.
Snap heads and points	6·43	5·49	5·96
Pan-heads and conical points	5·49	None made.	5·49
Countersunk heads and points..	5·14	4·91	5·02
Mean of the three	..	..	5·49

It thus appears that rivets with pan-heads and conical points have the advantage over both the other descriptions of riveting. The only exception to this is found in the second set of the experiments with steel plates and rivets, but as only one experiment was made the result cannot be relied on. It also becomes evident that countersunk riveting causes much less friction than the other systems. On comparison, it will be seen that in nearly all cases steel plates and rivets give less friction than iron, the only exceptions being the cases of rivets with snap-heads and points, and those with countersunk heads and points, in the second sets of experiments. The former of these exceptions is scarcely worth notice, as the difference is so small. The use of larger rivets with the same pitch, &c., gives an increase in the friction, but no law of increase appears to be conformed to.

Although these experiments do not give any definite idea of the probable amount of friction which would result from the use of rivets having different diameters and pitch, they yet serve to show how much the strength of a riveted joint is increased by the friction caused by the contraction of the rivets. Some engineers have been inclined to prefer putting in the rivets cold on account of the supposed injury done to the strength of the iron or steel in the neighbourhood of the hole, by its being brought to a great heat by the red-hot rivet, and then often cooled suddenly. The preceding experiments on the value of the friction obtained by hot-riveting seem to show that the additional strength thus secured in a joint, either wholly or partly, counterbalances the loss of strength (if any*) in the material around the hole from the cause in question, and consequently justify the almost universal practice of shipbuilders of working their rivets hot. It is true that a certain amount of contraction would be developed by what is called cold-riveting; but its amount would probably be much less than in the case of hot-riveting, and in the absence of experiments, it is not possible to assign its value.

In his prefatory remarks to the description of the experiments on friction Mr. Clark observes, "we have seen that in riveting "two plates together to resist a tensile strain, the sectional area "of the rivets should be equal to that of the plates themselves, "if we depend solely on the shearing of the rivet; but as rivets "are usually closed in a red-hot state, it is evident that the short-"ening of the rivet as it cools down must tend to draw together "the plates united, and before they can slip on each other the "friction thus induced must be overcome simultaneously with "the shearing of the rivet itself; *hence the value of the rivet is* "*greater than the value determined above by the amount of friction* "*produced by its contraction in cooling.*" The value here referred to as "determined above" is the shearing strength found by Mr. Clark's experiments described previously. It will be obvious, however, from the remarks made with respect to the experiments conducted by Mr. Doyne, and at Chatham on the shearing strengths of rivets, that the values thus obtained, although less than the shearing strengths found by Mr. Clark's experiments on bar-iron, include

* We say "if any," because experiments of Mr. Kirkaldy with steel plates appear to show that there is no such loss. At page 71 of his useful volume, he says of certain experiments, that they are "sufficient to show that the hot rivets do not reduce the strength by affecting the hardness of the plate." It would seem, however, that these experiments were made with plates which had been toughened in oil.

friction also. It would consequently be improper to arrange the fastenings of a wrought-iron structure on the assumption that the shearing strengths of the rivets (determined by experiments on rivet-bars), and the friction of the surfaces, might be treated as acting conjointly but independently. In the investigations on riveted work which are given in this chapter we shall therefore assume that friction is included in the values which are employed for the shearing strengths of the rivets.

There are passages in Mr. Fairbairn's works which agree with this view. In his 'Useful Information for Engineers' he says:— "From these facts it is evident that the rivets cannot add to the "strength of the plate, their object being to keep the two surfaces "of the lap in contact, and being headed on both sides the plates "are brought into very close union by the contraction or cooling "of the rivets after they are closed. It may be said that the "pressure or adhesion of the two surfaces of the plates would add "to the strength; but this is not found to be the case to any "great extent, as in almost every instance the experiments indicate "the resistance to be in the ratio of their sectional areas, or "nearly so."

It may be of interest to consider how far the shearing strength of the rivet is actually reduced in working and cooling. We are enabled to do this, approximately, if we assume Mr. Clark's results to hold with respect to the shearing strength of rivet-iron, the Chatham experiments to hold with respect to the shearing strength of rivets when in position, and the Pembroke experiments to hold with respect to friction. Taking first the case of a ¾-inch rivet, we find from Mr. Clark's data that the bar-iron would require a force of 19·52 tons for a double shear, and from the Chatham experiments we have 18 tons as the double shearing force of a ¾-inch rivet, including friction. The mean value of the friction caused by a ¾-inch rivet is found from the Pembroke experiments to be 4·6 tons, and hence it seems fair to consider that the double shearing strength of the rivet amounts to about 13½ tons, or about 6 tons less than the double shearing strength of the bar from which it was made. Considering similarly the case of a 1-inch rivet, we have 34·7 tons as the strength of the bar for a double shear, and 32 tons as the double shearing force of the rivet, including friction, the mean value of which is found to be 5·6 tons. The double shearing strength of the rivet is thus about 26·4 tons, showing a reduction from the strength of the bar of about 8 tons. As far as these two cases go the reduction is thus about proportional to the diameters of the rivets, but it would, of course, be improper to infer from two instances only that this is the general law of reduction. It is important to observe, however, that what we conjecture to be the principal cause of this loss of shearing strength in the finished rivet—viz. the interior stress of the iron due to the contraction—would obviously be proportional to the diameter.

We nevertheless give these two cases rather as indications of the fact that a great reduction really exists, than as measures of its amount. In addition to the loss of strength from the cause just mentioned, there is probably a further reduction due to the manipulation which the rivet has to undergo in being manufactured and put in. The whole question, including the dependence of the value of the rivet upon the extent of the frictional surface per rivet, obviously requires further experimental investigation.*

We propose to conclude this chapter with a brief consideration of the principles which should regulate the arrangement of butt-fastenings. We have previously illustrated (in Chapter X.) most of the modes of riveting butts which have been practised, and need not repeat the descriptions there given, but will proceed at once to state the experimental facts which have been determined with regard to a few modes of fastening. The amount of experimental knowledge which we possess on this subject is extremely limited, and is principally due to the researches of Mr. Fairbairn which are recorded in the Philosophical Transactions for 1850, and since published in the first series of his 'Useful Information for Engineers.' Mr. Clark gives some of the results of the experiments on riveted work made during the construction of the Britannia and

---

* The following is a theoretical and not unreasonable view of the matter :—The shearing strength and the tensile strength of rivet-iron are nearly the same, and consequently the strain required to break a rivet is about the same whatever be its direction. When, therefore, two plates are riveted together, and a force is applied along the plates to separate them, the force tending to rupture the rivets is the resultant of the tension upon them and the strain tending to shear them. These two forces are at right angles to one another; hence, if B is the breaking strain of the rivet, T its tension, and S the shearing strain, the rivet will break when the following equation is satisfied :—

$$B^2 = S^2 + T^2,$$

$$\text{or when } S = \sqrt{B^2 - T^2}.$$

Now, taking friction into account, and putting F for its co-efficient, and therefore FT for its amount; also calling the strength of the joint P, the equation of rupture is,

$$P = S + FT = \sqrt{B^2 - T^2} + FT.$$

Conway bridges, but does not enter into the subject at any length. In the *Transactions* of the Institution of Naval Architects for 1860, an account is published of experiments conducted by Mr. Mumford by the direction of Lloyd's Committee in order to test some arrangements of butt-fastenings, and it will be remembered that the conclusions deduced from those experiments were given in Chapter X. It will be sufficient, therefore, to state here with respect to the latter experiments, that it was considered that single riveting was proved to be too weak for the butt-fastenings of iron ships, and that double chain-riveted butts were stronger than double zigzag-riveted butts.* Mr. Fairbairn's experiments were made with single-riveted and double-riveted joints, and the final results at which he arrives are as follows:—

Taking the strength of the plate as	100
The strength of the double riveted joint would be	70
And the strength of the single riveted joint	56

It should be stated that the strength of the plate here taken is that of a cross section where it is *unpunched*, and that Mr. Fairbairn gives the following relative values of the strength of the plate through a row of rivet-holes in the joint, and of single and double riveted joints respectively:—

Taking the strength of the plate as	100
The strength of the double-riveted joint would be	97
And the strength of the single-riveted joint	76

One very important fact to be borne in mind is that the total sectional strength of the rivets in the lap, or on one side of the butt, should be made equal to the sectional strength of the plate taken through a line of rivet-holes. The rule on which calculations for the pitch and number of rivets in a butt or lap joint are usually made for Admiralty work is based on this consideration. It will be observed that this equality of strength is not fully attained in the plan employed by Mr. Fairbairn and others in bridge construction, and recommended by some writers for ship construction, viz., "that the collective *areas* of the rivets should be "equal to the sectional *area* of the plate taken through the line "of rivets." This principle is based upon the supposition that

* In 1855 a set of experiments was made at Woolwich dockyard in order to test the strengths of riveted joints, as compared with joints welded on Mr. Bertram's plan. The information given with respect to these experiments is wanting in detail, and the tabulated results show great discrepancies in the quality of the iron in the plates, so that it is impossible to put them to any practical use.

the shearing strength of the finished rivet is about equal to the breaking strength of the punched plate — a supposition which the Admiralty experiments do not fully bear out. Those experiments give a somewhat higher value to the strength per square inch of the finished rivet than to the punched plate.

In the use made by Mr. Fairbairn of the deductions from his experiments one feature claims especial notice, viz., the statement that the comparison of the strength of the joint with the strength of the unpunched plate may be safely taken as the standard value of joints. This comparison would be strictly correct in the case of a plate-tie which is unpierced except at the butts and ends, and requires to be specially strengthened at these parts. But in a ship, and most other wrought-iron structures, where the plates are necessarily pierced by holes which receive the fastenings of the frames and other stiffeners, it would evidently be worse than useless to make the butts as strong as the unpierced plates; for in order to secure uniformity of strength, it is only necessary to make the butts as strong as the unavoidably weakest section of the plates. The assumption that it is desirable to bring up the strength of the butt to the strength of the unpunched plate has been the basis on which most novel proposals for butt-fastenings have rested, but, as we shall see more clearly in the following examples, the present practice of iron shipbuilders supplies ample strength if the fastenings are well arranged. Mr. Fairbairn, however, in advocating the adoption of quadruple chain-riveting for the butts of outside plating, says, "it is to be hoped that vessels "will not in future be built at a loss of one-third of the longi- "tudinal tenacity as at present."* As he speaks immediately before this of his knowledge of the custom of double riveting the butts, and expresses his opinion that this system of fastening is comparatively weak, it seems fair to conclude that he has applied the comparison made by him between the strength of the unpunched plate and the double-riveted joint to an actual ship, without taking into account the lines of unavoidable weakness caused by the holes which receive the fastenings of the outside plating to the frames. This is obviously incorrect, for, as remarked above, in providing for the due strength of a single strake, the

* In a paper on "The Strength of Iron Ships," in the '*Transactions* of the Institution of Naval Architects for 1860;' also printed in the second series of 'Useful Information for Engineers.'

strength of the butts need only be brought up to the strength of the weakened sections of the plates, in order to preserve uniformity of strength. If the maximum strength of the plating were required to be attained, it might be considered desirable to increase the strength of the weakened sections at the frames by the use of broad liners such as are now fitted at bulkheads, and then the strength of the butts would require to be brought up to the strength of the unpunched plates. But as vessels are at present constructed, it would be a waste of both materials and workmanship to make the strength of the butts greater than that of the weakened sections. These remarks upon Mr. Fairbairn's statements are not intended to detract in the least from the undoubted value of his experiments, and the general correctness of his deductions, but to point out what, in our estimation, is an error into which he, in common with many other writers on the subject, has fallen, in urging the adoption of treble and quadruple systems of riveting in iron ship construction. We do so because we hold it to be of great importance that shipbuilders should not be urged into the use of unnecessary iron, or of superfluous labour.

Coming now to the consideration of the manner in which butt-fastenings should be arranged, it may be well to commence with the simplest case, that of a plate-tie which is unpierced except at the butts and ends. In this case, which will serve as an introduction to shipbuilding examples, it is desirable, as previously remarked, to make the strength of the butt approximate as closely as possible to the strength of a section of the unpunched plate. By a proper proportioning and disposition of the fastenings this result can be very nearly attained. In a paper on 'Wrought-iron Beams' in the *Proceedings* of the Institution of Civil Engineers for 1855, Mr. Barton makes the following remarks:—"There is no "necessity for any serious loss of strength if a judicious mode "of connecting the "bars is adopted. The "author devised the "method illustrated "in Fig. 244, and "tested it severely by "employing it in con- "necting bands for lifting stone instead of using chains for the

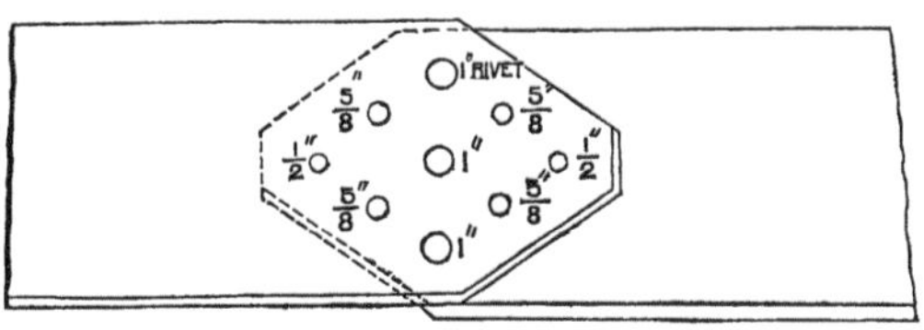

Fig. 244.

" purpose; and he found that when fracture took place it occurred, " either at some distance from the joint, or where the small rivet-" hole was placed, showing that the full strength of the section, " with the exception of this small rivet-hole was obtained by this " arrangement." In this case the plates are lap-jointed, but in the following case of the fastenings of a plate-tie (which my friend and assistant Mr. Barnaby worked out in detail), the plates are butt-jointed, although the principle on which both arrangements are based, is that the butt shall be as strong as the unpunched plate, with the exception of one rivet-hole. The investigation of the latter case is given in the fuller detail, because the method of calculation here employed is identical in character with that of the investigations of the cases taken from actual practice which follow. The plates united to form the tie are $\frac{10}{16}$ inch thick and 24 inches wide, with double butt-straps each $\frac{9}{16}$ inch thick, riveted with 1-inch rivets arranged as shown by *a* in Fig. 245, it being obvious that the use of extra thickness in the butt-straps must in this case be resorted to, because the strength of the straps through the line of holes next the butt has to be made equal to the strength of the plate through the single rivet-hole. The ordinary rule observed in shipbuilding is carried into effect here, all the rivet-holes being a diameter clear of the edges and butts. The tensile strength of the unpunched plate is assumed to be 22 tons per square inch of section, and it hence follows that we have—

$$\text{Breaking strength of the unpunched tie} = 24'' \times \tfrac{10}{16}'' \times 22 \text{ tons} = 330 \text{ tons.}$$

The butt may be fractured by breaking either the plate, or the butt-straps, and shearing the rivets. There are altogether ten modes of fracture which we propose to examine, commencing with those in which the *plate* is broken, observing that although the plates or straps might break in other ways, these ten modes appear sufficient for the present investigation, as they apparently comprise all the weakest cases. In these investigations we shall take the double shearing strength of a 1-inch rivet at 32 tons, this value being obtained from the data previously given in this chapter. The simplest mode of fracture is that illustrated by *b* in Fig. 245 where the plate has been broken through the single rivet-hole. As there is only this one hole in the breadth of the plate it will be fair to assume that the iron in the line of fracture retains its full strength of 22 tons per square inch. The effective breadth of the plate is reduced by the rivet-hole to 23 inches, and we consequently have for Mode I.:—

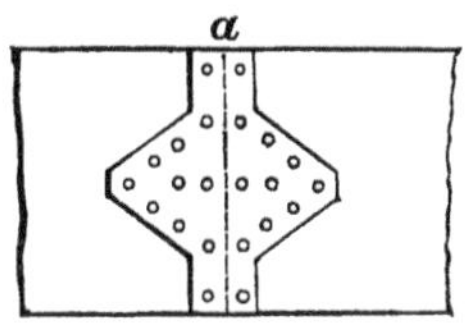

Breaking strength $= 23'' \times \frac{10}{16}'' \times 22$ tons $= 316$ tons.

A second mode of fracture is shown by *c* in Fig. 245, where the plate has been broken across two rivet-holes, and the single rivet

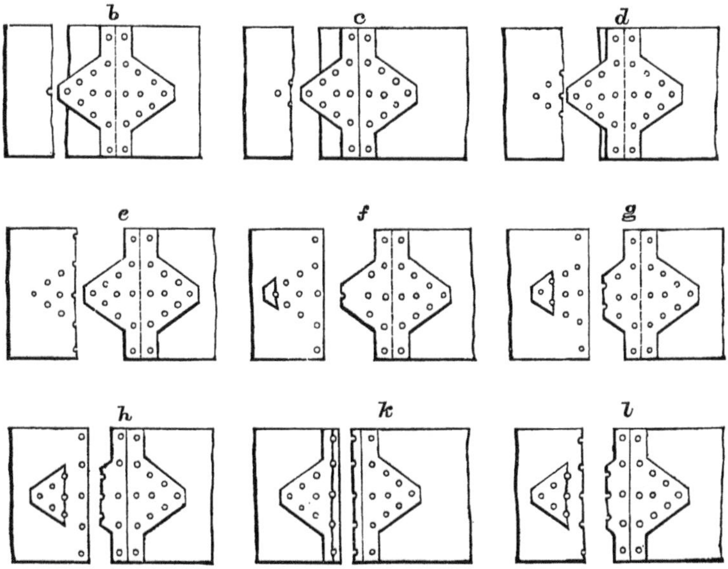

Fig. 245.

has been sheared twice. In this case also it may be fairly assumed that the tensile strength of the iron in the line of fracture is almost unchanged by the punching of the two holes. The effective breadth of the plate is reduced to 22 inches by the two rivet-holes, and we thus obtain for Mode II.:—

Breaking strength of plate $= 22'' \times \frac{10}{16}'' \times 22$ tons	$=$ 303 tons.
Added for double shear of one rivet	$=$ 32 ,,
Total breaking strength..	$=$ 335 ,,

A third mode of fracture is given in *d* in Fig. 245, where the plate has been broken through three rivet-holes, and three rivets have been sheared twice. In this case the tensile strength of the iron in the line of fracture may be considered to have been reduced by punching the holes to 20 tons per square inch. The effective breadth of the plate is 21 inches, and we have for Mode III.:—

Breaking strength of plate = 21″ × $\frac{10}{16}$″ × 20 tons = 263 tons.
Added for double shear of three rivets .. .. .. = 96 ,,

Total breaking strength .. .. = 359 ,,

A fourth mode of fracture is illustrated by *e* in Fig. 245, where the plate has been broken through the row of rivet-holes nearest the butt, and the remaining six rivets on that side of the butt have been sheared twice. Here, as the pitch of the rivets is about four diameters, it will be proper to take 18 tons as the tensile strength of the iron in the line of fracture. The effective breadth of the plate is reduced to 19 inches, and we obtain for Mode IV.:—

Breaking strength of plate = 19″ × $\frac{10}{16}$″ × 18 tons = 214 tons.
Added for double shear of six rivets .. .. .. = 192 ,,

Total breaking strength .. .. = 406 ,,

A fifth mode of fracture consists in shearing twice the eleven rivets on one side of the butt, and this gives for Mode V.:—

Breaking strength = 11 × 32 tons = 352 tons.

Before proceeding to consider the other cases of fracture in which the *straps* are broken across, it may be well to state that we shall assume 18 tons per square inch to be the tensile strength of the iron in all the lines of fracture, the breadth of the straps being proportioned in such a manner as to bring all the rivets within a diameter of the edges, as before described. A sixth mode of fracture is illustrated by *f* in Fig. 245, where the straps have been broken across the single rivet-hole, and the remaining ten rivets on that side of the butt have been sheared twice. Remembering that there are double straps each $\frac{9}{16}$ inch thick, and that the effective breadth of the straps along the line of fracture is 2 inches, we obtain for Mode VI.:—

Breaking strength of straps = 2 × 2″ × $\frac{9}{16}$″ × 18 tons = 41 tons.
Added for double shear of ten rivets .. .. = 320 ,,

Total breaking strength .. .. = 361

A seventh mode of fracture is shown by *g* in Fig. 245, where the straps have been broken through two rivet-holes, and the eight rivets between the fracture and the butt have been sheared twice. The total breadth of the strap at this part is 8 inches, and its effective breadth is consequently 6 inches, thus giving for Mode VII.:—

Breaking strength of straps = 2 × 6″ × $\frac{9}{16}$″ × 18 tons = 122 tons.
Added for double shear of eight rivets .. .. .. = 256 ,,

Total breaking strength .. .. = 378 ,,

An eighth mode of fracture is given in *h* Fig. 245, where the straps have been broken through three rivet-holes and the five rivets nearest the butt have been sheared twice. The total breadth of the strap is here 12 inches, and the effective breadth 9 inches; we thus obtain for Mode VIII.:—

Breaking strength of straps $= 2 \times 9'' \times \frac{9}{16}'' \times 18$ tons	= 182 tons.
Added for double shear of five rivets	= 160 ,,
Total breaking strength	= 342 ,,

Another mode of fracture is shown by *k* in Fig. 245, where the straps have been broken through the five holes nearest the butt. The effective breadth of the strap is here 19 inches, and we obtain for Mode IX.:—

$$\text{Breaking strength} = 2 \times 19'' \times \tfrac{9}{16}'' \times 18 \text{ tons} = 385 \text{ tons.}$$

The remaining mode of fracture is illustrated by *l* in Fig. 245, where the straps have been broken as in Mode VIII., and the plate has been broken through the line of holes nearest the butt. We thus have for Mode X.:—

Breaking strength of straps, as in Mode VIII.	= 182 tons.
,, ,, plate, as in Mode IV. ..	= 214 ,,
Total breaking strength	= 396 ,,

It will be seen from these results that in all the various modes of fracture, except the first, the breaking strength is greater than the strength of the unpunched tie-plate, and that the strength of the butt is, consequently, less than the strength of the tie by one rivet hole only.

Passing now from the consideration of plate-ties, we proceed to investigate a few cases of butt fastenings taken from actual practice. The first case is illustrated by Fig. 246, which represents the manner in which the butt of a stringer-plate *a* (36 by $\frac{5}{8}$ inches) is secured in an iron-clad frigate. In this ship the beams are covered with $\frac{1}{4}$ inch plating, marked *b* in the sketch, and the stringer *a* is worked upon it. The butt straps are double, each being $\frac{5}{16}$ inch thick, the upper strap coming directly upon the stringer and extending across the whole width, and the lower

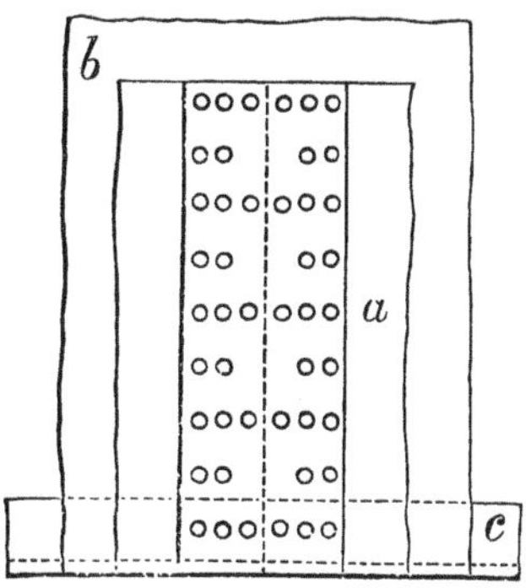

Fig. 246.

one being worked below the $\frac{1}{4}$ inch plating and extending from the edge of the angle-iron *c* to the edge of the stringer. The rivets used are ¾-inch, those in the outer rows having a pitch of about five and a half diameters, and the intermediate rivets being omitted in the rows nearest the butts. This arrangement is adopted in order to allow the edges of the strap to be efficiently caulked, and yet to keep the shearing strength of the rivets from becoming much greater than is required. The weakest section of the stringer is that through a line of rivet and deck-fastening holes in wake of the beams, and the strength of the butt should be made, as nearly as possible, equal to the strength of the stringer at the beams. As there would be about nine holes for rivets and deck-fastenings * in each beam, and an equal number of rivets is required in the row of rivet-holes furthest from the butt in order to make a watertight joint, the equality of strength is at once obtained, so far as the stringer-plate is concerned. It is still necessary, however, to examine how the strength of the fastenings is proportioned to the strength of the stringer. The effective breadth of the stringer through the line of rivet-holes furthest from the butt is reduced to 29¼ inches, and the effective sectional area to 18¼ sq. inches. Assuming, as before, the strength of the punched plate to be 18 tons per sq. inch, this gives a breaking strength for the stringer-plate of 328½ tons. There are twenty-three rivets on each side of the butt, and taking the double shear of a ¾-inch rivet at 18 tons, the strength of the butt fastenings will equal 414 tons. It appears therefore that the fastenings are considerably stronger than the plate in this case, the result being principally due to the additional strength obtained by the use of double butt straps. The pitch of the rivets in the outer rows is fixed by the consideration that the edge of the strap is to be made watertight, but if it were considered desirable to reduce the shearing strength of the butt fastenings, some of the rivets in the middle rows might be omitted; and, in point of fact, double-riveting would have been nearly sufficient. It will be obvious that in this case there is no necessity for calculating the breaking strengths corresponding to any other modes of fracture, as they would require either the plate or the straps to be broken through a line of rivet-holes, and a row of rivets to be sheared, thus giving greater breaking strengths than either

* Our remarks in Chapter IX. will show that we do not consider this a good arrangement, as the deck fastenings should be brought out up on the deck plating between the beams.

of the preceding calculations. It will also appear that since the united thickness of the butt straps equals the thickness of the stringer, that fracture will always take place across the plate rather than across the straps; we shall revert to this matter further on. One other feature of this arrangement claims attention, viz., the working of double straps at the stringer butt, when the double shear of the rivets might be secured by working a single strap upon the stringer, and making the $\frac{1}{4}$-inch plating serve as the lower strap. It is evident, however, that the single strap would require to be thicker than that now fitted in order to give sufficient strength; and that if the $\frac{1}{4}$-inch plating were made to serve as a strap to the stringer-butt its own strength and efficiency would be lost, and the full benefits which now result from the use of a stringer that is butt-strapped independently would not be attained.

The method which should be pursued in arranging the butt fastenings of a stringer may be briefly indicated here. Having settled the pitch of the rivets in the beams, the area of the unavoidably weakest section adjacent to the butt becomes known. The diameter of the rivets is, of course, determined from the thickness of the plate in accordance with the tables previously given. The number of rivets required on each side of the butt may then be determined, from the consideration that the shearing strength of the rivets should at least equal the breaking strength of the stringer at the beam. Supposing, for instance, that W tons is the breaking strength of the stringer at the beam, and S is the shearing strength of each rivet in the butt fastenings (either for a single or a double shear, according as single or double butt-straps are employed), then the minimum number of rivets required will be $\frac{W}{S}$. In placing the rivets care must be taken not to have the section of the stringer through the row of rivet-holes farthest from the butt weaker than the section at the beam. Ordinary stringer-butts, of course, do not require to be caulked, and can consequently be either double or treble riveted according to the number of rivets required and the breadth of the stringer. If, as in the case given in Fig. 246, the butt has to be caulked, the rivets nearest the edges of the straps must be spaced close enough for watertight work; but this is a special case as before stated. In determining the thickness of the butt-strap which is required, it is only necessary to avoid any arrangement which would allow the butt to separate by the fracture of the strap rather than by the breaking of the plate. In considering this point it is, of course, necessary to take account of the manner in which the shearing strength of the rivets in the butt acts, in more efficiently assisting either the plate or the strap. In the case illustrated in Fig. 246, it will be remarked that the butt might be separated by breaking the stringer across the line of rivet-holes farthest from the butts without shearing any rivets; whereas if the straps broke through this row of holes the other two rows of rivets would have to be sheared before separation took place. Hence the united thickness of the straps in this case might be made less than the thickness of the stringer without reducing the strength of the butt. The converse is true with respect to the butt fastening shown in Fig. 152, p. 206, where the strap is required to be thicker than the plate, in order to give the proper strength. This will be obvious to the reader if he follows out a similar mode of reasoning to that given above. In practice, single butt-straps to stringers and tie-plates are usually of the same thickness as the plates they connect, and when double butt-straps are employed, each strap is about $\frac{1}{16}$ inch thicker than the half-thickness of the plates. The foregoing considerations show, however, that in arranging work it is very desirable that the special circumstances of each particular case should be carefully regarded.

The preceding method of investigation is also applicable to deck tie-plates, clamp or spirketing plates, and other of the longitudinal pieces of a ship's frame, where it is only necessary to consider the strength of a single tie. But when it is desired to properly arrange the butt fastenings of outside or deck plating, it becomes necessary to take into consideration, not merely the butted plate, but some portion of the adjacent plating, and to investigate the amount of strength obtained from the edge riveting in the neighbourhood of the butt. The portion of the adjacent strakes of plating which has to be included in the calculation of the breaking strengths, varies, of course, with the shift of butts adopted. For example, with two passing strakes between consecutive butts in the same transverse section, it would be necessary to take into account the strength of the strake on each side of the butted strake; and with four passing strakes between consecutive butts, two strakes on each side of the butted strake would require to be included in the calculations. It may be remarked here that, until recently, it has been the general practice to consider the arrangement of butt fastenings as requiring to be determined from the consideration of the butted strake only; but, as we shall see from the following examples, this is not the correct method. It will be observed also that the arguments for the adoption of treble and quadruple riveted butts have been based upon experiments made with two plates either lap or butt jointed, which cannot be considered as applicable to assemblages of plating where the butts are carefully shifted. These considerations, added to those previously stated with respect to the unavoidable lines of weakness at the frames and beams, serve to show that double riveting gives sufficient strength to the butts of outside and deck plating, when the ordinary shifts of butts are followed. The following examples of the method of calculation which, in our opinion, fairly includes all the required considerations, and may be safely applied in practice, are taken from the outside plating of one of the iron-clad frigates of the navy, and of a first-class ocean mail steam-ship. The first case is illustrated in Fig. 153, p. 207, and is taken from the 'Hercules,' as built. The sketch shows an inside view of four strakes of the bottom plating, those marked *a* and *c* being outside strakes, and those marked *b* and *d* inside strakes. All the plates are $\frac{7}{8}$ inch thick, and the rivets are 1 inch in diameter. The butt-straps to outside strakes are $\frac{7}{8}$-inch thick, and those to inside strakes are $\frac{13}{16}$ inch. The diagonal shift of butts is adopted, and there are consequently two passing strakes between consecutive butts in the same frame space. The butts A B of the outside strake *c*, and C D of the inside strake *b* are those of which the strengths of the fastenings are to be investigated. As before remarked, it will be necessary in making the calculations to take into account the strength of the strake on each side of the butted strake. Thus in investigating the proportionate strength of the butt A B, it will be necessary to consider the strakes *b*, *c*, and *d*; and in dealing with the butt C D the strakes *a*, *b*, and *c* must be taken into account. The strakes *a* and *c* are each 3 feet 6 inches broad, the strake *b* is 4 feet 1 inch broad, and *d* is 4 feet. The frame marked *w* in the sketch is a watertight plate-frame, and consequently the pitch of the rivets which secure the bottom plating to it is less than that of the rivets in the bracket frames *f*, *f*. It is evident, therefore, that the weakest sections of the plating would be at the frames such as *w*, unless special means were used to compensate for the loss of strength resulting from the closeness of the rivets. This compensation can be made by the use of broad liner pieces under the frames *w*, similar to those marked *e, e*, which serve as butt-straps to the weakened sections in the manner previously explained when describing bulkhead connections. By this means the weakest sections of the plating are brought into the lines of rivet-holes in wake of the bracket frames *f, f*, at which the liners are not designed to add to the strength of the section. As previously remarked, the strength of the butt must be at least as great as the strength of the section through the line of rivet-holes at the frames *f, f*; or, in other words, the section through the butt must be at least as strong as the weakest section, in order that uniformity of strength may be preserved. In considering the modes of fracture which are possible to the three plates that have to be taken in connection with each butt, we may for convenience arrange them in the following order:—

I. The plates may be broken down through the line of rivet-holes at the bracket frame nearest to the butt of which the fastenings are under consideration. The supposed lines of fracture in each case are marked E F.

II. The adjacent plates to that which is butted may break along the line E F as before, and the butt fastenings on one side of the butt, together with the edge fastenings between the butt and the line E F, may be sheared.

III. The second mode of fracture may be varied by the butt-strap being broken down the line of rivet-holes nearest the butt, instead of the fastenings on one side of the butt being sheared.

IV. The second mode of fracture may also be varied by breaking the butted plate down through the row of rivet-holes marked K L, and shearing the edge fastenings between the lines E F and K L.

V. The plates on each side of the butted plate, and the butt-strap may be supposed to break along the line G H, which passes down through a line of rivet-holes in the butt.

VI. The preceding mode of fracture may be varied by the butt fastenings on one side of the butt being sheared instead of the butt-strap being broken along the line G H.

The tensile strength of the unpunched plate will be assumed to be 22 tons per sq. inch, and that of the punched plate for all except the two last modes of fracture will be taken at 18 tons. In the last two modes there are so few holes in the lines of fracture that it will be fair to assume that no allowance need be made for the reduction caused by punching. The single shearing strength of a 1-inch rivet is taken at $17\frac{7}{9}$ tons. All these values have been ascertained to be the average strengths by the experiments previously referred to.

We will first calculate the breaking strengths of the butt A B and the plates *b*, *c*, and *d* for the supposed modes of fracture.

MODE I.—Total breadth of plates, *b*, *c*, and *d* . = 139 inches.
Rivet-holes in line E F . . . . . = 27 ,,

∴ Effective breadth . . . . = 112 ,,

Breaking strength = $112'' \times \frac{7}{8}'' \times 18$ tons . . = 1764 tons.

MODE II.—Total breadth of plates *b* and *d* . . = 97 inches.
Rivet-holes in line of fracture . . = 19 ,,

∴ Effective breadth . . . = 78 ,,

Strength of plates *b* and *d* = $78'' \times \frac{7}{8}'' \times 18$ tons = 1228 tons.
Added for shear of rivets = $36 \times 17\frac{7}{9}$ tons . . = 640 ,,

Total breaking strength . . . . = 1868 ,,

MODE III.—Total breadth of strap to *c* . . . . . = 31 inches.
Rivet-holes in line of fracture . . . = 8 ,,

∴ Effective breadth . . . . . = 23 ,,

Strength of the strap = $23'' \times \frac{7}{8}'' \times 18$ tons . . = 362 tons.
Strength of *b* and *d*, as above . . . . = 1228 ,,
Added for shear of rivets in edges = $20 \times 17\frac{7}{9}$ tons = 355 ,,

Total breaking strength . . . . = 1945 ,,

MODE IV.—Total breadth of plate *c* . . . . . = 42 inches.
Rivet-holes in line K L. . . . . . = 12 ,,

∴ Effective breadth . . . . . = 30 ,,

Strength of plate *c* = $30'' \times \frac{7}{8}'' \times 18$ tons . . = 472 tons.
Strength of *b* and *d*, as above . . . . = 1228 ,,
Added for shear of rivets in edges = $12 \times 17\frac{7}{9}$ tons = 213 ,,

Total breaking strength . . . . = 1913 ,,

MODE V.—Total breadth of plates *b* and *d* . . = 97 inches.
Rivet-holes in line of fracture . . . . = 8 ,,

∴ Effective breadth . . . . . = 89 ,,

Strength of plates *b* and *d* = $89'' \times \frac{7}{8}'' \times 22$ tons = 1713 tons.
Strength of strap to *c* (as in Mode III.) . . . . = 362 ,,

Total breaking strength . . . . = 2075 ,,

MODE VI.—Strength of *b* and *d*, as above . . . . = 1713 tons.
Added for shear of butt fastenings = 16 × $17\frac{3}{4}$ tons = 284 ,,

Total breaking strength . . . . = 1997 ,,

For the butt C D, and the plates *a*, *b*, and *c* we find the following values for the breaking strengths:—

MODE I.—Total breadth of plates *a*, *b*, and *c* . . = 133 inches.
Rivet-holes in line E F . . . . . . = 26 ,,

∴ Effective breadth . . . . . = 107 ,,

Breaking strength . = 107″ × $\frac{7}{8}$″ × 18 tons . . = 1685 tons.

MODE II.—Total breadth of plates *a* and *c* . . . = 84 inches.
Rivet-holes in line of fracture . . . = 16 ,,

∴ Effective breadth . . . . . . = 68 ,,

Strength of plates *a* and *c* = 68″ × $\frac{7}{8}$″ × 18 tons . = 1071 tons.
Added for shear of rivets = 44 × $17\frac{3}{4}$ tons . . . = 782 ,,

Total breaking strength . . . . = 1853 ,,

MODE III.—Total breadth of strap to *b* . . . . = 49 inches.
Rivet-holes in line of fracture . . . = 12 ,,

∴ Effective breadth . . . . . . = 37 ,,

Strength of strap to *b* = 37″ × $\frac{13}{16}$″ × 18 tons . . = 541 tons.
Strength of *a* and *c*, as above . . . . = 1071 ,,
Added for shear of rivets in edges = 20 × $17\frac{3}{4}$ tons = 355 ,,

Total breaking strength . . . . = 1967 ,,

MODE IV.—Total breadth of plate *b* . . . . . . = 49 inches.
Rivet-holes in line K L . . . . . . = 12 ,,

∴ Effective breadth . . . . . . = 37 ,,

Strength of plate *b* = 37″ × $\frac{7}{8}$″ × 18 tons . . . = 583 tons.
Strength of *a* and *c*, as above . . . . = 1071 ,,
Added for shear of rivets in edges = 12 × $17\frac{3}{4}$ tons = 213 ,,

Total breaking strength . . . . = 1867

MODE V.—Total breadth of plates *a* and *c* . . . = 84 inches.
Rivet-holes in line of fracture . . . = 10 ,,

∴ Effective breadth . . . . . . = 74 ,,

Strength of plates *a* and *c* = 74″ × $\frac{7}{8}$″ × 22 tons = 1424 tons.
Strength of strap to *b* (as in Mode III.) . . . . = 541 ,,

Total breaking strength . . . . = 1965

MODE VI.—Strength of *a* and *c*, as above . . . . = 1424 tons.
Added for shear of butt fastenings = 24 × $17\frac{3}{4}$ tons = 427 ,,

Total breaking strength . . . . = 1851

If the strengths at the lines E F are taken as unity for each butt, we may put the results of the preceding calculations in the following form :—

	Butt A B.	Butt C D.
Breaking strength by Mode II. . .	1·059	1·099
,, ,, III. . .	1·102	1·167
,, ,, IV. . .	1·084	1·108
,, ,, V. . .	1·176	1·166
,, ,, VI. . .	1·132	1·098

From this table it can be readily seen exactly how much stronger the butt fastenings are than the section of the plating through the line E F. The above may be regarded as a well-arranged system of fastening, as it makes the plating at the butt stronger, under every circumstance of fracture, than the weakest section of the plating; but at the same time does not give, in any case, a very great margin of strength in favour of the fastenings, nor involve excessive weight or labour.

The second example of this method of calculation is based upon the sketch of bottom plating given in Fig. 247, which is taken from the Cunard liner 'Samaria,' lately built by Messrs. Thomson of Glasgow. This case may be regarded as a favourable instance of the highest class of merchant-ship construction (excepting in so far as the use of one passing strake only is concerned), but must not be taken as an illustration of the average arrangement of iron ships, as the care here taken in arranging the fastenings is very often almost entirely wanting. In order to avoid repetition, the possible lines of fracture are marked with the same letters in this sketch as in Fig. 153, p. 207. Here the strakes *a* and *c* are inside strakes, and *b* and *d* are outside strakes.

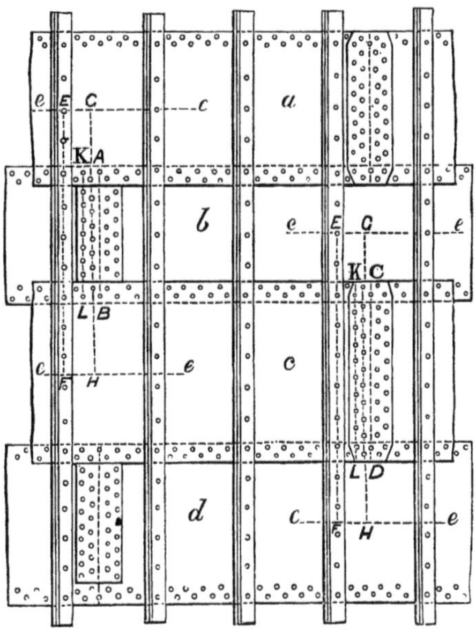

Fig. 247

The plates and butt-straps are $\frac{3}{4}$ inch thick, and the rivets are $\frac{7}{8}$ inch in diameter. The breadth of the strake *a* is 3 feet, that of *b* 2 feet 8 inches, that of *c* 3 feet 6 inches, and that of *d* 3 feet 1 inch. The brick arrangement of butts is adopted, and there is consequently only one passing strake between two consecutive butts in the same frame space. It will be necessary, therefore, to take into account the strength of one-half the breadth of the strake on each side of the butted strake in making the calculations of the breaking strengths. The lines marked *e e* are drawn at the middle of the various strakes, and it will be observed that fracture is supposed to begin at them. The same values will be employed for the strengths of punched and unpunched plates as were used in the preceding calculation, and the single shearing strength of a $\frac{7}{8}$-inch rivet will be taken at $13\frac{2}{3}$ tons. The modes of fracture being arranged and numbered as before, we obtain the following breaking strengths for the butt A B and the strake *b*, together with the strakes *a* and *c* out to the lines *e e*.

MODE I.—Total length of the line E F . . . . . . = 71 inches.
Rivet-holes in " " . . . . . = $9\frac{3}{16}$,,

∴ Effective length . . . . = $61\frac{13}{16}$,,
Breaking strength = $61\frac{13}{16}'' \times \frac{3}{4}'' \times 18$ tons . . . = 834 tons.

MODE II.—Total length of the line of fracture of *a* and *c* = 39 inches.
Rivet-holes in " " " = $4\frac{13}{16}$,,

∴ Effective length . . . . = $34\frac{3}{16}$,,
Strength of *a* and *c* = $34\frac{3}{16}'' \times \frac{3}{4}'' \times 18$ tons . . . . = 462 tons.
Added for shear of rivets = $19 \times 13\frac{2}{3}$ tons . . . . . . = 260 ,,

Total breaking strength . . . . = 722 ,,

MODE III.—Total breadth of strap to *b* . . . . = 22 inches.
Rivet-holes in line of fracture . . . . . . = $5\frac{1}{4}$,,

∴ Effective length . . . . = $16\frac{3}{4}$,,
Strength of the strap = $16\frac{3}{4}'' \times \frac{3}{4}'' \times 18$ tons . . . . . = 226 tons.
Strength of *a* and *c*, as above . . . . = 462 ,,
Added for shear of rivets in edges = $6 \times 13\frac{2}{3}$ tons . . . = 82 ,,

Total breaking strength . . . . = 770 ,,

MODE IV.—Total breadth of plate *b* . . . . = 32 inches.
Rivet-holes in line K L . . . . = $7\frac{7}{8}$,,

∴ Effective breadth . . . . = $24\frac{1}{8}$,,
Strength of plate *b* = $24\frac{1}{8}'' \times \frac{3}{4}'' \times 18$ tons . . . . . . = 326 tons.
Strength of *a* and *c*, as above . . . . = 462 ,,
Added for shear of rivets in edges = $2 \times 13\frac{2}{3}$ tons. . . . = 27 ,,

Total breaking strength . . . . = 815 ,,

MODE V.—Total length of line of fracture of a and c = 39 inches.
Rivet-holes in " " " = $1\frac{3}{4}$ "

∴ Effective length . . . . = $37\frac{1}{4}$ "
Strength of a and c = $37\frac{1}{4}'' \times \frac{3}{4}'' \times 22$ tons . . . . . = 615 tons.
Strength of strap to b (as in Mode III.). . . . . . . = 226 "

Total breaking strength . . . . = 841 "

MODE VI.—Strength of a and c, as above . . . . = 615 tons.
Added for shear of butt fastenings = $13 \times 13\frac{2}{3}$ tons . . = 178 "

Total breaking strength . . . . = 793 "

For the butt CD and the plates b, c, and d, we obtain the following breaking strengths:—

MODE I.—Total length of the line E F . . . . . . = $76\frac{1}{2}$ inches.
Rivet-holes in " " . . . . . . = $10\frac{1}{16}$ "

∴ Effective length . . . . = $66\frac{7}{16}$ "
Breaking strength = $66\frac{7}{16}'' \times \frac{3}{4}'' \times 18$ tons . . . . = 897 tons.

MODE II.—Total length of the line of fracture of b and d = $34\frac{1}{2}$ inches.
Rivet-holes in " " " = $4\frac{13}{16}$ "

∴ Effective length . . . . = $29\frac{11}{16}$ "

Strength of b and d = $29\frac{11}{16}'' \times \frac{3}{4}'' \times 18$ tons . . . . . . = 401 tons.
Added for shear of rivets = $27 \times 13\frac{2}{3}$ tons . . . . . . = 369 "

Total breaking strength . . . . = 770 "

MODE III.—Total breadth of strap to c . . . . = 42 inches.
Rivet-holes in line of fracture . . . . . . = $9\frac{5}{8}$ "

∴ Effective breadth . . . . = $32\frac{3}{8}$ "

Strength of strap = $32\frac{3}{8}'' \times \frac{3}{4}'' \times 18$ tons . . . . = 437 tons.
Strength of b and d, as above . . . . = 401 "
Added for shear of rivets in edges = $6 \times 13\frac{2}{3}$ tons . . . = 82 "

Total breaking strength . . . . = 920 "

MODE IV.—Total breadth of plate c . . . . = 42 inches.
Rivet-holes in line K L . . . . = $8\frac{3}{4}$ "

∴ Effective breadth . . . . = $33\frac{1}{4}$ "

Strength of plate c = $33\frac{1}{4}'' \times \frac{3}{4}'' \times 18$ tons . . . . . . = 449 tons.
Strength of b and d, as above . . . . = 401 "
Added for shear of rivets in edges = $2 \times 13\frac{2}{3}$ tons . . . = 27 "

877 "

MODE V.—Total length of line of fracture of b and d . . = $34\frac{1}{2}$ inches.
Rivet-holes in " " " . . = $1\frac{3}{4}$ "

∴ Effective length . . . . = $32\frac{3}{4}$ "
Strength of b and d = $32\frac{3}{4}'' \times \frac{3}{4}'' \times 22$ tons . . . . . . = 540 tons.
Strength of strap to c, (as in Mode III.) . . . . = 437 "

977 "

MODE VI.—Strength of b and d as above . . . . = 540 tons.
Added for shear of butt fastenings = 21 × $13\frac{2}{3}$ tons . . = 287 ,,

827 ,,

Assuming as before that the strengths of the unavoidably weak sections E F are taken as unity for each butt, we arrive at the following results :—

	Butt A B.	Butt C D
Breaking strength by Mode II. . .	·865	·858
,, ,, III. . .	·923	1·026
,, ,, IV. . .	·977	·978
,, ,, V. . .	1·008	1·089
,, ,, VI. . .	·950	·922

From this table it will be seen that the butt fastenings in this ship are, on the whole, fairly arranged, but it will be remarked that for both the butts of the inside and outside strakes there are modes of fracture for which the breaking strengths are more than one-tenth less than those required to separate the plates along the unavoidable weak sections E F. The strength of the ship is thus made less than it would be if the butt fastenings were brought up to a strength a little above that along the lines E F, which result might be arrived at by increasing the number of the rivets in both the butts, and the thickness of the strap to the butt A B; or, which would be much better, by avoiding the brick fashion shift of butts, and increasing the number of passing strakes as in the 'Hercules.'

It may be well here to call attention to the fact that in the preceding calculations we have neglected the shearing strength of the rivets in the frames adjacent to the butts, which would tend to increase the breaking strengths corresponding to some other modes of fracture. We have only taken account of the fastenings of the plating to the frames in so far as the plating is weakened by the holes punched to receive these fastenings; but it must be remembered that the breaking strengths of those modes where the lines of fracture cross transverse frames, will be increased by the assistance rendered by the shearing strength of the rivets in the frames, or the breaking across of the frames themselves. In the modes of fracture taken in the preceding calculations the breaking lines do not cross the frames, and consequently no account is taken of the strength of the frames or their fastenings. A further small correction should also in strictness be made on account of the reduction which the plates undergo in countersinking the holes; and it would be desirable in many cases to include this reduction in similar calculations.

In cases such as the preceding, where the butt fastenings of plates forming portions of assemblages of plating have to be arranged, the following order of procedure may be observed advantageously. The shift of butts, sizes and pitches of rivets in the frames and plate edges, and breadths of plating having been determined on, the strength of the unavoidably weakest section may be readily found. As the number of rivets in the edges of the plates between the frame and the butt is known, their shearing strength may be calculated, and hence the number of rivets may be found which is required on each side of the butt in order to bring the total breaking strength by Mode II. up to a little more than the strength of the section at the frames. Fracture by Mode II. is selected as the basis of the calculation, as it will be seen from the preceding tables to give the least total breaking strengths. When the number of rivets has been determined their positions have to be fixed, and will be governed by the breadth of the strake and the number of rivets required. In a butt of outside plating it is, of course, always necessary to have the rows of rivets next the butt spaced for watertight work, in order to obtain a good caulk of the butt joint. In deck plating, when the work is made watertight, the edges of the straps are caulked, and consequently the rows of rivets farthest from the butt are placed so as to make a watertight joint. These and other considerations having been taken into account and the positions of the butt fastenings settled, it becomes necessary to try whether the arrangement gives sufficient breaking strengths by the other modes of fracture. It will be obvious that a very fair approximation to the provision of the necessary breaking strength by Mode V. may be made by having a rather less number of rivet-holes in the lines G H than in the lines E F. With respect to the thickness of butt straps required, we need only refer to the remarks previously made with respect to stringers and tie-plates, and to call attention to the ordinary practice of shipbuilders as described in Chapter X. No general investigation or expression can be given which would much facilitate the process of arranging the fastenings, but it is hoped that by the aid of the foregoing explanations and examples the reader may be able to work out the details of any case that may occur in practice.

It was stated at the commencement of this chapter that the amount of experimental knowledge at present possessed with respect to the best forms and proportions for riveted work is extremely limited. The results of most of the experiments which have been made have been given in a condensed form in the preceding pages, and their applicability to actual practice in shipbuilding has been discussed. The calculations which have been given are of a simple, practical character, and are founded on, what in our opinion are, the most reliable experiments, both as regards the mode in which they have been conducted, and their resemblance to the actual proportions, pitches, &c., used in practice. The method of calculation is, confessedly, an approximate one, but being founded on experimental facts and conducted on intelligible principles, it tends to make the arrangement of butt fastenings sufficiently accurate for all practical purposes at present. While it has been considered necessary to point out what are thought to be errors in the views put forward by previous writers on iron construction, it is desired to give the strongest testimony to the value of the experiments and information which their works afford, of which we have availed ourselves so largely. We need only add a further statement of the opinions previously expressed with respect to the necessity and desirability of a complete set of experiments on the various kinds of riveted work, and a hope that such experiments may shortly be made.

In order to afford as detailed information as can be possibly required, with respect to the sizes and pitch of rivets used in the construction of the large iron-clad frigates of the Royal Navy, we have given the following table of the particulars of the riveting of the 'Hercules.'

TABLE of the SIZES and PITCHES of RIVETS employed in the 'HERCULES.'

Description of work.	Thickness of iron.	Breadth of lap.	Pitch of rivets.	Size of rivets.
	inches.	inches.	inches.	inches.
Flat keel-plates	Outer $1\frac{1}{8}$, Inner 1	..	$5\frac{1}{2}$	$1\frac{3}{8}$ & $1\frac{1}{4}$
Butt straps to ditto *	$1\frac{3}{8}$	20	4	$1\frac{1}{4}$
Keel angle-irons to flat keel plates	$6\times6\times1$ to 1 & $1\frac{1}{8}$ plate	..	6	$1\frac{3}{8}$
Ditto to vertical keel ..	ditto to $\frac{3}{4}$,,	..	6	$1\frac{3}{8}$

* Treble chain riveted.

TABLE of SIZES and PITCHES of RIVETS—*continued.*

Description of work.	Thickness of iron.	Breadth of lap.	Pitch of rivets.	Size of rivets.
	inches.	inches.	inches.	inches.
Butt straps to vertical keel (double) *	$\frac{7}{16}$	18	4	$1\frac{1}{8}$
Angle-irons on the upper edge of vertical keel to the keel-plate	$3\frac{1}{2} \times 3\frac{1}{2} \times \frac{9}{16}$ to $\frac{3}{4}$ plate	..	5 to $5\frac{1}{2}$	$\frac{7}{8}$
Ditto to the inner bottom	ditto to $\frac{1}{2}$,,	..	5 to $5\frac{1}{2}$	$\frac{7}{8}$
Transverse frames to short frame angle-irons	$5\frac{1}{2} \times 4 \times \frac{9}{16}$ to $\frac{7}{16}$,,	..	6 to $6\frac{1}{2}$	$\frac{7}{8}$
Watertight frames to ditto ..	$4 \times 3\frac{1}{2} \times \frac{9}{16}$ to $\frac{7}{16}$,,	..	$4\frac{1}{2}$	$\frac{7}{8}$
Transverse frames to vertical keel	$5\frac{1}{2} \times 4 \times \frac{9}{16}$ to $\frac{3}{4}$,,	..	$5\frac{1}{2}$	$1\frac{1}{8}$
Ditto to longitudinals ..	$3\frac{1}{2} \times 3\frac{1}{2} \times \frac{7}{16}$ to $\frac{1}{2}$ & $\frac{7}{16}$,,	..	$5\frac{1}{2}$	$\frac{7}{8}$
Ditto to continuous transverse angle-irons	$5\frac{1}{2} \times 3\frac{1}{2} \times \frac{7}{16}$ to $\frac{7}{16}$,,	..	5	$\frac{7}{8}$
Outside bottom plating to the frames	$5\frac{1}{2} \times 4 \times \frac{9}{16}$ to 1 ,, ditto to $\frac{7}{8}$, $\frac{13}{16}$, & $\frac{3}{4}$,,	.. ..	$6\frac{1}{2}$ $6\frac{1}{2}$	$1\frac{1}{8}$ 1
Edges of the outside bottom plating †	$1\frac{1}{8}$ to 1 ,,	7	$5\frac{1}{2}$	$1\frac{1}{4}$
	1 ,, $\frac{7}{8}$,,	$6\frac{1}{2}$	5	$1\frac{1}{8}$
	$\frac{7}{8}$,, $\frac{7}{8}$,,	$6\frac{1}{2}$	4 to 5	$1\frac{1}{8}$
	$\frac{7}{8}$,, $\frac{13}{16}$,,	$5\frac{1}{2}$	4 to 5	1
	$\frac{13}{16}$,, $\frac{13}{16}$,,	$5\frac{1}{2}$	4 to 5	1
	$\frac{13}{16}$,, $\frac{3}{4}$,,	$5\frac{1}{2}$	4 to 5	1
	$\frac{3}{4}$,, $\frac{3}{4}$,,	$5\frac{1}{2}$	4 to 5	1
Transverse frames behind armour	$3\frac{1}{2} \times 3\frac{1}{2} \times \frac{1}{2}$ (double) to $10 \times 3\frac{1}{2} \times \frac{1}{2}$ frame	..	6	$\frac{7}{8}$
Longitudinals to angle-irons on outer edges	$4 \times 3\frac{1}{2} \times \frac{9}{16}$ to $\frac{1}{2}$ & $\frac{7}{16}$ plate	..	5 to 6	$\frac{3}{4}$
Ditto to ditto on inner ditto	$3 \times 3 \times \frac{1}{2}$ to ditto ,,	..	5 to 6	$\frac{3}{4}$
Butt straps to longitudinals ‡	$\frac{1}{4}$ to $\frac{7}{16}$ plate, $\frac{5}{16}$ to $\frac{1}{2}$,,	$8\frac{1}{2}$	$3\frac{1}{2}$	$\frac{3}{4}$
Ditto to watertight ditto ‡	$\frac{5}{16}$ to $\frac{1}{2}$,,	$8\frac{1}{2}$	$3\frac{1}{2}$	$\frac{3}{4}$
Inner bottom to continuous transverse frames	$5\frac{1}{2} \times 3\frac{1}{2} \times \frac{7}{16}$ to $\frac{1}{2}$ & $\frac{7}{16}$,,	..	6	$\frac{3}{4}$
Ditto to edge strips ..	$\frac{1}{2}$ to $\frac{1}{2}$ plate, $\frac{7}{16}$ to $\frac{7}{16}$,,	5	4	$\frac{3}{4}$
Ditto to butt straps § ..	ditto	9	3	$\frac{3}{4}$
Transverse bulkheads to edge strips	$4\frac{1}{2} \times 2\frac{1}{4} \times \frac{1}{2}$ T-iron to $\frac{1}{2}$, $\frac{7}{16}$ & $\frac{3}{8}$,,	$4\frac{1}{2}$	$3\frac{1}{2}$ to $4\frac{1}{2}$	$\frac{3}{4}$
Ditto to butt straps ‖ ..	Same thickness as the plates.	$5\frac{1}{2}$	$3\frac{1}{2}$	$\frac{3}{4}$
Ditto to stiffeners	$3\frac{1}{2} \times 3 \times \frac{9}{16}$ to $\frac{1}{2}$, $\frac{7}{16}$, & $\frac{3}{8}$ plate	..	5 to 6	$\frac{3}{4}$
Ditto to inner bottom ..	$3\frac{1}{2} \times 3\frac{1}{2} \times \frac{1}{2}$ to $\frac{1}{2}$ & $\frac{7}{16}$,,	..	$3\frac{1}{2}$ to $4\frac{1}{2}$	$\frac{7}{8}$
Ditto to vertical portion of the inner bottom	$4 \times 4 \times \frac{5}{8}$ to $\frac{1}{2}$ & $\frac{3}{4}$,,	..	$3\frac{1}{2}$ to 4	1
Ditto to frames before and abaft double bottom	$\frac{7}{16}$ to $\frac{7}{16}$,,	$3\frac{1}{2}$	4	$\frac{3}{4}$
Wing passage bulkheads to edge strips	$\frac{1}{2}$ to $\frac{3}{4}$,, $\frac{1}{2}$ to $\frac{1}{2}$,,	6 $4\frac{1}{2}$	5 $3\frac{1}{2}$	1 $\frac{3}{4}$
Ditto to butt straps § ..	$\frac{1}{2}$ to $\frac{3}{4}$,, $\frac{1}{2}$ to $\frac{1}{2}$,,	$11\frac{1}{2}$ $8\frac{1}{2}$	$3\frac{1}{2}$ to 4 3	1 $\frac{3}{4}$

* Treble-chain riveted. † The butt fastenings have been described in Chapter X.
‡ Double butt-straps, double-chain riveted. § Double-chain riveted. ‖ Single-riveted.

TABLE of SIZES and PITCHES of RIVETS—*continued.*

Description of work.	Thickness of iron.	Breadth. of lap.	Pitch of rivets.	Size of rivets.
	inches.	inches.	inches.	inches.
Wing passage bulkheads to stiffeners	$3\frac{1}{2} \times 3\frac{1}{2} \times \frac{1}{2}$ to $\frac{1}{2}$ & $\frac{3}{4}$ plate	..	6	$\frac{3}{4}$ & $\frac{7}{8}$
Lower-deck stringer-plate to beams	$\frac{1}{2}$	36	$6\frac{1}{2}$	$\frac{3}{4}$
Do. to butt straps * ..	$\frac{3}{8}$	$12\frac{1}{4}$	$3\frac{1}{2}$ to 4	$\frac{3}{4}$
Ditto to strips on the intercostal plates between the frames	$\frac{1}{2}$	$4\frac{1}{4}$	4	$\frac{3}{4}$
Main-deck beams	$\frac{1}{2}$ plate to $\frac{1}{2}$ angle-iron	..	5	$\frac{3}{4}$
Main-deck stringer to gutter angle-irons	ditto to $\frac{5}{8}$ & $\frac{1}{2}$,,	..	$3\frac{1}{2}$	$\frac{3}{4}$
Ditto to beams	$\frac{1}{2}$	36	$7\frac{1}{2}$	$\frac{3}{4}$
Ditto to butt straps * ..	$\frac{1}{2}$	$12\frac{1}{4}$ & $8\frac{1}{2}$	4	$\frac{3}{4}$
Plating on main-deck ($\frac{1}{2}$-inch) to beams	$\frac{1}{2}$	..	$7\frac{1}{2}$	$\frac{3}{4}$
Ditto ditto to edge strips	$\frac{1}{2}$	$4\frac{3}{4}$	$4\frac{1}{2}$	$\frac{3}{4}$
Ditto ditto to butt straps †	$\frac{1}{2}$	$8\frac{1}{2}$	4	$\frac{3}{4}$
Ditto ($\frac{3}{8}$-inch) to beams ..	$\frac{3}{8}$	..	$5\frac{1}{2}$	$\frac{5}{8}$
Ditto ditto to edge strips	$\frac{3}{8}$	4	4	$\frac{5}{8}$
Ditto ditto to butt straps ‡	$\frac{3}{8}$	7 & 4	$3\frac{1}{2}$	$\frac{5}{8}$
Upper-deck stringer to gutter angle-irons	$\frac{1}{2}$ plate to $\frac{5}{8}$ angle-iron	..	4	$\frac{3}{4}$
Ditto to beams	$\frac{1}{2}$	36	$7\frac{1}{2}$	$\frac{3}{4}$
Ditto to butt straps ‖ ..	$\frac{1}{2}$	$12\frac{1}{2}$	$3\frac{1}{4}$	$\frac{3}{4}$
Ditto steel plating ($\frac{1}{2}$-in.) to beams	$\frac{1}{2}$	..	$7\frac{1}{2}$	$\frac{3}{4}$
Ditto ditto to edge strips	$\frac{1}{2}$	$4\frac{3}{4}$	$4\frac{1}{2}$	$\frac{3}{4}$
Ditto ditto to butt straps¶	$\frac{1}{2}$	$8\frac{1}{2}$	$3\frac{1}{2}$	$\frac{3}{4}$
Ditto steel plating ($\frac{3}{8}$-in.) to beams	$\frac{3}{8}$	..	$7\frac{1}{2}$	$\frac{5}{8}$
Ditto ditto to edge strips	$\frac{3}{8}$	4	$3\frac{3}{4}$	$\frac{5}{8}$
Ditto ditto to butt straps¶	$\frac{3}{8}$	$7\frac{1}{4}$	$2\frac{7}{8}$	$\frac{5}{8}$
Ditto steel plating ($\frac{1}{4}$-in.) to beams	$\frac{1}{4}$	..	$7\frac{1}{2}$	$\frac{1}{2}$
Ditto ditto to edge strips	$\frac{1}{4}$	$3\frac{1}{4}$	3	$\frac{1}{2}$
Ditto ditto to butt straps ¶	$\frac{1}{4}$	$5\frac{3}{4}$	$2\frac{1}{4}$	$\frac{1}{2}$
Magazine bulkheads to stiffeners	$3 \times 3 \times \frac{1}{2}$ to $\frac{7}{16}$ plate	..	5 to 6	$\frac{3}{4}$
Ditto to edge strips ..	$\frac{7}{16}$	$4\frac{1}{2}$	$3\frac{1}{2}$	$\frac{3}{4}$
Ditto to butt straps ** ..	$\frac{7}{16}$	$4\frac{1}{2}$	$3\frac{1}{2}$	$\frac{3}{4}$
Plating on platforms to beams	$\frac{3}{8}$	..	$3\frac{1}{2}$	$\frac{5}{8}$
Ditto to edge strips ..	$\frac{3}{8}$	4	$3\frac{1}{2}$	$\frac{5}{8}$
Ditto to butt straps † ..	$\frac{3}{8}$	7	3	$\frac{5}{8}$
Shaft passage bulkheads to stiffeners	$3 \times 3\frac{1}{2} \times \frac{1}{2}$ to $\frac{3}{8}$ plate	..	6	$\frac{1}{2}$
Ditto to edge strips ..	$\frac{3}{8}$	$3\frac{1}{4}$	4	$\frac{1}{2}$
Ditto to butt straps ** ..	$\frac{3}{8}$	$3\frac{1}{4}$	4	$\frac{1}{2}$
Bilge-keel angle-irons to bottom plating ††	$5 \times 4 \times \frac{9}{16}$ to $\frac{7}{8}$ & $\frac{13}{16}$ plates	..	5	1

* Double butt strap, riveted as in Fig. 246, page 365. Main deck stringer double riveted in the battery. † Double riveted. ‡ Double and single riveted. ‖ Treble chain riveted. ¶ Double chain riveted. ** Single riveted. †† Tap-rivets clenched on the inside.

TABLE of SIZES and PITCHES of RIVETS—*continued.*

Description of work.	Thickness of iron.	Breadth of cap.	Pitch of rivets.	Size of rivets.
	inches.	inches.	inches.	inches.
Armour shelf:—				
Angle-iron on outer edge to shelf plate	5 × 5 × ⅝ to ⅞ plate	..	5½	1⅛
Ditto to bottom plating	ditto to 13/16 ,,	..	5½	1⅛
Ditto on inner edge to shelf plate	ditto to ⅞ ,,	..	5½	1⅛
Ditto ditto to skin plating	ditto to ¾ ,, (two thicknesses)	..	5½	1⅛
Covering plate to shelf	¾	17	5	1
Plating behind armour to frames	3½ × 3½ × ½ to two thicknesses of ¾ plate	..	6 to 7	1
Ditto to edge strips ..	¾	6¼	5 to 6	1
Ditto to butt straps* ..	¾	16½	5	1
Ditto the two thicknesses to each other	¾	..	4 to 5½	1
Ditto to longitudinal girders	12 × 3½ × ½ to two thicknesses of ¾ plates	..	5	1
Plating above armour (before and abaft the battery) to frames	7 × 3½ × 7/16 or 4 × 3½ × 9/16 to ½ plate	..	5½	¾
Ditto to edge strips ..	½	5	5	¾
Ditto to butt straps † ..	½	9	3	¾
Ditto to armour plates ‡	½ plate to armour	7½ or 5½	4	⅞
Platform beams	½ plate to ⅜ or 7/16 angle-iron	..	6	⅝ & ½
Lower deck beams	½ plate to ½ angle-iron	..	6	¾ & ⅝

* There are edge-strips to the outer thickness only; the butts are treble-riveted for both thicknesses, the rivets being arranged so as to clear the armour-bolts. † Double riveted.
‡ Double-tap riveted at the edges, and treble at the butts.

We will conclude the present chapter with a few remarks upon the use of steel rivets in shipbuilding.

It is not without reason that great distrust of steel rivets is at present felt by shipbuilders. They undoubtedly require very careful manufacture and still more careful working, and in spite of both they have in many cases proved treacherous. Mr. Barber has stated publicly that "when steel rivets have been used, the heads "have been known to fly off when the vessel has bumped against "a pier-head, or been suddenly struck by a barge or a heavy "floating body." No case of this kind has come under our own notice, but we have frequently known steel rivet heads to fly off during the building of the ship, with the jar occasioned by knocking down other rivets in the same frame space, or by the caulking; and when there has been occasion to cut out rivets of this material during repairs, the heads have cracked off with a readiness quite unknown in good iron-riveted work. It is not intended by this to imply that the heads of iron rivets never fly off at all; on the contrary it is well known that the heads of small countersunk iron-rivets frequently fly off when over-hammered; but steel rivets have hitherto been found much more liable to this defect.

The following table gives the results of the tensile strength tests of the rivet steel used in some parts of the 'Penelope' and 'Inconstant' at Pembroke Dockyard.

Diameter.	No. of specimen.	Proof strain.	Breaking strain.	Average.
inch.	No.	tons. cwts. qrs. lbs.	tons. cwts. qrs. lbs.	tons. cwts. qrs. lbs.
½	1	6 9 2 6	6 15 0 0	
,,	2	,,	6 16 0 14	
,,	3	,,	6 13 0 0	6 15 3 9
,,	4	,,	6 12 3 14	34·59 tons per sq. inch.
,,	5	,,	6 18 3 0	
,,	6	,,	6 19 1 0	
⅝	1	10 2 1 26	11 15 2 0	
,,	2	,,	12 9 1 0	
,,	3	,,	11 9 2 14	11 17 0 0
,,	4	,,	11 13 2 14	38·64 tons per sq. inch.
,,	5	,,	12 7 2 0	
,,	6	,,	11 6 2 9	
¾	1	14 11 2 2	15 18 0 14	
,,	2	,,	15 12 3 14	
,,	3	,,	15 1 0 0	15 5 1 9
,,	4	,,	14 19 2 14	34·57 tons per sq. inch.
,,	5	,,	14 18 1 14	
,,	6	,,	15 2 0 0	
				Average 35·93 tons per sq. inch.

The rivet-steel employed by Mr. Kirkaldy in his experiments appears to have been about equal to the best of this steel, as he gives the mean breaking strength as 38·59 tons per sq. inch, and the mean shearing strength as 28·48 tons; the shearing strength being, therefore, about 26 per cent. less than the tensile strength. Mr. Kirkaldy's experiments with this rivet-steel appear to have proved the fact, which might have been inferred from the relation between the tensile and shearing strengths of the material, that, in steel, larger rivets are required for a given thickness of plate than are required in iron. For example, in experimenting with riveted plates, ·19-inch thick connected by rivets ·43-inch in diameter, several experiments failed from the shearing of the rivets, although they were larger than those used for iron plates of the same thickness.

The chief things to be attended to in working steel rivets are, first, to heat them sufficiently and yet to avoid raising them above a cherry-red heat; and, secondly, to knock them down and finish them off as quickly as possible. When the rivet has not been sufficiently heated, or the riveters have not been expert, they have had great trouble in cutting off the burr, and in doing so have often broken away part of the countersunk point with the burr. On the other hand, if the rivets are heated much above a cherry-red heat, they cannot be properly knocked down, as they waste away under the blows of the hammer. If great care is not taken, the rivet may be over-heated to an extent not sufficient to prevent its being knocked down, but sufficient to greatly deteriorate the quality of the finished rivet. It is advantageous to have plain knockdown or conical points to steel rivets in preference to snap-points, as a burnt or over-heated rivet is then more easily detected by the crack round the edges.

In comparing the advantages and disadvantages of steel rivets for steel work, there are other important considerations to be borne in mind, viz., that it is impossible, even under the best supervision, to ensure that every rivet shall be brought to its proper heat, and have its point knocked down before it has become cold; and that under these circumstances, the efficiency of the rivet is dependent, in a great measure, on the rivet-boy, who cannot be expected to show a large amount of judgment in the matter. It is to these considerations, in all probability, that the present prohibition of the use of steel rivets by Lloyd's Committee is due; and although we have felt bound, in the Government service, to make a limited and guarded use of steel in the form of rivets as well as of plates, we are not at all disposed to complain of the course which the Committee has taken.

CHAPTER XVIII.

ON TESTING IRON AND STEEL.*

THE quality of the iron and steel employed for shipbuilding purposes is obviously of primary importance, and the establishment of tests for these materials possesses corresponding interest. Lloyd's Rules merely provide that iron shall be of good malleable quality, capable of bearing a longitudinal strain of 20 tons per square inch, the material being guaranteed to the extent of having the manufacturer's name or trade-mark stamped upon it. The same rule applies to steel, but neither the minimum nor the maximum strain for it is fixed. The surveyors of the Committee exercise their own judgment and discretion in testing these materials. The Liverpool Committee's surveyors do the same, but their instructions direct that all iron plates (steel not being yet provided for in the Liverpool Rules) shall be of the best quality, branded with the maker's name, tough and malleable, the sheared edges to be free from rip, the surface free from flaws and blisters, and the punching reasonably free from cracks upon the convex side. The absolute mean breaking-strain must be 20 tons per square inch of the original section, and 24 tons per square inch of the broken section; and all brittle or coarsely crystalline iron has to be rejected. Angle-irons have to be free from veins and cracked holes, and rivet-iron has to be free from cracks and veins when laid up and finished.

These instructions point clearly enough to what is, we believe, the practice of the surveyors of both Committees, viz., that of limiting the tests of material chiefly to a general examination of plates and angle-irons, to tests of tensile strength, a careful scrutiny of the punchings, and a more or less close observation of the parts of the ship as the work proceeds.

In Her Majesty's Service, and in the case of ships built for that service by private shipbuilders, the materials undergo a more searching and rigid examination. The iron is supplied from the

* Part of this chapter was read to the Association of Foremen Engineers in the form of a Paper by the Author.

manufacturers subject to the following tests, which are carried out in the same manner, as nearly as possible, in the various establishments both public and private under the supervision of Admiralty officers :—

PLATE IRON (FIRST CLASS).

B. B.

Tensile Strain per square inch	Lengthways	22 tons.
	Crossways	18 „

FORGE TEST (*Hot*).

All plates of the First Class, of one inch in thickness and under, should be of such ductility as to admit of bending hot, without fracture, to the following angles :—

Lengthways of the grain	125 degrees.
Across	90 „

FORGE TEST (*Cold*).

All plates of the First Class should admit of bending cold without fracture, as follows :—

With the Grain.

1 in.	and	$\frac{15}{16}$	of an inch in thickness	to an angle of	15	degrees.
$\frac{7}{8}$	„	$\frac{13}{16}$	„	„	20	„
$\frac{3}{4}$	„	$\frac{11}{16}$	„	„	25	„
$\frac{5}{8}$, $\frac{9}{16}$	„	$\frac{1}{2}$	„	„	35	„
$\frac{7}{16}$	„	$\frac{3}{8}$	„	„	50	„
$\frac{5}{16}$	„	$\frac{1}{4}$	„	„	70	„
$\frac{3}{16}$	„	under	„	„	90	„

Across the Grain.

1 in., $\frac{15}{16}$, $\frac{7}{8}$, and $\frac{13}{16}$	of an inch in thickness	to an angle of	5	degrees.
$\frac{3}{4}$ and $\frac{11}{16}$	„	„	10	„
$\frac{5}{8}$, $\frac{9}{16}$, and $\frac{1}{2}$	„	„	15	„
$\frac{7}{16}$ „ $\frac{3}{8}$	„	„	20	„
$\frac{5}{16}$ „ $\frac{1}{4}$	„	„	30	„
$\frac{3}{16}$ „ under	„	„	40	„

PLATE IRON (SECOND CLASS).

B.

Tensile Strain per square inch	Lengthways	20 tons.
	Crossways	17 „

FORGE TEST (*Hot*).

All Plates of the Second Class of one inch in thickness and under, should be of such ductility as to admit of bending hot, without fracture, to the following angles :—

Lengthways of the grain	90 degrees.
Across	60 „

FORGE TEST (*Cold*).

All Plates of the Second Class should admit of bending cold without fracture, as follows :—

With the Grain.

1 in. and	$\frac{15}{16}$	of an inch in thickness	to an angle of	10	degrees.
$\frac{7}{8}$,,	$\frac{13}{16}$	,,	,,	15	,,
$\frac{3}{4}$,,	$\frac{11}{16}$	,,	,,	20	,,
$\frac{5}{8}$, $\frac{9}{16}$,,	$\frac{1}{2}$	,,	,,	30	,,
$\frac{7}{16}$,,	$\frac{3}{8}$	,,	,,	45	,,
$\frac{5}{16}$,,	$\frac{1}{4}$	,,	,,	55	,,
$\frac{3}{16}$,,	under	,,	,,	75	,,

Across the Grain.

$\frac{3}{4}$ in. and	$\frac{11}{16}$	of an inch in thickness	to an angle of	5	degrees.
$\frac{5}{8}$, $\frac{9}{16}$, ,,	$\frac{1}{2}$	,,	,,	10	,,
$\frac{7}{16}$,,	$\frac{3}{8}$	,,	,,	15	,,
$\frac{5}{16}$,,	$\frac{1}{4}$	,,	,,	20	,,
$\frac{3}{16}$,,	under	,,	,,	30	,,

Plates, both hot and cold, should be tested on a cast-iron slab, having a fair surface, with an edge at right angles, the corner being rounded off with a radius of $\frac{1}{2}$ an inch.

The portion of plate tested, for both hot and cold tests, is to be 4 feet in length, across the grain; and the full width of the plate, with the grain.

The plate should be bent at a distance of from 3 to 6 inches from the edge.

It is intended that all the Iron shall stand the Forge Tests herein named, when taken in four-feet lengths, across the grain; and the whole width of the plate, along the grain, whenever it may be necessary to try so large a piece; but a smaller sample will generally answer every purpose.

All plates to be free from lamination and injurious surface defects.

One plate to be taken indiscriminately for testing from every thickness of plate, sent in per invoice, provided they do not exceed fifty in number. If above that number, one for every additional fifty, or portion of fifty.

Where plates of several thicknesses are invoiced together, and there are but few plates of any one thickness, a separate test for plates of each thickness need not be made; but no lot of plates of any one thickness must be rejected before one of that lot has been tested.

Before describing in detail the manner in which these tests are practically applied, it is desirable to mention that we have found it necessary, at the Admiralty, to add to the above conditions a rigid stipulation with reference to the weight of iron materials. Since the introduction of armour-plated ships, and in view of the urgent necessity for carrying the utmost weights of armour and armament without exceeding a defined draught of water, it has become of great importance that the maximum weight of the angle-irons, plates, &c., entering into the construction of such vessels should be known with certainty; and for this reason the Admiralty have laid down the rule that the actual weight shall not exceed the weight due by calculation to the nominal dimensions. At the same time it is, of course, important that the actual weight should not fall sufficiently below the due weight to considerably reduce the strength; and the lower limit is consequently fixed at 5 per cent. below the calculated weight for plates and angle-irons of $\frac{1}{2}$ inch thick and upwards; and at 10 per cent. for those below $\frac{1}{2}$ inch. The standard weight of iron plates is taken at 10 lbs. per square foot for every $\frac{1}{4}$ inch in thickness, or 480 lbs. per cubic foot. In the case of angle-iron, it is presumed that the weight per foot of length should equal the weight of a plate a foot long of the same nominal thickness as the angle-iron, and of a breadth equal to the sum of the breadths of the angle-iron flanges, less the thickness; thus a piece of 3 by $3\frac{1}{2}$ by $\frac{1}{2}$ inches angle-iron, 1 foot long, would be equal in weight to a piece of plate 1 foot long, 6 inches broad, and $\frac{1}{2}$ inch thick; its maximum weight would consequently be 10 lbs., and its minimum weight $9\frac{1}{2}$ lbs.

It has not been found in practice that any great difficulty exists in conforming to these conditions in the manufacture of plates and angle-irons. In cases where the thickness was very small, some difficulty was experienced when the limit of 5 per cent. was fixed for all thicknesses, and therefore it was made 10 per cent. for all thicknesses below $\frac{1}{2}$ inch, as previously stated. That the difficulty must increase as the thickness diminishes is obvious from the fact that the percentage of weight and thickness being the same in all cases, the actual limits must continually approach equality as the thickness is reduced. In a 1-inch plate, for example, the maximum thickness allowed is 1 inch, and the minimum $\frac{19}{20}$ inch, the limits of thickness being $\frac{1}{20}$ inch apart. In a $\frac{1}{4}$-inch plate, if the maximum thickness allowed is $\frac{1}{4}$ inch, or $\frac{20}{80}$ inch, and the minimum $\frac{19}{80}$ inch, the limits are only $\frac{1}{80}$ inch apart. It is clear, therefore, that it must be very difficult to conform strictly to the rule formerly laid down in the manufacture of very thin plates. As a matter of fact, however, this consideration is of but little moment, as the quantity of thin iron employed in shipbuilding is small, and it is not any great disadvantage to waive, as is now done, the application of the 5 per cent. rule in these exceptional cases, substituting a limit of 10 per cent. for it. In iron plate and angle-iron of the thickness ordinarily used in shipbuilding the manufacturers can, with proper care, readily conform to the 5 per cent. limits assigned, and as a rule do conform to them. Such difficulties as are found to occur, usually exist where the manufacturer is paid by weight, and when the inducement naturally is to pass the superior limit. Where private shipbuilders make their own iron for ships building for the Admiralty this inducement does not exist. The judicious enforcement of the rules, however, soon results in the removal of all grounds of complaint in this respect. In order to furnish examples of the extent to which iron and steel plates have been found to vary from the prescribed weights when proper care has not been taken in the manufacture, we have given the following tables, which are records of the dimensions and weights of plates actually supplied to the dockyards by manufacturers.

IRON PLATES.

No. of Plates.	Length.		Breadth.		Thickness.	Actual Weight.			Estimated Weight.		
	ft.	in.	ft.	in.	16ths of in.	cwts.	qrs.	lbs.	cwts.	qrs.	lbs.
2	12	2	1	0	6	3	2	2	3	1	1
2	13	0	1	2¼	8	5	3	4	5	2	1
1	12	1	3	11½	6	6	3	12	6	1	17
1	13	0	3	11½	,,	7	2	4	6	3	16
1	13	6	3	9½	,,	7	1	8	6	3	12
1	19	6	1	2¼	8	4	1	23	4	0	15
1	21	9	1	2¼	,,	5	0	4	4	2	12
1	20	2	1	2¼	,,	4	2	4	4	1	3
1	19	0	1	2¼	,,	4	2	16	4	0	3
1	12	3	1	2¼	,,	3	0	4	2	2	11
1	11	6	3	10	6	6	1	20	5	3	17
1	10	10	3	10	,,	5	3	20	5	2	7
2	9	7	3	10	,,	10	1	26	9	3	10
1	7	7½	3	11½	,,	4	2	0	4	0	5
1	12	6	3	11½	,,	7	0	20	6	2	14
1	17	2	1	0	,,	2	1	24	2	1	6
1	18	2	1	0⅛	,,	2	2	4	2	1	23
2	17	10	1	0⅛	,,	5	1	4	4	3	8
1	13	6	0	11¾	,,	1	3	10	1	3	2
1	15	0	0	10¼	,,	1	3	12	1	2	24
2	6	0	3	5½	,,	5	3	10	5	2	6
1	8	7	3	9	,,	4	3	0	4	1	7
1	12	3	3	8	,,	6	2	22	6	0	2
2	12	2	3	9½	,,	12	3	12	12	1	12
1	7	7½	3	11½	,,	4	2	4	4	0	5
1	6	9½	2	1½	,,	2	0	14	1	3	21
1	5	2	3	7½	,,	2	2	4	2	2	1
1	4	3	2	10	12	3	1	12	3	0	25
2	8	11½	2	9½	,,	14	1	4	13	1	16
1	12	8	2	3	,,	8	1	4	7	2	15
1	12	11	2	1½	,,	7	3	14	7	1	11
2	4	9½	2	11½	,,	7	3	18	7	2	10
1	10	11	3	3	,,	9	3	14	9	2	0
1	9	4½	2	11	,,	7	3	4	7	1	8
1	8	11½	2	9½	,,	7	0	20	6	2	12
1	5	0	2	9½	,,	3	3	14	3	2	27
1	15	0	3	6	9	10	3	12	10	2	5
5	15	0	2	0	5	17	0	24	16	2	27
1	15	0	3	6	13	15	1	8	15	0	26
1	15	0	2	6	10	8	2	4	8	1	13
52						263	0	1	247	2	8

STEEL PLATES.

No. of Plates.	Length.		Breadth.		Thickness.	Actual Weight.			Estimated Weight.		
	ft.	in.	ft.	in.	16ths of in.	cwts.	qrs.	lbs.	cwts.	qrs.	lbs.
1	13	5	3	11	6	7	3	8	7	0	4
2	10	11	4	0	,,	12	3	24	11	2	22
2	13	4	3	11¾	,,	15	0	20	14	0	24
2	12	4	3	11¼	,,	14	0	20	13	0	11
2	10	6	3	1⅜	,,	9	0	24	8	3	6
1	13	0	3	10½	,,	7	2	8	6	3	0
2	7	8	3	0	5	5	2	8	5	0	15
2	10	2½	2	4	,,	5	2	24	5	1	9
2	14	0	3	11½	6	15	2	24	14	3	9
2	8	2	3	11¼	,,	9	1	16	8	2	14
1	11	9½	3	10¼	,,	6	2	4	6	0	10
2	8	8½	3	9	,,	9	1	8	8	3	0
2	13	0	3	11½	,,	14	2	8	13	3	14
1	12	6	3	10	,,	6	3	24	6	1	24
1	12	4½	2	4	5	3	2	4	3	1	0
3	10	1	2	4	,,	8	1	4	7	3	24
1	4	0	3	0	6	1	3	20	1	2	14
1	7	0	4	2	,,	4	0	20	3	3	17
2	14	0½	2	7	5	8	3	0	8	0	10
1	9	2	4	2	6	5	2	12	5	0	14
1	12	5	3	8	,,	6	3	4	6	0	10
1	12	0	3	9¼	,,	6	3	24	6	0	7
1	12	2½	3	7½	,,	6	2	12	5	3	20
2	12	4½	3	11¼	,,	14	1	4	13	0	16
1	13	5	3	11	,,	8	1	20	7	0	4
1	13	0	3	10½	,,	8	2	0	6	3	0
40						224	2	8	205	2	18

We now proceed to consider in detail the means by which the various tests are applied.

When parcels or lots of iron plates are delivered into the building yard they are spread out and examined with the object of first ascertaining if the manufacturer's name and the brand of quality are duly stamped upon each plate, and then of searching for surface defects, such as blisters, flaws, laminations, or bad places caused by dirt or cinders getting between the rolls during the rolling of the plates, any one of which, if considerable, would cause the overseer to reject the plate. This surface examination being completed, each plate is raised from the ground, and, being either hung by one edge or otherwise suitably supported, is tapped over with a small hammer; if it everywhere gives out a clear ringing sound the plate is considered to be solid, but if a heavy and dull sound be given out, it is presumed that the plate is laminated or otherwise defective. If this test is audibly decisive against the plate, it is at once rejected; but if the quality of the plate appears doubtful, a further test is resorted to. This consists in supporting the plate at its four corners, strewing the upper surface with sand, and lightly tapping over the under side; wherever the plate is sound the sand will be driven up off the plate by each tap of the hammer, but if it is blistered or laminated at any place the sand will not there be moved.

The plates which the foregoing tests show to be satisfactory are next carefully measured and weighed, their actual weights being compared with those due to their dimensions and specified thickness. They are weighed in lots not exceeding 5 tons, and if found to be either above the maximum or below the minimum limit, are rejected.

The next step is to test the tensile strength of the plates. For this test, and for the hot and cold forge tests likewise, plates are taken from the number of those which have thus far proved satisfactory, in the proportion of one to every fifty plates. If there are many more than fifty but not a hundred plates of the same thickness, the lot is divided into two, and a plate is taken indiscriminately from each. If there are several thicknesses of plates in the same invoice the same system is carried out, but those of each thickness are treated as a separate lot—unless their number be small. From the plates thus indiscriminately taken two pieces are cut out, one with and the other across the grain.

Fig. 248.

In some cases four pieces are cut for tensile tests, two with and two across the grain. The pieces cut out are carefully trimmed and filed down to the form shown in Fig. 248, having a parallel breadth for not less than 6 inches, and a breaking section of not less than one square inch—except in the case of thin plates, where the latter condition is waived, as the breadth of a piece of thin plate of one square inch section would be too great to break without tearing. In the Royal Dockyards these sample pieces for testing are never punched out, under existing orders, but are cut from the plate by sawing or drilling. Wherever punching is resorted to for the purpose, as it often is in private establishments, it is very necessary to leave a surplus width for planing and filing down, in order that the part to be broken may not be weakened by the injury which the iron sustains in immediate proximity to a punched hole. In the case of butt-straps, it is not, of course, always possible to have a parallel length of so much as 6 inches, as the fibre runs with the width of the strap. If in filing down the sample to its width the exact square inch of section is not secured, the actual section is carefully gauged and measured to two places of decimals, and the proper allowance is made. Two centre-punch marks are struck upon each sample at 6 inches apart, for the purpose of showing the elongation that takes place under the strain.

The most usual form of machine employed for testing the samples of plate is constructed on the principle illustrated by Fig. 249. It consists, as will be seen, of a steelyard of which the long arm carries the breaking weights, and the short arm pulls upon the sample to be broken. In the machine shown in the sketch the fulcrum is suspended from a hydraulic machine, *a*, set in motion by hand, and the lower end of the sample, *s*, is connected with a base plate. Should a considerable amount of elongation take place before the sample is fractured the fulcrum can be moved up by means of the hydraulic machine and the

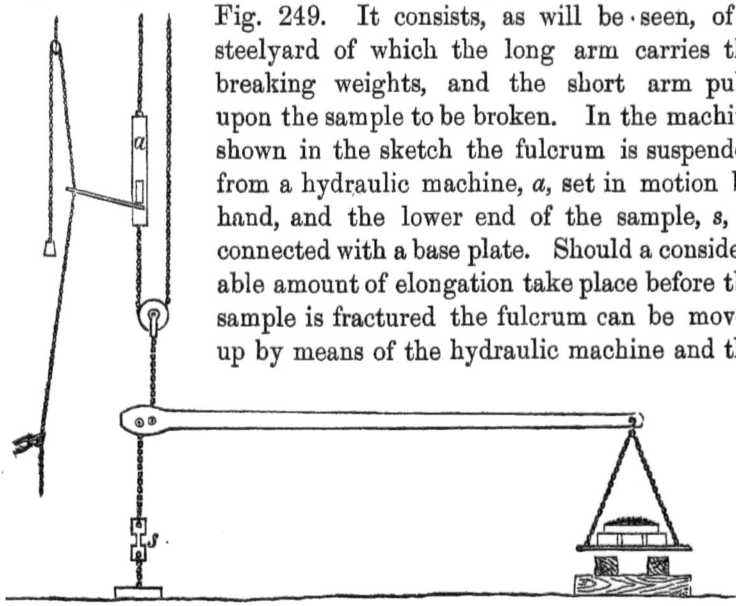

Fig. 249.

steelyard be always kept in a horizontal position. In some testing machines the fulcrum of the steelyard is fixed, and the lower end of the sample is shackled to, and fixed in position by a bolt which is screwed into a base plate, and which can consequently be tightened up at pleasure should a large amount of stretching throw the lever out of a horizontal position. When this occurs the weight is taken by means of a winch while the sample is readjusted. In other machines a small hydraulic press is substituted for this screw bolt, and answers the same purpose. The scale suspended at the extremity of the long arm serves to support the weight by means of which the tensile strain is partly produced. At first sufficient weights are placed on the scale to give a strain of about 10 tons per square inch of the sample's section, this weight being gradually increased by about a ton at a time until a strain of about 18 to 19 tons is reached if the sample be tested with the grain, or about 16 tons if it be tested across the grain. After these tensions are obtained, small weights, such as rivets, &c., are very gradually added, the lever being kept as nearly level as is practicable notwithstanding the increasing lift put upon the fulcrum by the hydraulic, or the stretching of the sample if the fulcrum is fixed. This process goes on until the sample breaks. When this happens the contents of the scale are carefully weighed, and the breaking strain is calculated from the result and recorded in a "test-book" kept for the purpose, together with a statement of the elongation and the appearance of the fracture. In order to illustrate the arrangement of this record, we have given opposite a specimen page, taken from the test-book of an Admiralty surveyor. It will be observed that the tabular form also includes columns giving the percentage above or below the estimated weight, and the particulars of the hot and cold forge tests to be described hereafter. In some test-books two additional columns are used besides those given in this table, in order to record the percentage of crystal in the fracture and the angles to which the samples are bent in the hot-forge tests, instead of giving this information in the column headed "Remarks." The date of the trial, and the name of the manufacturer of the iron are also recorded.

The calculation of the actual strain exerted upon a sample in a machine such as that illustrated by Fig. 249 is, of course, a simple enough operation, but to prevent mistakes it may be well to give a few words of explanation. The tensile strain put upon the sample is obviously composed of three parts:—

I. The strain put upon it by the leverage of the steelyard itself.

II. The strain produced by the weight of the scale-pan.

III. The strain produced by the weights in the scale.

For example, in a certain machine the lengths of the arms of the steelyard are in the proportion of 28 to 1, and it is found that a weight of 9215 lbs. suspended from the short arm will exactly balance the steelyard, without the scale. The weight of the scale is 227 lbs., and if W lbs. be the weight placed in the scale when the sample breaks, we shall obviously have:—

The strain of the steel-yard in lbs.	.. .. =	9215
,, scale pan in lbs.	.. .. =	227×28
,, weight in the scale in lbs.	=	$W \times 28$

SPECIMEN PAGE of TEST-BOOK, kept by ADMIRALTY SURVEYORS.

Description of Article.	Dimensions of samples.		Actual breaking strain of samples.		Breaking strain per sq. in.		Proof strain per sq. in.		Elongation in six inches.	Quantity represented by test.	Quantity rejected.	Percentage above or below estimated weight.	Angles bent cold.	Angles required.	Remarks.
	Size.	Area.	With the grain.	Across the grain.	With the grain.	Across the grain.	With the grain.	Across the grain.					With and across grain.	With and across grain.	
	inches.	inches.	tons.	tons.	tons.	tons.	tons.	tons.	inches.				degrees.	degrees.	
$\frac{9}{16}$ in. bottom-plates.	1·77 × ·545	·96465	22	19¼	22·8	19·95	22	18	with ·84 across ·35	30 plates	..	·78 below	with 44 across 15	with 35 across 15	Smithery tests quite satisfactory, the iron being of a grey colour, and fine-grained.
Ditto	1·75 × ·55	·9625	20½	17	21·29	17·66	22	18	with ·695 across ·3	29 plates	29 plates	..	broke broke	with 35 across 15	Rejected, having failed both in the cold smithery tests and tensile tests.
Frame angle iron, 9 × 3½ × $\frac{7}{16}$ ins.	..	..	..	..	..	..	..	..	..	9½ tons	9½ tons	..	..	..	The whole of this iron was rejected for not standing the smithery tests; it was seamy, and of very inferior quality.
$\frac{3}{8}$ Floor plates	2·73 × ·36	·9828	19¾	18·27	20·009	18·59	20	17	with ·28 across ·152	4 tons	..	3·25 below	with 45 across 17	with 45 across 15	Smithery tests satisfactory.
Frame angle-iron, 9 × 3¼ × $\frac{7}{16}$ ins.	2·04 × ·5	1·02	24¼	..	23·77	..	22	..	·18	11 tons	..	1·39 below	Folded together.	..	The grain of the iron slightly opened, but on the whole the smithery tests were satisfactory; no seams visible.
$\frac{7}{16}$ in. longitudinals.	2·38 × ·41	·9758	21½	18½	22·03	18·95	20	17	with ·88 across ·15	26 plates	..	1·75 below	with 45 across 16	with 45 across 15	Grey iron, fine grained, and of good quality; the smithery tests satisfactory.
$\frac{7}{16}$ in. bottom-plates, and $\frac{7}{16}$ in. butt-straps.	2·31 × ·43	·9933	24¼	18¾	24·41	18·87	22	18	with ·815 across ·12	29 plates 2¼ tons of butt straps.	..	1·8 below	with 52 across 22	with 50 across 20	These samples had a slight crystalline appearance, but the smithery tests were satisfactory.

If the area of the sample's section be A square inches and the breaking strain be B tons per square inch we obtain B A for the total breaking strain in tons. Hence we have the equation—

$$B\,A = \frac{9215 + (227 + W)\,28}{2240}$$

$$B = \frac{9215 + (227 + W)\,28}{2240 \times A}$$

If the breaking strain per square inch of the iron deduced from the breaking strain of the sample does not come up to the required standard, the lot of plates represented by the sample is rejected; but it should be understood that before the surveyor makes the foregoing tests, he sees the testing machine tried, and ascertains for himself the true amount of the weight that has to be allowed for the leverage of the machine and scale-pan. In some yards the lever can be balanced by means of a sliding weight, and then the strain on the sample is caused only by the weights in the scale.

The tensile strength of angle-iron is tested in exactly the same manner as that of plate-iron, except that it can only be tried with the grain, the flanges usually being too narrow to admit of a piece being cut out of sufficient size to test the strength across the grain. The piece with the grain is cut from one of the flanges, planed and prepared, as far as possible, like the plate samples, and broken by the same machine. All angle-iron is required to stand a strain of 22 tons per square inch, and if the breaking strain falls below this the lot is rejected. Owing, however, to the continued rolling of angle-iron in one direction into great lengths, a high longitudinal strength is usually developed in it, the larger bars generally standing from 25 to 26 tons per square inch, while the smaller exceed 30 tons.

Rivet-iron is sometimes subjected to a tensile test by the surveyors, especially when the rivets are manufactured on the establishment and a piece of the bar-iron from which they are made can be obtained. The mode of testing is similar to that described above.

We have next to consider the forge-tests to which plate-iron, angle-iron, rivets, and armour-bolts are subjected. The object of these tests is to ascertain the fitness of the iron for undergoing the various bending, twisting, hammering, and other more or less distressing processes to which the material is subjected, both hot and cold, while being worked up into the various parts of a ship.

We have already described the extent to which iron plates under test have to be bent, both hot and cold, and also given the sizes of the pieces thus tested, and the degree of sharpness of the angle of the cast-iron block or slab over which they are bent. As a matter of fact, the full sizes prescribed for the samples thus tested have not, for various reasons, been insisted upon in all cases, especially when there have been good independent reasons for having confidence in the quality of the material. Under such circumstances the Admiralty are content to test pieces of plate about 2 feet by 18 inches, or even less, resorting to the trial of pieces of the full width of the plate and 4 feet long when this seemed desirable. The sample pieces cut from the plate, after having their edges planed, are secured one by one to the cast-iron slab, about 3 or 4 inches from its edge, and are then bent down by moderate blows from a large hammer. The surveyor has to exercise great care in attending to this operation, for the result may be greatly affected by humouring and coaxing on the part of the hammer-man. By striking the iron in the direction of the fibre the workman can make an inferior iron bend with less symptoms of distress than a better iron may exhibit when used more roughly. The same leniency may be shown to the iron by bending it under a steady pressure instead of by blows. The blows should, therefore, be delivered not too lightly, and about square to the surface, and the first signs of fracture should be observed and recorded. In order to measure the angle through which the sample has been bent, it is usual to remove it from the slab. Should it be found that the required angle has not been obtained, the piece of plate is replaced on the slab, and the operation of bending is continued. Particular care is required, in order to ensure that the workman does not place the sample so that it shall project further over the edge of the slab than it did during the first part of the operation; because if this is done the sample may be bent through an additional angle without any further strain being brought upon the material—in fact, the required angle is then obtained by virtually increasing the radius of the corner of the slab on which the sample is bent, by means of the additional projection given to the plate. The surveyor has, therefore, to take every precaution to ensure the sample being replaced in, as nearly as possible, its original position after the measurement of the angle has been made, before he allows the bending to be completed. By careful piling of the iron—as the author has ascertained by many experiments—plates can be produced of almost equal ductility both lengthwise and crosswise of the plate; but as they are usually manufactured the fibre runs chiefly with the length of the plate, and therefore the strength and ductility are generally greater lengthwise than crosswise. The tensile and forge tests, in fact, require this. It is for this reason, however, all the more necessary to pay particular attention to the transverse tests; and experience shows that where samples broken under the tensile strains have exhibited a granulated structure with large grains or crystals, the samples bent cold across the grain rarely stand.* When, on the contrary, the fracture of the samples torn asunder by tensile strain exhibits a fine fibrous structure of light grey appearance, the iron is generally of good quality and stands the cold test well.

In carrying out the hot tests, pieces of similar dimensions are employed. They are heated until they assume an orange colour, and are then bent down to the prescribed angles in the same way as in the cold test. Great care is taken that the samples are not over-heated in the fire. The results of both the hot and cold forge tests are recorded in the surveyor's test-book in the manner shown by the table given on p. 393.

The forge tests to which angle-iron is subjected are usually hot tests only. The amount of rolling which iron in this form undergoes, generally secures, as we have already said, ample longitudinal strength, and the important thing to be further ascertained is the quality of the iron to withstand the necessary smithing operations, which are often very distressing in ship-work. For this purpose short pieces (say from 12 to 24 inches long) of the angle-iron are heated and treated as follows:—One piece has the flanges closed together with the hammer as shown by *a*, Fig. 250; another piece is opened out and hammered flat as in *b*; another piece, or this same

* The expression "across the grain," or "across the fibre," in this connection has sometimes been misunderstood. It has been said that as the fibre or grain is supposed to run, and usually does run, chiefly along the plate, this fibre or grain must be broken across when the plate is broken across, and that when the plate is fractured along its length, the fibre or grain must also be broken *along* and not *across*. There is no doubt much force in these representations; still it is obviously convenient to speak of breaking the iron across the grain when a narrow piece of plate cut off the end of a plate is broken across, and this is the sense in which the phrase is used in the Admiralty instructions.

piece, has the flanges turned back upon themselves as in *c;* and another piece has the flanges turned inwards as in *d.* Both the last-named tests are well calculated to try thoroughly the quality of the iron, and to develop indications of any "reediness," or looseness of structure that it may possess. Sometimes the hot tests are conducted in the manner illustrated in Fig. 251, a piece of angle-iron is cut nearly in two as shown by *a;* one-half is then heated and beaten out flat as in *b;* and the end is then doubled over as in *c.* The other half of the bar is then heated, and the flanges are curled inward or outward as the surveyor directs. These tests being satisfactorily past, the lot to which the tested piece belongs is usually accepted, but in some cases a piece is first broken across cold to exhibit a fractured section.

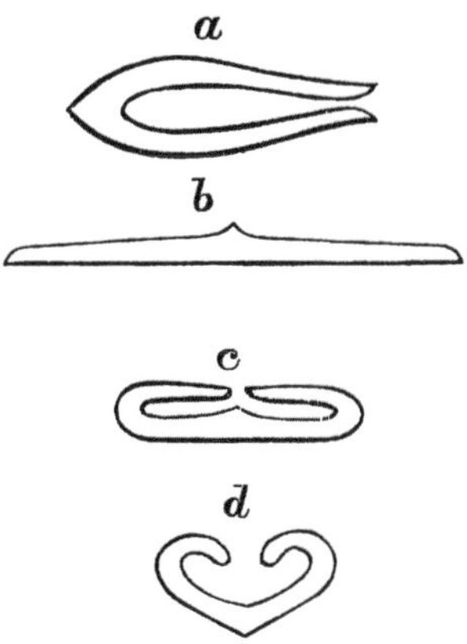

Fig. 250.

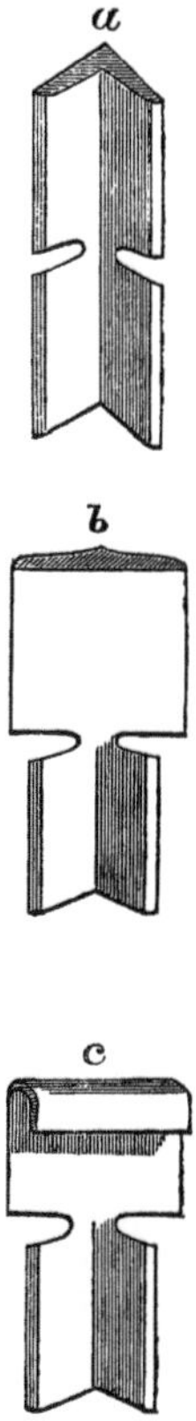

Fig. 251.

Rivets are first examined as to the correctness of their shape according to the pattern, and having been found correct are duly heated. The head is then flattened out very thin, as shown by *a* Fig. 252, and the iron should stand this test without cracking at the edges. The rivet shank is then heated and flattened out, and a punch is driven through it, bringing it to the form shown by *b.* No cracking should take place in the neighbourhood of the punched hole. Another rivet is bent cold under a hydraulic press or hammer to the form given in *c,* and should stand this without fracture. Another rivet is nicked on one side and similarly bent, that the nature of the fracture may be observed. When a piece of the bar from which the rivets are made can be procured, it is well to bend it double when cold, and then to bend the doubled portion a second time at right angles

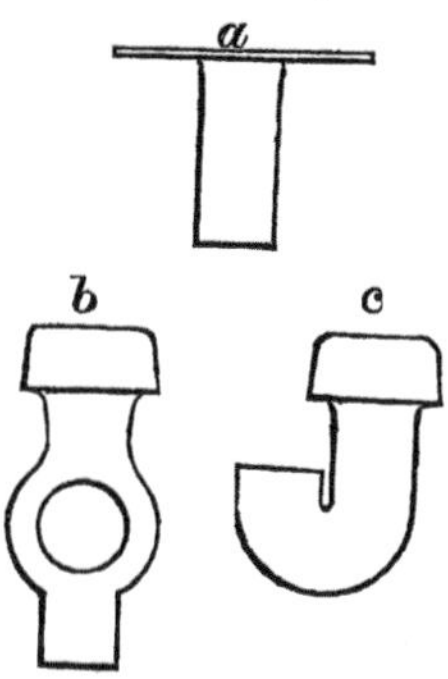

Fig. 252.

to the former bend. It is also well to nick a piece of the iron on one side with a chisel, and to bend it slowly over an anvil until it breaks; and having nicked a second piece all round, to break it suddenly by means of a smart blow. In the latter case the section should show a fine crystalline appearance, the crystals being very minute; in the former case it should present a very fibrous appearance, the fibres being fine and silky.

Armour-plate bolts are made of the best Lowmoor, Bowling, or other highest class iron, and are tested by being doubled cold under the hammer. In some cases this test is replaced by the following:—A length of the bar iron from which the bolts are made is taken, and, the ends being supported, a heavy weight is let fall upon the middle, thus bending the bar down very suddenly and severely testing the iron. It is also usual to test the tensile strength and ductility of the bar, the breaking strengths per square inch being recorded both for the original and for the reduced sections. As a very satisfactory example of the ordinary tests for armour bolts we give the following particulars of the trials recently made of the bars used in manufacturing the fastenings of one of the iron-clads of the Royal Navy. A bar of $2\frac{1}{2}$-inch bolt-iron broke under a strain of 113 tons 18 cwt., giving a strength of 23·2 tons per square inch of the original section; but before fracture took place the bar was elongated 4 inches on a length of 2 feet $10\frac{1}{2}$ inches, and the diameter of the breaking section was only $1\frac{15}{16}$ inches, thus giving a strength of 38·63 tons per square inch of the reduced section. A bar of $1\frac{3}{4}$-inch iron was also tested, and after stretching 3 inches on a length of 2 feet 10 inches, it broke on a strain of 54 tons 10 cwt., the diameter of the breaking section being $1\frac{1}{4}$ inches. This gives a strength of 22·66 tons per square inch of the original section, and of 44·41 tons per square inch of the reduced section. In both cases the forge tests were very satisfactory.

Recently experiments have been made to test the strength of armour bolts of various kinds under impact, at the Atlas Works (Sir J. Brown and Co.), and the Cyclops Works (Messrs. Cammell and Co.), Sheffield, and at the Millwall Ironworks. It has been considered desirable to make the experiments resemble actual practice as much as possible, and for this purpose the bolts have been inserted into holes in armour plates, their heads being countersunk in the ordinary manner, and the nuts have been hove up on their ends underneath a block of iron which is so arranged as to transmit the strain caused by impact directly to the nuts, as it would be if a ship's side were struck by projectiles. The bolts were tested by letting fall heavy weights from heights varying from 20 to 30 feet. The results were carefully observed, and are recorded in the Report of the Special Committee on the Gibraltar Shield. In some cases the bolt heads were drawn through the armour plate, and in others the bolts were fractured; but the amount of "work" applied to the bolt in each case being given, affords a comparative measure of the work done in breaking it. The account of the experiments given in the Report is well worth a careful study, but the subject is referred to in this connection only on account of its relation to a new mode of testing armour bolts, and consequently will not be treated at greater length.

We now come to speak of the Admiralty tests for steel plates, angle-bars, &c. The test of weight is identical with that already described for iron, the slight difference existing between the weight of iron and steel not being recognised in this test. The tensile and forge tests which the material has to withstand are as follows:—

		Tons.
Tensile Strain per square inch	Lengthways	33
	Crossways	30

The tensile strength is in no case to exceed 40 tons per square inch.

FORGE TEST (*Hot*).

All plates of one inch in thickness and under, should be of such ductility as to admit of bending hot, without fracture, to the following angles:—

	Degrees.
Lengthways of the grain	140
Across the grain	110

FORGE TEST (*Cold*).

All plates should admit of bending cold, without fracture, as follows:

With the Grain.	Degrees.	*Across the Grain.*	Degrees.
1 inch in thickness to an angle of	30	1 inch in thickness to an angle of	20
7/8 ,, ,, ,,	40	7/8 ,, ,, ,,	25
3/4 ,, ,, ,,	50	3/4 ,, ,, ,,	30
5/8 ,, ,, ,,	60	5/8 ,, ,, ,,	35
1/2 ,, ,, ,,	70	1/2 ,, ,, ,,	40
7/16 ,, ,, ,,	75	7/16 ,, ,, ,,	50
3/8 ,, ,, ,,	80	3/8 ,, ,, ,,	60
5/16 ,, ,, ,,	85	5/16 ,, ,, ,,	65
1/4 ,, and under ,,	90	1/4 inch and under ,,	70

Plates, both hot and cold, should be tested on a cast-iron slab, having a fair surface, with an edge at right angles, the corner being rounded off with a radius of ½ an inch.

The portion of plate tested, for both hot and cold tests, is to be 4 feet in length, across the grain, and the full width of the plate, with the grain, except in cases where the Officers are satisfied with a smaller sample.

The plate should be bent at a distance of from 3 to 6 inches from the edge.

All plates to be free from lamination and injurious surface defects.

One plate to be taken for testing from every invoice, provided the number of plates does not exceed fifty. If above that number, one for every additional fifty, or portion of fifty. Plates may be received, but may not be rejected, without a trial of every thickness on the invoice.

The pieces of plate cut out for testing are to be of parallel width from end to end, or at least 6 inches of length.

The edges should be drilled or sawn, and not punched, in cutting the sample from the plate.

These tests are carried out in the same way for steel plates as for iron plates, but a few additional precautions have to be taken in order to arrive at correct results. For example, in testing a sample of which the tensile strength exceeds the maximum strength allowed, care must be taken that the person manipulating the weights does not press heavily upon them as he places them in the scale, and thus cause a considerable increase in the strain brought upon the sample above that due to the weights in the scale. This result may also be attained by allowing the strain to be brought upon the sample *suddenly* after the steelyard has been readjusted, in order to bring it back to the horizontal position from which it has been moved by the elongation of the sample. This, of course, need only be guarded against in testing machines where the fulcrum is fixed, as in machines similar to that shown in Fig. 249 the fulcrum can be gradually elevated as the sample elongates. In other machines, however, it is necessary to bring the strain upon the sample very gradually after the readjustment of the steelyard is completed, or else the apparent breaking strength of the sample will be less than its real strength. The tests for angle-bars are the same in both cases, observing that angle steels are sometimes further tested by spreading a piece out flat, heating it, punching a small hole in one corner, and enlarging this hole to about 2 inches diameter, which the material is expected to stand without cracking round the hole. Steel rivets are also tested in the same way as iron rivets.

It will be observed that in the description of the tests to be applied to steel, it is provided that the pieces of plate cut out for testing are to be of parallel width for at least 6 inches in length as shown in Fig. 248. This provision, which is also carried out in testing iron plates, is of great importance. Before it was uniformly enforced by the Admiralty, it had been the practice on some occasions to reduce the sample to the required section by means of a circular arc as shown

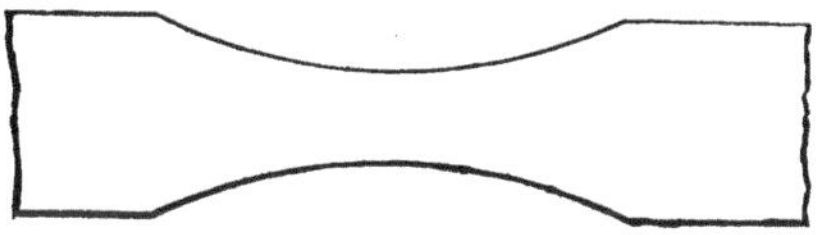

Fig. 253.

by Fig. 253. But this arrangement, by limiting the possibility of fracture to one part only, obviously gives a weak sample unfair chances of passing the tests, as will appear from the table on the next page, of the results of a series of trials of samples of steel cut from

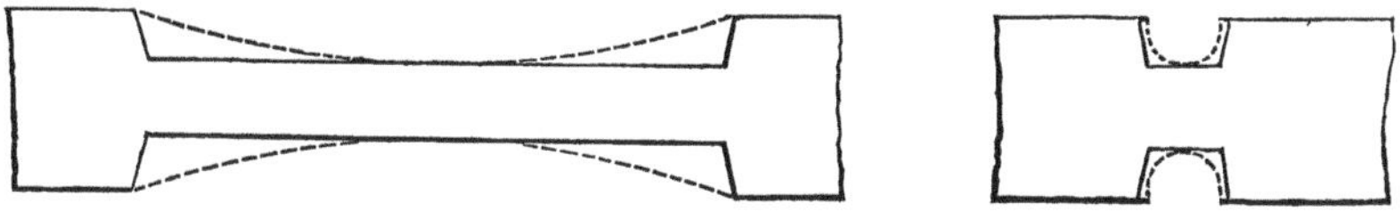

Fig. 254.

the same plates and reduced in width as shown in Fig. 254. Eight cases were taken, each case being tested by three experiments for each mode of reduction. The breaking section in each case was reduced to a breadth of 1 inch, the reduction from the extreme breadth of the samples ($2\frac{1}{4}$ inches) being made in one set of plates by circular arcs, and in the other set by parallel reductions. The lengths of the reduced parts were varied from 8 inches to 1 inch by successive deductions of 1 inch, and the sketches in Fig. 254 show the extreme cases of the longest and shortest reductions respectively. The forms of the circular arcs are shown in ticked lines, and those of the parallel reduction by drawn lines.

It will here be seen that throughout these experiments, which were conducted with extreme care, the same material broke at a less strain when trimmed down to a parallel breadth for a considerable distance, than when reduced to the same breadth at one place only. By comparing the samples reduced to a parallel breadth for a length of 8 inches with those similarly reduced for a length of 1 inch only, it will be seen that the apparent strength rose from an average of $19\frac{1}{4}$ tons to an average of $21\frac{3}{4}$ tons; or if we compare the former with the case of the shortest circular reduction we find it increasing from $19\frac{1}{4}$ tons to an average of about $23\frac{3}{4}$ tons. These experimental facts illustrate the importance of properly preparing the samples of steel plates for tensile tests in order that reliable results may be obtained for the compa-

TESTS of STEEL PLATES, with CIRCULAR and PARALLEL REDUCTIONS.

No. of Trial.	Length of reduction.	Dimensions.		Sectional Area.	Temperature of testing-house.		Breaking strain.						Breaking-strain per sq. inch.		Elongation in 4 inches.		Elongation in 3 inches.		Elongation in 2 inches.		Remarks.	
		Thickness.	Width.		Parallel.	Circular.	Parallel.			Circular.			Parallel.	Circular.	Parallel.	Circular.	Parallel.	Circular.	Parallel.	Circular.	Parallel.	Circular.
	inches.	inches.	inches.	inches.	degrees.	degrees.	tons.	cwt.	qr.	tons.	cwt.	qr.	tons.	tons.	16ths.	16ths.	16ths.	16ths.	16ths.	16ths.		
1	8	·51	1·01	·5151	70	70	19	15	0	21	12	2	38·342	41·982	4	7	..	..	..	..	Broke at one of the angles	Broke at centre.
2	,,	·51	1·01	·5151	69	69	17	15	0	21	5	0	34·459	41·254	3	7	..	..	..	..	,,	,,
3	,,	·51	1·01	·5151	68	68	20	7	2	21	5	0	39·555	41·254	6	9	..	..	..	..	,,	,,
1	7	·51	1·0	·5100	65	65	20	10	0	21	17	2	40·196	42·892	7	7	..	..	..	..	,,	,,
2	,,	·51	1·0	·5100	65	65	20	15	0	21	7	2	40·686	41·910	9	6	..	..	..	..	Broke at centre .	,,
3	,,	·51	1·0	·5100	65	65	19	17	2	21	15	0	38·970	42·647	4	8	..	..	..	..	,, angle .	,,
1	6	·51	1·01	·5151	65	65	20	17	2	21	2	2	40·526	41·010	12	3	..	..	..	..	,, near centre	,,
2	,,	·51	1·01	·5151	65	65	21	2	2	22	0	0	41·010	42·710	15	7	..	..	..	..	,, at centre .	,,
3	,,	·51	1·01	·5151	65	65	21	2	2	21	17	2	41·010	42·467	14	7	..	..	..	..	,, near centre	,,
1	5	·51	1·01	·5151	65	66	19	17	2	22	2	2	38·584	42·952	3	6	..	..	..	..	,, at angle .	,,
2	,,	·51	1·01	·5151	65	66	18	5	0	21	12	2	35·430	41·982	3	7	..	..	..	..	,,	,,
3	,,	·51	1·01	·5151	65	66	19	7	2	21	12	2	37·614	41·982	4	7	..	..	..	..	,,	,,
1	4	·51	1·01	·5151	66	66	21	7	2	22	12	2	41·496	43·923	6	6	..	..	..	..	,,	,,
2	,,	·51	1·01	·5151	66	66	20	15	0	22	5	0	40·283	43·195	5	5	..	..	..	..	,,	,,
3	,,	·51	1·01	·5151	66	66	21	2	2	22	15	0	41·010	44·166	8	6	..	..	..	..	,,	,,
1	3	·5	1·0	·50	72	72	20	15	0	21	0	0	41·500	42·000	10	2	..	..	..	..	,, at centre .	Broke near centre.
2	,,	·5	1·0	·50	72	70	19	15	0	21	17	2	39·500	43·750	6	5	..	..	..	..	,, angle .	,, at centre.
3	,,	·5	1·0	·50	72	70	19	17	2	22	0	0	39·750	44·000	11	3	..	..	..	..	,, centre .	,,
1	2	·51	1·01	·5151	67	68	21	7	2	23	10	0	41·496	45·622	..	..	8	4	..	..	,,	,,
2	,,	·51	1·01	·5151	67	69	21	17	2	23	7	2	42·467	45·379	..	..	5	3	..	..	Broke at angle .	Broke near centre.
3	,,	·51	1·01	·5151	68	69	20	15	0	21	5	0	40·283	41·254	..	..	9	4	..	..	,, centre .	,,
1	1	·51	1·0	·51	71	71	22	0	0	23	15	0	43·157	46·568	..	..	..	..	4	3	,, near centre	,,
2	,,	·51	1·0	·51	71	71	21	17	2	24	2	2	42·893	47·303	..	..	..	..	2	2	,, at angle .	,,
3	,,	·51	1·0	·51	71	71	21	10	0	23	12	2	42·156	46·323	..	..	..	..	2	1	,,	,,

rison of the strength of different samples. Similar experiments have also been made with iron plates, which show that the same care is necessary in the preparation of the samples, and that a shorter reduction, or a circular reduction, will give higher results than those obtainable with a longer, or a parallel reduction respectively.

It may not be uninteresting if, before concluding this chapter, we give a brief account of the mode of testing armour plates approved by the Admiralty and practised at Portsmouth. The sample plate to be tested is fixed on wood backing, and a 68-pounder gun is brought to within 30 feet of it, loaded with 13 lbs. of powder, and a certain number of ordinary cast-iron shot are fired at any selected part of the plate. The number of shot fired varies with the thickness of the plate—thus for a 10-inch plate 13 shots would be fired—the rule observed in taking aim being that successive shots shall overlap each other. Careful observations are made of the cracks, and the damage done to the back of the plate, and according to the amount of the injury the plate is placed in classification, three classes of plates being accepted.

It is not our intention to discuss here the various chemical processes which are in use for testing the quantity of carbon, phosphorus, sulphur, &c., in pig-iron, as this subject is fully treated in Dr. Percy's well-known and invaluable work on 'Iron and Steel.' It is worthy of observation, however, that Professor Eggertz's beautiful system of colour testing, is now largely used in the iron and steel works of this country; and looking to the extent to which the quality of steel is affected by variations in the proportions of carbon contained in it, we have no doubt that this and similar tests will become much more general as the employment of steel is extended.

CHAPTER XIX.

LLOYD'S AND THE LIVERPOOL RULES FOR IRON SHIPBUILDING.

IN the previous chapters of this work frequent reference has been made to the rules laid down by Lloyd's Committee, and also by the Liverpool Underwriters, for the regulation of the details of iron shipbuilding. If this volume were of an historical, and not of a practical nature, we should here trace with pleasure the extent to which this comparatively modern art has been fostered by these regulations, and more especially by the enlightened exertions of Lloyd's Committee,* and of Messrs. Martin and Ritchie, its chief shipbuilding officers. But as this would be inconsistent with the object of our work, we shall confine ourselves to such a reference to the subject as will suffice to illustrate its practical aspect.†

We give in an Appendix the latest edition of the revised rules

* Under the very able chairmanship of Mr. Chapman.

† In the fourth volume of the '*Transactions* of the Institution of Naval Architects,' Lloyd's revised rules were printed, and prefaced by an interesting introduction from the pen of Mr. Ritchie, from which we extract the following remarks:—
"In the year 1838, the Committee of Lloyd's Register classed the first iron vessel "that had ever been guaranteed as 'fit for the safe conveyance of dry and perishable "'cargoes.' This barque was appropriately named the *Ironsides;* she was 271 tons "register, was built at Liverpool, by Messrs. Jackson and Gordon, for Messrs. "Cairns & Co. She was launched on 17th October, 1838, and was built of angle-iron "and plate-iron, much in the same manner as iron ships are now built. She performed "her first voyage to Rio de Janeiro and back without damaging any cargo. She was "classed by the Committee of Lloyd's Register after careful consideration on the "15th November, 1838, as 'built of iron,' with no letter. From this date until 1844 "the Committee of the Register Book continued to class iron ships by the designation "'built of iron,' but with no letter. From 1844 till 1854 the Committee improved "the classification by marking them A 1, 'built of iron,' but this character was "limited to six years. However, before the termination of this six years' classifica- "tion the number of iron ships had so increased, and the demands for some kind of "higher class, based on admitted rules, was so general, that the Committee of Lloyd's "Register considered it necessary to form a code of rules for their own guidance, and "for that of the builders of iron ships; but having in vain solicited the assistance "and concurrence of the iron shipbuilders throughout the country to form such rules, "they in 1854 appointed a special Committee, composed of the members of their "Board most conversant with shipbuilding, who, with the assistance of their own "surveyors (by collating the scantlings of iron ships already classed) compiled their "celebrated Rules and Table of Scantlings for ships of six, nine, and twelve years' "grade, and which, when confirmed by the General Committee, was prefaced as "follows:—

of Lloyd's Committee, but the following comparisons are based upon the Rules for 1867, which differ from those of the present

---

'Considering that iron shipbuilding is yet in its infancy, and that there are 'no well understood general rules for building iron ships, the Committee have 'not deemed it desirable to frame a scheme compelling the adoption of a parti- 'cular form or mode of construction, but that certain general requirements should 'be put forward having for their basis thickness of plates and substance of frame, 'showing a *minimum in each particular* to entitle ships to the character A for a 'period of years; subject, however, to certain periodical surveys; and also to a 'continuation of such character, should their state and condition justify it on 'subsequent examination. For the purpose of attaining this object, the follow- 'ing rules and the accompanying table of dimensions have been formed.'

"These rules the Committee amended in 1857 by extending the room and space of "the frames in twelve-year ships from 16 inches to 18 inches, but increasing the "thickness of the plating in all grades one-sixteenth of an inch throughout. And "in 1861 the rules and tables of scantlings were *limited* to vessels whose extreme "length did *not* exceed *seven* times their extreme breadth, or *ten* times their depth of "hold. At the time the Committee drew up the first rules, in 1854, they felt that a "classification of six, nine, and twelve years, although it might approach the truth "as to the probable comparative durability of the various kinds of timber of which "wooden ships so classed were allowed by their rules to be built, yet these characters "could not correctly indicate the durability of vessels built of metal, which only "deteriorated by the wasting of the surfaces, and whose durability depended on "different laws from that of timber. But it was considered that these rules of "classing would serve until more experience was gained, not only on the durability "of iron when subjected to the continued action of sea-water and the corrosive "products of various cargoes, but also on unascertained points in the construction of "iron ships which could not be premised from the most complete knowledge of "wooden ships. And the result has been that the experience gained during the "following nine years induced the Committee in 1863 to reconsider these rules; and "with this view they submitted certain preliminary suggestions on the subject to the "now numerous iron shipbuilding establishments throughout the country, and to "their own intelligent and experienced staff of surveyors. In reply they received "valuable suggestions from twenty-four shipbuilding firms, and from twenty-eight "surveyors, and these were in the first place referred to the two principal surveyors, "Messrs. Martin and Ritchie, and their opinions having been modified and perfected "by the Special Committee were finally reported on to the General Committee, who "decided that the designation of class by *years* should be abolished, and the mono- "grams A B C be for the future used in their places to designate the grada- "tions of probable durability in such vessels, and indicate the length of time which "might intervene between the special surveys upon which such vessels should retain "their characters dependent on their continued efficiency. And in addition to this "important change the rules for the construction of iron ships have been altered in "the following particulars. The tonnage which is to regulate the strength of the "various parts of the vessel is to be that *below* the *tonnage* deck *only*, and not "to include the tonnage represented by a poop, or forecastle, or of a spar-deck "when intended for cabin accommodation alone. The skeleton of all ships, that is "to say, the keel, stern, sternpost, floors, frames, beams, keelsons, and stringers "of the A B C class are to be the same, but the thickness of the outside "plating may be various."

"The space of the frames in all grades is increased from 18 to 21 inches, and "under certain conditions to 24 inches, and consequently the spacing of the beams is "increased; and further to reduce the weight aloft, the plates of the beams, and the "outside plating from the lower part of the sheer-strakes, to three-fifths the depth of "holds above the upper side of keel, may be thinner; and also to relieve the ends of year only as respects some regulations for the measurement of tonnage. In order to illustrate the improvements which experience has shown to be necessary we will here mention the principal differences of detail between the rules of 1862 and the rules as now enforced; observing that while in 1862 the table of dimensions was limited in its application to those ships whose length on upper deck did not exceed seven times their breadth, or ten times their depth of hold, provision is now made in an additional section in the rules for ships which exceed these limits.

The sizes of keel, stem, and sternpost, are in both years regulated by the tonnage, and are the same for all grades. The only alteration from their dimensions of 1862 is a slight increase in the moulding for all vessels under 600 tons, the siding remaining the same. In 1862 hollow plate keels alone were mentioned, the thickness required being stated. Both hollow, and flat plate keels are now mentioned, the thickness for both being the same as that required for hollow keels in 1862 and the minimum breadth required being also stated.

The dimensions required for the frames are exactly the same in both years, as are also the requirements that wherever the frames are butted they shall have 4 feet lengths of corresponding angle-iron fitted back to back to cover and support the butts, or if welded the welds shall be perfect, with not less than 4 feet shift. The spacing of the frames in 1862 was 18 inches from centre to centre for all grades of classification; it is now increased to 21 inches, but if double frames be fitted extending from the keel to the upper part of the bilge for half the vessel's length amidships, the spacing may be further increased to 23 inches in vessels under 1000 tons, and to 24 inches in vessels of 1000 tons and upwards. In 1862 there was a table of thicknesses of floor-plates for each of

---

"the ship from unnecessary weight, the floor plates and the outside plating aloft "may be reduced for one quarter the length of the ship, forward and aft. The "breadth of stringer-plates is to be increased in proportion to the length of the "vessel, and rules are added for increasing the strength of the sheer-strakes, beam-"stringers, and bilge keelsons, in ships *exceeding* in length *seven* times their breadth "or *ten* times their depth. And it is now made requisite that the iron, of which all "ships shall hereafter be built, must be capable of sustaining a tensile strain of 20 "tons per inch.

"It should be borne in mind that, although the mode of constructing iron ships, "primarily intended by these rules, is the original and ordinary one of vertical frames "and longitudinal plating, the Committee do not hesitate to admit into the Register "Book, and into the same classes, vessels otherwise constructed if of equal strength; "and have classed ships with longitudinal frames, or with diagonal frames, and "many with double or cellular bottoms for water-ballast, &c."

the three grades, viz., 12 years, 9 years, and 6 years grade; they are now reduced to one table for all grades, this table corresponding almost identically with that for the 9 years grade in 1862. In 1862 the depth of the floor-plate at the middle line was proportional to the vessel's depth; it now depends upon the depth and breadth conjointly. The two rules would correspond in those ships of which the depth is two-thirds the breadth. In 1862 the floor-plates were required to extend beyond the bilge keelsons, and not to be less in depth at the bilge keelsons than the moulding of the frames; the floor-plates are now required to extend to a perpendicular height up the bilges of twice the depth of the floors amidships from the upper side of the keel at the middle-line, and not to be less moulded at their heads than the moulding of the frames. A reduction is now allowed in the thickness of the floor-plates at the extremities of the ship; in 1862 no such reduction was allowed. The sizes of the reversed angle-irons are the same in both years. The rules of 1862 require double angle-irons in the way of all keelsons; the present rules require them in way of all keelsons and stringers in hold. The height to which the reversed angle-iron is carried on the different frames is the same in both years, but a slight alteration has taken place in the wording of the rules;—wherever in 1862 the reversed angle-iron was required to extend to the middle or lower deck beam-stringer, it is now required to extend to above the middle or the lower deck beam-stringer angle-iron, and the butts of the reversed angle-irons are to be secured with butt-straps.

Middle line keelsons, if of single plate standing on the top of the floors, are required in both years to be in thickness the same as the garboard strake, and in depth two-thirds the depth of the floors, and to have double angle-irons at top and bottom, but the angle-irons are somewhat larger now than in 1862 for ships under 400 tons; above this they are the same. If box keelsons be adopted the rules in both years require the plates to be the same thickness as the floor-plates, and to be in depth two-thirds the depth of the floors; now the breadth is also given, viz., two-thirds of the depth. In both years the intercostal middle-line keelson is required to be of the same thickness as the floor-plates, and riveted to vertical angle-iron on all the floor-plates at each end. In 1862 the keelson was required to extend from the upper edge of the keel to above the upper edge of the floors sufficiently high to be riveted between double angle-irons extending all fore and aft; it is now required to extend from the upper edge of the keel to above the upper edge of the floors sufficiently high to be riveted to bulb iron bars of the same strength as the beams; or to extend only to the top of the floors, deeper bulb iron bars or bar of other form but equal in strength, being let down and riveted to it; in each case the bulb iron is to be fitted between double angle-irons extending all fore and aft, and riveted also to double reversed angle-irons on the top of the floors. Several additional forms of keelson not mentioned in 1862 are also now authorized. With reference to bilge keelsons the requirements are the same in both years, viz., bilge keelsons at the lower turn of the bilge in all vessels, double angle-iron stringers at the upper turn of the bilge in all ships of 500 tons and upwards, and intercostal side keelsons in ships of 1000 tons and upwards. Keelsons in both years are required to be continuous and not stopped at bulkheads.

Several alterations have been made in the rules for outside plating. In 1862 the plates were required to be not less than 9 feet in length; they are now required to be not less than 5 spaces of frames, the fore and after hoods being exceptions in both cases. In 1862 the minimum shift of butt was one frame space, it is now two spaces of frames. In 1862 the relative thickness of the plating was regulated as follows:—the distance from the keel to the sheerstrake was divided into three parts, viz., the garboard strakes, from the garboard to the upper part of the bilge, and from the upper part of the bilge to the sheerstrake, each of these portions being, for the most part, covered with plating $\frac{1}{16}$ inch thinner than that in the portion next below it, and the sheerstrake being of the same thickness as the plating in the middle portion. The upper portion, viz., from the upper part of the bilge to the sheerstrake, is now subdivided into two portions, the plating in the uppermost of these being reduced in most classes of ships by $\frac{1}{16}$ inch, the sheerstrake being as before, of the same thickness as the plating in the portion next above the garboard strake. The thickness of the plating differs also according to the grade in which the ship is to be classed. In 1862 ships of the 9 years grade were to have plating $\frac{1}{16}$ inch less in thickness, and ships for the 6 years grade $\frac{2}{16}$ inch less in thickness than that required for the highest or 12 years grade. Now the A and B grades correspond nearly to the 12 years and 9 years grades respectively of 1862, except in ships of 1000 tons and upwards where there is generally a reduction of $\frac{1}{16}$-inch or $\frac{2}{16}$-inch in the plating. The breadths of the garboard and the sheerstrake are now given; in 1862 they were not given. In 1862 no reduction in thickness at the extremities was allowed, except in the sheerstrake and the strake next below it which could be reduced $\frac{1}{16}$ inch in vessels of 1000 tons and under, and $\frac{2}{16}$ inch in vessels above 1000 tons, for one-quarter of the vessel's length from each end. The plating may now be reduced $\frac{1}{16}$ inch in vessels under 1200 tons, and $\frac{2}{16}$ inch in vessels of 1200 tons and upwards, for one-quarter of the length from each end, from the upper edge of the sheerstrake down to a perpendicular height above the upper side of the keel of three-fifths the depth of hold. In screw vessels, however, no reduction is to be made in the plating abaft, below the lower part of the rudder trunk. The rules in both years recommend that the sheerstrake be an outside strake, and require that all butt straps shall be fitted with the fibre in the same direction as the fibre of the plate. In the rules of both years reductions are allowed in raised quarter-decks, poops and forecastles, and upper decks in vessels with three decks, and in spar decks, in the parts and according to the proportions stated hereafter.

As regards beams, the rule in 1862 required the beams to be formed of bulb or any other approved iron plates, with angle-irons riveted on the upper edge; the rule now requires the beams to be made of H-iron, T-bulb-iron, or bulb-plate with double angle-irons riveted on the upper edge. The present rule also states that where no deck is laid the angle-irons may be of the same size as the reversed angle-irons on the frames. With reference to the spacing of the beams, detailed information will be given in the subsequent part of this chapter.

The rules relating to riveting are the same in both years, but the amount of lap required in the outside plating has been increased from that required in 1862, viz., from 5 to $5\frac{1}{2}$ diameters for double riveting, and from 3 to $3\frac{1}{4}$ diameters for single riveting. If double riveting is adopted where single is allowed both rules allow a reduction of $\frac{1}{16}$ inch in the diameter of the rivet provided that in no case the diameter be less than $\frac{5}{8}$ inch. The present rule requires the holes to be punched from the faying surfaces of the plates.

There is no difference between the rules of 1862 and those now enforced as regards the construction of bulkheads; the thickness of the plating, the dimensions and spacing of the stiffeners, and the connection to the side remaining the same. The rules of 1862, however, require all ships to have a watertight bulkhead at each end; they now require only the foremost one (or collision bulkhead) for sailing ships. The rules of 1862 require a sluice cock, or valve to be fitted at the limbers on each side of the middle line, at each watertight bulkhead, to be worked from the deck above: the rules now require the same when a pump is not fitted to each compartment.

In 1862 there was no provision made for double bottoms. At present a vessel is noted as having a "Double Bottom" if the inner bottom is carried forward to the fore bulkhead as usually fitted, and to an equal distance from the after part of the ship, and constructed as required by rules; or a ship may be noted as "Part Double Bottom" provided the inner bottom extends to at least one-half the length.

The thickness of the deck planking is the same in both years; but while in 1862 planks over 6 inches wide were to have two bolts in each plank in every beam, one of which might be a short screw bolt, both bolts must now be put through in all planks over 8 inches wide. The rules of 1862 state that if the planksheers and waterways are of wood the material must not be inferior in quality to that required for wood ships of the same grade. The rules now state that the waterways if of wood are to be fastened with screw bolts, with nuts on the under side of the stringer plates.

In both years the stringer plates are required to be of the same thickness as the floor plates. In 1862 the breadth was required to be not less than three times the depth of the beam; it is now as follows, viz., for the upper deck stringer in vessels with one or two decks, or the middle deck stringer in vessels with three decks, one inch for every seven feet of the vessel's entire length, for half her length amidships, tapering to three fourths the midship breadth at the extremities. Stringer-plates on the beams below the above mentioned decks may be reduced in width to three-fourths the midship breadth above named, being continued right fore and aft, and riveted by an angle-iron to the reversed angle-irons on the frames. Now, as in 1862, it is required that all upper deck-stringers, and those of the middle deck in vessels with three decks, shall be fitted home and riveted to the outside plating by means of an angle-iron, and that the middle-deck stringer shall have an extra angle-iron riveted to the reversed angle-irons and to the stringers. The rules of both years require all vessels to have tie plates ranging all fore and aft upon each side of the hatchways on each tier of beams, to be half the breadth of the stringer-plates, and of the same thickness. The rules of 1862 require plates of the same dimensions to be fitted from side to side diagonally where practicable. The rules at present require diagonal tie-plates to be fitted from side to side on the upper and middle decks in vessels with three decks, and on the upper deck in vessels with one or two decks, wherever the arrangements of the deck will permit them. Hatchways and mast holes must now be properly framed, the latter having mast partners at each tier of beams, except the orlop beams, the plating of which is not to be less in thickness than that required for stringers, and the united breadths of the plates are not to be less than three times the diameter of the mast. At the decks where the mast is wedged, an angle-iron of the same size as the main frames is to be riveted to the plates round the mast holes. There was no corresponding rule in 1862.

The rules of 1862 required that the plans of all vessels intended to be built, of which the length, measured from the fore part of stem to the after side of post on the range of the upper deck, exceeded seven times their extreme breadth or ten times their depth in hold, should be submitted to the Committee for approval, with full particulars of the arrangements for giving the vessel sufficient additional longitudinal strength, either by doubling or thickening the sheerstrake, and increasing the size of the stringer-plates, or otherwise. In the new rules the additional strength required is stated for ships exceeding the above-mentioned limits, and includes the strengthening of sheer-strakes, stringer-plates, and keelsons in hold, as will be seen further on in this chapter.

The rules of both years require the rudder to be made of best hammered iron, those of 1862 also requiring that where practicable the rudder shall ship and unship without docking. The size of the main piece at the heel is the same in both years, but the head is now enlarged in all ships.

The rules now, as in 1862, require that vessels intended for classification shall be surveyed five times as follows, viz.:—1st. On the several parts of the frame when in place and before the plating is wrought; 2nd. On the plating during the process of riveting; 3rd. When the beams are in and fastened and before the decks are laid; 4th. When the ship is complete, but before the plating is finally coated or cemented; 5th. After the ship is launched and equipped.

The rules laid down by the Liverpool Underwriters which were first promulgated in 1862, also exercise considerable influence upon the construction of many iron ships and therefore deserve consideration. We have given the latest edition of these rules in the Appendix, and here propose to briefly enumerate the alterations which have been made in them since their first issue, and then to state the principal differences which exist between Lloyd's and the Liverpool rules as now enforced.

The system of classification of the Liverpool rules has remained almost unaltered since their promulgation, but the times of the periodical surveys have been slightly changed. Thus in 1862 a ship built under survey was required to be thoroughly surveyed once in every three years, and a ship not built under survey once in every two years, provided they were in this country; and all vessels not built under inspection but surveyed not less than four years after the date of launching, were required to be subsequently surveyed once in every three years. Under the present regulations all ships classed for 20 years are to be surveyed once in every four years; ships built under survey and classed under 20 years are to be surveyed once in every three years; and ships not built under survey and classed under 20 years are to be surveyed once in every two years. The periods for which ships are classed vary from 10 to 20 years. The rules as published in 1862 contained the scantlings of ships intended to be classed for 20 years, and this arrangement has been conformed to in the present rules, but some particulars are given as to the reductions permitted in steamers intended to be classed in the lower grades.

Bar, side-bar, and flat-plate keels have always been recognised in these rules, but in the latest issue information has been given with respect to the butt-strapping of the centre plate of a side-bar arrangement, and of the flat-plate keelsons usually adopted in connection with it, which was not supplied in the rules for 1862. The regulation with respect to the row of rivets connecting the side bars with the centre plate has also been added, as well as that which states that the double reversed scarphing angle-iron is to be riveted to the flat-plate keelson in addition to its being riveted to the floors. The angle-irons connecting the floors with the centre plate have been increased in size from the dimensions of the reversed angle-irons to those of the frames. The regulations with respect to bar and flat-plate keels remain unaltered.

The rules having reference to the sizes and construction of the frames, the arrangements of their scarphing angle-irons, and the fitting and ending of the reversed angle-irons, are almost identical in the two editions, the only point requiring notice being that in ships having two tiers of beams, and not exceeding 16 feet depth of hold, the reversed frames are now ended at the upper bilge stringer and the gunwale alternately, instead of at the main and lower deck stringers as was formerly required. The spacing of the frames is unchanged, except that now in ships of 1000 tons the frames may be 24 inches apart for one-fifth of the vessel's length from each end, which is a provision not made in the rules of 1862. The depth and thickness of the floors which were originally given have been retained, but intermediate dimensions have been introduced in the table. A reduction of $\frac{1}{16}$ inch is now allowed in the thickness of the plates for one-fifth of the vessel's length from each end in floors which exceed $\frac{6}{16}$-inch in thickness. It is also provided that the depth of the floors shall be increased in spar-deck ships. The two latter provisions have been added in the present rules. The rule which governs the depth will be given hereafter.

In the rules for 1862 the only middle-line keelsons recognised were box-keelsons, double centre plate with top and bottom plates, and upright or single centre-plate keelsons. In the present rules fuller information is given with respect to these three keelsons, and in addition intercostal middle-line keelsons are admitted for ships of 1200 tons and under. The particulars of butt-strapping for keelsons now given were not furnished by the rule of 1862. The regulations of 1862 with respect to the hold-stringers are nearly identical with the present rules, but it will be seen that the depth of the lower hold now governs the position of the stringer, whereas in 1862 depth in hold was the measurement on which the rule was for the most part based. One very important alteration has been made in the present rules from those of 1862, viz., the omission of the hold-stringer then required in ships with a depth of hold exceeding 23 feet, which was constructed of the same form as the middle-line keelson, but formed of plates having one-half the thickness and two-thirds the depth of those used at the middle line.* Centre keelsons may now be reduced at their ends

* "There is no doubt that this stringer is generally in the neutral part of the ship's " side. This is inevitable. It is not designed for other than lateral strains, such as

to two-thirds the sectional area amidships, as explained hereafter. This was not allowed by the rules of 1862. Both rules require the longitudinal ties to be kept up through the bulkheads.

Very important alterations have been made in the rules with respect to outside plating. In 1862 the thickness of plating was governed by the length of the ship or the depth in hold, and the plating was of uniform thickness from the garboard strake to the sheer strake. Now, the thickness is based upon the depth in hold, the ship's length being disregarded, and variations in the thickness are given for the different grades of classification for 20, 18, and 16 years respectively. The thicknesses now given for the higher grade are identical with those of 1862, but intermediate dimensions are introduced. The same regulations as to length of plates and shift of butts are retained, but it is now required that in vessels of 1200 tons and over, three strakes of plates in the bilges shall be increased $\frac{1}{16}$ inch in thickness over half the vessel's length amidships. Other rules are now given for the plating of ships of excessive proportions which were not given in the rules of 1862. A reduction of about $\frac{1}{16}$ inch was allowed in the thickness of outside plating forward and aft in the rules for 1862, but in the present rule the reduction allowed is stated to be about one-sixth of the total thickness. In both cases the taper commences at one-fifth of the vessel's length from each end. The reductions now allowed for poops, forecastle, &c., are identical with those previously specified.

As regards beams the dimensions given in 1862 have been retained in the present rules, but intermediate breadths of vessels have been introduced in the table. It is also now provided that for one-fifth of the vessel's length from each end the thickness of the beams may be reduced $\frac{1}{16}$ inch. This is an addition to the rules of 1862. The regulations for the spacing of beams are nearly identical for the two years, and the same may be said of the requirements for hatch, forecastle, and poop beams, and the dimensions of beam-knees. One addition made in the present rules

" result from working in a rolling sea, lying in tiers in docks, and especially for " working in and out of docks. Yet, in vessels of greater depth than those into which " it is first introduced, it will be higher above the said axis, and to that degree " becomes available for longitudinal strain. It is generally of a limited length as " compared with the other stringers in the smaller vessels into which it is put, seldom " being carried into the ends."—*Mr. John Price*, Chief Surveyor of the Liverpool Committee, Sunderland, in a discussion at the Institution of Engineers in Scotland.

is that lower-deck beams are recommended to be one-eighth of the depth deeper, besides being $\frac{1}{16}$ inch thicker than the upper-deck beams, and in cases where the scantlings of the lower-deck beams are increased a proportional reduction is permitted in the upper-deck beams. Solid flanged beams are now recognised in the rules, but were not mentioned in the rules for 1862.

The present regulations with respect to the diameters, spacing, and arrangement of rivets are almost identical with those of 1862, the only important additions being that the particulars of treble-riveted butts are now given, and that the work is now specified in which the rivets are required to be conically formed under the head, whereas in the former rules the requirement was made for all work. The maximum thickness of the rivet-head is now fixed at two-thirds the diameter. No change has been made with respect to the quality of iron and the character of the workmanship, except the omission in the present rules of the regulation formerly made that drifting unfair holes would be considered bad work.

Important alterations have been made in the rule with regard to bulkheads. In 1862 only a collision bulkhead was required, but it was added that two years additional would be granted to ships which were thoroughly bulkheaded. In the present rules the collision bulkhead is still required for all ships, and for steamers bulkheads must be fitted at each end of the engine and boiler space, and at the fore end of the shaft-tube. The paragraph with respect to two years additional being granted has been omitted. In other respects the rules for the two years are nearly identical.

The regulations at present enforced with regard to decks only differ from those of 1862 in allowing a thickness of upper deck of 3½ inches in vessels up to 700 tons instead of 500 tons, and fixing the thickness in larger vessels at not less than 4 inches, without adding the remark made in the rules of 1862 that the decks may be made 4½ inches with advantage. The paragraph with respect to spar-deck vessels has been added to the present rules.

With regard to beam-stringers and deck-ties the rules for the two years differ considerably. The present table is much more elaborate than that previously given, although nearly all the dimensions given in 1862 are still retained. The reduction now allowed in the hold and orlop beam stringers at the ends was not allowed in 1862. The regulation with respect to upper-deck stringers of ships of and over twelve depths in length have also been added in the present rules. In the rules of 1862 no mention was made of the continuous stringer angle-iron now required to be fitted inside the frames, nor was any rule given for the arrangement of the upper-deck stringer in ships having a break at the after part of the deck, as is now the case. In other respects the rules agree.

The present regulations with regard to excessive proportions differ from those of 1862 in omitting the requirements which were formerly made for vessels of 10 and 11 depths in length. For vessels of 12, 13, and 14 depths in length the regulations are almost identical, but while the present rules require that ships above 12 depths in length and exceeding 1500 tons shall have the plating from the keel to the upper part of the bilge increased $\frac{1}{16}$ inch in thickness for one-half the vessel's length amidships, the rules of 1862 only required one strake above, and one strake below the bilge to be doubled for half their length amidships. The regulation now made with respect to the working a thicker plate instead of doubling the sheer-strake was not given in the rules of 1862. The notes attached to Tables 2 and 5 have been added in the present rules, and include further regulations with respect to excessive proportions.

No alterations have been made in the rules with regard to windlasses, rudders, masts, spars and sails, and painting and cementing.

In the present rules two additional paragraphs are given having respect to the extension of the character of vessels, and the form of survey for the extension of class. Rules are also given for composite ships, but these do not fall within the limits of this work.

Having thus briefly reviewed the alterations made in the Liverpool rules since their first appearance, we proceed to indicate the principal differences between Lloyd's and the Liverpool rules as at present enforced. The nomenclature adopted in the Liverpool rules differs somewhat from that of Lloyd's rules. In a ship with two decks they are denominated in Lloyd's rules upper and lower deck respectively; in the Liverpool rules they are called main and lower deck respectively. If a vessel has three decks, they are upper, middle, and lower decks in Lloyd's, and upper, main, and lower decks in the Liverpool rules. If the floors are divided at the middle line by a vertical plate, this plate is termed in Lloyd's rules a centre-plate keelson, or a centre through-plate keelson, according as it is only of the depth of the floors, or extends above or below them; in the Liverpool rules it is denominated a centre-plate keel. The classification is also differently arranged under the two sets of rules. The *gross tonnage* is the basis of Lloyd's scantlings, subject to certain exceptions, while *dimensions* are the principal basis of the Liverpool rules.

Lloyd's Committee class all vessels A, subjecting them to special periodical survey, the vessels retaining their grade only so long as these surveys show them to be in a fit condition to carry dry and perishable cargoes to all parts of the world. Degrees of strength and probable durability are indicated thus, A_{A} A_{B} A_{C}.* A_{A} A_{B} denote vessels built in accordance with the rules. A_{C} denotes vessels considered entitled to the A character, but which have not been built in accordance with the rules. The Liverpool Committee propose to class ships on their general merits, for a specific period, having special reference to the quality of the materials, to the character of the workmanship, to the arrangement and size of the parts where the principal strains are experienced, and to the equipment. The latter Committee will class in red all vessels submitted to the inspection of their surveyors while building, for periods varying from ten to twenty years. The Committee will also class in black ships already built but not submitted to the inspection of their surveyors while building, for periods, varying according to their merits, from ten to twenty years. Lloyd's Committee require that vessels shall be surveyed every four, three, or two years respectively, according as they are classed A_{A}, A_{B}, or A_{C}. The Liverpool Committee require that vessels having certificates for twenty years shall be surveyed once in every four years, and that vessels having certificates for less than twenty years shall be surveyed once in every three or two years respectively, according as the vessel has a red or black certificate. Lloyd's rules require the whole of the iron to be of good malleable quality, capable of bearing a longitudinal strain of 20 tons per square inch, and to have the manufacturer's trade-mark, or his name and the place where made, legibly stamped upon it in two places. The Liverpool rules require all iron to be tough and malleable and branded "best" with maker's name, and to bear an absolute mean breaking strain of 20 tons per square inch, or 24 tons per square inch of broken section, the rule thus taking ac-

* See foot-note, p. 404.

count of the ductility of the iron. Both rules specify that the iron shall exhibit the ordinary properties of good material, and also that the workmanship shall be of the best description and well executed.

The following is a detailed comparison of the two sets of rules, framed chiefly with reference to the differences that exist between them:—

The Liverpool rule states that the form of keel must be either that of the centre plate or that of a bar; Lloyd's rule mentions the same forms. When bar-keels are adopted, the scantling in both rules is regulated by the tonnage of the vessel, but the Liverpool rule requires a larger scantling than Lloyd's for ships under 1500 tons, the same for ships from 1500 to 2500 tons, and smaller for ships above 2500 tons. When the garboard strakes are thicker than those required by the rule, Lloyd's rule allows a proportionate reduction in the thickness of the keel, but the garboard strakes must in this case extend to the bottom of the keel. In both rules the thickness of the centre plate depends upon the tonnage, but a somewhat thicker plate is required by the Liverpool rules than by Lloyd's. The Liverpool rules state that the butts of the centre plate are to be secured by double butt-straps treble chain-riveted, each strap being two-thirds the thickness of the centre plate; Lloyd's rules give no directions on this point. Lloyd's rule requires the side bars to be of such a thickness as, together with the centre plate, shall make up the thickness of a solid keel; the Liverpool rule gives the sizes for ships of various tonnages; but although put in a different form the rules are nearly identical. If the centre plate extend only to the top of the floors, both rules require a flat keelson-plate to be worked on the top of the floors, riveted to double reversed angle-irons on the upper edges of the floors, and to the centre plate by short fore and aft angle-irons underneath. The Liverpool rule requires plates of less thickness than Lloyd's, but of greater width, except in ships under 200 tons where the breadth is greatest as required by Lloyd's. The angle-irons on the upper edge of the centre plate are largest as required by Lloyd's rule. If the centre plate extends above the floors to form a keelson, both rules require that it shall be riveted by two fore and aft angle-irons to two flat keelson-plates, one on each side, each being of the same thickness and half the width of the single plate mentioned above: the size of the angle-irons required by the two rules is about the same as regards sectional area, but they are thickest in the Liverpool rules. The Liverpool rules state that the butts of the flat keelson-plates are to be double chain-riveted, the butt-straps being worked on the upper side of the plate.

When flat keel-plates are used, both rules require the centre plate to be connected to them by two continuous angle-irons, the sectional area of the angle-irons required by the two rules being about the same for similar ships. Lloyd's rules, as we have seen, give the breadth and thickness of hollow or flat keels; the Liverpool rules do not.

Both rules require stems and stern-posts to be of the same dimensions as bar-keels. The Liverpool rule requires that propeller posts shall be double the thickness of, and the same breadth as, bar-keels; also that the feet of the stem and stern-post shall be extended to form part of the keel for a distance not less than four and a half feet; the rule also allows that the stem may be gradually reduced in sectional area one-fourth, from the load-line upwards. Lloyd's rule requires the stern-posts and the after end of the keel in steam-vessels to be double the thickness of, or twice the sectional area of, the adjoining length of the keel (the siding in no case to be less than that given for bar-keels); and to be tapered into the adjoining length of the keel. Lloyd's rules require the scarphs of keel, stem, and stern-posts to be in length eight times the thickness given for bar-keels. The Liverpool rule gives the actual lengths for ships of various tonnages, but the two rules are nearly identical.

The sizes of the frames and reversed angle-irons are regulated in Lloyd's rules by the tonnage, and in the Liverpool rules by the depth of hold, but for well-proportioned ships the sectional areas required by the two rules are nearly identical. When frames are butted, Lloyd's rules require that the covering piece shall be not less than four feet long. The Liverpool rules require that they shall be four feet long in vessels up to 900 tons, and six feet long in vessels over 900 tons, but the pieces to connect the heels of the frames across the middle-line, where bar-keels are used, are to be not less than four feet long. The Liverpool rules require the reversed angle-irons to extend to the upper part of the bilge and the gunwale alternately in vessels of under 12 feet depth of hold, while Lloyd's rules require no reversed angle-irons above the upper part of the bilge in vessels under 300 tons. In ships above 300 tons the rules are nearly identical. Lloyd's rules now require, as we have seen, double reversed angle-irons in the way of all keelsons and hold-stringers, while the Liverpool rules require them in the way of all keelsons, hold and beam stringers, except where great closing bevil would be necessary in the double angle-iron. The maximum limit allowed for the spacing of the frames by the Liverpool rule under any conditions is 21 inches from centre to centre for ships under 1000 tons; while by Lloyd's rule this may be increased to 23 inches under the conditions already specified. For ships of 1000 tons and upwards both rules allow the spacing throughout the ship to be increased to 24 inches under certain conditions, and the Liverpool rules allow of this spacing being adopted for one-fifth the vessel's length from each end without any double frames being fitted. With reference to floor-plates, Lloyd's present rule, as previously stated, makes the depth of floor-plate at the middle line to depend upon the breadth and depth of the ship, viz., the depth in inches to be two-fifths the sum of the breadth and depth; the Liverpool rule makes it to depend upon the breadth alone. The depths at the middle line are given for various breadths of ship obtained evidently from the following rule:—Depth in inches equal to two-thirds the breadth of the ship. The rules would correspond for those ships whose depth was two-thirds their breadth. For spar-deck ships, the Liverpool rules also require an increase of $\frac{3}{8}$ inch in the depth of the floor-plate for each foot of height of the spar-deck. The thickness is by Lloyd's rules made to depend upon the tonnage, but by the Liverpool rules on the breadth only. A table of thicknesses is given in each case, the Liverpool rule requiring a somewhat less thickness of floor-plate than Lloyd's for ordinary proportioned ships. Both rules allow a reduction in the thickness of the floor-plates at the extremities of the ship. Lloyd's rules require the outer ends of the floor-plates to be carried up the bilge to a perpendicular height of twice the depth of the floors amidships from the upper side of the keel; the Liverpool rules say they must be carried well up into the bilge, and be half the centre depth at the lower turn of the bilge. Lloyd's rule gives no intermediate moulding.

Both rules require for middle line keelsons either a box keelson, a single plate with double angle-iron at the top and bottom, or an intercostal middle line keelson with a bulb iron bar on the upper edge riveted to two fore and aft angle-irons; the Liverpool rule also recognises a double centre plate keelson with top and bottom plates, and limits the application of the intercostal middle line keelson to vessels of 1200 tons and under. A deeper and thicker keelson is required by the Liverpool rules than by Lloyd's. The angle-irons are similar in size. In box keelsons the thickness is nearly the same in both rules. The particulars of the butt strapping of keelson work are given in the Liverpool rules but not in Lloyd's.

The Liverpool rule requires all vessels to have double angle-iron stringers both at the upper and the lower turn of the bilge; Lloyd's rule requires the lower one only for vessels under 500 tons. Lloyd's rule requires, as already stated, a side intercostal keelson to be fitted to all vessels of 1000 tons and upwards as far forward and aft as practicable; the Liverpool rule requires the same for all vessels over 32 feet beam, to extend for two-thirds of the vessel's length when practicable. The Liverpool rule requires in vessels of 15 feet depth of the lower hold an extra stringer between the upper bilge stringer and the lower deck beams, and in vessels above 16 feet, but under 18 feet in depth, a bulb-iron of the size of the lower-deck beams to be secured between the angle-irons forming the side stringers, and a bulb-iron to be riveted between the angle-irons on the upper or the lower bilge stringer. The sectional area of the centre keelsons at the extremities may be reduced to two-thirds the area amidships, the reduction only extending to the heel of the fore and main masts, and not exceeding in length one-third the length of the vessel when taken together. Both rules require keelsons, stringers, &c., to be continuous as far as practicable.

Both rules now require outside plates to be not less in length than five spaces of frames. The thickness of the plating is, in the Liverpool rules, made to depend upon the depth of hold of the ship; in Lloyd's, upon the tonnage. The Liverpool rule, in most cases, provides for a thinner plating than Lloyd's. In the Liverpool rule the plating is arranged in four divisions, the garboards, bilge and bottom, sides, and sheer-strakes. There are also three classes of ships 20, 18, and 16 years respectively for which the thickness is varied. The thickness of plating between the garboard and sheer-strakes is uniform in vessels of the highest class, but for the 18 and 16 year class the side plating is $\frac{1}{16}$ inch thinner than the bilge and bottom plating. For vessels of 1200 tons and upwards special regulations are made as before stated. Lloyd's rules divide the distance from the garboard strake to the sheer-strake, as we have seen, into four parts, the plating in each part being generally $\frac{1}{16}$ inch thinner than that in the part next below it, that in the lowest being $\frac{1}{16}$ inch thinner than the garboard strake. A reduction in thickness is allowed in the plating at the extremities in all ships by the Liverpool rules, and over a portion of the plating by Lloyd's rule. Butt straps are required by both rules to be of the same thickness as the plates they connect, and no butts in adjoining strakes may be nearer than two spaces of frames. The Liverpool rule requires a shift of 3 feet between the butts of the upper-deck stringer and of the sheer-strake. There is no rule in Lloyd's for this. Lloyd's rule gives the breadth of the sheer-strakes and the garboard strakes, the Liverpool rule does not. When the united lengths of the poop and forecastle do not exceed three-fifths of the entire length of the upper-deck, Lloyd's rules allows a reduction of one-fourth from the thickness which would be required for the same parts in the range of the upper-deck in ships with two decks, the outside plating, beams, stringer-plates upon beams, angle-irons on stringer-plates, and flat of deck; in raised quarter-decks a reduction of one-fifth in the same parts is allowed. The Liverpool rules allow the following reductions—the sides of the poop and forecastle to be one-third lighter than the shell plates amidships; the poop beams to be one-third, and the forecastle beams one-fourth lighter than the scantling given in the table for beams; the poop and forecastle stringers to be one-third lighter than lower-deck stringers. In vessels with three decks, Lloyd's rules allow a reduction of one-sixth from the dimensions given for such parts in the range of the upper-deck in vessels with two decks, in the scantling of beams, the flat of deck, and the plating, but not in the dimensions of the sheer-strakes. In the way of spar decks Lloyd's rules allow a similar reduction of one-fourth in all beams, stringers, plating, and flat of deck. The Liverpool rules also allow a reduction of one-sixth in the scantling of the material above the main deck of spar deck vessels.

With regard to beams, Lloyd's rule requires them to be in depth $\frac{1}{4}$ inch for every foot of the length of the midship beam, and to be in thickness $\frac{1}{16}$ inch for every inch in depth. The Liverpool rule gives the dimensions of the beams and of the angle-irons for the upper edge for various breadths of ship, but these dimensions are obtained by the same rule as Lloyd's, and therefore for these breadths the rules are identical. Lloyd's rule requires the two sides of the angle-iron on the edges of the beam to be not less in breadth than three-fourths the depth of the plate, and to be in thickness $\frac{1}{16}$ inch for every inch of the two sides of the angle-iron. The Liverpool rule allows the beams to be reduced $\frac{1}{16}$ inch in thickness for one-fifth of the vessels length from each end. The Liverpool rule requires a beam on alternate frames at all main decks. Lloyd's rule requires a beam on alternate frames at the upper (or main) deck in vessels with one or two decks, and on the upper and middle (or main) decks in vessels with three decks. Vessels of 12 feet to 13 feet depth according to Lloyd's, or 11 feet to 13 feet by the Liverpool rule, are to have lower-deck (or hold) beams, at least one to every eighth frame. Vessels from 13 feet to 15 feet deep are required by both rules to have lower-deck beams on every fourth frame, and vessels of 15 feet to 18 feet are to have lower-deck beams on every second and fourth frame alternately. Vessels of 18 feet and upwards are required by both rules to have lower-deck beams on every alternate frame. Vessels over 24 feet in depth are required by Lloyd's to have orlop beams on every sixth frame; the Liverpool rule requires the same in vessels over 17 feet in depth of lower hold. In vessels over 18 feet in depth of the lower hold, the Liverpool rule requires orlop beams on every fourth frame. Both rules require stringer-plates on orlop deck beams. The Liverpool rule requires the lower-deck beams to be $\frac{1}{16}$ inch thicker, and one-eighth of the depth deeper than the upper deck beams.

With reference to riveting, the Liverpool rule requires a larger rivet than Lloyd's rule for a given thickness of plate in nearly every instance; there are two thicknesses, however, viz., $\frac{6}{16}$ inch and $\frac{11}{16}$ inch for which the rules correspond, and one, viz., $\frac{5}{16}$ inch for which Lloyd's rule requires the larger rivet. Lloyd's rules require the rivet points to be slightly convex; the Liverpool rules say they must be perfectly fair with the surface of the plating, except the keel rivets which are to project slightly. The Liverpool rule requires the rivets to be laid up round the head. Lloyd's rule requires a lap of $5\frac{1}{2}$ times the diameter of the rivet for double riveting, and $3\frac{1}{4}$ times for single riveting. The Liverpool rule requires the same lap for double riveting of the edges, but a breadth of 6 diameters for butts. The comparison between the two systems of riveting is further discussed in Chapter XVII.

Both rules require a collision bulkhead to be fitted forward in all ships, and in steam ships a similar bulkhead must be fitted aft in addition to the engine-room bulkheads. The Liverpool rules require bulkheads to be stiffened on both sides with angle-irons 4 feet apart, one set horizontal and the other vertical. Lloyd's rules require but one set of smaller angle-irons placed vertically, and two feet six inches apart. The thickness required for the plating of bulkheads is about the same in both rules, $\frac{1}{4}$ inch being the minimum. The required connection to the side is the same in both rules.

The Liverpool rules make no mention of double bottoms. Lloyd's, as we have seen, now make provision for them.

Lloyd's rule requires the ceiling to be not less than $1\frac{1}{2}$ inches or greater than 3 inches in thickness. The Liverpool rule gives no dimensions, but states that the ceiling in the flat of hold is to be laid in hatches.

The thickness of the wood deck as required by the Liverpool rule is $\frac{1}{2}$ inch more than is required by Lloyd's, for all vessels under 1000 tons. For ships above 1000 tons the rules agree. The Liverpool rule requires lower decks of ships above 500 tons to be 3 inches thick, and in decks laid with East India teak it allows a reduction of one-sixth in the thickness.

With reference to stringer plates, the Liverpool rule requires a much wider stringer than Lloyd's rule, the disproportion being greatest in the smallest ships, and decreasing with the size, till in the largest ships the sizes approximate to an equality. The thickness required by both rules is about the same, but in some places the Liverpool stringers are thinner than Lloyd's. The angle-irons required by Lloyd's rule are about the same for the smaller ships, and greater for the larger ships than is required by the Liverpool rule. Lloyd's rules allow the principal deck stringer to taper for one-quarter the length of the vessel at each end, to three-quarters its breadth in midships. The Liverpool rule allows the main deck stringer to be reduced one-fourth in width and one-sixteenth inch in thickness, the tapering to commence at one-fifth the length of the vessel from the ends. Special regulations are made for ships exceeding 12 depths in length, or having a break in the after end of the upper deck. Lloyd's rule requires the lower deck and orlop stringers to be three-fourths the midship breadth of the principal deck stringer, the breadth to be continued all fore and aft. The Liverpool rules allow a reduction not exceeding one-third, to be made in the width of the stringers on the orlop deck beams; and Lloyd's rules also allow a reduction in the width of the hold beam stringers, it being provided in each case that such reduction must be fully compensated for. The Liverpool rules allow the thickness of these stringers to be reduced $\frac{1}{16}$ inch for one-fifth of their length from each end.

With regard to the tie-plates, Lloyd's rule gives them narrower than the Liverpool rule, except for the smallest ships. The Liverpool rule requires double angle-irons on the upper side of the main deck tie-plate. Lloyd's rule, as we have seen, requires in addition that diagonal tie-plates shall be fitted; the Liverpool rules make no mention of diagonal tie-plates.

We have also seen what Lloyd's arrangements are for hatchways and mast holes; the Liverpool rule requires mast partners at decks where wedged to be plated over twice the width of the hole cut out of them, and to take three beams in length.

In ships of which the depth is less than five-eighths the breadth, the Liverpool rule requires the tie-plates and stringers to be increased in strength; Lloyd's rule requires for ships over ten depths in length, a thicker sheer-strake or a doubling strake 9 inches broad, increased to 12 inches if the ship exceeds eleven depths in length. For vessels over twelve depths in length Lloyd's rule requires a thicker sheer-strake (or a doubling strake 18 inches broad), the principal deck stringer to be increased in thickness or width, and a bulb plate riveted between the double angle-irons at the lower part of the bilges. The Liverpool rule requires that such vessels shall have the sheer-strake doubled for half the length. In ships above thirteen and not exceeding fourteen depths in length Lloyd's rule requires the sheer-strake to be doubled over its whole breadth. The Liverpool rule requires that in such vessels the sheer-strakes shall be doubled for two-thirds of the vessel's length amidships, and the sheer-strakes, upper-deck stringers, and ties treble-riveted for half the length amidships. In ships above 1000 tons the main-deck stringers and the sheer-strakes must be treble-riveted for half the length amidships. All vessels of 1200 tons and above, are to have three strakes of plates at the bilges increased $\frac{1}{16}$ inch in thickness for half the length amidships, and in all ships of 1500 tons and above, and exceeding twelve depths in length, a further addition of $\frac{1}{16}$ inch is to be made in the thickness of all the bilge and bottom plating for a similar length. Ships of and over thirteen depths in length are to have $\frac{1}{16}$ inch in thickness added to the tabulated thicknesses of plating in addition to the previous requirements, excepting the bilge and bottom plating of ships of 1500 tons which have been increased in accordance with the preceding regulation. In place of doubling the sheer-strake a single thickness of plate, one and a half times the thickness of the sheer-strake, may be worked. In ships above 1000 tons the main-deck stringers and the sheer-strakes must be treble-riveted for half the vessel's length amidships. Vessels exceeding twelve depths in length and more than 1500 tons are required to have the plating from the keel to the upper part of the bilge increased $\frac{1}{16}$ inch in thickness for half the vessel's length amidships.

In both rules the size of the main piece of the rudder is regulated by the tonnage. In general the size as required by Lloyd's would be greater than that required by the Liverpool rule, except in the case of the rudder heads in steam-ships which would be largest by the Liverpool rule.

The Liverpool rule states that all the surfaces of iron ships are to be properly painted with good oil paint, and that cement is to be laid in the bottom so as to cover the frames and rivet-heads, to be raised in the centre to the height of the limber holes, and extend to the upper part of the bilges. There is no special rule on this subject in Lloyd's.

The Liverpool rules do not name periods at which ships building for classification shall be surveyed; but they state that their surveyors are to have free access at all times to vessels which hold a certificate from the Committee, and the same rule probably holds in the case of vessels building for certification.

From the foregoing comparison of the two sets of rules it will be seen that considerable differences still exist, not only between the scantlings enforced by the two Committees, but also between the principles upon which the arrangements laid down have been determined; and it is not at all surprising that in both respects each set of rules has been subjected to the criticism of builders, who not unfrequently have to produce ships for owners who desire to class their ships under either Committee, or under both. An able and elaborate discussion upon this subject took place in 1866-67 at the

Scottish Shipbuilders' Association (now incorporated with the Institution of Engineers in Scotland) with the view of bringing about an assimilation of the two systems, but not, as has been seen, with a thoroughly satisfactory result. It was shown, however, from evidence furnished during the debate, that both Committees were in the habit of making limited concessions to owners and builders who desire to class under both sets of rules; and the last edition of the Liverpool rules undoubtedly affords many evidences of a substantial approximation to the rules of Lloyd's, while still retaining characteristic differences. Owing to the descriptive character of the present work we shall not here enter upon a fuller investigation of this large and important question; but we may do so in a future volume.

It is only necessary to add, that Lloyd's Committee alone have issued rules and regulations for ships built of steel. They have been previously given in Chap. XVI., but for convenience it may be here stated that they are to the following effect:—That ships built of steel of approved quality, under special survey, will be classed in the Register Book with the notation "experimental" against their characters. In all cases, however, the specifications for the ships must be submitted to the Committee for approval. A reduction will be allowed in the thickness of the plates, frames, &c., of ships built of steel, not exceeding one-fourth from that prescribed for iron ships. In no case, however, are the rivets to be made of steel, nor will any reduction be allowed in the sizes of rivets from those prescribed for ships of the same tonnage, built of iron. In other respects the rules for the construction of iron ships will apply equally to ships built of steel.

CHAPTER XX.

SYSTEMS OF WORK.

IN this chapter we propose to give a brief outline of the general methods of proceeding with the work of building iron ships practised on the Mersey, the Clyde, the Thames, and the Tyne, and in the Royal dockyards. The descriptions given are based upon the methods practised by some of the principal firms on each river; but it will be obvious that, as most shipbuilders have peculiar methods of performing some portion of the work, it would be impossible to frame any general description which would include all these special cases. The information here brought together may, we think, be relied on as affording the means of comparing generally the different systems of conducting the work. We shall have occasion to indicate the advantages claimed for, and the disadvantages urged against, the various modes of building, and in order to avoid repetition, we will in the first place describe the system of shipbuilding usually practised on the Mersey, and afterwards point out in succession the principal differences between that system and the systems pursued in the other yards enumerated above. The precedence has been given to the Mersey system on account of the fact that iron shipbuilding was first extensively practised on that river.

THE MERSEY SYSTEM.

The order in which the work is usually conducted on the Mersey is as follows:—A model of the ship on a scale of ½ inch to a foot is prepared immediately after the drawings have been received, and on the model the general arrangement of the edges and butts of plating, the directions of the longitudinal work, deck-lines, &c., are drawn. That no confusion may occur in ordering the plates from the manufacturers, and that a correct account may be given to the workmen, it is customary to mark the strakes, in order, alphabetically, and to number the plates in each strake. The lengths of the plates used are regulated by the specification, averaging, as before stated, about 10 feet. The lengths of the frames, reversed angle-irons, &c., are taken from the body plan on the mould-loft floor. The dimensions, actual weights, and particulars of the results obtained by the testing of both plates and angle-irons are recorded in an order book, and in cases where the plates have a peculiar shape there is a rough sketch given of the form to which they must be brought by the manufacturer. A margin of about 1 inch in length and ½ inch in breadth is allowed in the dimensions recorded in the order book above the net dimensions of plates on the broadside of an iron ship; but forward and aft where there is considerable curvature and twist a greater margin is given. Floor plates are usually ordered to the required taper, and afterwards bent to the proper curves. When centre plate keelsons are adopted, each of the floors is in two separate pieces, as previously explained. In ships with bar keels each floor is usually made up of two pieces welded together, the welds of adjacent frames being placed on opposite sides of the middle line in order to give a good shift.

The laying-off of the ship is proceeded with simultaneously with the preparation of the model, and when it has been completed, the lines to which the angle-iron frames are to be bent are transferred to boards prepared for the purpose, and rased in. There are two boards, each being large enough to take the midship section of the ship, the fore body being transferred to one and the after body to the other. In order to show the lines more clearly, the upper surfaces of these boards are covered with a composition of lamp-black, size, and water. The name commonly given to these boards by the workmen is the "scrive" or "scriveing" boards, but we shall refer to them as the "blackboards" throughout the following description. In addition to the lines to the outside of the frames, the positions of plate edges, diagonals, level lines, heights of floors, beam-ends, &c. (which answer for bevilling spots), are marked upon the blackboards, which are then removed to a place appropriated for the purpose situated near the furnace in which the angle-irons are heated. The levelling blocks or bending slabs on which the frames are bent are made of cast iron, the upper surface being straight and out of winding, and perforated with holes placed at intervals of about 6 inches. The line to which the frame is to be bent is transferred from the blackboard to the slab by means of a soft iron bar, known as "set" iron, (about 1¼ by ⅜ inch) which is bent to the line on the board, has the bevilling spots, &c., marked on it, and is then removed to the slab, on which the curve is drawn and the spots are marked. Iron pins are then fixed in the holes in the slab which are near the outside of the line, and small blocks and wedges of iron are prepared to fill in the spaces between the pins and the curve. When the angle-iron is sufficiently heated it is drawn out of the furnace and placed on the slab, being gradually brought to the required curve and bevillings by means of levers, heavy hammers, iron wedges, &c. The bevillings are given out on a separate board, as in wood shipbuilding, and are applied to the back of the angle-iron. Care has to be taken in bringing the flanges to the correct bevilling to avoid striking too heavily, as the angle-iron, even when of good quality, is liable to open at the root under very heavy blows. The backs of the flanges are also liable to become hollowed while the bevilling is being performed unless special care is taken to keep them straight, which, it will be obvious, is an essential condition for good work, for otherwise the flange would not fit accurately against the plating.*

The bending and bevilling having been completed, the angle-iron is allowed to cool, and is then taken to the blackboard and tried to its curve, any unfairness or alteration of form which may exist being corrected. The plate edges and other stations before enumerated, are notched in on the frames, and the rivet holes for the fastenings of the outside plating are marked, the pitch varying from six to eight diameters, according to the space between the plate edges. The spacing of the rivets for the reversed angle-irons is regulated by the positions of the rivets in the outside plating,

---

* The above method is that ordinarily adopted in preparing the frames of an iron ship, but in some of the French dockyards (according to M. de Freminville's book, published in 1862), a different plan is followed. Instead of keeping one flange of the angle iron in a transverse plane and bevilling the other to fit against the outside plating, the French *twist* the angle iron itself, keeping the flanges at right angles to each other, and thus leave only the edge at the junction of the two flanges in the transverse plane. Professor Rankine proposed the same method in a paper published in the '*Transactions* of the Institution of Naval Architects for 1865,' being unacquainted, apparently, with the French system. The advantages claimed for twisting as compared with bevilling are, that the inner arm of the angle-iron stands everywhere at right angles to the plating, and thus gives greater support to it; that the material in the angle-iron is less damaged, and that the labour required is less. It would, of course, be necessary to bend the beam arms to fit against the frames, but this is thought to be no serious disadvantage, as the stringer stiffens the beam arms. This part of the subject has already been discussed in Chapter VIII. The floor plates would also require to be twisted.

the rule observed being that no two rivets shall come in the same transverse section of the angle-iron frame, as its strength would otherwise be seriously reduced. The average pitch of the rivets in the reversed bars also, is from six to eight diameters. The holes in the frames should be punched from the back in order that the countersink obtained by punching may assist in keeping the rivet in place. The holes in the frames which receive the rivets in the plate edges are generally drilled after the ship is framed and the plate edges faired and marked in. When the punching has been completed the frame is again tried to the curve on the board, and any alteration of form caused by punching is corrected.

While the frames are being prepared, the keel is proceeded with and temporarily put together on blocks alongside the slip or the dock where the ship is to be built. This course is adopted whatever may be the character of the keel, whether bar, side bar, or flat plate. When the keel is made ready, the frame stations are painted upon it, and it is taken to pieces and removed to the permanent blocks in the dock or slip, where it is put together again, and riveted up. The fore and after ends of the keel have been previously prepared so as to scarph with the stem and stern post respectively, and it is usual to make the moulds for the stem and stern post at as early a stage of the work as possible, in order that they may be forged, and that no time may be lost in waiting for their completion. The work amidships is often well advanced, however, before the sternpost is got in place. The details of the keel arrangements, and of the connections of the stem and sternpost with the keel, which are commonly adopted on the Mersey, are of an identical character with those previously described, and need not be repeated here. It may be stated, however, that it is usual to drill all holes in connection with bar keels, and in all work where three thicknesses come together the holes are drilled in the middle thickness and punched in the outer thicknesses in order to ensure their being well filled by the rivets.

While the frames and keel of the ship are in progress, beam moulds, with the round-up and lengths marked on them, are given out to the workmen to guide them in making the beams. The processes of bending and straightening the beams are performed by means of screw presses worked by hand, or by hydraulic presses, the beams being cold. In forming the beam knees the ends are the only parts put into the fire, and the plan adopted in nearly all instances is that illustrated by Fig. 99, page 146, the beam-arm being split up for a short distance, the lower part turned down, and a piece of plate welded in. The moulding of the frames determines the number of rows of rivets which may be employed in connecting the beam-knee with the frame, double or treble zigzag riveting being preferred for this purpose. In setting off the fastenings in the knees, templates are used. These templates are put in place at the ship, and the holes are arranged so that they may clear the holes in the other flanges of the frame angle-irons, two rivets being usually put in the upper part of the beam-arm above the line of the weld made in forming the knee. The templates are then removed to the beams, and the positions of the holes are transferred to the knees. After the holes have been drilled or punched in the knees, the beams are put in place, set fair to the beam-line, and fixed, and then the holes are drilled through the frames. It is usual to punch the holes for the deck fastenings indiscriminately in the flanges at the upper edge of the beams, without regard to the positions of the edges of the deck planking; this, of course, tends to make bad work, and should not be practised.

When the keel has been fixed in position on the permanent blocks, the process of framing is commenced, the frames amidships being first put up, and the work being continued forward and aft simultaneously. Before any frames are hoisted, staging is erected at the topsides, and the sheer or gunwale harpins are suspended from it, ready to receive the frames when raised in place. When raised, the frames are shored, stiffened by cross-spalls, and temporarily secured at the keel; when a considerable number has been put up the other harpins and ribbands are fixed in place and the frames faired. Stages are then made around the ship (without being secured to any part of her) at different heights for the purpose of proceeding with the plating, the latter operation being commenced as soon as the frames are set fair and fixed in place. In the mean time the floor plates are prepared from the lines got in on the blackboard, and, having been bent to form, are put in place, and have the holes for the fastenings to the frame angle-irons marked. They are then taken out of the ship, the holes in the upper edges for the fastenings of the reversed bars are set off, and all the holes are punched. The floors are then fixed in place and temporarily secured. The reversed bars are also prepared while these operations are proceeding, and are bent, bevilled, punched and faired in a similar manner to the frame angle-irons. In taking account of the holes for the rivets securing the reversed angle-iron to the frame and floor plate, it is usual to employ a light batten, which is bent around the line of holes after the frame and floor plate are fixed in place, and then transferred to the reversed bar. When the preparation of the reversed angle-irons is completed they are put in place, and the riveting up of both the floors and reversed bars to the frames proceeds simultaneously. It is a very common practice in ships with a continuous centre-plate keelson to run the reversed angle-irons across the middle line, as shown in Fig. 73, page 80, the butts of adjacent frames being on opposite sides of, and not less than 6 feet from the centre keelson. Before getting the beams in, it is usual to work a strake of plating at or near the beam ends, and to shore the ship at this part. The whole of the work connected with the construction of the vessel is thus progressing simultaneously. It should be stated, however, that when the ship is properly faired, the spots notched on the frames at the plate edges, heights of deck, &c., are faired through and corrected by means of battens, and the lines are then marked in on the frames.

In plating a ship, the inside strakes are first worked, and the position and shift of butts are made to correspond with the arrangements previously made on the model, the foreman in charge of the work usually having a duplicate of the model to guide him in regulating the plating. The diagonal shift of butts, illustrated by Fig. 134, page 190, is that generally preferred, but the shifts shown in Fig. 133, page 189, and Fig. 136, page 191, are also frequently employed. In Chapter X. we gave a brief description of the operation of plating, but it may be convenient here to more fully illustrate the details of the process, notwithstanding the fact that some of the statements previously made will be repeated. The lowest strake of the bottom plating is generally an inside strake, and is, in most cases, the first strake put on, the work being continued upwards, and the two inside strakes upon which an outside strake laps being fixed in place before it is worked. As soon as the inside strakes are riveted to the frames, the harpins which were originally placed in wake of the outside strakes are removed, and the ship is shored under the inside strakes, thus leaving the space free for working the outside strakes. In taking account of the bottom plating, templates are generally used, the most common form of template consisting of a light batten mould of which the outside dimensions are a little greater than those of the plates. Cross battens are fixed on the templates at intervals corresponding to the frame space, and when the templates are put in place at the ship, these battens cover the frames. For a plate of an inside strake it is only necessary to take account of the edges and butts on the battens forming the frame of the template, and of the positions of the rivet-holes in the frame angle-irons on the cross battens. The positions of the holes are marked upon the template by means of a wood plug with a hollow end, or a hollow cylinder, which is dipped in whiting, and put through the holes from the inside, thus marking the outlines. When the account has been taken the template is taken down and laid on the plate, and the positions of the lines and rivet-holes are transferred to it. The method of transferring the positions of the holes requires some notice, as in many cases bad work is caused by carelessness in this respect; full particulars of the operation will be given further on. The edges of the plate are next sheared, and the butts planed to the lines obtained from the template, after which the positions of the rivets in the edges and butts are set off. In setting off the edge riveting of inside strakes, templates are used which have the positions of the holes marked upon them. As the frame space and the pitch of the rivets in the edges are constant quantities, this mode of setting off the fastenings is a very good one, the only care required being to make the edge fastenings work in well with the butt fastenings of adjacent strakes. This can be readily done if two templates are employed, the first having the edge fastenings arranged to suit the frame spaces in which a butt comes, and the second being adapted to frame spaces in which there is no butt of the adjacent strakes. Templates are also used for setting off the butt fastenings. The centres of the holes are generally bored through the templates, and in order to transfer their positions to the plates, a sharp pointed centre punch is driven through the template. After the holes have been punched the plate is curved by passing it through the rolls, the proper curvature being secured by the use of section moulds made to the frames nearest the butts, the backs of the moulds being out of winding. In some portions of a ship's bottom the amount of curvature is so small as to render this operation unnecessary; but in other portions special care is required, as will be explained more fully hereafter. When the plate has been sheared, planed, punched, and curved as above described, it is put in place, and temporarily secured by means of cotters and pins.

In working a plate of an outside strake a similar template is used, and when put up in place, in addition to having the positions of the butts and edges of the plate, and of the rivet-holes in the frames marked upon it, account has to be taken of the rivet holes in the edges of the inside plates which it overlaps. In marking the positions of the holes upon the template the same method is adopted as is explained above for the plates of an inside strake. When the holes have been marked the template is removed and laid on the inside of the plate, and the holes are then transferred from the inside of the template to the inside of the plate. This is done by means of a "marker" or "reverser" of the form shown in Fig. 255. The end of the marker is forked, in order that it may

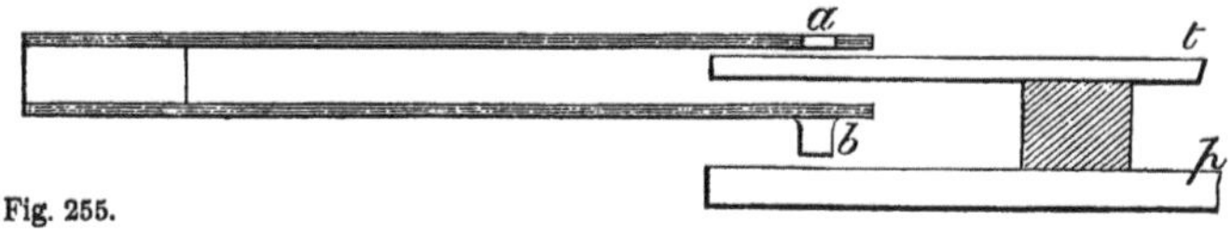

Fig. 255.

be put on over the edge of the template *t*, and have the hole *a* in the upper part brought exactly over the outline of the hole marked on the template. On the lower limb of the marker there is a projecting plug *b* vertically under the hole *a*, and the template *t* is chocked up at such a height above the plate *p* as to allow the lower part of the plug to just clear the surface of the plate when the hole *a* is brought well with the outlines marked on the template. When the marker has been placed in this position the workman presses down the plug *b* on the plate, and as the plug has been previously dipped in whiting it marks the position of the hole to be punched. Both the hole and the plug on the marker are, of course, of the same diameter as the rivets used. The plates of the outside strakes are punched from the inside on account of the fact that it is the faying surface; but the holes for the edge riveting in the inside strakes require, for a similar reason, to be punched from the outside, while the holes for the rivets securing all plates to the frames, and those for the butt fastenings, should always be punched from the inside. The remaining operations involved in the preparation of a plate of an outside strake, — punching, shearing, planing, &c.,—are identical with those described above for an inside strake, and when they have been completed, the plate is temporarily secured in place by cotters and pins. The liners between the frames and the outside strakes are fitted after the plates are prepared and fixed. Wooden templates are used in preparing the liners, being put in place in order to have the positions of the rivet-holes marked, and then transferred to the liners. The holes are punched in the liners. When the curvature of the frames is at all considerable the liners are bent to the form required. After their preparation is completed the liners are driven in between the plates and angle-irons by the workmen, and fixed in their proper positions.

With comparatively light plates, wood templates are often dispensed with, and the plates themselves are put up and marked, but the ordinary practice is that given above. In cases where a plate has a large amount of twist, such as boss plates, &c., special means are employed to ensure accuracy in taking account of it. The common plan is to take four iron rods about $\frac{5}{8}$ inch in diameter, to cut them to the lengths of the edges and butts of the plates, and to weld them at the corners. The frame thus formed is put up in place at the ship, and bent to the shape required to give a correct account for the plate. Short pieces of angle-iron are then bent to the curve of the frames, and a bed is formed which has these angle-irons for its transverse framing, their ends being placed well with the twist given by the iron frame. The plate is heated and bent to the form of the bed or brake, after which it is put up in place and fitted, the holes being drilled. All difficult twisted plates with considerable curvature are thus worked, and the iron in the plates requires to be of very superior quality in order to stand the bending. The various processes of marking, bending, punching, &c., are performed by workmen known as "platers," each being assisted by from 4 to 6 "helpers," the number of the latter being regulated by the weight of the plates, averaging about one man to every cwt.

The edges and butts of bottom plating are generally double-chain riveted, but in some cases treble-chain riveting is employed for butt-fastenings. The pitch of the rivets in the edges is from $3\frac{1}{2}$ to 4 diameters. It is usual to joggle the butt straps to the outside plates in order to receive the edge fastenings of the adjacent plates, as shown in the section given in Fig. 145, page 202. The liners to the outside strakes are of the breadth of the frame angle-irons, on all except the bulkhead frames, where they extend to the adjacent frames before and abaft. The object of this arrangement has been pointed out in Chapter XI.

While the plating is being proceeded with, the work in the interior of the ship is also advancing, the riveting of the reversed angle-irons and floor-plates being completed, the beams being got in and fastened, the deck and hold stringers being fitted and fastened, &c. The usual mode of fitting deck stringers consists in laying the plate upon the beams and taking account upon it of the curve of the edge, positions of scores, rivet holes, &c. As the weight of the stringer plates is not usually very great, and they are easily moved and placed, this is found to be the best mode of procedure, no moulds being required to be made. The transverse watertight bulkheads are also built in the ship, but are delayed as long as is consistent with the progress of the work, in order to allow free access to every part of the hold. The plates, butt straps, stiffeners, &c., of the various bulkheads are prepared, fitted, and punched outside the ship, in readiness for being put together, and when the work is sufficiently advanced they are put in place and riveted. The ordinary bulkhead connections described in Chapter XI. are generally adopted, and the outline of the bulkhead plating is obtained from the section marked on the blackboard. In ships having a double bottom it is usual to secure the bulkheads to deep plate frames (moulded about 9 inches); and these frames are bent to the form required, put in place, and secured to the inner skin by double angle-irons, before the bulkhead plating is put in place. The taking account, fitting, &c., of all these portions of the work is performed by platers, who have the assistance of helpers as before stated.

After working about three-fourths of the outside plating of a ship, men are set to work at closing up the joints, riming out unfair holes, &c., preparatory to the riveting being commenced. The riveting is generally done by piece-work, a set of riveters being composed of two riveters, a holder-up, and two boys. The rivets used in the outside plating are of a conical form under the head, and the heads of all the rivets in the ship are laid up. Care is taken that the holes are well filled, and the points of the rivets flush with the surface of the plates. All rivets are examined by the foreman of riveters before being paid for, and in order to try if the surfaces of the plates are brought close, and to test the tightness of the rivets, the following course is adopted:—Rivets are marked in different parts of the ship, and the rivet on each side of a marked rivet is cut out. Screw-bolts of the size of the rivets are then placed in the holes and hove up as tightly as possible, in order to try if the rivet between them can be loosened, which will only be the case if the work was not properly drawn together when the rivets were put in. Riveting machines are not used in most of the principal yards on the Mersey. When the riveting has been advanced to some extent the edges (when not planed) are chipped fair and cleaned, and then the caulking of the edges and butts is commenced. In caulking a lap-joint the edge of the plate is first fullered with a tool, then it is split with another tool, and lastly the splitting tool is reversed, and the split part of the edge is driven in against the plate which it overlaps, as shown by *a* in Fig. 256. The caulking of a butt-joint differs from that of a lap-joint in requiring the butts to be first chipped smooth, then split on both sides of the butt, and afterwards fullered off, as shown by *b* in Fig. 256. The closeness of the joints is tested before they are caulked, by trying to insert a sharp instrument (usually a thin steel blade) at various parts. The caulking being found satisfactory, a painter follows, and thus marks the work as complete, while oxidation of the finished portions is prevented. At Liverpool it is customary to give the bottoms of iron ships three good coats of red lead paint before launching, the rivet-heads being cemented flush before the second coat is applied.

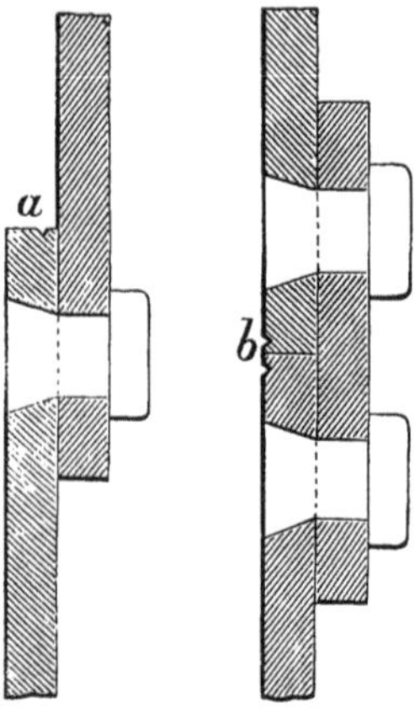

Fig. 256.

THE CLYDE SYSTEM.

We now come to the description of the method of conducting the work of shipbuilding pursued on the Clyde. The laying-off of the ship having been completed, the whole of the lines representing the outside of frames, the inside of floors, the positions of side keelsons, edges of outside plating, deck heights, diagonals, &c., are, in most cases, transferred from the mould-loft floor to a similar floor in the workshop near the bending slabs. Battens are used for the purpose of transferring the lines and stations, and in order to facilitate the work, the fore and after bodies are drawn on different parts of the floor. The plan of having a fixed floor instead of portable boards (such as are used on the Mersey, and sometimes employed on the Clyde) is thought to ensure greater accuracy in the preparation of the frames, and to render it perfectly safe to cut them to their lengths before they are put up. The arrangement of the butts of the outside plating, &c., is usually made on an expansion drawing prepared from the fair lines on the mould-loft floor, the disposition of the edges being made on the floor itself, and transferred to the expansion drawing. No model is used for this purpose, as is always the case on the Mersey. The same method is adopted on the Clyde as on the Mersey for distinguishing the strakes and plates, the former being marked alphabetically and the latter numbered. On the Clyde, however, all the dimensions are taken from the mould-loft floor, or the expansion drawing, and tabulated before the plates are ordered, and the positions of the butts are also recorded. It is usual to allow a margin of about 1 inch in length and from $\frac{1}{2}$ to $\frac{3}{4}$ inch in breadth above the net dimensions of the plates. A copy of the record is supplied to the foreman of the yard, so that he need not refer to the expansion drawing for the purpose of fixing the positions of the butts, &c. A similar arrangement is carried out with the deck stringer and tie plates, bilge stringers, keelsons, &c., and the frame and reversed angle-irons, the lengths of the latter being obtained from the body plan on the floor. The dimensions and particulars of the materials required are recorded in an order book. The diagonal shift of butts is usually adopted, but in many ships, on both the Mersey and the Clyde, the brick arrangement of butts is followed, of which a sketch is given in Fig. 133, page 189.

When the operation of transferring the lines and stations from the mould-loft floor to the floor in the workshop (or to the black-boards) has been completed, the preparation of the frames is commenced. An expansion batten is applied to the line on the floor representing the moulding edge of the frame, and all the holes for the rivets in the outside plating (except those in the edges) as well as those for the fastenings of the reversed bars, are marked on the batten. It is then laid on the angle-iron bar, and the positions of the holes are transferred from the batten to the angle-iron. The holes are punched previously to the frame being bent, and the operations of bending and bevelling are performed in a way similar to that previously described. After it has become cold the angle-iron is placed on the floor and tried to the line, any adjustment which may be necessary being made. The head and heel of the frame are then marked for cutting off, and the edges of outside plating, heights of decks, ribbands, &c., are notched in the bar.

While the frame angle-irons are being prepared, the floor-plates are moulded, bent, &c., at another part of the floor, and when ready are brought to the frames, and have the holes which have already been punched in the angle-irons marked upon them, and the holes for the fastenings of the reversed angle-irons, side keelsons, &c., set off, as well as the positions of the watercourses. They are then removed to the machine, and the holes are punched. It should be stated that the mode of demanding the floor-plates and arranging the welds of the pieces forming each floor is generally similar to that described for the Mersey system. In some yards, however, the floor-plates are made in three pieces, the welds on each side being placed at the lower turn of the bilge, and having no shift on adjacent frames. When the floors are thus made, only the floor-arms are bent to the required curves on the slabs, the centre piece of each floor being moulded and chipped to the slight curve required. The reversed bars are also taken account of and bent while the other work is advancing, the operations being conducted similarly to the corresponding operations described above for the frame angle-irons. The positions of the holes in the reversed angle-irons are transferred from the floors and frame angle-irons either by means of expansion battens, or by laying the reversed bars upon the frames and floor-plates, when fixed in position on the floor. The holes in the fore and aft flanges of the reversed bars to receive the fastenings of the ceiling are often punched before the bending is performed, but the holes in the other flanges are never punched until after the operation of bending is completed.

During the time occupied in preparing the frames the keel-work is being proceeded with. Bar-keels are most commonly adopted, and it is usual to put them together on the blocks, to rivet the scarphs, and to prepare and fix the stem and stern-posts, before any frames are raised. No difficulty is experienced in carrying on the work in this order as the stems are simply continuations of the bar-keels, and the sternposts are very simple forgings. The keel, &c., having been fixed, and the frames, reversed bars, and floor-plates got ready, the putting together of the framing is begun. The frames are laid across the keel near their stations, and the riveting up is readily performed.

The beams are usually made of bulb-iron, with double angle-irons riveted to the upper edge, and the common mode of forming the knees is that given in Fig. 100, p. 146, the end of the beam being turned down, and a piece of plate welded on to form the upper corner. It is considered that this mode requires much less time for its performance than the method given in Fig. 99, p. 146. When beams of the Butterley pattern, or other beams with a solid upper flange are employed, the beam-knee is formed by splitting the end, turning down the lower part, and welding in a plate, in the ordinary manner. Machine-riveting is sometimes employed for beam-work, care being taken that the work is properly closed and the holes made fair before the rivets are put in. The holes for the deck fastenings are usually drilled by hand after the beams are in place, as this method is considered to be less expensive, and to ensure a good disposition of the fastenings in relation to the edges of the planks. In wake of stringer and tie plates the deck fastenings are brought upon the plates, and clear of the beam-flanges. The bending of the beams is usually performed by Rennie's machine, the beams being worked cold, but in some cases the beams are heated and bent on the slabs. This operation, together with the forming of the knees, and the riveting of the beam angle-irons, is conducted simultaneously with the preparation of the frame and reversed angle-irons, and of the floor-plates. The beams are cut to the lengths and bevellings given on a board prepared for that purpose, the length in no case being taken from the ship. The positions of the fastenings are set off on both the frames and the beam-knees by means of templates having the holes marked on them. which are prepared by the draughtsman. All the holes are usually punched in both frames and beam-knees. In some yards, instead of giving out the lengths on the beam-board, the beams are brought to the frames at the same time as the floor-plates, and the frame angle-irons, floor-plates, and beams having been fixed in their proper positions on the floor, the stations of the rivet-holes, and the lengths of the beams are marked-by from the frames, the holes in the latter having been punched previously. Should the vessel not exceed from 500 to 600 tons burthen, the beams are always riveted to the frames before they are raised, and the same course is often adopted in ships of 900 tons burthen. For large ships it is a common practice to rivet either the upper or the main deck beams to the frames before they are erected, and to put in the remaining tiers of beams after the frames have been raised. The reason for this difference is, that for large ships the weight of the beams added to that of the frames would render it necessary to have an additional number of purchases, in order to prevent an undue strain being brought upon any part of the frame while it is being raised; and even if the power of the tackles were sufficiently great, there would still be the risk of straining the frames by the surging of the ropes. In all cases the frames are laid across the keel and riveted together before being erected, and in many instances the beams are attached to them also, as above stated. This is considered to give greater facilities for performing the riveting than are afforded by the contracted spaces left between the frames after they are raised.

The transverse bulkheads are put together and fitted to the frames before they are erected, the fitting and punching of the plates and stiffening bars being performed either at the ship's side or in the workshop, and the various parts being marked and numbered in order to enable them to be readily put together in the ship, when the work is sufficiently advanced. It is a very common practice on the Clyde to arrange the plates in the bulkheads with the greatest length vertical.

The sheer and bilge harpins having been firmly shored and secured in their respective positions, and the riveting of the frames having been completed, the frames are raised, each frame being horned and plumbed in order to ensure the correctness of its position before it is secured. Immediately after the frames have been erected, the other ribbands and harpins are put up and the frames are faired. The work in the interior of the ship is then commenced, such as centre and side keelsons, intercostals, &c. The stringer-bars for the several decks are set and bevelled during the time that the frames are being hoisted and fixed, and the moulds which are made for this purpose are preserved for cutting the stringer-plates and trimming the wood waterway. Sometimes the stringer-plates themselves are put in place and marked, this mode being always adopted when the plate has to be scored out between the frames. The work on the various decks is also proceeded with as soon as the framing of the ship is sufficiently advanced. In raising the frames different modes of procedure are adopted, some builders commencing amidships and working forward and aft simultaneously, while others begin at one extremity of the ship and work either forward or aft as the case may be; the latter plan is that most commonly adopted. The stern framing and plating abaft the transom frame are very often fitted together on the ground, the holes marked and punched, butt-straps prepared, &c., and afterwards put together and riveted up in place. In some small vessels the whole of the stern work has been riveted together on the ground before being hoisted. Rounded gunwales are very generally adopted at the stern, the frames being turned inwards and connected with the poop-beam on the transom-frame. This makes a very strong and simple arrangement of the stern framing.

The working of the bottom plating is commenced as soon as all the frames are faired and fixed, and the hold keelsons, beams, deck-stringers, &c., put in place and riveted. The methods of work are usually of an identical character with those adopted on the Mersey. The templates in common use are formed of light wood battens, like those previously described, and are used in a similar manner. Sometimes templates formed of light strip-iron are used for taking the length and breadth account of plates, and in other yards iron templates are also employed in marking the positions of the holes, movable pieces of zinc being attached to the edge and cross bars of the template for the purpose. Some builders employ a patent template consisting of a sheet of fine wire-gauze attached to a square frame of strip-iron. The merit of this template is thought to consist in dispensing with the use of a reverser; for when the positions of the holes have been marked upon the template in the usual manner they can be transferred to the plate by means of a stiff brush (of circular section) which penetrates the openings in the gauze, and marks the holes upon the plate. In very many cases, however, templates are altogether dispensed with, and the plates themselves put up in place and marked, this practice having been followed in some yards with plates of which the weight was half a ton.

The edges and butts of the plates from the garboard out to the round of the bilge are usually double riveted, the rivets being placed either zigzag or chain fashion; the latter arrangement is now preferred, especially for butts, for which it is sometimes adopted when the edges are double-zigzag riveted. From the bilge to the sheer-strake the edges are often single riveted and the butts double riveted. The sheer-strake is double riveted throughout. Below the light water-line the rivet-points are left well rounded, but above this line they are flush with the surface of the plates. The heads of all rivets are laid-up in watertight work. A "set" of riveters consists of 2 riveters, 1 holder-up, and 1 boy in most cases, but sometimes 2 boys are employed. The processes of riveting aud caulking are conducted similarly to those previously described.

From the preceding remarks it will be seen that the principal differences between the Clyde and Mersey systems of shipbuilding consist, in the substitution on the Clyde of an expansion drawing for a model in taking account of the plating and disposing the butts; the use of a fixed floor instead of portable boards on which the curves for the frames, &c., are drawn; and the modes of preparing the frame and reversed angle-irons, and the floor-plates, forming the beam-knees, and putting the frames and beams together. In other respects the modes of conducting the work at the two ports are almost identical. The shipbuilders who adopt the Mersey system (with more or less important variations) consider that by erecting the frame angle-irons and fairing them by harpins and ribbands previously to the floor-plates and reversed angle-irons being attached, and then riveting these plates and angle-irons in place, while the frame angle-irons are firmly secured to the harpins and ribbands, they avoid uncertainty as to the fairness of the frames. They are also of opinion that their mode of procedure admits of the work being proved and corrected at different stages before finally riveting it together, and allows the different portions of the work, framing, plating, &c., to be carried on simultaneously, thus diminishing the chances of error, and the time required for the construction of the ship. It is also urged that the practice of punching the holes in angle-irons before they are bent is objectionable, either on account of the iron breaking in wake of the holes, or of the holes being made oval by the bending, especially where sharp curves are required. To these objections the advocates of the Clyde system reply that the holes in the angle-irons can be punched with greater ease before the bars are bent, and that with careful workmanship the breaking of the angle-iron or distortion of the holes may be avoided; while the readjustment of the bars to the curves, which has to be performed when, as on the Mersey, the holes are punched after the bars are bent, is rendered unnecessary. They are also of opinion that with good supervision, and the employment of skilled workmen, the Clyde system can be carried out without the occurrence of serious errors, and that vessels can be built more quickly and cheaply by this method. There is, doubtless, a considerable amount of truth in the remarks made by the advocates of both systems, and there are advantages and disadvantages attaching to both methods; but it must, we think, be admitted that if it were possible to rely upon every operation being accurately performed, both in laying-off the vessel, and in fitting and riveting the work, the Clyde system would be the better of the two. To what extent this correctness is attained in practice, is a question the answer to which depends, almost entirely, on the character of the workmanship, and the strictness of the supervision.

THE THAMES SYSTEM.

We next propose to give a brief outline of the method of conducting the work practised in the principal yards on the Thames. In its general character the Thames system of shipbuilding approaches the Mersey system, but there are some important differences which we will presently point out. The arrangement of the framing and plating is made on a half-block model of the ship, to a scale of $\frac{1}{4}$ inch to a foot, on which are drawn the edges and butts of the outside plating, the butts of stringers, &c. The strakes and plates are distinguished, as on the Mersey, by being marked alphabetically and numbered, respectively. The dimensions of the plates are usually written on the model. Simultaneously with the preparation of the model the laying-off of the vessel is proceeded with, and as soon as the body-plan is completed the widths of the plates are checked with the widths obtained from the model, previously to the iron being ordered. A margin of about 1 inch in length and from $\frac{1}{2}$ to 1 inch in breadth is allowed above the net dimensions in preparing the specifications for the plates. The girths of the sections are also taken in order to guide the draughtsman in demanding the angle-iron for the frames and reversed bars. The particulars of the plates and angle-irons required are recorded in an order book, in the same manner as on the Mersey. The moulds which are used to guide the workmen in bending the frames, are generally made on the principle of a skeleton or spider mould, so that one mould serves for several sections. It will be seen that this constitutes a very important difference from the plan pursued on the Mersey of having portable boards as previously described. In getting in the curve to which a frame is to be bent, the mould is laid upon the bending slab, and spots are transferred from the mould with a piece of chalk, from six to eight sections being transferred to the slab at one time. The angle-irons, having been heated, are bent to the curves given by the chalk-marks, the operations of bending and bevelling being conducted in the manner before described.

Another skeleton mould is prepared on the mould-loft floor, and lines representing the upper and lower edges of the floors are marked upon it. If the vessel has a bar-keel the mould must be made to take the whole breadth of the floor; but if a centre-plate keelson is adopted, a half-mould only will be required. This also differs from the Mersey system, in which the lines for the upper edges of the floors are got in upon the blackboards.

The setting off and punching of the holes in the frames are conducted in exactly the same manner as on the Mersey, but the holes which come in the plate-edges are usually punched by a "bear" after the ship is in frame, instead of being drilled, as they are on the Mersey. After the holes have been punched, the frames are again placed on the bending slabs, and readjusted to the curves. This operation does not require the reheating of any angle-irons, except those of the largest size.

While the frame angle-irons are being bent, the keel-work is being arranged and proceeded with, either in a workshop or some other convenient place, where the various parts are fitted together, and temporarily secured. When the preparation has been completed, the keel is taken to pieces and removed to the permanent blocks in the slip or dock, where it is finally put together and has the stations of the frames marked upon it from a room-and-space batten. Staging is then erected around the slip or dock, being hung from standards, and a gunwale harpin is prepared, put in place, shored, and stiffened by cross-spalls at about every sixth frame. The stations of all the frames are marked on this harpin in the usual manner. The frame angle-irons are next hoisted into place, fixed at their respective stations, and secured at the heads and heels. When a sufficient number has been put in place they are regulated at the bilges, and made fair by other ribbands and harpins which are secured to the frames by screw-bolts and plates.

The account for the floor-plates is next taken, the plates being usually sheared to the taper given by the floor-mould and then bent on the slabs, after which they are put into place, and have the rivet-holes in the frames marked on them. The reversed bars are also bent, and the positions of the rivet-holes in the frames and floor-plates are transferred to them by means of a batten on which there are secured sliding pieces of zinc, which can be made to coincide with the holes when the batten is bent to the required curve. As the punching of the holes is completed, the floor-plates and reversed bars are put in place and riveted, the fairness of the frames being ensured by the harpins and ribbands previously fixed. A person known as a "liner" generally acts in conjunction with the foreman in charge of the work, his duties consisting principally in seeing that all measurements are set off at the ship in accordance with the instructions furnished from the mould-loft, and that all lines required for plating are correctly transferred from the block-model to the frames; he also superintends the getting in of all sheer-lines.

The beams are prepared by means of a beam-mould on which the lengths are marked, and the bevellings of the beam-knees are usually given on a beam-arm board. Sometimes the beam-arm board is dispensed with, and the bevellings of the knees are then marked upon the beam-mould itself; but in this case the breadth of the mould should not be less than the depth of the beam in order to afford good guidance in taking the bevellings. In some cases the beams of small vessels are bent cold to their round-up, but in others the beams are heated before being bent. The mode of forming the beam-knees generally adopted is similar to that practised on the Mersey, but some builders prefer the Clyde method shown in Fig. 100, p. 146. The rivet-holes for the fastenings in the beam-knees are generally punched in the frames before they are raised, and drilled through the knees after the beams are in place; sometimes, however, the beam-knees are first punched by means of a "bear," and the holes in the frames drilled after the beams are in place. Double-zigzag riveting is generally employed in beam-knees. The holes for the deck-fastenings are either drilled or "beared" in the beam-flanges after the beams are in place.

It is usual on the Thames to lay-off the midship portion of the ship first, and to give out moulds for the frames, &c., of that part, so that the work of building may be proceeded with while the extremities of the ship are being laid-off and the moulds prepared. Without, however, waiting for the framing to be completed, the platers commence fitting the outside plating, decks, stringers, &c., amidships. The process of plating is conducted similarly to that described for the Mersey, and the shifts of butts adopted are the same. The arrangements of the riveting of the edges and butts of the outside plating are generally in accordance with Lloyd's Rules. The ordinary batten template is very commonly employed in taking account of the plates, but there are two other descriptions of templates sometimes used, which require to be noticed. The first has been patented, and consists of a frame formed of light longitudinal battens with cross battens attached, the pins which secure the ends of the cross battens to the longitudinal battens being capable of motion in slots, so that the cross battens can be placed at a considerable inclination, and a moderate amount of curvature can thus be given to the longitudinal battens. By this and other arrangements, the edge of the template can be brought to the curve of the plate to be taken account of, and the cross battens nearest the ends can be fixed to the bevels of the butts. In order to take account of the holes in the frames and edges, pieces of zinc with holes cut in them are attached to the cross and longitudinal battens, and can be fixed directly over the holes which are to be transferred, when the template is in place at the ship. By this arrangement no reverser is required, as the holes can be marked directly upon the plate. The great difference between this and the ordinary form of template consists in the fact that the curve of the edge of the plate is taken off, and that the positions of the holes are taken account of by means of the moveable pieces of zinc. The second kind of template is a modification of the first, as a somewhat similar mode is adopted for taking account of the curvature of the plate-edges, but the positions of the holes are marked upon the battens, as upon the ordinary template, and transferred to the plate by means of a reverser. The latter form of template is now most generally used, only it is usual to have the longitudinal or edge-strips of wood, and the cross-bars of very thin strip-iron.

The bulkheads are, upon the Thames, prepared outside the ship, as is done on the Mersey and Clyde, but in this case it is usual to fit the plates next the ship's side, by means of a template which is made to the form of the sides after the frames are fixed in place. When it is thought that the work is sufficiently advanced, the bulkheads are put together in their respective positions and riveted.

While this work is progressing in the amidship portion of the vessel the framing of the extremities is being proceeded with; the stem and sternpost are put in place; the beams, stringers, &c., are fixed and fastened, and the outside plating is gradually completed. The remainder of the work of riveting, caulking, &c., is conducted similarly to that previously described.

It will thus be seen that the principal difference between the Mersey and Thames systems of shipbuilding consists in the use of frame and floor moulds on the Thames, instead of the blackboards employed on the Mersey. It should, however, be stated, that some of the shipbuilders on the Thames use a large blackboard, to which one side only of each body plan is transferred from the mould-loft floor. The curves to which the frames are bent, are transferred from the board to the slab by means of iron templates or set-irons instead of wooden moulds. The reason why only one side of each body plan is drawn on the blackboard instead of both sides as is done on the Mersey, is that on the Thames the floor-plates are fitted in place at the ship after the frames are fixed, and the board is used simply to regulate the bending of the frames and putting the bulkheads together; while on the Mersey the floor-plates are moulded from the board, and consequently the complete sections are required.

THE TYNE SYSTEM.

The system of shipbuilding practised on the Tyne may be regarded as a combination of the Clyde and Mersey systems, as it includes some of the more important features of each. The arrangement of the butts and edges of outside plating, &c., is drawn on a model of the ship made on a scale of $\frac{1}{4}$ inch to a foot. After the ship has been laid-off on the mould-loft floor, the breadths of the strakes of plating are transferred from the model to the floor, and the lines of the plate edges are faired; so that the correct breadths of the plates being obtained from the body plan, and their lengths being taken from the model, they can be ordered from the makers. The margin allowed above the net dimensions of the plates in preparing the specification is about 1 inch in length and $\frac{1}{2}$ inch in breadth for the plating near the extremities of a ship, and $\frac{1}{2}$ inch in length and $\frac{1}{4}$ inch in breadth for the plating amidships; all the particulars of form, dimensions, &c., are recorded in an order-book. The same mode of distinguishing the various plates and strakes is adopted as is employed on the Mersey, the latter being lettered and the former numbered. An expansion drawing of the plating is prepared from the model, and given to the foreman in order to guide him in placing the butts. Deck plans are also prepared showing the position of the stringer-plates and angle-irons, tie-plates, &c., the butts of this work being always shifted with the butts of the outside plating. This practice is also adopted at all the other ports, and, as before remarked, is essential to the proper construction of an iron ship. After the disposition of the butts and edges, and the laying-off of the ship has been completed, the lines and stations are transferred to a fixed floor or board near the bending slabs, the method of transferring the lines being conducted in a manner similar to that pursued on the Clyde. The process of bending and bevelling the frames is conducted similarly to that on the Mersey, a piece of "set" iron being used in transferring the curves from the board to the slab. When the frames have cooled they are tried to the lines on the board; if any adjustment is necessary it is performed, and the heights of decks, positions of plate edges, ribbands, &c., are all notched in upon them. The holes in the frames are then set off, and both the holes in the plate edges and those for the fastenings in the beam-knees, are punched before the frames are put up. It will be seen therefore that in the preparation of the frame angle-irons the Mersey system is followed, except in the use of a fixed floor (similar to that used on the Clyde), and the practice of punching the holes for the edge riveting of the plates, and those for the beam-arm fastenings before the ship is in frame.

During the time that the frame angle-irons are being punched the reversed angle-irons, floor-plates, and beams are being prepared. The reversed angle-irons are marked, bent, and punched in the manner previously described for the Mersey system, but the operations are performed before the frames are raised. The floors are prepared from the lines drawn on the board, are fitted to the frames, have the holes marked upon them, and are punched. The frame and reversed angle-irons and the floor-plates are riveted together at the side of the slip, near their stations in the ship, before the frames are raised, this modification of the Clyde system being adopted because it is considered that when the frames are laid across the keel, the keel work is liable to be made unfair while the riveting of the frames is being performed. It should be added, that the keel work is proceeded with while the frames are being prepared, and the stem and sternpost are also completed. When the scarphs, &c., of the keel have been fitted, the holes drilled or punched, and the other work performed, it is fixed in position on the blocks and riveted, the stem and sternpost also being put in place and secured.

In raising the frames it is usual to commence with the after or transom frame, and when this frame has been fixed in its true position and shored, the second harpin down from the top-side is put up. The after end of the harpin is secured to the transom frame, and its fore end is suspended from the standards which support the staging. The work of erecting the frames is then continued, the frame next the transom frame being first raised, and the framing being gradually brought forward. Cross-spalls are fixed to the alternate frames, the middle line being cut in each cross-spall. The stations of the frames are also marked on the harpin and on the keel, so that they can be accurately fixed in their proper positions, and the middle line can be plumbed down from the cross-spalls to the keel in the usual manner. When the number of frames put up has advanced the framing to the fore end of the piece of harpin first fixed, another piece is put up and the work is continued until all the frames are erected. It should be stated that in some cases the topside harpin, and another which comes just at the bilge, are put up before the frames are raised. When the framing is completed, the remaining rib-bands and harpins are put in place, and the frames are set fair.

The beams are, in the mean time, bent to their round-up (the bending being performed while the beams are cold), and the beam-knees are either welded to the central part of the beam, or formed by riveting on a piece of plate to the side of the beam after the end has been split and the lower part turned down, as shown in Fig. 101, p. 146. This mode of forming the beam-knee differs from the plans adopted on the Clyde and the Mersey. When the beams are made of bulb-iron with double angle-irons on the upper edge, the riveting is performed by hand, as it is considered that when the riveting machine is used, the round-up of the beam is increased while the riveting is being performed, and the beam has to be again put into the beam-bending machine and brought to its correct form. The holes for the deck fastenings are drilled in the beam flanges before the beams are put in place.

The bulkheads are lined off upon the board, and the account is there taken for fitting and punching the plates and stiffeners, but moulds are made at the ship to the bulkhead frames after they are in place. Moulds are also made at the ship for the stringer-plates and angle-irons as soon as the beams have been got in.

When the frames have been fixed and faired the work in the interior of the ship (keelsons, stringers, &c.) is at once begun, the beams are put in and fastened, and the work on the decks is proceeded with, simultaneously with the working of the outside plating. The lengths of the beams are always taken from the ship, and the holes in the frames for the knee fastenings are marked on the mould when in place, and thence transferred to the beam-knee. These holes are usually punched in the knees. This mode of taking the lengths differs from the methods adopted on both the Mersey and the Clyde. The operation of plating is conducted similarly to that previously described, except that instead of using a "reverser" to transfer the positions of the holes from the template to the plate, it is usual for the workman to drive a small centre punch through the template at the centre of each of the holes marked upon it, and thus to make an indent in the plate (on which the template is laid) which determines the centre of the hole to be punched. The position of the hole is then marked on the plate with a plug of which the end has been dipped in whiting. It will be obvious that this method is not as accurate as that previously described, for the workman has to trust to his eye both in determining the centre of the hole marked on the template, and in driving the punch squarely through the template, while in placing the plug so that the point thus fixed may be the centre of the hole marked on the plate there is a further chance of incorrectness. It should be added, that for plates not exceeding 5 cwt. in weight, the usual practice is to put them up in place, and to mark the positions of the holes directly upon them. In marking the holes on plates which have a considerable curvature, a "reverser" is sometimes employed.

The spacing of the rivets, shifts of butts, modes of punching, &c., are similar to those previously described, the lengths of the plates used varying from 8 feet to 10 feet 6 inches. Lloyd's Rules are usually conformed to in the riveting of the butts and edges, chain riveting being generally adopted. In all watertight work the rivet-heads are laid up, and all the rivet-points are made flush with the surface of the outside plating, as it is considered that they are more quickly oxidized if left rounded. In the keel work, however, the rivet-points are a little rounded. The operations of riveting and caulking are here performed in a manner similar to that described for the Mersey system, great care being taken in testing the closeness of the work and the tightness of the rivets before the caulking is begun.

The practice on the Tyne, like that on the Clyde, is to lay-off and expand the stern plating, to prepare the stern frames, and to fit the whole of the stern work together before putting it in place; and in some small vessels the whole of the stern framing and plating is riveted together in the workshop before being raised.

From this brief description it will be seen that the Tyne method differs somewhat from all the systems previously described, although, as before remarked, it resembles both the Mersey and Clyde systems in some particulars. As these resemblances and differences have been pointed out in passing, it is unnecessary to again state them here.

THE SYSTEM OF THE ROYAL DOCKYARDS.

The last system of shipbuilding which we propose to describe is that practised in the Royal dockyards, and it will be assumed that the bracket-plate arrangement is (as is now almost universally the case) adopted in the construction of the imaginary vessel of which we shall trace the progress; while, in order to represent the general practice as far as possible, we shall suppose her to be armour plated. Directly the drawings are received the laying off is proceeded with, and, as soon as the midship section has been got in upon the mould-loft floor, demands are prepared from it for the framing and plating of that portion of the length amidships, of which the transverse form does not differ materially from that of the midship section. By this means a supply of materials is ensured by the time that the laying off is completed, and the work of building can be at once commenced. In the mean time a model of the ship is prepared on a scale of $\frac{1}{2}$ inch to a foot, and the positions of the edges and butts of the bottom plating and armour plates, the longitudinal frames, deck heights, and transverse frames are marked upon it. It is found desirable to have the model pivoted at the ends in order to give facilities for drawing these lines upon it. The disposition of the butts of the flat and vertical keel plates, keel angle-irons, and gutter plate is first arranged on a separate drawing, and the demands for the plates and angle-irons are made from it. Other expansion drawings are also made from the model, one of which shows the arrangement of the bottom plating up to the armour shelf, and another that of the skin-plating behind armour. The lines for the edges of the bottom plating are first determined on the model, and the longitudinal frames are made to follow the plate lines, so that the holes for the fastenings of the continuous angle-irons on the outer edges of the longitudinals may be brought, as nearly as possible, to the centre of the strake of plating. In arranging the butts of the bottom plating and of the longitudinal plates and angle-irons, regard is had to the positions of the butts of the keel work, previously determined on, and care is taken that the butts of the longitudinal framing are well shifted with each other, and with the butts of the outside plating. A drawing is prepared showing the arrangement of the butts of the longitudinals, similar to that given in Fig. 93, p. 130. The diagonal disposition of butts is now followed for the bottom plating, there being two passing strakes between consecutive butts. The edges of the plating and stations of the longitudinals are transferred from the model to the body plan on the floor, and, the lines having been faired, the laps of the plating are marked. In demanding the plates for the bottom the breadths are taken from the body plan. For plates in the midship part of the ship the allowance made over the net length is about 1 inch, and over the breadth from $\frac{1}{4}$ to $\frac{3}{8}$ inch. Care is taken to allow for the curve in the edges and the bevelling of the butts in plates with a considerable amount of twist. The longitudinal plates and angle-irons are also demanded from the dimensions taken from the floor, the breadth of the longitudinals being decreased towards the extremities, as previously explained. In tapering the longitudinals it is usual to reduce the breadths in such a manner as to give sufficient depth at the extremities of the double bottom to allow men to enter for the purpose of making repairs or painting.

The moulding of the short transverse plate and bracket frames is, of course, regulated by the breadths of the longitudinals, and when these have been determined and the inside lines of the frames faired, the dimensions of the plates and angle-irons can be obtained from the floor, and the demands prepared. It is usual to make an expansion drawing of the continuous transverse frames and the deep frames behind armour, showing the positions of the scarphs and butts, similar to Fig. 94, p. 131, and from the lengths taken from the floor in preparing this expansion the angle-irons are demanded. An expansion drawing is also prepared of the inner bottom, the disposition of the butts and edges is made upon it, the butts being shifted with those of the longitudinals and the bottom plating, and the demands for the plates are made out from the dimensions thus obtained. A similar course is followed with the plating and angle-irons in the wing passage bulkheads.

In arranging the plating behind armour it is first necessary to fix the positions of the butts and edges of the armour plates, and this is usually done on a separate expansion drawing. The edges of the armour plates being fixed, the positions of the longitudinal girders behind armour are known, and these determine the positions of the edges of the inner thickness of skin-plating, as the edge fastenings are made to work in as fastenings in the girders. The edges of the outer thickness of skin-plating are shifted from those of the inner thickness and kept clear of the armour bolts. The butts of both thicknesses are well shifted with each other and with the butts of the armour. A sketch showing the character of this disposition is given in Fig. 137, p. 192.

An expansion drawing is also prepared showing the disposition of the light plating above the armour belt in the unprotected portions of the ship. The dimensions for this expansion are taken from the floor, and the demand for the plating and light angle-iron frames is prepared in a manner similar to that described above. As the ship advances dispositions and demands have to be made also for the plates and angle-irons in bulkheads, engine and boiler bearers, rudder work, &c., as well as demands for beams and the materials required for the various decks. Records of all demands are kept in an order-book, together with estimates of the weights of plates and angle-irons.

The moulds for the stem and sternpost are prepared at as early a stage of the work as possible, in order to give time for the manufacture, the sternpost especially being in many cases a cause of delay.

On this account the engineer's drawing, showing the height of screw-shaft, &c., is required at an early stage of the work.

As soon as the materials have been received the preparation of the flat and vertical keel plates, and transverse and longitudinal framing is commenced. Sectional moulds are given out from the loft to guide the workmen in flanging the flat keel plates, and for the plates forward and aft where there is a considerable amount of twist, the sectional moulds are connected by light battens, in order that their correct application may be ensured. The flat keel plates are flanged under a hydraulic press, having been first heated in a furnace placed near the press. In some private yards the flanging is performed by special plate-bending machines. The putting together of the keel work is conducted in the manner explained in Chap. VII. (p. 123), the foreman in charge of the work being guided by the expansion drawing prepared at the mould-loft. In the construction of most of the ships built in the Royal dockyards, it has been necessary (on account of the slip or dock being either occupied or under repair) to fit the keel together on temporary blocks in the workshops or by the side of the dock. In such cases it is also usual to rivet up the flat keel plates in such lengths as can be conveniently removed to the permanent blocks.

The short transverse plate and bracket frames are prepared from moulds which give the curves of the inner and outer edges, the bevelling of the ends, and the moulding of the brackets. Amidships one mould will, of course, serve for several frames, but in general a separate mould is made for each frame, and is accompanied by a bevelling board by which the preparation of the short frame angle-irons is regulated. The laps of the bottom plating are marked upon the moulds, and the joggles for the continuous longitudinal angle-irons are cut out. In putting a bracket frame together, the brackets are moulded and cut to shape, the short frame angle-iron is bent and bevelled, and the holes for the rivets securing it to the brackets are punched, their positions having been set off so as to clear the holes in the other flange; these latter holes, which receive the fastenings of the bottom plating, are for the most part drilled, a few being punched before the frame is put together in order to allow the bottom plating to be secured when first put up. The brackets are then put in position on the angle-iron and the holes are marked and punched. The holes for the rivets in the upper edges and ends of the bracket plates are set off upon them and punched, and the short connecting angle-irons are marked from the brackets, punched, and temporarily secured by cotters and pins. In a watertight frame, the frame angle-irons are forged staple fashion to the form given by the mould, the holes are set off and punched in the angle-iron, and being marked on the plate, are punched in it also. A similar course is followed with the lightened plate frames. The frame angle-irons and short connecting bars are riveted to the brackets and plates before the frames are put in place, machine riveting being generally adopted.

The longitudinals are prepared from moulds given out from the mould-loft, on which the scores for the continuous transverse angle-irons are marked or cut out, and the positions of the butts of the continuous longitudinal angle-irons, and of the bottom plating are marked, together with the stations of the transverse frames. The holes for the fastenings in the continuous and short angle-irons on the edges of the longitudinal, in the short connecting angle-irons on the transverse bracket and plate frames, and in the butts of the longitudinal plates themselves, are then set off and punched, and the frames are ready to go in place. A separate mould is made for each length of the longitudinals. The continuous angle-irons on the outer edges are bent cold to the curves required in the midship part of the ship; but forward and aft they require to be heated and bent on the slabs to the curves given by the moulds for the longitudinals. The holes in both flanges of these angle-irons are punched before the bars are put in place. The short angle-irons on the inner edges are taken account of, put in place on the longitudinals, and are riveted, after the longitudinals are fixed in the ship.

Separate moulds are also prepared for the continuous transverse angle-irons, and for the deep frames behind armour. The moulds are usually made so that one edge shall give the curve of the frame adjacent to that given by the other edge, one mould thus serving for two frames. The edges of plating, positions of butts and scarphs, &c., are marked upon these moulds, and spiling lines with check measurements are given out with them, together with spread battens showing the proper breadths at the heights of the longitudinals and decks, so that accuracy may be ensured if the moulds should warp. When all the brackets or plate frames corresponding to any section have been prepared as far up as the longitudinal next below the armour shelf, they are fixed in their proper relative position on the floor of the workshop, and, allowance being made for the longitudinals, the correctness of the form of the section is tested by means of the spread battens and moulds. The continuous transverse frames are bent in the usual manner on the slabs, and completed as will be described hereafter. The deep reversed frames behind armour are bent and bevelled, and the holes are set off and punched in the outer edge of the transverse flange to receive the fastenings of the double angle-irons. These angle-irons are bent and bevelled, and, being brought to the frames, have the holes marked upon them. They are next taken to the press and punched, and then fixed in place on the frames. The riveting machine is used in preparing these frames, and care is taken in setting off the holes to avoid bringing them into the same sections of the angle-iron with the holes which receive the fastenings of the skin-plating behind armour.

The process of framing is commenced as soon as a portion of the keel has been fixed on the blocks, and the riveting and caulking have been sufficiently advanced. The stations of the transverse frames are marked upon the vertical keel plate from a batten given out from the mould-loft. As soon as these operations are completed a tier of short transverse frames is put up amidships, and temporarily secured to the vertical keel, the heads being fixed to a ribband which is afterwards put up and shored. When this has been done the fitting of the plates of the lowest longitudinal is proceeded with, they having been previously prepared from the moulds given out from the mould-loft as described above. When put in place the longitudinals are temporarily secured by cotters and pins, the butt straps are prepared, and the continuous angle-irons on the outer edges are fixed. A portion of the length of the lowest longitudinal having been completed, another tier of transverse frames is put up, and a ribband is fixed and shored near their heads; then another longitudinal is fitted and fixed; and so on until the longitudinal is reached which forms the upper boundary of the double bottom, and is usually situated at the foot of the wing passage bulkhead. This longitudinal has to be made watertight, and the manner in which this has been accomplished has been previously illustrated in Fig. 84, p. 114, and Fig. 88, p. 125. Previously to completing this watertight work, the frames behind armour have to be hoisted in and the continuous transverse angle-irons put in place. The latter are in some cases put in and have the holes in the brackets and plates marked upon them, the butts fitted and the holes set off for the fastenings, and are then taken out and have the holes punched. In other cases the holes have been drilled in place, but the former plan is thought to be the cheaper, and is that adopted in recent ships. In getting the frames behind armour into position it is usual to put up one or two at each end of a length of the sheer ribband, and to secure the ribband to the inside of the frames in order to avoid having to hoist the frames in over the ribband, as would require to be done if it were put on the outside. When the ribband has been fixed, the other frames which come upon it are put in, brought to their stations, and secured. The fairing of a portion of the framing is then completed, cross-spalls being fitted to every fourth or fifth frame, and ribbands being put up on the outside of the frames. The lower ends of the vertical frames are scarphed with the continuous transverse angle-irons, and it is usual to punch the holes for the fastenings in either the frame or the angle-iron before it is put in, and to drill them through the unpunched thickness in place.

Between the armour shelf and the longitudinal next below it the transverse framing is formed by lightened plates with angle-irons on the edges, as shown in Plates 4 and 5. These frames are prepared from moulds made in the mould-loft, and are completed, with the exception of cutting the heels, in the same manner as the other short transverse frames. In order, however, to secure accuracy in the armour shelf-line, the moulds are put up in place, and a fair line is got around the ship by means of battens, the lower ends of the frames being afterwards cut to the lengths thus obtained.

While the framing amidships has been thus advancing, the keel is being extended both forward and aft, the transverse and longitudinal framing is being put in place in wake of it, and the preparation of the remainder of the framing is being proceeded with. Simultaneously with this the riveting up of the various parts of the frame and the connecting angle-irons is being performed, and, as soon as possible, the working of the bottom plating is commenced on the midship part where the framing is most advanced. The only points requiring notice with respect to the mode of plating adopted, are, that the harpins and ribbands on the bottom are always placed between the edge of an inside strake and a longitudinal, so that they need not be removed until the outside strakes of plating are worked; that the lines for the plate edges are got in upon the frames by a draughtsman from an account furnished from the mould-loft; and that a thin blackboard is used for taking account of the plates instead of a batten template. It has been previously explained that most of the holes in the frame angle-irons for the fastenings of the bottom plating are drilled in place, a few only of the holes between the plate edges being punched previously in order to allow the plating to be temporarily secured when first put up. The holes for the edge fastenings are always drilled in the frame angle-irons. The positions of the holes are transferred to the plates by means of reversers similar to that previously described. It is the practice, as far as possible, to complete the riveting of the framing and the bottom plating, together with the fitting of the drain pipes in the double bottom, before the inner skin is worked. The disposition made at the mould-loft is conformed to in working the inside plating, and the mode of taking account of the plates is very similar to that described for outside plating. The plating is flush-jointed both at the edges and butts, and the strips are worked below it. The holes for the fastenings are drilled in the continuous transverse frames, and punched in the plates. The riveting and caulking of the plating in both the inner and outer bottoms are performed in the manner before described.

The armour shelf having been completed for a portion of the length amidships, the working of the skin-plating behind armour and of the longitudinal girders is commenced. The lines for the plate edges are got in on the frames from an account furnished by the mould-loft draughtsman, and the holes for the fastenings are drilled in the frame angle-irons and longitudinal girders and punched in the plates. The disposition previously made is carried out by the foreman, who is guided by the expansion drawing, and the butt fastenings are arranged so as to clear the armour bolts. The taking account of the plates, punching, &c., are conducted similarly to the processes described above.

Simultaneously with this the beams are being put in, the deck-lines having been previously got in upon the frames. It is usual for the T-bulb and H-iron beams to be supplied to the dock-yards by the makers, with the knees formed and the proper round-up. For this purpose the makers are furnished with sketches of the beam having the figured dimensions marked upon them, with beam moulds giving the round-up of the decks, with check battens marked from the mould-loft floor in order to test the lengths, and with batten moulds showing the bevel of the beam-knees and the inside curves of the beam-arms. The usual allowance made over the true length taken from the floor is $\frac{3}{4}$ inch on each beam-arm, the additional length being allowed on the outer edges of the arm, and the true lengths taken from the floor being conformed to in making the moulds for the inside curves of the knees. When made beams are adopted, the plate welds are carefully shifted in adjacent beams, the beam webs are bent to their proper round-up on the slabs, and have the knees formed by splitting the ends and welding in pieces as shown in Fig. 99, p. 146, or by welding the knees on. The holes for the fastenings in the beam angle-irons are then set off and punched, the angle-irons are bent to the curves, brought to the beam-plates, have the holes marked, are taken to the press, and are punched, after which they are temporarily secured to the beam-plates by cotters and pins until the riveting is performed. The riveting of made beams is usually done by the machines. In taking account of the beams the lengths of the beams and bevellings of the knees given from the mould-loft are conformed to, being tested at the ship previously to cutting the beams. The outer edge of the beam-arm is accurately fitted against the transverse flange of one of the double angle-irons on the deep reversed frames, as shown in Plate 4. The holes for the fastenings in the beam-arms are usually set off and punched before the beams are put in, templates being used for setting off the fastenings; the holes in the frames are drilled after the beams are in place. The holes for the fastenings of deck planking and plating are always drilled in the beam flanges after the beams have been fixed. The deck planking is now fastened to the beams only in cases where there is no iron deck.

The work in the hold is also being proceeded with during this time. As soon as the inner bottom has been sufficiently advanced the bulkheads are fitted and fastened. A sketch is prepared for each bulkhead from the dimensions taken from the mould-loft floor, and on it the disposition of the plating and stiffeners is made. The plates and angle-irons required for the bulkheads are also demanded from these sketches. The bulkheads are fitted together outside the ship, the holes for the fastenings are marked and punched, the strips and stiffeners fitted, &c., and the various pieces marked in order to facilitate the putting together in place. In building the bulkheads in the ship the midship part is first put up, and the ends of the plates coming on the inner bottom are cut to the lengths and curves taken from the ship. The fitting of watertight doors, sluice valves, &c., can be proceeded with as soon as the riveting of the bulkheads is completed. The bulkhead connections are usually similar to those described for the 'Hercules' in Chap. XI.

The work on the different decks is commenced directly the deck framing is completed for a portion of the length. Plans of the decks are prepared from the mould-loft floor, and the dispositions of the butts and edges of the deck stringers and plating are made upon them, the demands for plates and angle-irons also being prepared from these drawings. No moulds are used in fitting the stringer-plates, except in places where great care is needed, as for instance where the frames behind armour are run up through the stringer which comes upon the upper edge of the armour belt, as described in p. 122. In nearly all instances the stringer-plates themselves are put in place and marked, and this course is thought to be both cheaper and more expeditious. The deck plating is also laid upon the beams, and the holes for the fastenings to the beam flanges are marked upon it, after which the plates are removed to the press, and the holes are punched. In recent ships the fastenings of the deck planking have been brought out upon the plating, clear of the beams, and it is usual to set off the holes upon the plates and to punch them, care being taken in setting them off to allow for the strakes of planking and to make good fastenings.

Meanwhile the armour plating of the midship portion of the ship is being performed, having been commenced as soon as the skin plating behind armour has been completed for a sufficient length. We shall give a full description of the operations connected with armour plating in another chapter; but it may be remarked here that the system followed in the dockyards of advancing the framing of the midship portion of the ship before that of the extremities, allows the armour plating of the broadside in wake of the central battery, and the construction of the armour bulkheads at the ends of the battery, to be proceeded with simultaneously with the framing of the bow and stern, which is in nearly all cases delayed by the necessity for allowing a considerable time for the manufacture of the stem and sternpost.

The framing and plating of the unprotected portions of the vessel above the armour belt are commenced as soon as possible, the mode of conducting the work requiring no special remarks as full particulars of the arrangements have been already given in Chap. VII. The remainder of the work in completing the framing and plating, putting on the armour in the belt, and finishing the bow and stern, is conducted in a manner similar to that described above. The various fittings in the hold, watertight flats, engine and boiler bearers, shaft passages, magazines, chain lockers, &c., and the works connected with the decks and topsides, the gunnery arrangements, ports, &c., are completed as the ship advances. The rudder and its fittings are generally prepared and fitted in place before the ship is launched.

From the preceding description it will be evident that no comparison can be made between this mode of conducting the work and the systems previously described, as they had reference to transversely framed ships while the iron-clads of the navy have combined transverse and longitudinal framing and a double bottom. It will be observed, however, that in respect of fitting the keel work on temporary blocks, riveting up the framing in place and proceeding with the plating while this is being done, getting in the beams, and some other details, the dockyard system approaches that adopted on the Mersey. It may be of interest to state that in the ships built for the navy by Messrs. Laird of Birkenhead the continuous transverse angle-irons and the deep frames behind armour have been prepared from blackboards in a manner similar to that described for ordinary transverse frames, only the double angle-irons have been riveted to the deep frames before they were put up. The short transverse frames have been prepared from wood batten-moulds, and the frame angle-irons and bracket plates have been fitted before being put in place. In the preparation of the keel work the ordinary course has been followed, the plates and angle-irons being fitted, punched or drilled, &c., on temporary blocks erected near the slip. The keel has then been removed to the permanent blocks and riveted, the general mode of proceeding with the framing of the vessel previously described (by first putting up and securing a tier of short transverse frames, then working a longitudinal, then fixing another tier of frames, and so on) being carried out. In the construction of the 'Audacious' and 'Invincible' Messrs. Napier of Glasgow have adopted a different course, as they first put up the continuous transverse angle-irons and faired them by ribbands worked upon the inside, and then brought on the bracket and plate frames, and the longitudinals.

CHAPTER XXI.

ARMOUR PLATING.

IT is proposed in this chapter to give an account of the ordinary modes of working and fastening armour-plates. The allied topics connected with the processes of the manufacture of armour-plates, and their distribution on the hulls of iron-clad ships would, while affording very interesting subjects for discussion, hardly fall within the limits of a treatise like the present, but may be more fully treated of in a future volume. Neither will any attempt be made to describe the almost innumerable projects which have been brought forward for improving the armour and the fastenings. The remarks made will, for the most part, be confined to the armour-plating of iron-built ships, and it will be assumed that the armour is supported by wood-backing worked upon the skin-plating carried by the vertical frames. While the information given will thus be of a strictly practical character, it has been considered desirable to afford the means of tracing the progress of armour-plating since its introduction into the ships of the Royal Navy, and for that purpose we have given at the end of this chapter a reprint of a paper by the Author of this work published in the *Transactions* of the Institution of Naval Architects for 1866, "On the 'Bellerophon,' 'Lord Warden,' and 'Hercules' Targets."

A disposition of the edges and butts of the armour-plating is usually made on the model of the ship on which the arrangement of the outside plating, &c., is drawn. In most cases the positions of the ports determine the positions of the butts of the strake of armour next below the port-sill, the butts being kept as clear of the ports as is possible. The arrangement of the butts in the other strakes of plating is regulated by those first determined on, the brick-fashion shift of butts being generally followed, but in some ships special dispositions have been made. The butts of the armour are nearly always placed over the vertical frames, as they are considered to be best supported when so situated. In the Royal Dockyards it is customary to prepare an expansion drawing of the armour-plating, the dimensions being taken from the mould-loft floor. This drawing guides the foreman in charge of the work, and is of use to the draughtsman in recording the dimensions of the several plates. In demanding the plates the breadths are carefully checked from the lines on the floor, and where there is considerable curvature and twist in a plate, as for instance under the counter of a vessel, special means are adopted in order to obtain a correct account of the dimensions and form. The practice in the Government service in such cases is to make a set of moulds representing sections of the outside of the ship, and to fix them in their proper relative positions so as to obtain an accurate representation of the ship's side in wake of the plating to be taken account of. The slight expense of materials and workmanship thus involved is found to be more than compensated in the accuracy with which the plates can be specified for. Drawings of the various plates are prepared, showing their forms and having the figured dimensions, thickness, and estimated weights marked on them, together with the distinguishing letter and number by which each plate is known. These drawings are forwarded to the manufacturers, and accompanied by printed forms on which are given the numbers and particulars of the plates ordered. In preparing the specifications for armour plates at the extremities of a ship it is usual to allow a margin of about $\frac{3}{4}$ inch on each edge and butt above the net dimensions; but for the plates amidships the allowance made is not so great, as they have only a moderate amount of curvature and twist. The ordinary dimensions of the plates used are a length of from 15 to 16 feet and a breadth of from 3 to 4 feet, but the dimensions of many of the plates necessarily vary greatly from these.

It has been previously explained that the disposition of the butts and edges of the skin-plating behind armour is regulated by that of the armour-plating, and an example of such a disposition with two thicknesses of skin-plating has been given in Fig. 137, p. 192. With one thickness of skin-plating the disposition is, of course, more readily made, but in all cases the same considerations regulate the arrangements, viz., to keep the edges of the skin-plating clear of the armour-bolts, and to shift the edges and butts of the thicknesses with each other, and with those of the armour. The butts of the skin-plating are commonly placed in the middle of the frame space, and the armour-bolts pass through the lines of the butts, the butt-fastenings being arranged previously in such a manner as to clear the bolts. In some ships the butts have been placed as close to the frames as was compatible with the allowance of space required for the butt-straps, and the armour-bolts have been thus made to clear the straps; in others of the iron-clads the butts of the skin-plating have been placed on the frames; and in some ships the butts of the inner thickness of skin-plating have been brought upon the frames, and those of the outer thickness placed between the frames. In all recent ships longitudinal girders have been worked upon the skin-plating in wake of the armour, their positions being to some extent governed by those of the edges of the armour-plates, from which they are usually distant about 12 inches. They have an average spacing of about 2 feet between the girders, being about the same distance apart as the vertical 10-inch frames inside the skin-plating. It has been previously remarked that the positions of these girders determines the disposition of the edges of the inner thickness of the skin-plating, as the edge fastenings are made to work in as fastenings in the girders. The wood backing is worked between the girders, there being usually two strakes of backing between each pair of girders. This is now the usual arrangement of the protected portion of the side of an iron-clad ship, but it will be remembered that in the earlier ships there are two thicknesses of wood backing with two continuous longitudinal girders (as shown in the section of the 'Warrior' given in Plate 3), and that in the 'Northumberland' class there is only one thickness of backing and no longitudinal girders. In both these cases it was considered very desirable to arrange the edges of the strakes of wood backing so that they might clear the armour-bolts, in order to prevent the liability to leakage through the holes which would otherwise have existed. With the present arrangement the same result is accomplished, but in ordinary cases no care is needed in arranging the breadths of the strakes of backing to clear the bolts, as they generally fall about 3 inches outside a girder and are consequently the same distance clear of the edge of a strake of backing.

Having thus briefly sketched the process of disposing the butts and edges of the armour and skin-plating, wood backing, &c., we pass to the illustration of the practical operations connected with armour-plating. In the preceding chapter an account has been given of the manner in which the skin-plating and the longitudinal girders are usually prepared and secured. When these operations are completed, the positions of the armour-bolts are marked upon the plating, so that all fastenings may be kept clear of them. Amidships the positions of the bolts can be easily set off as there is comparatively little curvature in the side, and the positions relatively to the girders are known. But at the bow and stern greater care is required, and the ordinary method adopted is to make a light batten-mould to the dimensions of the plate, to set off upon the mould the positions of the bolts, and, having put the mould in place at the ship with its outer surface in the position which the outside of the plate will occupy when it is fixed, to square in the positions of the bolts from the mould to the skin-plating. As the midship framing is nearly always advanced considerably beyond that of the extremities, the working of the wood backing and the armour-plating is usually begun amidships, the lowest strake of armour resting upon the shelf being the first put up, and the work being continued upwards and toward the extremities simultaneously, as the construction of the ship progresses. When the working of the skin-plating and girders is sufficiently advanced, the fitting of the wood backing is commenced. In some of the earlier ships the rivets in the skin-plating were snap-pointed, and great care was taken to accurately fit the backing over the rivet-points; but in some of the later iron-clads the rivet-points have been countersunk, and consequently the labour and expense of fitting the backing over them have been avoided. In all cases great care is required in order to make the backing bed fairly upon the surface of the plating, and fit well upon the edge-strips and stringers. All the faying surfaces of the backing are thickly coated with red lead, waterproof glue, or some other approved material, and all the joints are caulked and made watertight after the backing bolts have been driven. The ordinary fastenings in the backing consist of through-bolts with deck-heads sunk into the wood, secured upon the inside of the skin-plating by nuts hove up on screw-threads cut on the ends of the bolts. In order to prevent leakage through the bolt-holes, hempen grummets saturated with paint are placed between the nuts and the plating, and are compressed and made to fit tightly around the bolt by the heaving up of the nuts. The backing bolts are usually disposed so that there may be one in every frame space, and they are placed on opposite edges of the strake alternately, care being taken to keep them clear of the armour-bolts.

The backing having been fitted, bolted, and caulked, is made fair, and the preparations for taking account of the plating are commenced. It may be remarked here, that in general the plates are supplied to the shipbuilder planed to the sizes given in the demand or specification; but when the builders of the ship make their own armour-plates, they often leave the edges rough until the plates are to be worked, when they are lined to about 1 inch greater length, and ½ inch greater breadth than the net dimensions, and are either planed or slotted down. In order to render the following description of the method of taking account of an armour-plate as clear as possible, it will be assumed that the plate is of the ordinary dimensions, and that the curve and twist required are about the average amount, say 9 inches curvature lengthwise, or "bend," and 2 inches crosswise, or "dish," with a corresponding amount of twist. It will also be supposed that the ordinary case met with in practice is that under consideration, one of the butts and one of the edges of the plate requiring to be fitted against plates already in place. In taking account of such a plate five moulds would be made—two of these would be sectional moulds made to the butts, and two others would be sectional moulds made to the edges of the plate to be fitted; the remaining one would consist simply of three battens nailed together and fitted against the butt and edge of the plates already in place. The edges of the sectional moulds are all fitted to the backing, either close to the plates already fixed, or along the lines got in for the plate-edges and butts. In some cases the backs of all four moulds are planed out of winding in order to obtain the means of checking their correct application to the plate; but only the backs of the sectional moulds made at the butts are usually so planed. The ends of the moulds are cut to the bevelling of the butts and edges respectively, and marks are put upon them to distinguish the lower from the upper, and the fore from the after ends. It is also usual to mark a straight line upon the longitudinal sectional moulds before they are removed from the ship's side, in order to check their accuracy before they are used, as the light mould stuff is very liable to warp by exposure; this precaution is especially requisite when the back edges of the moulds are not out of winding. If the plate were a "shutter-in" of a strake both butts would require to be fitted against plates already in place, and a light batten-mould would be made for the purpose. If the plate were the last to be put on a broadside it would, most probably, require to be fitted to the edges and butts of four plates, and a batten-mould would be made to the outline of the space it had to fill. In both these cases the four sectional moulds would be made as described above.

Before being bent, the plates are heated in reverbatory furnaces constructed in such a manner as to prevent the flame from impinging upon and damaging the plate-edges. In order to allow the plates to be moved easily, it is usual to lay them upon low iron carriages which are made to travel on rails and can be readily drawn out of, or pushed into the furnaces; in some cases these carriages form a portion of the furnace bottom. In most workshops there are special arrangements made for conveying the plates from the furnaces to the bending machines or the cradles; and the latter are usually placed as near as possible to the furnaces in order to reduce as much as possible the time and cost of the conveyance of the plates. Unless proper attention is paid to all the details of the process of heating the plates it is impossible to obtain the ductility and softness necessary for bending. The conclusion arrived at by the Iron Plate Committee was, that from good red heat to bright red were the safest limits of temperature, and that with proper precautions the quality of the iron will be rather improved than otherwise by the heating of the plates, on account of the annealing effect produced. Great care needs always to be taken that the heating is done gradually, in order to secure greater uniformity of temperature throughout the mass than would otherwise be obtained; and the temperature of the furnace must never be raised so high as to injure the iron. It is usual to keep the furnace-door closed during the whole time occupied in heating a plate, and to allow the damper to remain in one position. The time which the plate is kept in the furnace depends, of course, upon its thickness and upon the temperature of the furnace when the plate is put in. Under ordinary circumstances, from 3 to 5 hours may be taken as a fair average of the time occupied by the operation, but in every instance the appearance of the plate is that which guides the workmen in determining when it is sufficiently heated. It should be added that the work of taking account of the plate and making the moulds, is usually performed during the time that the heating is being proceeded with.

There are two modes of bending armour-plates now in general

use; the first by hydraulic pressure, and the second by what is known as the "wedge and tup" or "cradle" system. When the first method is adopted, powerful hydraulic presses are employed, and the curvature and twist required in the plates are given by means of cast-iron blocks, known as "packing," the sectional forms of these blocks being such as to approximate to the curves to which the plates have to be bent. When placed in the press, the lower surface of the plate rests upon a cast-iron slab carried by the piston, the upper part of the slab being curved concavely to the form to which the plate is to be brought. The packing is then piled upon the plate and the upper surface of the blocks being pressed against the framing at the upper part of the machine, when the piston is forced upwards, the plate is gradually bent down until it fits against the curved edge of the slab beneath it. It may be remarked that the outer surface of the plate is that which is generally placed lowest in the press, and that the packing for plates of ordinary curves and twists can be selected from a stock kept at hand, those required in extreme cases being the only ones specially prepared.

The cradle system of bending is that most generally adopted both in the private and Royal yards. A sketch of a cradle with an

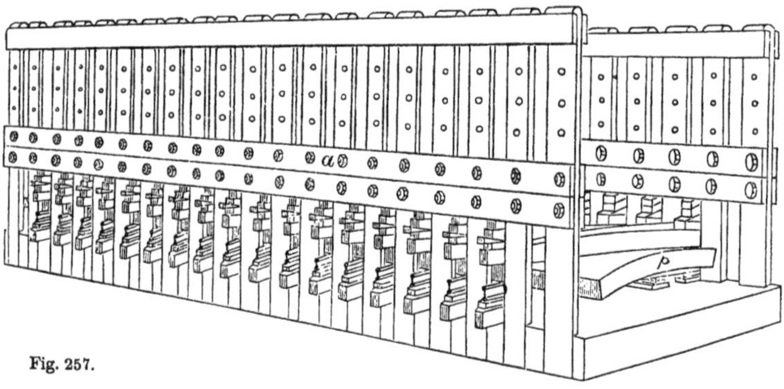

Fig. 257.

armour plate fixed in it is given in Fig. 257. The sides are formed of vertical bars of which the lower ends are secured to a cast-iron slab, and the upper ends are stiffened by longitudinal plates. The vertical bars are pierced by numerous holes, and the longitudinal bars marked *a* can be shifted up and down and secured in the most convenient position. The cradle is placed near the heating furnaces in most yards, and arrangements are made to facilitate the conveyance of the plates from the furnaces to the cradle, as remarked above. In preparing the cradle for bending a plate it is first necessary to determine the position in which the plate is to be placed. Two bars are then procured, of which the upper edges fit the curves of the sectional moulds made at the butts of the plate. These bars or packings are usually about 6 by $2\frac{1}{2}$ inches, and are of sufficient length to span the breadth of the cradle. A large number of them is kept in stock, and it is possible, in most cases, to select bars which will serve the purpose; but when this cannot be done the bars are bent to the required shape. The two bars selected to fit the sectional moulds are next put in the positions which the ends of the plates will occupy in the cradle, the moulds are applied upon them, the back edges brought out of winding, and the bars are wedged up as shown in the sketch. The longitudinal sectional moulds made to the plate edges are then applied upon the cross bars in the proper positions (determined by the other sectional moulds), and the remainder of the cross packing bars can then be put in place and wedged, their upper edges being kept close to the edges of the longitudinal moulds, and the curves of the various bars being so adjusted as to give a fair surface on which to bend the plate. These preparations having been completed, and the plate being sufficiently heated, it is drawn out of the furnace and conveyed to the cradle, where it is placed in the position previously determined. Two strips of plate-iron (about 5 inches wide and 1 inch thick) are then placed upon the armour plate near the edges, and the upper cross bars shown in the sketch are put in place; as their lower edges bear upon the strips, the heated plate, *p*, is preserved from being indented by them when the workmen drive in the wedges shown between the cross bars and the plates marked *a*. By means of these wedges the plate is gradually forced down upon the bed formed by the cross bars, beginning at the middle of the length, and working out towards the ends. Care is always required to bring the plate well down on each cross bar, as it would otherwise require considerable adjustment afterwards. The whole of the operation must be performed as quickly as possible, as the difficulty of bending the plate increases very rapidly with the diminution of the temperature. It is worthy of remark that the inside of the plate is that bearing upon the packing bars, as this is a point on which the advocates of the cradle system lay some stress.

With respect to the comparative merits of these two modes of bending, various opinions are entertained. For ordinary curves and twists it appears that the bending can be performed more cheaply by hydraulic pressure, when the operation is conducted by men who have mastered the working of the machine; but for plates having considerable curvature and twist the cradle system is generally preferred. The latter mode of bending is also more readily understood by the workmen, and less is left to their skill and judgment than is the case with the hydraulic press. It is also worthy of remark that in this, as in many other instances, the men are sometimes prejudiced against the use of the press on account of the reduction made by its adoption in the amount of manual labour required.

The work of bending can be done much more expeditiously with the cradle, but the plates generally require to be adjusted at the press after cooling, and this considerably reduces the difference between the times occupied by the two methods; in some instances no adjustment is required, but these cases are exceptional. When plates are bent by hydraulic pressure they leave the press finished. In order to afford a more definite idea of the times occupied in bending a plate by the two methods, the following examples are given, the particulars being taken from actual practice. A plate 15 feet 6 inches long, 3 feet 6 inches broad, and 5½ inches thick, to which an ordinary amount of curvature and twist is to be given, would require 8 workmen, including a leading man, to be engaged for 10 hours in bending it by the press; whereas it might be bent in the cradle in from 15 to 25 minutes by 16 men and a leading man, and would afterwards occupy the men at the press about 6 hours in its adjustment. An allusion was made above to the effects supposed to be produced on the material by the two methods of bending. The advocates of the cradle system state, that as the wedging is all done on the upper surface of the plate, the curvature and twist are obtained almost entirely by the compression of the material between the centre of the thickness of the plate (or the neutral axis) and the lower surface. They consider that by this method the extension of the material on the other side of the neutral axis is considerably less than it would be if the pressure were applied on the inside of the plate, as is done in the press; and think that a deterioration in the material in the outer portion of the plate is of much greater importance than any which may occur in the inner portion. While there may be some truth in this view, especially in the case of plates which have to be bent to sharp curves, it hardly appears to have sufficient weight to lead to the preference of one mode of bending before the other. The principal reasons which determine the practice of various ship-builders are rather those connected with the time occupied and the cost of workmanship involved in bending, than any injury done to the material in the armour plates.

Before leaving this part of the subject, it may be well to call special attention to the necessity which exists for the exercise of the greatest care in ensuring accuracy of form in the plates before they are removed from the press; for, should they not be found to fit the ship's side when put in place, it would be necessary to take them back to the press to readjust them, and in all probability the edges would require to be re-planed, as an alteration in the curvature would affect their bevelling.

After being bent to the required form, the plate is lined to the proper shape by means of the batten-mould previously described. The positions of the armour bolts are then marked upon the plate, it is taken to the machine shop, the edges and butts are planed, and the holes drilled and countersunk. In some cases the edges and butts of the plates have been chipped and filed by hand, but this can hardly fail to be a more expensive and less satisfactory mode of performing the work than by the use of planing machines. When there would be considerable curvature in the length of a plate, if the edges were made to coincide exactly with the lines got in upon the model, it is usual to plane the edge in two straight lengths which are so arranged as to approach the curve as nearly as possible, and to bring the angle which is thus formed in the edge, directly over or under the butt of the plates in the adjacent strake. In a few of the earlier iron-clads the edges of the armour plates were tongued and grooved (as shown in Plate 3), as it was thought probable that greater resistance would thus be offered to the turning up of the edges when the plates were struck, and that a stronger connection would be made. The experiments conducted by the Iron Plate Committee showed, however, that this was a mistake, and that the arrangement rather caused mischief than otherwise. In later ships therefore the plate edges are plain, as shown in the section of the 'Bellerophon' in Plate 4.

When the preparation of the armour plate is completed it is

conveyed to the ship and put in place. If the butts and edges are found to fit against those of the plates already worked, and to coincide with the lines got in for the breadth and length, the bolt holes are bored through the wood backing and drilled through the skin plating. The plate is then taken down and any little fairing of the wood backing which may be required is performed. The holes in the skin-plating are rimed through from the inside in order that they may be a little larger than the bolts, so that the screw-threads may not be injured when the bolts are driven.

This is the mode of conducting the work most commonly practised; but in some yards it is the custom to first set off the positions of the armour bolts upon the skin-plating, after having lined in the plate edges, &c., and then to drill a small hole (about ½ inch) through the skin. The backing is next worked and the holes previously drilled are continued out through it, care being taken to make them square to the side. The mould or template for the armour plate is then put in place, the positions of the bolt-holes are marked upon it and transferred to the plate. When the holes have been drilled and countersunk, and the edges and butts planed, the plate is put up and the holes are rimed to their full size in the backing and skin-plating. Any slight inaccuracy which may occur in marking or transferring the centres of the holes is corrected by the process of riming; and it is thought that this method ensures a greater exactness in the positions of the holes in the skin-plating relatively to the fastenings. The method adopted in taking the account for bending the plate is similar to that described above.

A thick coating of tar, paint, waterproof glue, or some other material, is put on over the surface of the backing, and, the faying side of the plate having also been coated, it is put up in place and temporarily secured by shores while the fastenings are being driven. It may be added that waterproof glue is now used for coating the backing and the plates of all the ships of the Royal Navy, and that it is usual either to slightly warm the plates, or to thin down the glue, in order to increase the adhesion of the plates to the side.

Having thus briefly illustrated the common methods of taking the account of, bending, and fitting armour plates, we turn to the consideration of the modes of fastening. In nearly all cases the fastenings consist of conical-headed bolts countersunk in the plates. Great objections have been made to this arrangement on account of the reduction in the strength of the plates supposed to be caused by the bolt-holes. The Iron Plate Committee consider, however, that the experiments made at Shoeburyness prove that with soft iron, such as is preferred for armour plates, the reduction in strength is not at all serious. This experimental proof, of course, removes the basis on which have rested most of the ingenious but expensive propositions for attaching armour without the use of bolts passing through the plates. The two methods of fastening now in use in ships such as we are considering, which have the armour worked upon wood backing, are screw-bolting and through-bolting.* Screw-bolting has been almost universally adopted in the French iron-clads, of which the greater number are wood-built, and has also been employed in some ships built for foreign Governments in this country, among others in the iron-built Italian turret-ship 'Affondatore.' On the French bolts the screw-thread is raised above the shank, as shown in Fig. 258 instead of being cut into it

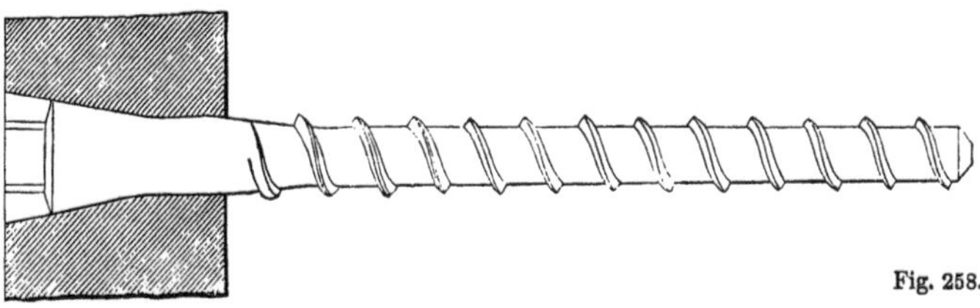

Fig. 258.

in the usual manner; the heads are countersunk in the armour, and the points are screwed into the wood timbering or backing. In order to allow the bolts to be screwed in, square projecting portions are formed upon the heads upon which spanners can be fitted, and when the bolts have been hove up, the countersunk holes in the armour are cemented flush with the surface. No screw armour bolts are used in the ships of the Royal Navy except in places where through-bolting is impracticable (as, for instance, in wake of waterways and other work in the interior), and the screw-bolts used are of a different pattern from the French, having a much

* In the construction of the iron-clads of the United States Navy, where the armour has been made up of several thicknesses, the plates are connected together by means of through-rivets, and the fastenings to the side consist either of through-bolts clenched upon rings inside the ship, or of conical-pointed bolts driven partly through the timbering of the side. These arrangements are of a special character, and are, therefore, only mentioned in passing, not having been adopted in this country or in France.

finer thread. A trial was made at Shoeburyness in 1864 upon what is known as the "small plate" target, of which the fastenings were composed of screw-bolts on the French pattern. In their Report the Committee stated that this mode of fastening proved a most marked success, and recommended that further experiments should be made in order to test the general applicability of this mode of fastening. The great objections to screw-bolting are thought to consist in the facts that the bolts are torn out of the wood by a less strain than is required to fracture them; that there would be great difficulty experienced in getting the bolts out after they had been in place for a considerable time; and that in an iron-built ship the plates would be simply secured to the wood backing.

Through-bolting is the kind of fastening generally employed in this country, and undoubtedly gives greater strength to the structure than screw-bolting, as it better combines the various parts. A sketch of an armour bolt, similar to those used in the ships of the Navy, is given in Fig. 259. The head is countersunk about half through the armour, the ordinary rule observed being as follows:—Depth of countersink equal to $1\frac{1}{8}$ times the diameter of the bolt, and the diameter of the head at the surface of the plate, equal to $1\frac{1}{4}$ times the depth of the countersink. Upon the end of the bolt a comparatively fine thread is cut, it having been found that bolts with fine threads will stand much better than those with coarse threads, as would be anticipated from the greater sectional area of the bolts left in wake of the thread when it is fine. There are, of course, limits to the fineness, on account of the liability to strip the thread which is experienced when it becomes very shallow. Threads cut in the lathe, or chased, are preferred to those made by a die, as they can be run out gradually, or made to "die out." The part of the bolt on which the screw-thread is cut is usually made about $\frac{1}{8}$ inch less in diameter than the shank, in order to give facilities for driving the bolt without enlarging the hole in the wood backing

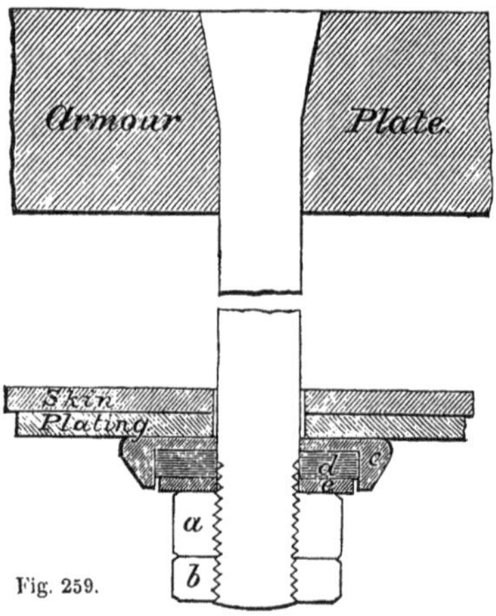

Fig. 259.

to such an extent as to render the fastenings comparatively slack when driven. There are two nuts on each bolt similar to those marked *a* and *b* in the sketch, the latter being intended to serve as a lock or preventer nut to *a*, and to secure it from being loosened by the impact of projectiles. It will be remarked that between the nut *a* and the skin-plating there is interposed an arrangement known as the "elastic cup-washer." This has been introduced since the construction of the earlier iron-clads in which the nuts were screwed up on simple iron plate-washers. Experience has shown that unless some elastic substance is placed between the nuts and the skin-plating, the ordinary through-bolts are very liable to be broken under the great strain suddenly brought upon them when the side is struck. In nearly all cases fracture has taken place in the screw-thread, where the weakest section of the bolt is found, and the nuts and bolt-ends have been scattered on the inside, thus greatly increasing the probability of casualties occurring. The elastic cup-washers have been introduced for the purpose of supplying sufficient elasticity to enable the material in the bolts to withstand these sudden strains, and the results of the experiments made at Shoeburyness prove that the object aimed at has been satisfactorily attained. Before describing this kind of washer, it may be well to state that other washers, formed of cork and other materials, have been tried, but have not been found as efficient. The cup-washers are arranged as follows:—A hexagonal wrought-iron cup, *c*, is put on over the point of the bolt in the same manner as a common plate-washer; within this cup and around the bolt a washer, *d*, of vulcanized india-rubber is fitted, its thickness being rather less than the depth of the cup; upon the washer *d* a plate-washer, *e*, is placed, and when the nuts are screwed up it forms a loosely-fitting cover to the cup-washer *c*, thus protecting the india-rubber washer *d* from the action of the atmosphere, and forming a base on which the nut *a* bears. The washer *c* serves to prevent the lateral extension of the india-rubber when it is compressed by the heaving up of the nut *a*, and the hexagonal shape of the cup is adopted in order to prevent the india-rubber washer from rotating during the process of screwing up. The elasticity of the washer *d* becomes available when the side is struck and prevents the bolt breaking, while the elastic force due to the compression of the india-rubber, is, so to speak, stored in the washer, and tends to keep the nuts tight even when they have been shaken by the blows on the side. Another advantage of this arrangement is, that when the bolts are not driven exactly square to the skin-plating the india-rubber readily accommodates itself to the inclination of the bolt, and forms a fair as well as a yielding surface on which the nut bears by means of the washer *e*. The strain is thus made very nearly uniform at all parts of the nut, a result which it is hardly possible to attain with the common plate-washers, even with the most careful workmanship. It may be of interest to state, that experiments made on the 'Bellerophon' target at Shoeburyness show that with elastic cup-washers the preventer nuts similar to *b* might be dispensed with; in practice, however, these nuts are usually fitted, as the strength of the fastenings is undoubtedly increased by this means.

Before describing the manner in which through-armour bolts are made watertight, it may be well to call attention to a few facts connected with the operation of driving the bolts. It is usual to give a "drift" to the bolts of from $\frac{1}{8}$ inch to $\frac{1}{4}$ inch, that is, to have the diameter of the holes in the wood backing less than that of the bolts by these amounts. It has been previously explained that the diameter of the bolt in wake of the screw-thread is made less than that of the other part of the shank in order that the latter may fit tightly. In driving the bolts it is usual to employ a "monkey," similar to that shown in Fig. 260, which generally consists of an

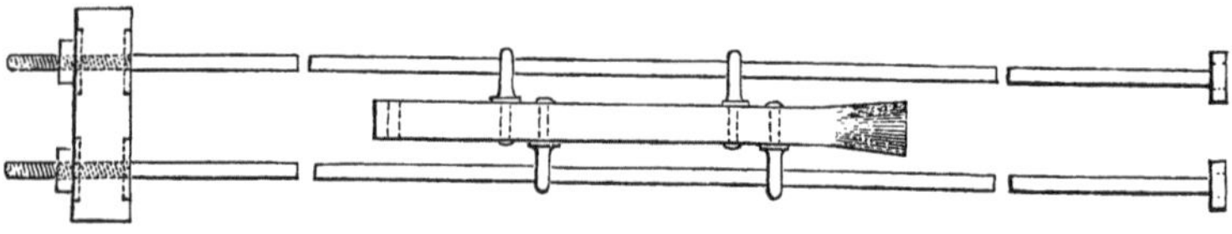

Fig. 260.

iron bar about 40 inches long by 3 inches square, made to traverse on two guide rods by the workmen pulling on ropes attached to each side. This machine is employed also in driving all the longest bolts in a wood ship, and is found to strike a smarter blow than a maul with less manual labour. After a bolt has been driven, a hempen grummet saturated with red lead is threaded on the point upon the skin-plating, the washers are put on, and the nuts screwed up. The pressure thus brought upon the grummet forces it closely against the plating and around the bolt; the elasticity of the india-rubber washer afterwards tending to keep it in position and to keep the bolt-holes watertight, even when the nuts have been loosened. These hempen grummets are in general use in

H. M. Service, but if it were deemed preferable they might be dispensed with, the india-rubber washer being formed as shown in Fig. 261. The hole in the cup-washer would then require to be enlarged to admit of the introduction of the lower part of the washer, marked *a a*, and when the compressive strain caused by screwing up the nuts was brought upon the washer, the latter would be pressed against the bolt and plating, and the hole made watertight. No special appliances are used for screwing up the nuts, but the workmen simply employ a spanner with a moderate length of leverage, as this is considered to ensure sufficient tightness.

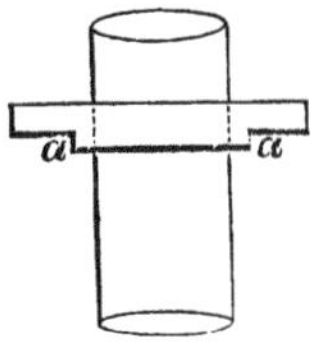

Fig. 261.

The preceding description applies to the ordinary through-armour bolt, but there is another kind of bolt, proposed by Major Palliser, which well deserves consideration. The great feature of the proposal consists in reducing the diameter of the shank of the bolt for a portion of the length to about an equal size with the smallest diameter in wake of the screw-thread. An ordinary plate-washer is fitted under the nut, and the screw-threads on the bolts are much finer than those usually employed. This plan is an application of the well-known principle that a screwed bolt is much less liable to break under a suddenly applied strain, if a portion of the shank is reduced to an equal sectional area with the iron left uncut at the thread. By this reduction a comparatively long space of uniform strength with the unavoidable weak section of the bolt in wake of the thread is provided, and the bolt-stretching freely in this space performs an amount of work equivalent to the suddenly-applied strain. If, on the other hand, the bolt shank is of uniform diameter, and a screw-thread is cut on a portion of the length, there is scarcely any elongation, and the bolt usually breaks off at the thread, unless some arrangement, such as an elastic washer, is made to prevent it.* Major Palliser has made numerous

* Mr. Chalmers has proposed another kind of armour bolt which would be in accordance with the principle above stated. The bolt would be made of two half-round pieces of iron with a piece welded in between them for the head, and a thin piece (about ½ inch thick) welded in at the point in wake of the screw-thread. A slot would thus be left at the centre of the bolt extending from the under side of the head to the bottom of the screw, and this it is proposed to fill in with wood. Other proposals have been made having the same object in view. Mr. Crampton proposes to turn down the bolt at several portions of the shank, and thus allow it to fill the hole at many parts while the weakened sectious would admit of elongation taking place. Mr. Paget, Mr. Hughes of the Millwall Ironworks, and Mr. Parsons, have proposed to reduce the sectional area of a portion of the length by boring out a hole in the centre.

experiments with bolts in order to test this principle, and a record of the results of those experiments, and of the trials of this kind of fastening made at Shoeburyness, will be found in the *Transactions* of the Institution of Naval Architects for 1867. Experiments have also been made at Chatham dockyard with reference to Major Palliser's proposal, the particulars of which are given in a paper by Mr. Barnaby in those *Transactions* for 1866. The following are the results deduced from the latter experiments:—

1. That iron bolts of good quality and of uniform diameter, subjected to a steadily increasing strain, before breaking elongate about one-fifth of their original length.
2. If the diameter is not uniform, but is decreased through a portion of the length, then the reduced part elongates about one-fifth of its length before breaking, and the larger portion scarcely stretches at all.
3. If this reduced part is very short, as in the thread of a screw, the strain required to break the bolt is the same per square inch of the unstretched section as in the previous cases, but there is scarcely any elongation before rupture.
4. If the whole length of the bolt is made to the reduced diameter of the screw-thread, so that the thread projects from the bolt, the breaking strain (gradually applied) is the same as before; but, as the bolt will stretch one-fifth of its length before breaking, it becomes thereby less liable to rupture by a sudden blow, because the work done in producing rupture is proportional to the work performed, *i.e.* to the weight or strain applied multiplied by the elongation.

From the preceding remarks it will be seen that Major Palliser's plan would prevent the bolt from being broken under the jarring strains suddenly brought upon the nuts by the impact of projectiles, by means of the elongation of the bolt itself. The bolt would thus be permanently increased in length, and consequently reduced in sectional area in the stretched part, although it would not be broken. With the elastic washer arrangement the requisite elasticity is given to the fastenings without altering the lengths or the sectional areas of the bolts, and the fitting of new washers after an action would restore the efficiency of the bolts; whereas bolts on Major Palliser's plan would either require to be replaced or be reduced in strength. The proposed kind of bolt has the practical disadvantage of being made to fill the holes and to be watertight, only by some special arrangement. The method of doing this proposed by Major Palliser consists in putting on a metallic casing around the reduced portion of the bolt, in order to bring it up to a uniform diameter throughout. This plan is, of course, attended with a certain amount of expense, and requires great care to be exercised in order to ensure its success. Watertightness is not required in the fastenings of armour plating on land fortifications, and this kind of bolt can therefore be employed without being cased. It has been used in the shields for Gibraltar and Malta, but on the occasion of the first trial the bolts gave way to such an extent as to lead the Committee to consider it unfair to the structure to proceed with the trial. Major Palliser attributes the failure of the bolts to the method of manufacture, as he considers that the iron in the head of a bolt is considerably injured by being "upset" or "jumped" in order to form the head. It should be stated also that a part of the original proposal made by Major Palliser consisted in the manufacture of the bolts by drawing them down under a hammer from bars of the diameter required for the heads, this plan being adopted in order to avoid upsetting the iron in the heads. The opinions of those who have superintended the manufacture of armour bolts are, however, in most cases opposed to this view, and it is considered by these gentlemen that experimental proof has been obtained that the ordinary mode of forming the heads does not materially, if at all, weaken the bolts. It may be interesting to add, that before the second trial was made on the Gibraltar shield new bolts manufactured on Major Palliser's plan were substituted for those which had failed, and other changes were made. Full particulars of these will be found in the Report of the Special Committee, who consider that the trial proved that the alterations adopted had considerably assisted the bolts, and had effected a marked improvement. It seems, however, that this improvement was partly due to the precautions taken against the bolts being sheared, to the adoption of a different quality of iron in the manufacture, and to the lengthening of the reduced portion of the bolt-shank. Further trials are required in order to decide the comparative merits of the Palliser bolt, and the ordinary bolt with the elastic cup-washer; but at present it is considered preferable to adopt the latter in the armour fastenings of the ships of the Royal Navy.

The question of the proper diameters of armour bolts required for different thicknesses of plating is still very unsettled. At first the bolts and nuts employed were comparatively small, and the fastenings were few in number; but the trials conducted by the Iron Plate Committee have shown the desirability of increasing both the size and number of the fastenings. In order to afford information of the present practice of H. M. Service on both these points, the following particulars of the armour fastenings of the 'Hercules' are given. The 6-inch plates are secured with $2\frac{3}{4}$-inch bolts, the diameter of the heads being $3\frac{1}{2}$ inches; and the 8-inch and 9-inch plates are fastened with 3-inch bolts with heads having a diameter of $4\frac{1}{4}$ inches. The holes in the plates to receive the bolt-heads are countersunk one-half through the plates; the nuts, &c., are of the ordinary pattern described above, with elastic washers. The bolts are 9 inches in from the plate edges, and 1 foot in from the butts; but in order to give additional support to the butts, another bolt is introduced at about 16 inches from each butt on the centre line of the plate. There are two bolts in each plate between every two frames, and as the frame space is 2 feet, this is also the distance between the fastenings on each edge. The centre of the plate is thus left unsecured and unpierced, the arrangement of the fastenings being such as best resists the tendency of the plate edges to turn outwards when the side is struck.

Reference has been made to the small-plate target in which the French screw-bolt fastenings were tried by the Iron Plate Committee. The plates used were smaller than those employed in this country (about 5 feet 9 inches by 2 feet 6 inches), their thicknesses being $4\frac{3}{4}$ inches and $5\frac{9}{10}$ inches respectively. The bolts were only $1\frac{1}{2}$ inch diameter, and were arranged very differently from those of the 'Hercules.' They were placed in three rows, one of which came on the centre line of the plate, and the other two 5 inches in from the edges. The bolts in the outer rows were opposite each other, and those in the middle row were placed midway between them. The trial was considered favourable to this mode of fastening; but, as remarked above, further experiments were thought necessary in order to determine whether it was generally applicable, and it is not without serious drawbacks for iron-built armour-clads. Even in the case of wood-built vessels there will probably be great difficulty in getting these screw-bolts out after they have rusted and wasted in the timber.

In the preceding part of this chapter the various processes connected with the preparation and fixing of armour plating have been fully described, and it need only be added, in conclusion, that after the bolts have been driven the edges of the plates are caulked in the same manner as the butt joints of bottom plating, illustrated by Fig. 256, p. 438. The bolt-heads are also chipped fair with the surface of the plates, and any slight projections at the edges and butts of adjacent plates are chiselled away. As soon as these operations are completed, it is customary to lay on a good coat of paint in order to preserve the surface of the plating from oxidation.

ON THE 'BELLEROPHON,' 'LORD WARDEN,' AND 'HERCULES' TARGETS.*

THE object I have in view in this Paper is to place before the Institution a simple record of facts concerning the latest adopted forms of naval structures intended to resist shot and shell. I propose to describe the manner in which the targets above named were constructed, and the principal reasons why their several forms of construction were successively adopted. My sole object in submitting the Paper is to contribute definite, and perhaps useful information to the Institution, believing, as the originators of the Institution always have believed, that the bringing together of exact facts, and the record of actual experience, are among the best and most valuable functions of a scientific body like ours.

If we neglect the smaller iron-clad vessels of our navy, such as the 'Scorpion,' 'Wyvern,' 'Enterprise,' 'Favourite,' and some other vessels of like tonnage, and if we neglect also the wooden frigates which are plated with 4½-inch armour only—and we may without disadvantage neglect all these in studying the development of our iron-clad fleet with a view to future action—we may say that five different, or rather modified, systems of construction have been successively adopted in Her Majesty's iron-plated ships, these systems being known by the names of the five ships, 'Warrior,' 'Minotaur,' 'Bellerophon,' 'Lord Warden,' and 'Hercules.' We might, in strictness, neglect the 'Lord Warden' system, as it is essentially associated with wooden hulls, the construction of which, for receiving armour-plating and powerful engines, will no doubt soon cease in this country, if it has not already ceased. But as there are features of construction which are not without interest in the 'Lord Warden' system, I have given it a place in the present Paper.

Permit me, however, first to remind you of the features of construction adopted in the 'Warrior's' and 'Minotaur's' sides.

Neglecting the small floating batteries built during the Russian war, the 'Warrior' was the first iron-clad ship designed in this country. In constructing her side to resist shot and shell, her designers had before them a most difficult and obscure problem to solve, and it is gratifying to know that they

* From the *Transactions* of the Institution of Naval Architects for 1866, vol. vii.

solved it with marked success. I am of opinion, that it could not then by possibility have been better or more happily solved; for the 'Warrior' target remained unequalled long after other minds had been brought to bear upon the problem, and after other targets had been built in the expectation of surpassing the original structure.

The 'Warrior's' side and the 'Warrior' target (for the two structures are alike), were composed, as most of you doubtless know, of a series of vertical iron frames 10 inches deep, and about 2 feet apart, covered with a skin-plating nine-sixteenths of an inch thick, against which were placed two thicknesses of teak, the inner 10 inches thick, and the outer 8 inches, the whole being faced with a 4½-inch armour plate. A section of the target is given in Fig. 262.

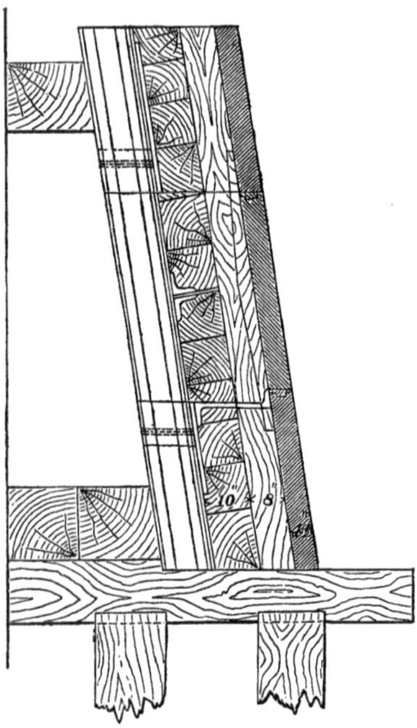

Fig 262.

There were also embodied in this target two subordinate, but nevertheless important, components or features, which I request you particularly to observe, viz., the double skin-plating above and below the line of ports, and the external stringers upon the iron frames between the ports. These portions of the structure, which I shall refer to again hereafter, were no doubt introduced with a very careful regard, both to the structural strength of the hull, and to its resisting power when struck by shot and shell.

I regret that I have not time to detail the nature and results of the trials to which this well-constructed target was subjected on more than one occasion. I must however state, that it was signally superior to all others that had been tried up to March, 1862; and I must also express my strong conviction that the 'Warrior' target was not, as some have supposed, a lucky hap-hazard combination of parts, or a mere attempt to imitate a wooden ship's side in connection with an iron hull; but, on the contrary, a highly skilful and scientific construction, carefully designed in view of the objects which had to be accomplished. It was, and is, and will remain, a remarkable illustration of the ability which is brought to bear in this country upon even the most novel and difficult mechanical problems as soon as their solution is felt to be necessary.

The 'Minotaur' target differed from the 'Warrior' mainly in the reduction of its wood backing, and in an increase of equivalent weight in the armour. A single layer of 9-inch teak and armour plates 5½ inches thick were used in this, the frames and skin-plating remaining about the same. A section of this target is given in Fig. 263. For a long time it was supposed that this target had proved much inferior to that of the 'Warrior,' and there were not wanting persons to publicly, and strongly and repeatedly, censure the departure that had been made from the 'Warrior' system. I must confess that I was never able to join in that censure myself, and when it became my duty to consider, with the Controller of the Navy and his officers, how the 'Bellerophon' might best be built in this respect, we ventured to adhere to the reduced thickness of wood backing and the increased thickness of armour, notwithstanding the outcry against them. I am happy to be able to state what, perhaps, many gentlemen present may not yet have heard (for it is ill news that flies apace, and not good news), viz.,—that all the gloomy and disparaging comparisons which were drawn between the 'Warrior' and 'Minotaur' targets have recently proved to be in error, it having been discovered that what is known as "2 A" powder was used with two out of three rounds of 150 lbs. cast-iron spherical shot, which were fired from the 10½-inch gun, at the 'Minotaur' target, the effect of using this powder having been to raise the striking velocity of the shot from 1,620 feet to 1,744 feet per second. The change in the powder was made (I know not how or why) immediately after the first round, and invalidated all the comparisons that were made in and after the report of the trial. The 'Minotaur,' 'Agincourt,' and 'Northumberland' are now known to possess much greater strength than has been supposed, and are in all probability at least equal to the 'Warrior' in that respect. When the great cost of these large ships and the time which has been required for building them are considered, it must be highly satisfactory to the country to learn that no mistake was made in designing their armour, and that they are really as stout and strong as their designers proposed.

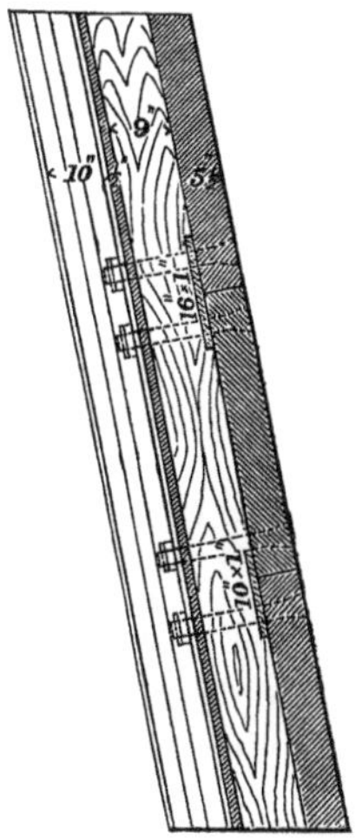

Fig. 263.

I have now to describe to you the 'Bellerophon' target, of which a section is shown by Fig. 264; and in order to make the principles of its construction clear, I must mention the two points in reference to which the 'Warrior' and 'Minotaur' targets appeared to me susceptible of improvement. It seemed, first, that a great addition to the general stability and strength of the structure might be secured if the strong vertical iron frames of the ship were crossed horizontally by other frames of approximately equal strength, and spaced like the vertical frames; and, secondly, that the risk of shot or shell passing through the structure, between the frames, would be greatly reduced, and the resistance of the frames much more effectually elicited, wherever a shot or shell might strike, if the skin of the ship were considerably thickened. In other words, it appeared highly desirable to extend, throughout the entire structure, that double skin-plating, and those external frames or stringers, which had already been introduced, as we saw a minute ago, in the weakened portions of the 'Warrior' target. These features constitute the characteristic merits—for

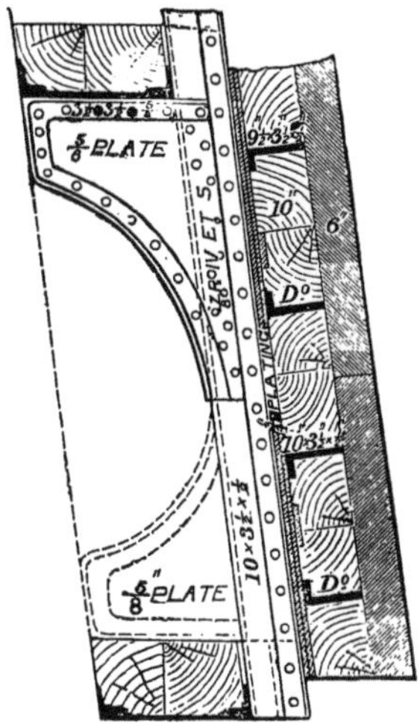

Fig. 264.

they proved on trial to be merits—of the 'Bellerophon' target; and it is a pleasure to me, and not by any means a subject of regret, to know that the germs of these improvements may be traced in the structure designed by my predecessors. By virtue of these we secure many important objects. The combined horizontal and vertical 10-inch frames, connected by the doubled skin of ¾-inch iron, constitute an enormously strong and rigid structure, eminently well adapted to sustain the armour under all circumstances, while both the doubled skin and the external stringers (to which we fitted butt-straps in the 'Bellerophon' herself), increase the longitudinal strength of the ship to a most unusual extent.

It will complete the general description of the 'Bellerophon' target when I state that the armour was 6 inches thick, and the teak 10 inches; and that instead of forming the external frames or stringers of a plate and two angle-irons, as was done in the 'Warrior,' we formed them of one large angle-iron 10 by 3½ inches.

You are now in a position to understand the true reasons that existed for riveting external stringers to the outside of the 'Bellerophon's' skin-plating, and you cannot fail to see how little the adoption of that arrangement had to do with the notion of giving direct support to the armour plates. I mention this because it has been supposed, and stated publicly on many occasions, that these edge plates were adopted in imitation of a quite different system, and with a view of rigidly backing up the armour. This, however, is wholly a mistake; for much as I, for one, should like to banish the teak from our iron-clads, and to make their hulls of iron throughout, I am of opinion that a rigid iron backing has many disadvantages. In fact, so far were we from valuing these edge plates as direct armour supports, that we caused them to be reduced in depth behind one of the plates of the target, and to a large extent in the ship also, expressly in order to keep them from too immediate contact with the armour; and we did so because it appeared undesirable to bring the force of a blow so directly and fully upon that portion of the hull proper of the ship which is immediately in front of the shot, as these plates would otherwise tend to bring it, especially if placed closer together. We put armour upon a ship to protect the hull, which we require to preserve from the blow as effectually as possible. A very rigid backing, in direct contact with the skin of the ship, must obviously transfer much of the shock of a shot to that skin; whereas a moderately yielding backing allows the force to expend itself upon the armour which is put there to receive it, and thus protects the skin from its violence. This is a very important point, and one upon which too hasty opinions may easily be formed. I have given the most careful consideration to the matter, and have seen many corroborations of the soundness of the views here expressed. There is one test which is easily applied, but which is usually applied in a manner the very reverse of what it should be. It is this: wherever you see an armour plate that is supported by close rigid edge-plates struck upon a line of support, you will find that the armour plate is comparatively but little injured, and on removing it from its backing you will find that the edge plate has scored more or less deeply into the back of the armour plate. Now what does this point to? To the fact that the edge plate has been driven back with violence upon that which supports it, viz., that very skin of the ship which you desire to preserve intact. If the external frames or stringers of the 'Bellerophon' had been situated within a few inches of each other, I should have considered this circumstance so serious as to destroy all prospect of success in carrying out the plan; but with the frames 2 feet apart it is not so, as the isolated edge plate of ½-inch iron buckles up under the blow before it can injure the skin. I will only add on this head that in expressing the foregoing views I am not neglecting the consideration that closely situated edge plates must tend greatly to distribute the blow: I am well aware of that fact: but the answer to it is that the time allowed for distributing the force is very short, and that so far as they distribute it at all, they distribute the blow upon and over the skin of the ship, which we wish to preserve, and take it for that purpose from the armour, which is employed expressly to receive it.

I now pass to the consideration of the 'Lord Warden' target, shown in section in Fig. 265, concerning which a very few words will suffice. In the first instance the 'Lord Warden' was to have been an ordinary wooden frigate's side, plated with 4½-inch armour plates, and therefore affording but little scope for even attempts at improvement. There was, however, room for one change which appeared desirable, viz., the solidification of the frame in the wake of the armour. I am aware that there are disadvantages attending this; but the advantage, in the case of an armour-plated ship, appears to me to far outweigh them. The advantage consists in this, that by the simple device of filling the timbers in solid—for which purpose no very great weight of timber is necessary—you at once more than treble the depth of the side through which a shot or shell must penetrate in order to let water into the ship. If you will consider the case of an ordinary wooden iron-clad frigate, between wind and water, you will observe that, so soon as you have penetrated the armour and outer planking, the water is perfectly free to pass down through the openings of your frames, and sink the ship, even before the frames themselves are materially injured. But fill in the openings, and thus make the frame solid, and you at once become protected from such a catastrophe until the entire solid side—plating, outer planking, frames, and inner planking—are all successively pierced. I consider this to be an ample reason for solidifying the frames in the wake of the armour-plating of wooden frigates; and that was accordingly done in the case of the 'Lord Warden,' and of her sister ship the 'Lord Clyde.' Had this not been done, there would have been but 9 inches of plank between the inside of the ship and the armour; but with the fillings fitted, we have a thickness of about 2 feet 7 inches of solid timber behind the armour.

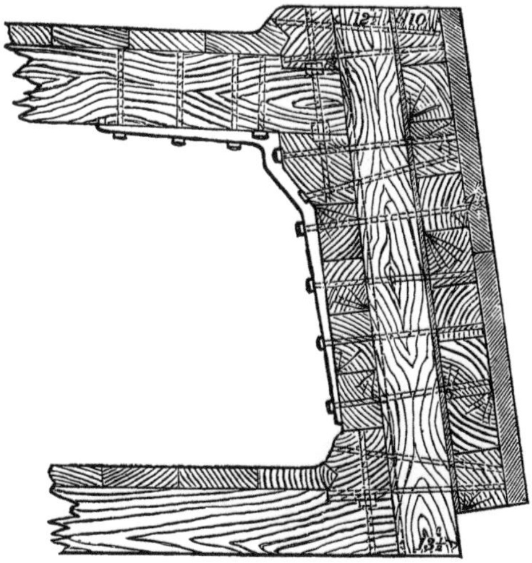

Fig. 265.

After the armour for these ships was ordered, and during the progress of the hulls, it became highly desirable to add, if practicable, to the strength of the side in the wake of the battery deck, and we found that by modifying other weights, we could carry an extra $1\frac{1}{2}$ inch of iron plating, for a depth of 10 feet entirely around the ship. The questions then arose:—in what form, and in what manner can this additional iron be best applied?—observing that, as the armour was already manufactured, the thickness of the solid plates could not be increased. Should the extra plate be put inside the main armour plate and in contact with it; or on the frames of the ship behind the planking; or on the frames of the ship, but inside; or inside of the inner planking? In my opinion there was but one proper place for it, viz., where it was placed, between the frames and outer planking of the ship. I am not aware that any one ever thought it wise to place it outside of that planking, in immediate contact with the thick armour; but there were some, I believe, who considered that it would have been better to place it inside of everything, on the inner planking. But how such a notion can have possessed any mind I cannot imagine; for consider what would be the consequence of placing an inch-and-a-half plate in that position! A blow from a shot capable of penetrating the outer armour and the wooden side, would obviously be free to strip such a plate from its fastenings, and drive it bodily, or in fragments, across the deck among the crew, to their great destruction. There would be nothing but the bolt-points to prevent this. Of course, the case would be very different if the inner plate could be supported by iron frames; but the use of these was in this instance out of the question, as the whole weight at our disposal was insufficient for both frames and plating, and there were also other objections to the use of iron frames, in addition to the thick wooden side.

There was really but little choice, therefore, of a position for this additional plate. Its obvious and natural place was upon the outside of the frames of the ship, and this position possessed an important advantage, which I will briefly explain.

A ship's side has to resist both shot and shell, and one of the most important things to guard against is, the explosion of a shell within the wood backing. When the 'Warrior' was designed it was very improbable that any shell would pass completely through the $4\frac{1}{2}$-inch armour and explode in the backing, and therefore a great depth of wood backing was unobjectionable on this ground; but as Mr. Whitworth, Sir William Armstrong, and others, improved their shells, this contingency became probable, and hence it appeared to me most important so to adjust the thicknesses of the backing and armour, that a shell, which was large and powerful enough to pass through the armour, should be too large to bury itself within the backing, and explode there. If this were not regarded, it is obvious that a single shell might strip several armour plates from a ship's side, and expose her to speedy destruction. In the 'Bellerophon' target this was carefully regarded, as it is not to be expected that a shell large enough to break through a 6-inch armour plate would be less in length than the thickness of the backing, viz., 10 inches: and the importance of the precaution was signally illustrated, for the shell that did penetrate the plate was stopped by the stout iron skin before it got within the armour, and consequently exploded harmlessly backward through the hole which it had made. By placing the extra iron plate of the 'Lord Warden' upon the frames of the ship, as before described, the same object is accomplished there also, so that a shell piercing the $4\frac{1}{2}$-inch plate may encounter the $1\frac{1}{2}$-inch plate before it has space to bury itself.

In the 'Hercules' target (illustrated in section by Fig. 266), which I will now briefly describe, the same principle of construction has been carried out. In the 'Hercules' herself, provision has been made for 9-inch armour at the water-line, and for wood backing, varying from 10 to 12 inches. It was considered that in this case also it was out of the question to suppose that a shell which could penetrate a 9-inch plate would be of less than 12 inches in length, and consequently in this case also it was presumed that the armour plate, the backing, and the skin of the ship must all three be pierced, before a shell could do any serious injury to the structure.

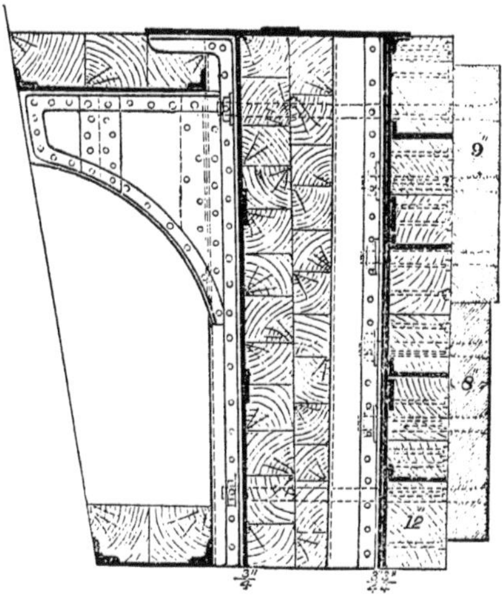
Fig. 266.

The arrangement of frames, skin-plating, external girders, and armour plates of the 'Hercules' target, resemble those of the 'Bellerophon' (although differing somewhat in dimensions), but there is a further source of strength in the former, which the 'Bellerophon' did not possess, viz., a second wood backing, supported by a second series of frames and skin-plates. This part of the construction was adopted, not for its supposed excellence as a shot-resisting arrangement in the abstract only, but as a means of making good use of the wing bulkheads. These wing bulkheads in the 'Warrior,' 'Bellerophon,' and other ships, serve the treble purpose of acting as a second line of defence against splinters and debris, of serving to stop the flow of water from an injured place in the side to the hold of the ship, and of enclosing a passage, by means of which access to the ship's side can be obtained conveniently, for effecting repairs, and for other purposes. In the case of the 'Hercules,' where we had to provide for resisting the shot of 20-ton guns, it was considered that such a bulkhead would not be required all the time the ship had to withstand guns of moderate size and power; and that, against larger guns, it would be a more effectual defence if placed nearer to the side, and made to support additional teak logs, placed between it and the skin proper of the ship. For this, and for other reasons, which I cannot dwell upon, the 'Hercules' target was constructed as you see it upon the drawing, and although it was actually penetrated at Shoeburyness, by two 600-lb. projectiles, fired with 100 lbs. of powder in each case, striking in succession upon nearly the same spot, it is no doubt proof to any single shot fired from any gun that exists in the world.

APPENDIX.

LLOYD'S REGISTER OF BRITISH AND FOREIGN SHIPPING.

RULES FOR THE BUILDING AND CLASSIFICATION OF SAILING AND STEAM VESSELS BUILT OF IRON.

ALL vessels will be classed A so long as on careful annual and periodical *special* surveys they are found to be in a fit and efficient condition to carry dry and perishable cargoes to and from all parts of the world.

Differences of construction, as regards thickness of plating, strength, and probable durability, &c., will be indicated by the letters A B and C placed inside the letter A,—thus, A(A) A(B) A(C).

A(A) A(B) will denote that the vessels have been built in accordance with, or equal to, the Rules and Table G.

A(C) will denote vessels which are considered entitled to the A character, but which have not been built in accordance with the Rules.

All vessels to be subject to occasional or annual survey when practicable.

To entitle ships to retain their respective characters in the Register Book, the following Special Surveys must be held periodically:—

SURVEY No. 1.—The vessel to be placed on blocks of sufficient height in a dry dock, or on ways; the limber boards, and ceiling equal to one strake fore and aft on both sides removed, with both surfaces of outside plating exposed.*

SURVEY No. 2.—The vessel to be placed on blocks of sufficient height in a dry dock, or on ways; the limber boards, and ceiling equal to *three* strakes fore and aft on both sides removed, with both surfaces of outside plating exposed.*

SURVEY No. 3 BY TWO SURVEYORS, ONE TO BE AN EXCLUSIVE OFFICER OF THE SOCIETY.—The vessel to be placed on blocks of sufficient height in a dry dock, or upon ways; proper stages to be made and the hold to be cleared,

* In cases where the inner surface of the bottom plating is coated with cement or asphalte, if a sufficient quantity of ceiling be removed to enable the coating to be carefully inspected, and tested by beating or chipping, and the coating be found sound and good and adhering satisfactorily to the iron, the removal of such coating will be dispensed with. Ships which have undergone the above examination will be noted in the Register Book thus (*s.s.No.*1-68), (*s.s.No.*2-68), (*s.s.No.*3-68); and if not submitted to such survey, will be liable to have their characters suspended.

the close ceiling in the hold to be removed, so that the rivets and plates of keel, and flat of bottom, may be thoroughly examined; coal bunkers of steam vessels to be cleared, the whole of the frames, stringers, hooks, floor plates, keelsons, engine and boiler bearers,* ends of beams, watertight bulkheads, rivets, and inner surface of the plating to be exposed;† all oxidation to be removed by being cut or beaten off the several parts above named, also from the outside plating, rivets, keel, stem, sternpost, and rudder, so as to completely lay bare all the surfaces of iron; the planksheers and waterways, if of wood, to be scraped bright: and when the vessel is so prepared, the Surveyors are to ascertain, by drilling, the thickness of the plating, also the condition of all the parts of iron above named, and of the planksheers, waterways, flat of decks and their fastenings; such parts as may be found defective, or less than three-fourths of the required substance by Rule, are to be removed and replaced with proper materials, equal in substance and quality to the original construction.

Whenever the bottom plating is to be cemented, a survey is to be held prior to the cement being laid.

Every ship classed A must be submitted to a *special periodical survey* every *four* years:—the first survey according to No. 1; the second according to No. 2; the third according to No. 3; and afterwards according to No. 1 and No. 3 *alternately* at intervals of four years.

Every ship classed B must be submitted to a special periodical survey every *three* years, as per Nos. 1, 2, and 3, afterwards Nos. 1 and 3.

Every ship classed C must be submitted to a special periodical survey every *two* years, as per Nos. 1, 2, and 3, and afterwards Nos. 1 and 3.

RULES FOR THE BUILDING OF IRON SHIPS.

Length, breadth, and depth.—The scantlings given in Table G are intended for ships, the length of which, measured from the fore part of the stem to the after part of the sternpost, on the range of the upper deck, does not exceed seven times their breadth, or ten times their depth of hold, taken from the upper part of floors to the top of the upper-deck beams. For ships which exceed these proportions, see Section 16.

Tonnage.—In flush-decked vessels having either one, two, or three decks (not being spar or awning decked), the tonnage under the upper deck, *without abatement of the tonnage of the space for the crew, or for the propelling power of steam vessels,* is to regulate all the scantlings of the hull, and also the equipment of the vessel.

In vessels having a *raised quarter deck,* or a poop, or top-gallant forecastle, or deck houses, or awning deck, or spar deck, the total tonnage below the tonnage deck is to regulate the scantlings of the hull; but the register tonnage, as cut on the main beam of sailing vessels and of steam vessels, *with the addition of the tonnage of the space required for propelling power,* is to

---

* Whenever the engines and boilers are taken out for repair, the engine and boiler bearers, with the floor-plates, keelsons, rivets, &c., under them may, at the request of the Owners, be surveyed in anticipation of the above rule.

† See note on previous page.

regulate the equipment, also the size of the main piece of rudder and windlass, and the keel and keelsons and their number, and the scantling of the stringer plates on the upper and lower deck beams, and the requirements as to double riveting.

1. *Quality of Iron and Workmanship. Maker's Name.*—The whole of the iron to be of good malleable quality, capable of bearing a longitudinal strain of twenty tons per square inch, and all plate, beam, and angle iron *to be legibly stamped in* TWO *places* with the manufacturer's trade mark, or his name and the place where made, which is also to be stated in the report of survey. The workmanship to be well executed, and submitted to the closest inspection before coating or painting: any brittle or inferior article to be rejected. (It is not intended to prevent the coating of the plates *inside* in the way of the frames.)

2. *Keel, Stem, Stern, and Propeller Posts.*—The keel, stem, stern, and propeller posts are to be either scarphed or welded together, and to be in size according to Table G (*see also* page 492); if scarphed, the length of scarphs to be eight times the thickness given in the table for keels; and the sternposts and after end of keel, for screw-propelled vessels, to be double the thickness of, or twice the sectional area of, the adjoining length of keel (but the siding in no case to be less than the thickness of the keel given in Table G), and to be tapered fair into the adjoining length of keel. Where the garboard-strakes are thicker than required by the Rules, and extend to the bottom of the keel, the thickness of the keel may be proportionably reduced, but such reduction not to exceed one-third of the requisitions of the Rule. Where the keel and keelsons are made of several thicknesses of plates, the plates that form the keel to be in thickness, taken together, the same as is required for a solid keel, as per Table G; and the butts of the several plates of which the keel is formed to be carefully shifted from each other, and from the butts of the garboard-strakes, which in all cases must also be shifted, so as not to be opposite, or nearer to each other than two spaces of frames. For thickness and breadth of hollow or flat-plate keels, see foot-note of Table G.

3. *Frames.*—The frames to be of the dimensions set forth in Table G; to be in as great lengths as possible, fitted close on to the upper edge of the keel, and in all cases to extend to the gunwale; and when butted on the keel (except when double frames, or centre through-plate keels, are adopted) and wherever elsewhere butted, to have not less than four-feet lengths of corresponding angle-iron fitted back to back to cover and support the butts and receive the plating. If welded together, the welds to be perfect, with not less than four-feet shifts.

Spacing.—If single frames be adopted, the space from centre to centre is not to exceed 21 inches all fore and aft; but provided an additional frame, for half the vessel's length amidships, be fitted at opposite sides of each floor-plate, across the keel, and extended to upper part of bilges and riveted through floor-plates and main frames, also through the outside plating as required for main frames, the space may be increased to 23 inches in ships under 1000 tons, and to 24 inches in ships of 1000 tons and upwards.

4. *Floor-Plates.*—The floor-plates to be in depth at middle line according to the following rule, viz.:—To the vessel's depth, measured from the top of keel to the top of upper or spar deck beams amidship, add the extreme breadth of the vessel; two-fifths of that sum *in inches*, to be the depth of the floor-plates at middle line; their thickness to be as given in Table G; but at each

end of the vessel, for one quarter of her length, they may be reduced in thickness one-sixteenth of an inch where the plates are less than ten-sixteenths, and two-sixteenths of an inch where the plates are ten-sixteenths and upwards. The floor-plates to extend up the bilges to a perpendicular height of twice the depth of floors amidships from upper side of keel at middle line, and not to be less moulded at their heads than the moulding of the frames. A floor-plate to be fitted and riveted to every frame, and to be extended across the middle line: but where a vertical centre plate is adopted at middle line, then the floor-plates are to be efficiently connected to it on each side by double vertical angle-irons.

Watercourses.—Watercourses are to be formed through all the floor-plates on each side of middle line, so as to allow water to reach the pumps freely.

5. *Reversed Angle-Iron.*—Reversed angle-iron on frames to be in size as per Table G. All vessels, of whatever size, to have reversed angle-iron riveted to every frame and floor-plate across the middle line to the height of upper part of bilges, and to have double reversed angle-iron in way of all keelsons and stringers in hold; and, in addition, all vessels of 300 tons and upwards to have reversed angle-iron extended from bilges to the upper-deck beam stringer on *alternate* frames, and vessels of 800 tons and upwards to have reversed angle-iron extended on *every* frame from bilges to above lower-deck or hold beam stringer angle-iron if the vessel has two decks or tiers of beams, and to above the height of middle-deck beam stringer angle-iron if the vessel has three decks or tiers of beams. The rivets for securing the reversed angle-iron to the frames and floor-plates to be in diameter equal to those specified in the Table for the outside plating, and not to exceed eight times their own diameter apart. Butts of reversed angle-iron to be secured with butt-straps.

6. *Middle Line Keelsons.* (*See also* page 492).—The middle line keelson, if of single plate, and standing above the floor-plates, to be of the same thickness as the garboard-strakes, and to be two-thirds of the depth of floor-plates, well fitted and riveted thereto; and an angle-iron of the size as per Table G, to be fitted on each side both on the top and the bottom, extending all fore and aft; the lower angle-irons to be riveted to the double reversed angle-irons on the top of floors. If a box keelson be adopted, it is to be formed with a foundation plate, the plating to be of the thickness as per Table G, the depth not to be less than two-thirds of the depth of floor-plates, and the breadth of the box two-thirds its depth.

If an intercostal middle line keelson be adopted, it is to be of the same thickness as the floor-plates (see page 493), and to be riveted to vertical angle-iron on all floor-plates at each end, the plates to extend from upper edge of keel to above the upper edge of floor-plates, sufficiently high to be riveted to bulb-iron bars, of the same strength as the beams, or to deeper bulb-iron bars let down, or bars of other form, but of equal strength, between double angle-irons, of the dimensions given in Table G, extending all fore and aft, and the said double angle-irons of keelson are to be riveted to double angle-irons on top of all floor-plates.

Where flat-plate keels are used, the intercostal keelson plates and centre through-plates to be fitted close down on and connected to the keel by double angle-irons of the dimensions given in Table G, riveted all fore and aft to the keel and keelson.

If the middle line keelson be formed of a centre through-plate, extending from the lower edge of the keel to the top of the floors, it must not be less in thickness than that required in Table G for intercostal keelsons. To strengthen the floor-plates transversely at their intersection at the middle line, in addition to the double vertical angle-iron riveted to their ends and to the centre plate keelson, there is to be a flat keelson plate, of the same thickness as the garboard-strakes, and not less than three-fourths the breadth given in Table G, riveted to double reverse angle-irons on the upper edge of floors, and to two fore and aft angle-irons on the upper edge of the centre through-plate of the keelson. But should the centre through-plate keelson be extended up above the upper edge of the floors, then it is to be riveted by two fore and aft angle-irons, of the size as per Table G, to two flat plates, one on each side of the middle line, to be well riveted to the double reverse angle-irons on the upper edge of the floors. In all cases the centre plate keelson to be extended to the stem and sternpost, and connected thereto where practicable.

7. *Bilge Keelsons.*—The bilge keelsons to be fitted and secured in an efficient manner, and to extend all fore and aft, and placed at lower turn of bilges, according to the form of the bottom.

Intercostal Side Keelsons.—In ships of 1000 tons and upwards, an intercostal keelson to be fitted on each side, as far forward and aft as practicable, and to be placed about midway between the middle line keelson and the bilge keelson, with double angle-iron riveted on the top of floor-plates. (*See also* page 492.)

Stringers.—All vessels of 500 tons and upwards to have fitted between the bilge keelsons and the hold beams, at the upper part of the turn of bilge, strong angle-irons, as stringers, extending all fore and aft, riveted back to back and to the reversed irons on the frames, the size of them not to be less than those used for the middle line keelson. (*See also* page 492.)

In all cases the middle line, side, and bilge keelsons, and *where practicable*, the stringers, are to be carried fore and aft, without being cut off at the bulkheads, the latter being made watertight around them; and where such parts of the ship are necessarily separated, the longitudinal strength to be efficiently maintained to the satisfaction of the Surveyor.

8. *Plating.*—No plates to be less in length than five spaces of frames, except the fore and after hoods. No butts of outside plating, in adjoining strakes, to be nearer each other than two spaces of frames. In vessels under 1200 tons, the plating may be reduced from the thickness shown in Table G, one-sixteenth of an inch forward and aft, for a distance not exceeding one quarter of the length of the vessel from each end, below the upper edge of main sheerstrake, down to a perpendicular height from upper side of keel of three-fifths the internal depth of hold; and in ships of 1200 tons and upwards, a reduction of two-sixteenths will be allowed; the plates next abaft and next afore the quarter length of the vessel, to be of an intermediate or graduated thickness, between that required in midship and the reduction allowed at the ends. In screw-propelled vessels, however, no reduction is to be made in the plating at the after end, below the lower part of the rudder trunk.

Butt Straps.—All plates are to be well fitted, and secured to the frames and to each other; the butts to be closely fitted by planing or otherwise, and to be united by butt-straps, of not less than the same thickness as the plates, and of sufficient breadth for riveting, as described hereafter, and to be fitted with the fibre of the iron in the same direction as the fibre of the plates to which they are riveted; the space between the plating and the frames to have solid filling or lining pieces, closely fitted in one length, and of the same breadth as the frames.

Sheerstrake.—It is recommended that in all cases the sheerstrake be an outside strake, so as to admit of the butt-straps or lining pieces being extended, in one piece, from the foreside of the frame next afore the butts to the aftside of the frame next abaft the butts, or to admit of doubling the sheerstrake where it may be required.—*For breadth of sheerstrake see foot-note in Table G.*

9. *Reductions allowed in raised Quarter decks, Poops, Forecastles, &c.*—In raised quarter decks, a reduction of one-fifth from the thickness required by the Table G for such parts in the range of the upper deck in ships with two decks will be allowed in the outside plating, beams, stringer-plates upon beams, angle-iron on stringer-plates, and flat of deck.

Poops and Top-gallant Forecastles.—In full poops and top-gallant forecastles a reduction of one-fourth from the dimensions required by the Table G for such parts in the range of the upper deck in ships with two decks will be allowed in the outside plating, beams, stringer-plates upon beams, angle-iron on stringer-plates, and flat of deck, but in no case need the outside plating exceed six-sixteenths of an inch in thickness. The united lengths of poop and forecastle are not to exceed three-fifths of the entire length of the upper deck.* All frames are to extend to the stringer-plates of poop and forecastle.

Where the poop or forecastle is constructed in a rounded form at the gunwale, the beams may be of plain angle-iron, not less in dimensions than the sizes required in Table G for the main frames; a beam to be properly riveted to every alternate main frame, with a scarph not less than four feet in length. The breast beams are to be double, and the rounded gunwale is to be plated and properly constructed in all respects to the satisfaction of the Surveyor.

In vessels with three decks (viz., upper, middle, and lower deck), a reduction of one-sixth from the dimensions given for such parts in the range of upper deck in ships with two decks will be allowed in the scantling of beams, flat of deck, and plating, *but not in the dimensions of sheerstrake.*

Spar Decks.—In vessels having three decks or tiers of beams, where the space under the upper deck is to be used only for the accommodation of crew and passengers, or to enclose the engine openings of steam vessels, the scantlings are to be regulated as defined at page 492. The total depth of hold in spar-decked ships must not exceed thirteen-sixteenths, nor be less than twelve-sixteenths of the ship's extreme breadth. In spar decks a reduction of one-fourth from the dimensions required by the Table G, for such parts in the range of the upper deck in ships with two decks, will be allowed in the dimensions of all beams and stringers, and thickness of plating, and flat of deck; but all frames are to extend to the stringer-plates of spar deck.

Deckhouses or other erections are allowed on spar decks, but only to the extent of one-tenth of the total superficial area of the spar deck, and are not to exceed seven feet in height. They are not to be placed nearer to either of the ends than one-fifth of the entire length of the vessel.

Vessels to which this rule applies, as regards an entire spar deck, will be noted in the Register Book thus:—"*Spar-decked.*"

10. *Beams.*—Beam-plates to be in depth one-quarter of an inch for every foot in length of the midship beams, and to be in thickness one-sixteenth of an inch for every inch in depth of the said beams, and to be made of H-iron,

* Parties desirous of making any alterations in the length or construction of *poops* and *forecastles*, may submit their plans for the Committee's consideration and approval.

T bulb-iron, or bulb plate with double angle-irons riveted on upper edge; the two sides of each of these angle-irons to be not less in breadth than three-fourths the depth of beam plate, and to be in thickness one-sixteenth of an inch for every inch of the two sides of the angle-iron; or the beams may be composed of any other approved form of beam iron, equal in strength. Where beams below the upper or middle deck (including orlop beams) have no deck laid upon them, the angle-irons on their upper edges are required to be of the dimensions of the angle-iron of the reverse frames. All beams to be well and efficiently connected or riveted to the frames, with bracket ends or knee-plates; each arm of knee-plates at ends of beams not to be less in length than twice and half the depth of beams, and to be in thickness equal to the beams. The beams to be placed over each other, and pillared where practicable.

Upper-deck beams in vessels with one or two tiers of beams, and the upper (or spar deck) and middle-deck beams in vessels with three tiers of beams, to be fastened to alternate frames.

Vessels of 12 feet and under 13 feet depth of hold, or where the tonnage (see page 492) exceeds 200 tons, shall be required to have as many hold beams as may be practicable or convenient, fastened to at least every eighth frame. Vessels not being of a depth to require hold beams are to have a double angle-iron stringer riveted to reverse frames extending all fore and aft about midway between bilge keelson and deck beams.

Vessels of 13 feet depth and under 15 feet, to have hold beams fastened to every fourth frame.

Vessels of 15 feet depth and under 18 feet, to have hold or lower-deck beams fastened to every second and fourth frame, alternately.

Vessels of 18 feet depth and above, to have hold or lower-deck beams fastened to every alternate frame, and the same number of middle-deck beams, where such are required.

All vessels having two decks (viz., upper and lower deck), and exceeding 24 feet in depth from the top of floor-plates to the upper side of upper-deck beams, and vessels with three decks (viz., upper, middle, and lower deck), and exceeding 24 feet in depth to the upper side of middle-deck beams, and where the depth from under side of lower-deck beams exceeds 15 feet, such ships to have orlop beams fastened to every sixth frame; also to have stringer-plates and angle-iron on their ends, all fore and aft, *equal in strength* to the requirement at Section 15; but, in the case of flush-deck ships, a depth of 25 feet will be allowed, provided the lower hold does not exceed 16 feet in depth from the under side of lower-deck beams. Should a house be constructed on such flush-deck ship for lodging crew or for store-room, the same not to extend within 10 feet of the sternpost.

When the spaces between beams exceed two spaces of frames, a knee or bracket plate is to be riveted to alternate frames and to the stringer-plate at underside.

Depth of Hold for Space of Beams.—For the arrangement of beams the depth of hold is to be measured amidship from the top of the floor-plates to the top of the upper-deck beams in vessels with two decks, and to the top of the middle-deck beams in vessels with three decks.

Where a deviation from the foregoing Rules as applying to beams takes place in way of engine-rooms or hatchways, or where no deck is intended to be laid, and the above-named spaces would materially interfere with the stowage of cargo, and where partial or entire bulkheads with horizontal stringers between them, or larger beams are substituted for ordinary beams in wider spaces, a sketch with all particulars must be submitted through the resident surveyor, for the Committee's consideration. The middle deck to be a perfect deck laid and caulked.

11. *Rivets and Riveting.*—The rivets to be of the best quality, and to be in diameter as per Table G. (*See also* page 492.) The rivet-holes to be regularly and equally spaced and carefully punched opposite each other from the faying surfaces in the laps and lining pieces or butt-straps, and to be countersunk all through the outer plating; the rivets not to be nearer to the butts or edges of the plating, lining pieces to butts, or of any angle-iron, than a space not less than their own diameter, and not to be further apart from each other than four times their diameter, or nearer than three times their diameter, and to be spaced through the frames and outside plating, and in reversed angle-iron, a distance equal to eight times their diameter apart. When riveted up they are completely to fill the holes, and their points or outer ends are to be round or convex, and not to be below the surface of the plating through which they are riveted. All vessels to have all edges or horizontal joints of outside plating double-riveted from the keel to the height of upper part of bilges, all fore and aft; but vessels of 700 *tons and above*, intended for the highest grade, are to have all edges or horizontal joints of outside plating double-riveted *throughout.** The stem, sternpost, keel, edges of garboard-strakes and sheerstrakes, and butts of outside plating,* and butts of floor-plates, breasthooks, transoms, and plates of beams, also butts of keelsons, stringers, shelf-plates, and all longitudinal ties, to be double-riveted in all vessels. The overlaps of plating, where double-riveting is required, not be less than five and a half times the diameter of the rivets; and where single-riveting is admitted, to be not less than three and a quarter times the diameter of the rivets. If double-riveting be adopted where single-riveting is allowed by the Rules, the diameter of the rivets may be reduced one-sixteenth of an inch below that prescribed by the Rules, provided that in no case the diameter be less than five-eighths of an inch. The butts and edges of outside plating to be truly fitted, carefully caulked, and made watertight.

12. *Bulkheads.*—Steamers, in addition to the engine-room bulkheads, to have two watertight bulkheads, built at a reasonable distance from the ends, to extend from the keel and outside plating to the upper deck in vessels with two decks, and to the middle deck in vessels with three decks (otherwise called "tonnage deck"); but the aftermost bulkhead will not be required to extend to this height if it be continued above the load water-line, and be connected to a watertight platform or deck of iron extending from its upper part entirely round the after part of the vessel, thus rendering the lower after body a watertight compartment. The bulkhead is to be made watertight where a screw shaft passes through. And in the construction of vessels propelled by machinery care must be taken that the engine and boiler bearers are properly constructed (and where they may interfere with the longitudinal strength of the vessel they must be extended a sufficient distance beyond the bulkheads of the engine and boiler space, to compensate for such interruption); and after the machinery and boilers are fitted, then as many hold or lower-deck beams are to be introduced as may be practicable; and knee or bracket

* The above requirement as regards double-riveting does not apply to poops or forecastles.

plates are to be added and riveted to the stringer-plates, and to alternate frames which have no beams in the said space; and the vessels are to be otherwise made secure where necessary in the engine-room to the satisfaction of the Surveyors. In sailing-ships the foremost or collision bulkhead only will be required. All plating of bulkheads to be of the thickness prescribed in Table G; and when fitted between two frames at each side of the vessel, to be strongly riveted through them; or if attached only to one frame, then to have brackets or knee-plates riveted horizontally against the side plating of the vessel and to the bulkheads, foreside and afterside alternately, near the middle of the outside plates, and to be strongly riveted thereto. Lining pieces between these frames and outside plating in way of bulkheads, are to be plates extending in one piece from the foreside of the frame afore, to the aftside of the frame abaft the bulkhead frames. The bulkheads to be supported vertically by angle-irons (of the dimensions given in Table G) not exceeding two feet six inches apart; and to be efficiently connected and riveted together and to the corresponding floors, beams of the several decks, and the frames. All such bulkheads to be caulked and made thoroughly watertight.

Sluice Cocks.—Where a pump is not fitted in each compartment, a sluice cock, or valve, is to be fitted at the limbers on each side of middle line, at each watertight bulkhead, so as to allow water to be shut off, or to reach the pumps when required; the same to be worked from the deck above.

DOUBLE BOTTOMS.—To entitle a vessel to be noted in the Register Book as having a "*Double Bottom,*" the inner or second bottom must be efficiently constructed, with the plating carried forward to the fore bulkhead, as usually fitted, and to an equal distance from the after end of the ship; the plating not to be less in thickness than that given in Table G for plating of bulkheads, excepting the flange plate, which must be one-sixteenth thicker. The double bottom must be efficiently connected to the outside plating and frames of the main body of the ship. The butts and edges may be single-riveted. "Man holes" must be constructed, or provision made for the removal of a portion of the plates so as to enable the inner surface of outside plating, the frames, floors, keelsons, and rivets to be thoroughly examined, and coated when required. The upper side of the plating must be protected with wood planking as ceiling.

Should a smaller portion of the ship be constructed as above, such ship may be marked "*Part Double Bottom,*" provided such portions extend to at least one-half of the length.

13. *Ceiling.*—The wood ceiling or lining is not to be less than 1½ inch, nor more than 3 inches in thickness in any case, and is to be so fastened to the reversed angle-irons or frames that it may be easily removed for survey and painting.

14. *Decks, Waterways, and Planksheers.*—The flat of upper deck to be fastened by screw-bolts from the upper side, with nuts at the under side of the angle-iron of the beams; where the planks exceed six inches in width there must be two bolts in each plank in every beam, one of which may be a short screw-bolt, provided the planks do not exceed eight inches in width, in which case both bolts must be put through. The waterways, if of wood, to be fastened with screw-bolts with nuts at under side of stringer-plates.

15. *Stringers on ends of Beams.*—All vessels to have stringer-plates (of the thickness given in Table G) upon the ends of each tier of beams. (*See also* page 492.) Those upon the ends of upper-deck beams in vessels with one or

two decks or tiers of beams, and on ends of middle-deck beams in vessels with three decks or tiers of beams, to be in width one inch for every seven feet of the vessel's entire length, for half her length amidship, and from thence to the ends of the vessel they may be gradually reduced to three-fourths the width amidship—in no case, however, is the width to be less than eighteen inches amidship. The stringer-plates are to be fitted home and riveted to the outside plating at all upper decks, and at the middle deck in vessels having three decks, with angle-iron of the dimensions given in Table G: the middle-deck stringer-plate to have an additional angle-iron extending all fore and aft inside of the frames, riveted to the reverse angle-iron on the frames, and to the stringer-plate. Stringer-plates on ends of beams below the upper deck in vessels with two decks, or below middle deck in vessels with three decks, may be reduced in width to three-fourths the midship breadth above named, this breadth is to be extended all fore and aft, and to have an angle-iron of the dimensions given in Table G, extending all fore and aft, riveted to the reverse angle-iron on the frames, and to the stringer-plates. In cases where no deck is laid, and the width of stringer-plate on ends of hold beams is objected to, it may be reduced, provided such reduction be fully compensated for. The objectionable practice of cutting through the stringer-plates for the admission of wood rough-tree stanchions will not be allowed.

Tie-plates.—All vessels to have tie-plates ranging all fore and aft upon each side of the hatchways *on each tier of beams*, and in addition thereto the beams of the upper and middle decks in three-decked or spar-decked ships, and of the upper deck in vessels of one or two decks, must have the tie-plates fitted from side to side diagonally, whenever the arrangements of the deck will admit of them; the tie-plates are to be in width once and a half the depth of beams, and of the thickness required for stringer-plates, and to be well riveted to each other, and to the beams, deck-hooks, and transoms; and all butts to be properly shifted. Upon hold beams where no deck is to be laid, or where tie-plates would interfere with stowage of cargo, an angle-iron of the dimensions given in Table G for angle-iron on beam stringers, placed at middle line, extending fore and aft wherever practicable, and well riveted to all beams, deck-hooks, and transoms, will be admitted in lieu thereof.

All hatchways and the mast-holes of sailing ships are to be properly framed to receive half beams where required, and the latter to have mast partners at each tier of beams (except at orlop beams) the plating of which is not to be less in thickness than is required for stringer-plates, and the united breadths of the plates not to be less than three times the diameter of the masts. The said plates are to be well riveted to each other, and to the beams, and angle-iron carlings; and at the decks where masts are to be wedged, an angle-iron of the dimensions required for the main frames of the ship is to be properly fitted and riveted to the plates round the mast-holes. The skylights and mast-holes of steam vessels must be properly secured to the satisfaction of the surveyors.

16. In the following cases additional longitudinal strength beyond that stated in Table G will be required, viz.:—

Ships above 10 *depths.*—Ships above 10, and not exceeding 11 depths in length, to have the main sheerstrake increased in thickness one-sixteenth of an inch amidships, for three-fourths the length of ship; or to have a doubling strake not less than 9 inches broad, for the same distance amidships.

Ships above 11 *depths.*—Ships above 11, and not exceeding 12 depths in length, to have the main sheerstrake increased in thickness two-sixteenths of an inch amidships, for three-fourths the length of ship; or to have a doubling strake not less than 12 inches broad, for the same distance amidships.

Ships above 12 *depths.*—Ships above 12, and not exceeding 13 depths in length, to have the main sheerstrake increased in thickness two-sixteenths of an inch amidships, for three-fourths the length of ship; or to have a doubling strake not less than 18 inches broad, for the same distance amidships; and the stringer-plate upon ends of upper-deck beams, in vessels with one or two decks, or on ends of middle-deck beams, in vessels with three decks, is to be increased two-sixteenths of an inch in thickness for half the ship's length amidships, or be proportionately increased in width for the same distance, and the vessels to have a bulb-plate of the dimensions required for beam plates, placed between and riveted to the double-angle iron keelson, at lower part of bilges, for half the length of the ship amidships.

In all the above cases, the doubling plate is not to be of less thickness than the strake next below the sheerstrake, and fitted at the upper edge of the sheer-strake.

Ships above 13 *depths.*—In ships above 13, and not exceeding 14 depths in length, the main sheerstrake to be doubled its entire breadth for three-fourths the length of ship amidships, the doubling is not to be of less thickness than the strake next below the sheerstrake and fitted upon the edge of the same, and to extend in one or two breadths of plating to the upper edge of sheer-strake. The stringer-plate on ends of beams and the bulb-plate between the angle-irons at bilges to be as is required in the preceding case.

Ships above 14 *depths, or* 7 *breadths.*—In cases of ships which exceed 14 depths or 7 breadths in length, the builders are to submit to the Committee, through the resident Surveyor, their plans for giving the vessel sufficient additional strength longitudinally. The depth for the foregoing purpose in spar-decked ships is to be taken from the under side of the "tonnage" or middle deck to the top of the floor-plates.

17. *Rudder.*—The main piece of rudder to be in size according to Table G, of the best hammered iron, and the plating of it to be carefully stayed and riveted. (*See* page 492.)

Windlass.—The windlass, *for all grades*, if of wood, is to be composed of the woods comprised in line No. 1 of the Table A; namely, English, African, and Live Oak; Adriatic, Italian, Spanish, Portuguese, and French Oak; East India Teak, Morung Saul, Greenheart, Morra, and Iron Bark.

18. *Surveys while building.*—Vessels intended for classification to be surveyed as follows, viz.:—

1st. On the several parts of the frame, when in place, and before the plating is wrought.

2nd. On the plating during the progress of riveting.

3rd. When the beams are in and fastened, and before the decks are laid.

4th. When the ship is complete, but before the plating is finally coated or cemented.

5th. And lastly, after the ship is launched and equipped.

For Equipment, see Sections 32, 71, 72, 73, 74, 75, and 76 of Lloyd's Regulations for Wood Ships and Table No. 22, which are not appropriate to this volume.

SHIPS NOT BUILT UNDER SURVEY.

19. In cases of ships not surveyed while building for which a character may be required, application must be made to the Committee in writing, who will direct a special examination to be made by two Surveyors of the Society (one of whom shall be an exclusive officer), for which purpose the vessel is to be placed on high blocks in a dry dock or upon ways; the hold to be cleared and proper stages made; the rivets and plating of keel, and flat of bottom thoroughly examined; the close ceiling in the hold to be removed, and coal bunkers of steam vessels to be cleared; the whole of the frames, stringers, hooks, floor-plates, keelsons, engine and boiler bearers, ends of beams, water-tight bulkheads, rivets, and inner surface of the plating exposed to view;* all oxidation to be removed by being cut or beaten off the several parts above named, also from the outside plating, rivets, keel, stem, sternpost, and rudder, so as to completely lay bare all the surfaces of iron; the planksheers and waterways, if of wood, to be scraped bright; and when the vessel is so prepared, the Surveyors are to ascertain, by drilling, the thickness of the plating, also the condition of all the parts of iron above named, and of the planksheers, waterways, flat of decks and their fastenings; and send a detailed report thereon, and on the dimensions and quality of the materials and workmanship, to the Committee, who will then assign the vessel such character as the facts may appear to them to warrant, and define the periodical Surveys to which they shall respectively be subjected; but in no such case will a higher character than A/B be allowed.

Mem.—The foregoing Rules have been framed for Iron Ships built with vertical frames and longitudinal plating. Parties desirous of constructing vessels varying from the Rules, must submit their plans with specifications, for approval.

RULES FOR THE SURVEY OF IRON SHIPS CLASSED FOR PERIODS OF YEARS.

All vessels to be subject to occasional or annual survey when practicable, and every third year to be specially surveyed in dry dock or laid on blocks; with both surfaces of outside plating exposed; † and whenever the engines or the boilers of iron steam ships are taken out, the vessel shall be submitted to a particular and special survey.

CONTINUATION OF IRON SHIPS TO THE CHARACTER A.

20. If, on the termination of the period of original designation, or if at any subsequent period, not exceeding one-half the number of years assigned

---

* In cases where the inner surface of the bottom plating is coated with cement or asphalte, if a sufficient quantity of ceiling be removed to enable the coating to be carefully inspected, and tested by beating or chipping, and the coating be found sound and good, and adhering satisfactorily to the iron, the removal of such coating will be dispensed with.

† In cases where the inner surface of the bottom plating is coated with cement or asphalte, if a sufficient quantity of ceiling be removed to enable the coating to be carefully inspected, and tested by beating or chipping, and the coating be found sound and good, and adhering satisfactorily to the iron, the removal of such coating will be dispensed with. Ships which have undergone the above examination will be noted in the Register Book thus (*t.s.*); and if not submitted to such triennial Survey, will be liable to have their character suspended.

originally, or on Restoration, an owner shall wish to have his ship remain or be replaced on the letter A, he is to send a written notice thereof to the Secretary, and the Committee shall then direct a special survey, as follows, to be held by not less than two competent persons, to be appointed by the Committee, one of them to be a Surveyor, the exclusive servant of the Society.

SURVEY.

The vessel to be placed on high blocks, in a dry dock, or upon ways, and proper stages to be made, so that the rivets and plates of keel, and flat of bottom, may be thoroughly examined; the whole of the ceiling or lining inside to be entirely removed; coal bunkers of steam vessels to be cleared, so as to expose the whole of the frames, stringers, hooks, floor-plates, keelsons, engine and boiler bearers, ends of beams, watertight bulkheads, rivets, and inner surface of the plating, to view; the hold to be cleared; all oxidation to be removed by being cut or beaten off the several parts above named, also from the outside plating, rivets, keel, stem, sternpost, and rudder, so as to completely lay bare all the surfaces of iron;* the planksheers and waterways, if of wood, to be scraped bright: and when the vessel is so prepared, the Surveyors are to ascertain, by drilling, the thickness of the plating, also the condition of all the parts of iron above named, and of the planksheers, waterways, flat of decks and their fastenings; and upon the Owner consenting to remove and replace with proper materials, equal in substance and quality to the original construction, such parts as may be found defective, or less than three-fourths of the required substance by Rule, such vessel, upon the repairs and efficiency being reported to the Committee, may be Continued on the letter A for a term of years not exceeding one-half the number of years assigned originally, or on Restoration, subject to occasional or annual survey when practicable. The period of Continuation will, upon all occasions, commence from the time the ship may have gone off the letter A, without regard to the date when the survey for this purpose may be held.

RESTORATION OF IRON SHIPS TO THE CHARACTER A.

21. If, *at any age of a vessel*, an Owner be desirous to have his ship Restored, such Restoration, on his application to the Committee, and consenting to the special survey hereinafter described, to be held by two Surveyors, one of whom shall be an exclusive servant of the Society, and performing the repairs thereby found requisite, will be granted for a period not exceeding two-thirds of the time originally assigned, the same to be calculated from the date of such repairs.

Survey and Requisites for Restoration.

The vessel to be placed on high blocks, in a dry dock, or upon ways, and proper stages to be made, so that the rivets and plates of keel, and flat of

* In cases where the inner surface of the bottom plating is coated with cement or asphalte, if a sufficient quantity of ceiling be removed to enable the coating to be carefully inspected, and tested by beating or chipping, and the coating be found sound and good, and adhering satisfactorily to the iron, the removal of such coating will be dispensed with. Ships which have undergone the above examination will be noted in the Register Book thus (*t. s.*); and if not submitted to such triennial Survey, will be liable to have their character suspended.

bottom, may be thoroughly examined; the whole of the ceiling or lining inside to be entirely removed; coal bunkers of steam vessels to be cleared, the boilers to be taken out, and also the engines (unless it shall be shown by previous survey that the removal is unnecessary), so as to expose the whole of the frames, stringers, hooks, floor-plates, keelsons, engine and boiler bearers, ends of beams, watertight bulkheads, rivets, and inner surface of the plating, to view; the hold to be cleared; all oxidation to be removed by being cut or beaten off the several parts above named, also from the outside plating, rivets, keel, stem, sternpost, and rudder, so as to completely lay bare all the surfaces of iron; * the planksheers and waterways, if of wood, to be entirely removed, and also the flat of upper deck, except under special circumstances, to be sanctioned by the Committee in each case: and when the vessel is so prepared, the Surveyors are to ascertain, by *drilling,* the thickness of the plating, also the condition of all the parts of iron above named, and of the beams and their fastenings; and upon the Owner consenting to remove such parts as may be found defective, or objected to, or less in thickness than hereinafter admitted for repairing such vessel, and replace them with proper materials equal in quality and substance to that required in the Table G for the nine years' grade in those originally classed 12 A, and equal in quality and substance to that required in the Table G for the six years' grade in vessels originally classed 9 A or 6 A, such vessel, upon the repairs and efficiency being reported to the Committee, may be restored to the letter A, for a term of years not exceeding two-thirds the number of years assigned originally, subject to occasional survey.

Iron ships which have been Restored under the foregoing Rule shall be entitled to Continuation thereon, subject to the same conditions of survey and examination as are prescribed for ships proposed to be Continued at the expiration of the period first assigned to them; but, in like manner, the term of such extended continuance to be limited to a period not exceeding one-half the number of years for which the ship may respectively have been restored, without reference to the period originally assigned to them.

22. Vessels not surveyed while building will be classed A from year to year only, but for a period not exceeding six years. (*See also* Section 19.)

23. On the expiration of the terms assigned to ships classed A, they will be liable to lapse (like ships built of wood).

24. One year will be added to the character of all ships of the A class built under a roof which shall project at each end beyond the length, and on each side beyond the breadth, a quantity equal to one-half the breadth of the vessel.

IRON SHIPS ALREADY CLASSED A 1.

Iron ships, built prior to the promulgation of the Rules will be allowed to remain in the Register Book classed A 1 from year to year, *subject to annual survey,* until the expiration of six years from their date of build, and then be examined to determine the period to which they may be entitled under the

---

* In cases where the inner surface of the bottom plating is coated with cement or asphalte, if a sufficient quantity of ceiling be removed to enable the coating to be carefully inspected, and tested by beating or chipping, and the coating be found sound and good, and adhering satisfactorily to the iron, the removal of such coating will be dispensed with.

Rules; and if, on such examination, it shall be found the ships are entitled to the 9 or 12 years' grade, it will be in the option of the owners either to adopt such period respectively, or continue the vessel A 1 from year to year, as above, until the expiration of the extended period; but if it shall be found that the term of years for which a vessel would have been entitled to remain on the A character has expired, she will be classed Æ, if entitled thereto, unless specially surveyed for Continuation or for Restoration.

By order of the Committee,

GEORGE B. SEYFANG, *Secretary.*

No. 2, *White Lion Court, Cornhill, London,*
1*st July,* 1868.

Table of Minimum Dimensions of Frames, Plating, Rivets, Keels, Keelsons,

All plates, and all beam and angle-iron, used in ships intended for classification, are to be stamped

Tonnage.	Keel, Stem, and Stern-post for all Grades.*	Distance of Frames from Moulding edge to Moulding edge all fore and aft for all Grades.	THICKNESS OF OUTSIDE PLATES.† Garboard Strakes* and Single Plate Middle Line Keelsons standing upon floors.		From the Garboard to the upper part of Bilge, and the Sheerstrakes.*		From upper part of bilge to a perpendicular height from upper side of keel of three-fifths the internal depth of hold, measured from the upper side of upper deck in all Ships, whether spar-decked or otherwise.		From three-fifths the depth of hold (measured from the upper side of upper deck in all Ships, whether spar-decked or otherwise) to lower edge of Sheerstrakes.	
			A	B	A	B	A	B	A	B
	Inches.	If single frames be adopted the space from centre to centre is not to exceed 21 inches, all fore and aft; but provided an additional frame, for half the vessel's length amidships, be fitted at opposite side of each floor-plate across the keel, and extended to upper part of bilges, and riveted through floor-plates and main frames, also through the outside plating, as required for main frames, the space may be increased to 23 inches in ships under 1000 tons, and to 24 inches in ships of 1000 tons and upwards.	16ths of an inch.	16ths of an inch.	16ths of an inch.	16ths of an inch.	16ths of an inch.	16ths of an inch.	16ths of an inch.	16ths of an inch.
100 and under 200	$6 \times 1\frac{1}{2}$		8	7	7	6	6	5	6	5
200 and under 300	$6\frac{1}{4} \times 2$		9	8	8	7	7	6	6	5
300 and under 400	$6\frac{1}{2} \times 2\frac{1}{4}$		10	9	9	8	8	7	7	6
400 and under 500	$6\frac{3}{4} \times 2\frac{1}{2}$		10	9	9	8	8	7	7	6
500 and under 600	$7 \times 2\frac{1}{2}$		11	10	10	9	9	8	8	7
600 and under 700	$7 \times 2\frac{3}{4}$		11	10	10	9	9	8	8	7
700 and under 800	$7\frac{1}{4} \times 2\frac{3}{4}$		12	11	11	10	10	9	9	8
800 and under 900	$7\frac{1}{2} \times 3$		12	11	11	10	10	9	9	8
900 and under 1000	8×3		13	12	12	11	11	10	10	9
1000 and under 1200	$8\frac{1}{2} \times 3$		13	12	12	11	11	10	10	9
1200 and under 1500	9×3		14	13	13	12	12	11	11	10
1500 and under 2000	10×3		14	13	13	12	12	11	11	10
2000 and under 2500	12×3		15	14	14	13	13	12	12	11
2500 and under 3000	$12 \times 3\frac{1}{4}$		15	14	14	13	13	12	12	11
3000 and under 3500	$12 \times 3\frac{1}{2}$		16	15	15	14	14	13	12	11

Mem.—The Scantlings of the above Table are intended for ships, the length of which, measured from the fore
their breadth or ten times their depth of hold, taken from the upper part of floors to the top of the upper-deck beams.

STEMS, STERNPOSTS, FLOOR PLATES, BEAMS,‡ BULKHEADS, STRINGERS, &c.

legibly in two places with the manufacturer's trade mark, or his name and the place where made.

§ Thickness of Stringer Plates upon Beams, Floor Plates, Hooks, Crutches, and Box or Intercostal Keelsons for all Grades.	Thickness of Plates for Bulkheads for all Grades.	FRAMES. Dimensions of Angle Iron for all Grades.	Dimensions of Reversed Angle Iron on Frames, Bulkheads, and Box Keelsons, for all Grades.	Dimensions of Angle Iron on Beam Stringers, or Keelsons, for all Grades.	RUDDER for all Grades. Diameter at the Head.	RUDDER for all Grades. Diameter at the Heel.	Thickness of Wood Flat of Upper Deck.	Main Piece of Windlass.	Tonnage.
16ths of an inch.	16ths of an inch.	Inches.	Inches.	Inches.	Inches.	Inches.	Inches.	Inches.	
5	4	$2\frac{1}{2} \times 2\frac{1}{2} \times \frac{6}{16}$	$2\frac{1}{4} \times 2\frac{1}{4} \times \frac{5}{16}$	$3 \times 3 \times \frac{6}{16}$	3	2	$2\frac{1}{2}$	14	100 and under 200
6	4	$3 \times 2\frac{1}{2} \times \frac{6}{16}$	$2\frac{1}{4} \times 2\frac{1}{4} \times \frac{5}{16}$	$3 \times 3 \times \frac{6}{16}$	$3\frac{1}{2}$	2	$2\frac{1}{2}$	15	200 and under 300
7	5	$3\frac{1}{4} \times 2\frac{3}{4} \times \frac{6}{16}$	$2\frac{1}{2} \times 2\frac{1}{2} \times \frac{5}{16}$	$3\frac{1}{2} \times 3 \times \frac{6}{16}$	$3\frac{3}{4}$	$2\frac{1}{4}$	3	16	300 and under 400
7	5	$3\frac{1}{2} \times 2\frac{3}{4} \times \frac{7}{16}$	$2\frac{3}{4} \times 2\frac{1}{2} \times \frac{6}{16}$	$4 \times 3 \times \frac{6}{16}$	$4\frac{1}{4}$	$2\frac{1}{2}$	3	17	400 and under 500
8	6	$3\frac{3}{4} \times 2\frac{3}{4} \times \frac{7}{16}$	$3 \times 2\frac{1}{2} \times \frac{6}{16}$	$4\frac{1}{4} \times 3\frac{1}{4} \times \frac{7}{16}$	$4\frac{1}{2}$	$2\frac{3}{4}$	$3\frac{1}{2}$	18	500 and under 600
8	6	$4 \times 3 \times \frac{7}{16}$	$3 \times 2\frac{3}{4} \times \frac{6}{16}$	$4\frac{1}{2} \times 3\frac{1}{2} \times \frac{7}{16}$	$4\frac{3}{4}$	$2\frac{3}{4}$	$3\frac{1}{2}$	19	600 and under 700
9	6	$4\frac{1}{4} \times 3 \times \frac{8}{16}$	$3 \times 2\frac{3}{4} \times \frac{7}{16}$	$4\frac{3}{4} \times 3\frac{3}{4} \times \frac{8}{16}$	5	3	$3\frac{1}{2}$	20	700 and under 800
9	6	$4\frac{1}{2} \times 3 \times \frac{8}{16}$	$3 \times 3 \times \frac{7}{16}$	$5 \times 4 \times \frac{8}{16}$	$5\frac{1}{4}$	3	$3\frac{1}{2}$	21	800 and under 900
10	7	$4\frac{3}{4} \times 3 \times \frac{8}{16}$	$3\frac{1}{4} \times 3 \times \frac{7}{16}$	$5 \times 4\frac{1}{4} \times \frac{9}{16}$	$5\frac{1}{2}$	3	$3\frac{1}{2}$	22	900 and under 1000
10	7	$5 \times 3 \times \frac{9}{16}$	$3\frac{1}{2} \times 3 \times \frac{8}{16}$	$5 \times 4\frac{1}{2} \times \frac{9}{16}$	$5\frac{3}{4}$	3	4	23	1000 and under 1200
11	7	$5 \times 3\frac{1}{2} \times \frac{9}{16}$	$3\frac{1}{2} \times 3 \times \frac{8}{16}$	$5\frac{1}{2} \times 4\frac{1}{2} \times \frac{9}{16}$	6	$3\frac{1}{4}$	4	24	1200 and under 1500
11	8	$5\frac{1}{2} \times 3\frac{1}{2} \times \frac{10}{16}$	$4 \times 3\frac{1}{2} \times \frac{8}{16}$	$6 \times 5 \times \frac{9}{16}$	$6\frac{1}{2}$	$3\frac{1}{2}$	4	25	1500 and under 2000
12	8	$6 \times 4 \times \frac{10}{16}$	$4\frac{1}{2} \times 3\frac{1}{2} \times \frac{9}{16}$	$6\frac{1}{2} \times 5\frac{1}{2} \times \frac{10}{16}$	$7\frac{1}{4}$	$3\frac{3}{4}$	4	27	2000 and under 2500
12	8	$6\frac{1}{4} \times 4 \times \frac{11}{16}$	$4\frac{1}{2} \times 3\frac{1}{2} \times \frac{10}{16}$	$6\frac{1}{2} \times 5\frac{1}{2} \times \frac{10}{16}$	$7\frac{3}{4}$	4	4	28	2500 and under 3000
12	9	$6\frac{1}{2} \times 4 \times \frac{11}{16}$	$4\frac{1}{2} \times 3\frac{1}{2} \times \frac{10}{16}$	$6\frac{1}{2} \times 5\frac{1}{2} \times \frac{10}{16}$	8	$4\frac{1}{2}$	4	30	3000 and under 3500

part of the stem to the after part of the sternpost, on the range of the upper deck, does not exceed seven times (*For ships which exceed these proportions, see Section 16. See also exceptions in Section 9.*)

‖ RIVETS. Diameter of Rivets required	$\frac{5}{8}$ of an Inch.			$\frac{3}{4}$ of an Inch.		
for Thickness of Plates	$\frac{5}{16}$	$\frac{6}{16}$	$\frac{7}{16}$	$\frac{8}{16}$	$\frac{9}{16}$	$\frac{10}{16}$

Notes to Table G.

* Hollow or flat-keel plates and garboard-strakes, and main sheerstrakes, are not to be less in breadth than as follows, viz.:—In ships under 500 tons, 2 ft.; in ships 500 and under 1000 tons, 2 ft. 6 in.; in ships 1000 tons and upwards, 3 ft. When hollow or flat-plate keels are adopted, their thickness should not be less than one and a half that of the garboard-strake. (*For keels of other forms, see Section* 2.)

† Plating.—No plates to be less in length than five spaces of frames, except the fore and after hoods. No butts of outside plating in adjoining strakes to be nearer each other than two spaces of frames. In vessels under 1200 tons the plating may be reduced from the thickness shown in Table, one-sixteenth of an inch forward and aft, for a distance not exceeding one quarter of the length of the vessel from each end below the upper edge of main sheerstrakes, down to a perpendicular height from upper side of keel of three-fifths the internal depth of hold, including the height of the spar-deck in spar-deck ships; and in ships of 1200 tons and upwards, a reduction of two-sixteenths will be allowed; the plates next abaft and next afore the quarter length of the vessel to be of an intermediate or graduated thickness, between that required amidships and the reduction allowed at the ends. In screw-propelled vessels, however, no reduction is to be made in the plating at the after end below the lower part of the rudder-trunk.

Butt Straps.—All plates are to be well fitted, and secured to the frames and to each other; the butts to be closely fitted by planing or otherwise, and to be united by butt-straps, of not less than the same thickness as the plates, and of sufficient breadth for riveting, as described hereafter, and to be fitted with the fibre of the iron in the same direction as the fibre of the plates to which they are riveted; the space between the plating and the frames to have solid filling or lining pieces, closely fitted in one length, and of the same breadth as the frames. It is recommended that in all cases the sheerstrake be an outside strake, so as to admit of the butt-straps or lining pieces being extended, in one piece, from the foreside of the frame next afore the butts to the aftside of the frame next abaft the butts, or to admit of doubling the sheerstrake where it may be required. (*For breadth of sheerstrake see note * above.*)

‡ Beams.—Beam-plates to be in depth one quarter of an inch for every foot in length of the midship beams, and to be in thickness one-sixteenth of an inch for every inch in depth of the said beams, and to be made of H iron, T bulb-iron, or bulb-plate with double angle-irons riveted on upper edge; the two sides of each of these angle-irons to be not less in breadth than three-fourths the depth of beam-plate, and to be in thickness one-sixteenth of an inch for every inch of the two sides of the angle-iron; or the beams may be composed of any other approved form of beam-iron equal in strength. Where beams below the upper or middle deck (including orlop beams) have no deck laid upon them, the angle-irons on their upper edges are required to be of the dimensions of the angle-iron of the reverse frames. All beams to be well and efficiently connected or riveted to the frames, with bracket ends or knee-plates; each arm of knee-plates at ends of beams not to be less in length than twice and a half the depth of beams, and to be in thickness equal to the beams. The beams to be placed over each other, and pillared where practicable.

§ Floor-Plates.—The floor-plates to be in depth at middle line according to the following rule, viz.:—To the vessel's depth, measured from the top of keel to the top of upper or spar-deck beams amidships, add the extreme breadth of the vessel; two-fifths of that sum *in inches*, to be the depth of the floor-plates at middle line—their thickness to be as given in Table; but at each end of the vessel, for one quarter of her length, they may be reduced in thickness one-sixteenth of an inch where the plates are less than ten-sixteenths, and two-sixteenths of an inch where the plates are ten-sixteenths and upwards. The floor-plates to extend up the bilges to a perpendicular height of twice the depth of floors amidships from upper side of keel at middle line, and not to be less moulded at their heads than the moulding of the frames. A floor-plate to be fitted and riveted to every frame; and to be extended across the middle line, but where a vertical centre plate is adopted at middle line, then the floor-plates are to be efficiently connected to it on each side by double vertical angle-irons. Watercourses are to be formed through all the floor-plates on each side of middle line, so as to allow water to reach the pumps freely.

Stringer and Tie-plates.—All vessels to have stringer-plates (of the thickness given in Table) upon the ends of each tier of beams. Those upon the ends of upper-deck beams in vessels with one or two decks or tiers of beams, and on ends of middle-deck beams in vessels with three decks or tiers of beams, to be in width one inch for every seven feet of the vessel's entire length, for half her length amidships, and from thence to the ends of the vessel they may be gradually reduced to three-fourths the width amidships —in no case, however, is the width to be less than eighteen inches amidships. The stringer-plates are to

$\frac{7}{8}$ of an Inch.			1 Inch.			Rivets to be $\frac{1}{8}$ of an inch larger in diameter in the stem, sternpost, and keel.
$\frac{11}{16}$	$\frac{12}{16}$	$\frac{13}{16}$	$\frac{14}{16}$	$\frac{15}{16}$	$\frac{16}{16}$	

be fitted home and riveted to the outside plating at all upper decks, and at the middle deck in vessels having three decks, with angle-iron of the dimensions given in Table: the middle-deck stringer-plate to have an additional angle-iron extending all fore and aft inside of the frames, riveted to the reverse angle-iron on the frames, and to the stringer-plate. Stringer-plates on ends of beams below the upper deck in vessels with two decks, or below middle deck in vessels with three decks, may be reduced in width to three-fourths the midship breadth above named, this breadth is to be extended all fore and aft, and to have an angle-iron of the dimensions given in Table, extending all fore and aft, riveted to the reverse angle-iron on the frames, and to the stringer-plates. In cases where no deck is laid, and the width of stringer-plate on ends of hold beams is objected to, it may be reduced provided such reduction be fully compensated for. The objectionable practice of cutting through the stringer-plates for the admission of wood rough-tree stanchions will not be allowed. All vessels to have tie-plates ranging all fore and aft upon each side of the hatchways *on each tier of beams*, and in addition thereto the beams of the upper and middle decks, in three-decked or spar-decked ships, and of the upper deck in vessels of one or two decks, must have the tie-plates fitted from side to side diagonally, wherever the arrangements of the deck will admit of them; the tie-plates are to be in width once and a half the depth of beams, and of the thickness required for stringer-plates, and to be well riveted to each other, and to the beams, deck-hooks, and transoms—and all butts to be properly shifted. Upon hold-beams where no deck is to be laid, or where tie-plates would interfere with stowage of cargo, an angle-iron of the dimensions given in Table for angle-iron on beam-stringers, placed at middle line, extending fore and aft wherever practicable, and well riveted to all beams, deck-hooks, and transoms, will be admitted in lieu thereof. All hatchways and the mast-holes of sailing ships are to be properly framed to receive half beams where required, and the latter to have mast partners at each tier of beams (except at orlop beams), the plating of which is not to be less in thickness than is required for stringer-plates, and the united breadths of the plates not to be less than three times the diameter of the masts. The said plates are to be well riveted to each other, and to the beams, and angle-iron carlings; and at the decks where masts are to be wedged, an angle-iron of the dimensions required for the main frames of the ship is to be properly fitted and riveted to the plates round the mast-holes. The mast-holes and sky-lights of steam vessels must be properly secured to the satisfaction of the Surveyors.

‖ RIVETS AND RIVETING.—The rivets to be of the best quality, and to be in diameter as per Table; the rivet-holes to be regularly and equally spaced and carefully punched opposite each other from the faying surfaces, in the laps and lining pieces or butt-straps; and to be countersunk all through the outer plating; the rivets not to be nearer to the butts or edges of the plating, lining pieces to butts, or of any angle-iron, than a space not less than their own diameter, and not to be further apart from each other than four times their diameter, or nearer than three times their diameter, and to be spaced through the frames and outside plating, and in reversed angle-iron, a distance equal to eight times their diameter apart. When riveted up they are completely to fill the holes, and their points or outer ends are to be round or convex, and not to be below the surface of the plating through which they are riveted. All vessels to have all edges or horizontal joints of outside plating double-riveted from the keel to the height of upper part of bilges, all fore and aft; but vessels of 700 *tons and above*, intended for the highest grade, are to have all edges or horizontal joints of outside plating double riveted *throughout*.* The stem, sternpost, keel, edges of garboard-strakes, and sheerstrakes, and butts of outside plating,* and butts of floor-plates, breast-hooks, transoms, and plates of beams, also butts of keelsons, stringers, shelf-plates, and all longitudinal ties, to be double-riveted in all vessels. The overlaps of plating, where double-riveting is required, not to be less than five and a half times the diameter of the rivets; and where single-riveting is admitted, to be not less in breadth than three and a quarter times the diameter of the rivets. If double-riveting be adopted where single-riveting is allowed by the Rules, the diameter of the rivets may be reduced one-sixteenth of an inch below that prescribed by the Rules, provided that in no case the diameter be less than five-eighths of an inch. The butts and edges of outside plating to be truly fitted, carefully caulked, and made watertight.

LLOYD'S REGISTER OF SHIPPING, LONDON,
2nd July, 1866.

---

* *The above requirement as regards double-riveting does not apply to poops and forecastles.*

LIVERPOOL UNDERWRITERS' REGISTRY FOR IRON VESSELS.

ESTABLISHED 1862.

CONDUCTED BY A JOINT COMMITTEE OF UNDERWRITERS, SHIPOWNERS, AND SHIPBUILDERS.

1. *Durability of Iron Ships.*—Experience has shown that Iron Ships are much more durable than was at first supposed. By the use of cement inside, and by careful attention to the outside coating, a well-constructed Iron Ship can be reckoned upon to last, in first-class condition, for a period of at least twenty years: wear and tear of equipment, and of the wood used in their construction, must in all cases be excepted.

2. *System of Classification.*—The Committee propose to class ships on their general merits, having special reference to the quality of the materials, to the character of the workmanship, to the arrangement and size of the parts where the principal strains are experienced, and to the equipment; a system of classification which is considered preferable to one based mainly on tables of scantling.

3. *Class—Red Certificate.*—The Committee, continuing the classification adopted by the Liverpool Underwriters' Association for the last six years, will class in RED, for periods varying from ten to twenty years, all Iron Vessels, whether Steamers or Sailing Ships, which have been, or may be, submitted to the inspection of the Surveyors of that Association or of this Committee during construction, and be built and completed to their satisfaction.

4. *Black Certificate.*—The Committee will also class in BLACK ships already built, but not inspected by their Surveyors while building, for periods varying according to their merits, from ten to twenty years from the date of launching, but subject to survey as hereafter mentioned.

5. *Renewal of Certificate.*—When the period originally assigned to a ship shall have expired, the Committee will renew her Certificate for such a period as they may consider her entitled to.

6. *Equipment.*—Character of equipment will be denoted by numerals, 1, 2.

7. *Surveys.*—A thorough Survey will be required once in every four years for vessels with a Certificate for twenty years, and once in every three years for vessels with a Red Certificate for less than twenty years. Vessels with a Black Certificate for less than twenty years to be surveyed every two years. When vessels are abroad at the time they become due for survey, they must be thoroughly examined on their return to the United Kingdom. The Surveyors are at all times to have free access to examine vessels holding a Certificate from this Committee; and in case of defects reported by them not being made good, the classification of the ship shall be revised.

8. *Reference in case of complaint.*—Any dispute shall be referred to three Engineers or Shipbuilders, one to be chosen by the Shipowner, one to be chosen by this Committee, and a third to act as umpire, to be chosen by the other two.

9. *Extension to other Ports.*—The Committee do not debar themselves from the co-operation of other Ports in this system of classification, should it meet

with the approval of shipowners and shipbuilders elsewhere; and they will hold themselves open to any arrangement which the requirements of the case may call for.

10. *Powers in regard to the above Clauses.*—The above clauses shall be subject to such alterations as the Committee may from time to time deem desirable.

TABLES OF SCANTLING, &c.

1. The following Tables indicate the character of the arrangements, the quality of the materials, the class of workmanship, and the least scantling which may be adopted in order to ensure the Twenty Years' Classification. The scantling for Ships differing from these Tables must be submitted to the Committee, who reserve to themselves the power of making such alterations in their Rules, and of permitting such deviations from them as they may see fit.

2. *Keels.*—The form of Keel may be either that of the centre plate or that of a bar. If a centre plate be adopted, the thickness to be as per Table 1, the butts to be secured by double butt-strips, each of a thickness equal to two-thirds that of centre plate, and to be treble-riveted; the upper part may form a centre keelson above the floors, with a horizontal plate on each side of a width not less than two-thirds the depth of floors at centre line and the same thickness as the floors. The butts of the horizontal plates to be double-riveted; the strips to be put on the upper side. Double angle-irons of size as per Table 5 must be riveted to the upper edge of the centre plate; the part below the floors should project to form a keel not less than the depth given for bar keels, and must have a double row of rivets to fasten the garboard-strake plates; also an intermediate row to hold the side plates which form the keel proper. In all cases when this kind of keel is adopted, the floor ends at centre line are to be riveted to the centre plate with double angle-irons of the size given for frames the full depth of the floors, the ends of these angles to be joggled over the reverse bar and frame and riveted through them.

Also at the upper edge of floors a scarphing angle-iron, the size of angles for reverse bars, is to be passed through centre plate and riveted to horizontal plate and to floors at both sides, and to scarph on each side, in vessels under 1000 tons 2 feet, and in vessels over 1000 tons 3 feet.

In vessels where the centre plate is finished flush with the tops of floors the horizontal plates must be in one width, and that width not less than one and one-third depth of floors at centre, and must be secured to centre plate by fore and aft angles underneath, of the size as per Table 2.

In vessels where the centre plate is secured to a flat garboard-strake plate or flat keel, it must be secured thereto by two continuous longitudinal angle-irons of the size per Table 5, the floors being notched at bottom corners to admit them.

Bar keels may be of any scantling not less than per Table 1.

3. *Stems and Sternposts.*—To be of size per Table 1. Stems above the load line may be gradually reduced one-fourth in sectional area. Propeller posts to be double the thickness and of the same breadth as bar keels. Rudder gudgeons to be forged on the sternposts.

The feet of stems and sternposts to be extended so as to form part of the keel, not less than four and a half feet long.

4. *Frames and Reverse Frames.*—All frames and reverse frames to be in

size as per Table 2, and spaced so as not to exceed 21 inches from centre to centre throughout in vessels under 1000 tons. In vessels of 1000 tons, and above, the frames may be spaced 24 inches from centre to centre for one-fifth the vessel's length from each end, or may be spaced throughout so as not to exceed 24 inches from centre to centre, provided a double frame of the same size as the frames be carried from the centre line to the upper turn of bilge, and be properly secured to floors and shell of vessel, for three-fifths the vessel's length amidships.

Lining pieces under frames to be in one length and thickness, and the breadth of the frame.

All frames should be in one length, but when butted they must have scarph pieces same size as frames, 4 feet long, in vessels up to 900 tons, and 6 feet long in vessels above 900 tons, with a good shift of butts. Lapping pieces to connect heels of frames across centre line, where bar keels are used, to be not less than 4 feet long, and of the same size as frames.

Reverse frames to be riveted on every frame, and to be carried to upper turn of bilge and gunwale, alternately, in vessels under 12 feet in depth of hold; in vessels with two tiers of beams, and not exceeding 16 feet depth of hold, to be carried to upper side of the upper bilge stringer and gunwale, alternately; and in vessels with three decks to the main and upper deck alternately.

Double reverse angle-irons to be fitted on the frames in way of all keelsons, hold and beam stringers. Where much closing bevel is required at the ends of vessels, the pieces requiring closing may be left out, and the keelsons or stringers fastened with double rivets to the single reverse frames only.

5. *Floors.*—Floors of a depth at centre and thickness as per Table 3, are to be riveted on every frame, and to be half the centre depth at lower turn of bilge, are to be carried well up into the bilge, and to be finished the depth of the moulding edge of the frames. A reduction of one-sixteenth of an inch in thickness may be made for one-fifth the vessel's length from each end in floors which exceed six-sixteenths of an inch in thickness.

In spar-deck vessels the floors are to be increased three-eighths of an inch in depth for each foot in height of this deck.

Limber holes are to be cut on each side in floors and intercostal plates, so as to allow the water to flow freely to the pumps.

All vessels to have breakwaters or wash plates fitted between floors.

6. *Beams.*—Beams to be of bulbed iron, with strongly bulbed lower edge, with double angles on top edge of size per Table 3, or of bulbed T-iron, or of any other approved form, to be in depth as per Table 3, and to be spaced as follows:—

All main decks to have a beam upon every alternate frame. Vessels 11 to 13 feet in depth must have one lower-deck beam on every eighth frame. Vessels from 13 to 15 feet in depth, one on every fourth frame. Vessels 15 to 18 feet in depth, one on every second and fourth frame alternately. Vessels from 18 to 24 feet in depth, one on every alternate frame. Vessels over 17 feet in depth of lower hold, to have orlop beams on every sixth frame. Vessels exceeding 18 feet depth of lower hold, to have orlop beams, one to every fourth frame. In all cases where orlop beams are used, a stringer, same size as the one on the lower deck, is to be riveted upon orlop beams and to reverse frames with angle-iron same size as lower-deck stringer.

It is recommended that lower-deck beams be made one-sixteenth of an inch thicker and one-eighth of the depth deeper than upper-deck beams. In cases where the scantling of the lower-deck beams is increased, a proportionate reduction will be permitted in the upper-deck beams.

Hatch beams and fore-and-afters in upper deck to be one-sixth deeper than upper-deck beams. Poop beams to be one-third and forecastle beams one-fourth lighter than upper-deck beams, and to be spaced over them. Beam knees to be 2½ times the depth of the beams.

When solid flanged beams are used, the section must be submitted to the Committee for approval.

7. *Stanchions for Beams.*—Stanchions of size per Table 4 to be fixed to every beam amidships for one-third of the vessel's length, and to alternate beams forward and aft, and to be secured to the beams under which they are placed, by at least two rivets. When a bulb-iron forms part of the centre line keelson, the hold stanchions must be made to stride the bulb, and be riveted through the angle-irons on the top of the floors.

8. *Plates.*—Thickness of plates to be as per Table 2. Sides of poop and forecastle may be one-third lighter than shell plates amidships, but need not exceed six-sixteenths. Gunwale or sheer strakes to be one-eighth of an inch thicker, and garboard-strakes to be one-sixteenth of an inch thicker than shell plates amidships.

No openings for side lights or ports to be cut in the sheer-strakes without compensation. Bulwark plates need not exceed five-sixteenths of an inch in thickness.

No plate to be less than five spaces of frames in length, with the exception of those at the extreme ends of the vessel.

When double strakes are used, the doubling plates are to be of the same thickness as the strake next adjoining.

Vessels of and over 1200 tons to have three strakes of plates in the bilges increased one-sixteenth in thickness over half the vessel's length amidships.

9. *Butts and Seams of Plating.*—All butts in garboard-strakes, shell plating, stringers, and scarphs of keels to be two clear spaces between frames apart. Butts in the garboard-strakes must not be opposite each other. Butts upon upper-deck stringer-plates must not be nearer than 3 feet to butts of sheer-strake.

Butt-strips to have grain of iron in the same direction, and to be of the same thickness as the plates which they connect together.

Butts to be closely fitted either by planing or jumping; when jumped, the ridge formed by jumping to be chiselled off the inside, in order that the butt-strips may fit closely. The ridge outside to be hammered into the joint.

All butts and seams to be chipped fair and caulked tight.

All seam and butt holes must be punched from the surfaces which are placed together so that the taper of the holes shall be in opposite directions.

The holes to be punched fair, and opposite each other; unfair holes will render the piece of work badly punched liable to rejection. Where holes cannot be truly punched, they must be drilled through fair. Every hole requiring to be countersunk must be countersunk quite through the plate.

Breadth of lap in seams for double-riveting to be 5½ times the diameter of the hole punched, and in the butts to be six diameters of the hole punched.

10. *Hold Keelsons and Stringers.*—The centre keelson, if standing above

floors, may be either box-form with top, bottom, and two side plates, the inside width of the box to be two-thirds the depth of the side plates, the angle-irons to be all outside, and of size per Table 5. Or double centre plate, with top and bottom plate. Width of top plate to be two-thirds the depth of centre plates. The depth and thickness of side plates for box keelsons, and of centre plates for centre-plate keelsons, also angle-irons for the latter to be in accordance with Table 5; or a single centre plate with top plate may be used provided the plates be one and a half times the thickness given in the Table; the vertical plate to be fitted with double butt-strips, and the butt-strips of the top plate to be treble-riveted.

Or a centre line intercostal keelson may be adopted in vessels of 1200 tons and under, provided the intercostal plates be of the thickness given in Table 1 for centre-plate keels, and each plate be secured to the floors, and each other by double angle-irons on their fore and after ends; of the size given in Table 2 for reverse frames; the intercostal plates to stand above the floors to a height equal to the largest flange of the angle-irons for keelsons, per Table 5, two of which angle-irons are to be riveted, one on either side, to the intercostal plates with a bulb-iron of the size required for lower-deck beams between them.

This keelson not to be adopted in vessels over ten depths in length without proportionate increase of strength; particulars to be submitted to the Committee.

An intercostal keelson is to be put at half floor, in vessels exceeding 32 feet beam, for two-thirds of the vessel's length where practicable, to be fastened to side of floors, and to project above floors to form a keelson, with double angle-irons riveted to the upper edge, of size as per Table 5. The half floor intercostal plates to be the same thickness as floors.

All vessels to have two bilge stringers, one at lower and one at upper turn of bilge, formed of double angle-irons of size per Table 5.

Vessels exceeding 15 feet in lower hold to have a side stringer of double angle-irons, size as per Table 5, between the upper bilge stringer and the lower-deck beams. Vessels exceeding 16 feet and under 18 feet in depth of lower hold to have a bulb-iron, size of lower-deck beams, secured between the angle-irons which form the side stringers, and to have a bulb-iron riveted between the angle-irons of either upper or lower bilge stringers for two-thirds of the vessel's length, same size as lower-deck beams.

All keelsons to extend fore and aft as far as practicable, and to be continued through the bulkheads, or if stopped at the bulkheads to be connected therewith to the satisfaction of the Surveyor.

The centre keelson may be reduced at ends to two-thirds sectional area, per Table 5; the reduction to extend from heel of fore and of mizen mast, and not to exceed one-third length of vessel when taken together.

11. *Beam Stringer and Deck Ties.*—Stringer plates are to be laid upon the ends of each tier of beams and riveted thereto through both beam angles, and are not to be less in width and thickness than per Table 5. Main-deck stringer plates may be reduced in width one-fourth at ends of vessel, and one-sixteenth of an inch in thickness; this reduction to begin at one-fifth the vessel's length from each end.

Hold and orlop beam-stringers may be reduced one-sixteenth of an inch in thickness for one-fifth of their length from each end.

In vessels of and over twelve depths in length, the upper-deck stringer and tie plates may be reduced as follows:—At each end for a distance equal to one-tenth of the vessel's length, two-sixteenths of an inch; for a further distance on each side, equal to one-fifth of vessel's length, one-sixteenth of an inch.

Stringer plates must in no case be reduced below six-sixteenths of an inch in thickness.

Stringer plates on upper deck, and in vessels with three decks on main and upper decks, to be fitted and riveted to shell plates with angles as per Table for keelsons. A continuous angle-iron of the size for stringers, per Table 5, to be fitted to inside of frames upon all beam stringers below upper deck, whether the stringers are carried to shell plates or not; and also on the upper-deck stringer when the frames are carried through the stringer to form bulwark stanchions. All stringer and tie plates to extend fore and aft where practicable, and not to be stopped at bulkheads.

If desired, stringers on orlop-deck beams may be diminished in width not exceeding one-third, if proportionately increased in thickness; angle-iron on gunwale stringer to be formed round the scupper holes, and if butted at scuppers to be otherwise strengthened.

Poop and topgallant forecastle stringers and ties may be one-third lighter than lower-deck stringers. When the upper deck, aft, is raised so as to form a "break," the deck ties and stringers are to be continued on the raised portion as they would have been had no "break" occurred in the deck line, and the deck stringers are to be made to scarph with the raised portion to the satisfaction of the Surveyors. The foremost plate of the "break" must project into and form part of the bulwark for half the length of the plate, so as to afford a scarph of 3 to 4 feet in length.

Tie plates, ranging all fore and aft, of size per Table 5, to be laid upon each tier of beams on both sides of hatches, and riveted to both angle-irons of the beams, and at ends of vessel to stringer plates. Ties on main deck to have double angle-irons, of size per Table 5, riveted on their upper sides, ranging all fore and aft, and with the tie plates riveted to beams and to stringer plates at ends of vessel.

On the hold beams of vessels under 600 tons, and on orlop-deck beams where no deck is laid, two angle-irons, back to back, on each side of hatchways, same size as for keelsons, and riveted through and through and to the beams, may be substituted for tie plates.

Mast partners at deck, where wedged, to be plated over twice the width of the hole cut out of them, and to take three beams in length.

12. *Bulkheads.*—One "Collision Bulkhead" must be placed at a reasonable distance from forward, subject to the approval of the Surveyor. Bulkheads to be stayed on both sides with angle-irons four feet apart, same size as frames, one set vertical and one set horizontal. In steamers a bulkhead must be fitted at each end of the engine and boiler space, and one at the fore end of the screw stern pipe.

All bulkheads to be well riveted either between a double frame, or if to a single one, to be well and strongly secured by brackets.

All lining pieces in way of bulkheads to extend from frame forward to frame abaft, and to be made perfectly tight.

Bulkheads to be one-third lighter than shell plates, and to be fitted with sluice valves or cocks, or to have a pump in each compartment.

13. *Rivets and Riveting.*—Rivets to be in accordance with Table 6.

All vessels to be double-riveted in bottom, bilges, and sheer-strake, and all vessels above 600 tons to be double-riveted throughout.

Vessels above 1000 tons to be treble-riveted in the main-deck stringers and sheer-strakes for half the vessel's length amidships.

All double or treble riveting in butts of plates to be in parallel rows, or what is termed chain riveting. All butts to be double or treble riveted, as per Rules.

Rivets to be four diameters apart, from centre to centre, longitudinally in seams and vertically in butts, except in the butts where treble-riveting is required, when the rivets in the row farthest from the butt may be spaced eight diameters apart, centre to centre. Rivets in framing to be eight times their diameter apart, from centre to centre, and to be of the size required for shell plating.

Rivets in bar keels, stems, and sternposts, to be one-eighth of an inch larger than in the butts of garboard-strakes.

Iron decks to have their butts treble-riveted amidships, for one-half the vessel's length.

Rivets in the seams of sheer and garboard strakes may be of the size required by Table 6 for the strakes of plating adjoining them, but the rivets in the butts of the sheer and garboard strakes must be of the size required by Table 6 for plates of equal thickness.

Rivets in shell plating, in deck ties and stringers, in centre plate and in flat keels and keelsons are to have their necks bevelled under their heads, so as to fill the countersink made in punching, and their heads must be no thicker than two-thirds the diameter of the rivet.

14. *Rudder, Rudder Heads and Pins.*—Rudder frames to be forged solid. Rudder heads and pins not to be less than the diameters given in Table 1. Rudder heads of screw steamers in all cases to be one inch larger in diameter than as given in Table 1.

15. *Quality of Iron and Character of Workmanship.*—All plates to be of the best quality (branded best, with maker's name), tough and malleable, the sheared edges to be free from rip, the surface free from flaws and blisters, and the punchings reasonably free from crack upon the convex side. The absolute mean breaking strain to be 20 tons per square inch. Brittle or coarsely crystalline iron to be rejected. The strain per square inch of broken section to be 24 tons.

Angle-irons to be free from veins and cracked holes. Care to be taken that the iron be not burned in the bending furnace. Rivet iron to be free from cracks and veins, when laid up and finished. Riveted seams, butts, frames, floors, keelsons, stringers, and ties to be laid up quite close, so as to prevent the introduction of the thinnest knife used for trying riveted work.

Each rivet to fill the hole and be laid up close round the head, and when finished to be flush and fair, neither projecting above nor sinking below the surface of the part through which it passes. Keel rivets must project slightly.

Details not specified to be approved by the Surveyors.

16. *Excessive Proportions.*—Vessels whose depth is less than five-eighths of their breadth to have their upper-deck longitudinal deck-ties and stringer-plates increased in width one-tenth for half the vessel's length amidships. When the depth is less than half the breadth, the plans to be submitted for the approval of the Committee.

Vessels exceeding 12 depths in length to have their sheer-strakes doubled for half the vessel's length amidships.

Vessels above 13 depths, and not exceeding 14 depths in length, to have their sheer-strakes doubled for two-thirds the vessel's length amidships, and to have the sheer-strakes and upper-deck stringers and ties treble-riveted for half the length amidships.

In place of doubling the sheer-strake, a single thickness of plate may be used, when practicable, provided it be once and a half the thickness of the plate required by Table 2, with treble-riveted butts and butt-strips one-sixteenth of an inch thicker than the plates which they connect together.

Plans of all vessels exceeding 14 depths in length to be submitted for the approval of the Committee.

17. Vessels above 12 depths in length and exceeding 1500 tons, to have the plating from keel to upper part of bilge increased one-sixteenth of an inch in thickness for half the vessel's length amidships.

Other scantlings which depend on excessive proportions are included in Tables 2 and 5.

18. The dimensions referred to above are to be taken as follows:—

Length, to be taken between the perpendiculars.

Breadth, to be the extreme breadth.

Depth of hold is to be taken from the upperside of floors to underside of main deck, in vessels of two tiers of beams, and to underside of spar-deck in vessels of more than two tiers of beams.

19. *Decks.*—For vessels up to 300 tons the upper decks are to be not less than 3 inches thick, thence to 700 tons, 3½ inches. Above 700 tons, decks are not to be less than 4 inches. Lower decks, in vessels above 500 tons, to be 3 inches.

All decks should be fastened with galvanized screw-bolts and nuts.

Poop decks should be fastened with galvanized wood screws.

Waterways, margin planks, and plank sheers to be made of teak or green-heart, or other approved wood.

Deck planks may be of pine, and must be free from sap, rot, and shakes, reasonably free from knots, and caulked perfectly watertight. Upper, fore-castle, main, poop, and quarterdeck planks not to be more than 6 inches wide.

Decks laid with East India teak may be one-sixth less in thickness.

Ceiling in flat of bottom to be laid in hatches.

Spar-deck Vessels.—Vessels whose depth is equal to or exceeds three-fourths of their breadth, and which are constructed with three tiers of beams, may be classed as spar-decked, and will be allowed a reduction of one-sixth in the scantling of material above the main deck.

For floors of spar-deck vessels, see Section 5.

20. *Windlasses.*—Windlass to be sufficiently strong, and firmly secured with bitts at decks. All windlass spindles to be in one length. Diameter of windlass spindles to be in accordance with Table 1.

21. *Masts, Spars, and Sails.*—All sailing and steam vessels to have a full and complete set of masts, yards, sails, &c.; sufficient in size and strength, and of the best quality. Vessels under 600 tons to have one spare lower yard, and one spare topmast; above 600 tons, to have a spare topsail yard in addition. All vessels to have one complete suit of sails, and sailing and auxiliary steam vessels at least one spare foresail, foretopsail, and foretopmast staysail.

Iron masts to be made according to Table 7.

Heel of the mast to be properly supported in all vessels.

Angle-irons when used, to run the whole length. Seams to be single-riveted. The butts in masts in way of wedging-deck to be treble-riveted, remainder double.

Yards of steel to be as per Table 7. The seams single-riveted, the butts for one-third the length in the middle treble-riveted, remainder double.

Angle-irons, when used, to run the whole length.

Yards of iron to be one-third heavier than steel yards.

22. *Painting and Cementing.*—All the surfaces to be properly painted with good oil paint.

To prevent wear from wash of bilge water in the bottoms of iron vessels, fresh Portland cement is to be laid on so as to cover the plates, frames, and rivet-heads. The cement is to be raised in the centre to the level of the limber holes, and to be taken up to upper part of bilge.

23. *Extension of Character.*—All vessels which may have completed the term of years originally assigned to them, and all vessels for which a class is sought, but which are of an age beyond the term for which their scantlings would have originally entitled them, may, by undergoing the following survey, obtain such class or extension of class as the Committee may, on the Report of Survey, consider their actual condition entitles them to.

Form of Survey for Extension of Class.—The vessel to be placed in dry dock, all ceiling to be removed, and both surfaces of plating and the wood-work of hull to be thoroughly exposed by chipping and scraping; in this condition an inspection is to be made, and holes bored in the plating, stringers, and floors, and in such other parts as may be directed, but in no case to be less than one hole for every ten tons register tonnage.

SCANTLING FOR STEAMERS.

The Committee having been requested to indicate the reductions from the Scantling for vessels of the Twenty Years' Class, which would be permitted for steamers intended to class for lower grades, publish the following for the guidance of owners and builders:—

STEAMERS TO CLASS EIGHTEEN YEARS.

30. The following reductions from the scantling for the Twenty Years' Class will be permitted for Steamers intended to class Eighteen Years:—

Plating.—A reduction of one-sixteenth of an inch in thickness in the sheer-strake, in the garboard-strake, and in the side plating.

Stringers on Beams.—A reduction of one-sixteenth of an inch in thickness, and of one-twelfth in width of the lower-beam stringers.

Ties on Beams.—One-third reduction in width and one-sixteenth of an inch in thickness.

Floors.—A slight reduction, either in moulding depth, or thickness.

Keelsons.—A reduction, not exceeding two-sixteenths of an inch, in the total thickness of the centre plate or plates, and a slight reduction in the angle-irons.

Bilge Keelsons.—A reduction of one-fifth in breadth of one flange of the angles, and a reduction of two-sixteenths of an inch in thickness.

STEAMERS TO CLASS SIXTEEN YEARS.

31. The following reductions from the Scantling for the Twenty Years' Class will be permitted for Steamers intended to class Sixteen Years :—

Plating.—A reduction of one-sixteenth of an inch in thickness in the sheer-strake and in the garboard-strake; a reduction of two-sixteenths in the side, and of one-sixteenth of an inch in the bottom and bilge plating.

Riveting.—Single-riveting permitted in seams of side plating from bilges upwards.

Stringers on Beams.—A reduction in width of one-twelfth in lower-beam stringers, and of one-sixteenth of an inch in thickness in both upper and lower beam stringers.

Ties on Beams.—One-third reduction in width and one-sixteenth of an inch in thickness.

Frames.—A slight reduction in size or thickness.

Floors.—A slight reduction either in moulding depth, or thickness.

Keelsons.—A reduction not exceeding two-sixteenths of an inch in the total thickness of centre plate or plates, and a slight reduction in the angle-irons.

Bilge Keelsons.—A reduction of one-fifth in breadth of one flange of the angles, and a reduction of two-sixteenths of an inch in thickness.

32. In the sixteen and eighteen years' classes, as in the twenty years' class, the Committee will take into consideration any compensation which may be offered for deviations from the approved scantling.

W. W. RUNDELL, *Secretary.*

Liverpool, September, 1866.

33. The above regulations will also apply to Sailing Vessels intended to class for these grades.

W. W. RUNDELL, *Secretary.*

Liverpool, September 10, 1867.

TABLE I.

Tonnage	250	500	750	1000	1500	2000	2500	3000
Thickness of Centre Plate Keels	$\frac{8}{16}$	$\frac{9}{16}$	$\frac{10}{16}$	$\frac{11}{16}$	$\frac{12}{16}$	$\frac{12}{16}$	$\frac{13}{16}$	$\frac{13}{16}$
Size of Side Plates for Centre Plate Keels ..	$6 \times \frac{12}{16}$	$7 \times \frac{12}{16}$	$8 \times \frac{13}{16}$	$9 \times \frac{14}{16}$	$10 \times 1\frac{1}{8}$	$12 \times 1\frac{1}{8}$	$12 \times 1\frac{1}{8}$	$13 \times 1\frac{1}{8}$
Size of Bar Keels, Stems and Sternposts, in inches	6×2	7×2	$8 \times 2\frac{1}{4}$	$9 \times 2\frac{1}{2}$	10×3	12×3	12×3	13×3
Length of Scarphs in Bar Keels, in inches	18	19	20	22	24	26	28	30
Diameter of Rudder Heads, in inches.. ..	$3\frac{1}{2}$	$4\frac{1}{2}$	5	6	$6\frac{1}{2}$	7	8	$8\frac{1}{2}$
Diameter of Rudder Heels and Pintles ..	2	$2\frac{1}{2}$	$2\frac{3}{4}$	3	$3\frac{1}{4}$	$3\frac{1}{2}$	4	$4\frac{1}{4}$
Diameter of Windlass Spindles	3	$3\frac{1}{2}$	$3\frac{3}{4}$	4	$4\frac{1}{2}$	5	$5\frac{1}{2}$	6

NOTE.—Rudder Heads in Screw Steamers, in all cases, to be one inch larger in diameter.

TABLE II.

Frames, Reverse Frames, and Plating.

Depth of Hold in Feet.	Sizes and Spacing of Frames and Reverse Frames, in Inches. Frames—all Grades.	Reverse Frames.	Space apart of Frames.	Thickness of Plating for Three-fifths Length of Vessel, amidships, in Sixteenths of an Inch. Sheer Strakes. 20 Years' Class.	Sheer Strakes. 18 Years' Class.	Sheer Strakes. 16 Years' Class.	Sides. 20 Years' Class.	Sides. 18 Years' Class.	Sides. 16 Years' Class.	Bilge and Bottom. 20 Years' Class.	Bilge and Bottom. 18 Years' Class.	Bilge and Bottom. 16 Years' Class.	Garboard. 20 Years' Class.	Garboard. 18 Years' Class.	Garboard. 16 Years' Class.	Depth of Hold in Feet.
	ths.	ths.														
5	2½ × 2½ × 5-16	2 × 2 × 4-16	Frames in vessels below 1000 tons are to be spaced 21 ins. apart from centre to centre, and in vessels of 1000 tons and above they may be spaced 24 inches apart. See *Sect. 4.*	8	7	7	6	5	4	6	6	5	7	6	6	5
10	3 × 2½ × 5-16	2½ × 2 × 4-16		8	7	7	6	5	4	6	6	5	7	6	6	10
12½	3 × 2½ × 6-16	2½ × 2½ × 4-16		9	8	8	7	6	5	7	7	6	8	7	7	12½
14	3 × 3 × 6-16	2½ × 2½ × 5-16		9	8	8	7	6	5	7	7	6	8	7	7	14
15	3 × 3 × 6-16	2½ × 2½ × 5-16		10	9	9	8	7	6	8	8	7	9	8	8	15
16	3½ × 3 × 6-16	2½ × 2½ × 6-16		10	9	9	8	7	6	8	8	7	9	8	8	16
17½	4 × 3 × 6-16	2½ × 2½ × 6-16		11	10	10	9	8	7	9	9	8	10	9	9	17½
18	4 × 3 × 7-16	3 × 2½ × 6-16		11	10	10	9	8	7	9	9	8	10	9	9	18
20	4½ × 3 × 8-16	3½ × 3 × 6-16		12	11	11	10	9	8	10	10	9	11	10	10	20
22	5 × 3 × 9-16	3½ × 3½ × 7-16		12	11	11	10	9	8	10	10	9	11	10	10	22
24	5 × 3 × 9-16	3½ × 3½ × 7-16		12	11	11	10	9	8	10	10	9	11	10	10	24
26	5½ × 3 × 9-16	4 × 3½ × 8-16		13	12	12	11	10	9	11	11	10	12	11	11	26
28	5½ × 3½ × 9-16	4 × 3½ × 8-16		13	12	12	11	10	9	11	11	10	12	11	11	28
30	6 × 3½ × 10-16	4½ × 3½ × 9-16		13	12	12	11	10	9	11	11	10	12	11	11	30

Note.—All vessels of 1200 tons and above are to have three strakes of plates at the bilges, increased one-sixteenth of an inch in thickness for half the vessel's length amidships; also, in all vessels of 1500 tons, and above, and exceeding 12 depths in length, a further sixteenth of an inch in thickness is to be added to all bilge and bottom plating for a similar length. All vessels of and over 13 depths in length are to have one-sixteenth of an inch added to the tabulated thickness of all shell plating, in addition to the previous requirements, excepting only so much of the bilge and bottom plating of vesssels of 1500 tons and above as have already been increased in thickness by the preceding Rules.

All tiers or strakes of plating, after continuing the prescribed thickness for the given length, may from this distance be reduced in thickness by a fair taper towards the ends of the vessel, and at the extreme ends a reduction of one-sixth from this thickness will be allowed.

For increase of sheer-strake see *Section* 16.

To face p. 521.

TABLE V.—BEAM STRINGERS, TIE

NOTE.—The

Length of Vessel in Feet.	Thickness of all Stringers, Deck Ties, and their Angles, in 16ths of an inch.	Width of Hold and Orlop beam Stringers.	Width of Hold and Orlop beam Ties.	9	10	10½	11	11½	12	12½	13
100	6	18	6½	30 7¾	25 6½	25 6½	25 6½	25 6½	25 6½	25 6½	....
105	6	18	6½	30 7¾	27½ 7	25 6½	25 6½	25 6½	25 6½	25 6½	25 6½
110	6	18	6½	*30 7¾	27½ 7	27½ 7	25 6½	25 6½	25 6½	25 6½	25 6½
115	6	18	6½	*30 7¾	30 7¾	27½ 7	27½ 7	25 6½	25 6½	25 6½	25 6½
120	7	20	7½	*36½ 10½	*31 9	31 9	28½ 8¼	28½ 8¼	26 7½	26 7½	26 7½
125	7	20	7½	*36½ 10½	*31 9	31 9	31 9	28½ 8¼	28½ 8¼	26 7½	26 7½
130	7	20½	9	....	*32½ 10¾	*32½ 10¾	32½ 10¾	32½ 10¾	29½ 9¾	29½ 9¾	27 9
135	7	20½	9	....	*38 12½	*32½ 10¾	*32½ 10¾	32½ 10¾	32½ 10¾	29½ 9¾	29½ 9¾
140	7	21	9½	....	....	*39 13½	*33½ 11½	*33½ 11½	33½ 11½	33½ 11½	31 10½
145	7	21	9½	....	....	*39 13½	*39 13½	*33½ 11½	*33½ 11½	33½ 11½	33½ 11½
150	8	21½	10	....	....	....	*39 13½	*39 13½	*33½ 11½	*33½ 11½	33½ 11½
155	8	21½	10	....	....	....	....	*40½ 14	*35 12	*35 12	35 12
160	8	22	10½	....	....	....	....	*42 14½	*42 14½	*36 12½	*36 12½
165	8	22	10½	....	....	....	....	....	*42 14½	*42 14½	*36 12½
170	8	22½	11	....	....	....	....	....	....	*43½ 15	*43½ 15
175	8	22½	11	....	....	....	....	....	....	....	*43½ 15
180	9	23	11½	....	....	....	....	....	....	....	*43½ 15
185	9	23	11½	....	....	....	....	....	....	....	....
190	9	24	12	....	....	....	....	....	....	....	....
195	9	24	12	....	....	....	....	....	....	....	....
200	9	25	12½	....	....	....	....	....	....	....	....
205	9	25	12½	....	....	....	....	....	....	....	....
210	9	26	13	....	....	....	....	....	....	....	....
215	9	26	13	....	....	....	....	....	....	....	....
220	10	27	13½	....	....	....	....	....	....	....	....
225	10	27	13½	....	....	....	....	....	....	....	....
230	10	28	14	....	....	....	....	....	....	....	....
235	10	28	14	....	....	....	....	....	....	....	....
240	10	29	14½	....	....	....	....	....	....	....	....
245	10	29	14½	....	....	....	....	....	....	....	....
250	10	30	15	....	....	....	....	....	....	....	....
255	10	30	15	....	....	....	....	....	....	....	....
260	10	30½	15½	....	....	....	....	....	....	....	....
265	10	30½	15½	....	....	....	....	....	....	....	....
270	11	31	16	....	....	....	....	....	....	....	....
275	11	31	16	....	....	....	....	....	....	....	....
280	11	31½	16	....	....	....	....	....	....	....	....
285	11	31½	16	....	....	....	....	....	....	....	....
290	11	32	16½	....	....	....	....	....	....	....	....
295	11	32	16½	....	....	....	....	....	....	....	....
300	11	32½	17	....	....	....	....	....	....	....	....
305	11	32½	17	....	....	....	....	....	....	....	....
310	11	33	17½	....	....	....	....	....	....	....	....
315	11	33	17½	....	....	....	....	....	....	....	....
320	12	33½	18	....	....	....	....	....	....	....	....
325	12	33½	18	....	....	....	....	....	....	....	....
330	12	34	18	....	....	....	....	....	....	....	....
340	12	34½	18	....	....	....	....	....	....	....	....
350	12	36	18	....	....	....	....	....	....	....	....
360	12	38	18	....	....	....	....	....	....	....	....
				9	**10**	**10½**	**11**	**11½**	**12**	**12½**	**13**

* All Stringers and Ties marked thus are allowed to be reduced in accordance with 11th section of the Ta
sixteenth of an inch in thickness of all Stringers. For the 16 or 18 years' grades a reduction in Ties on Beams c

PLATES, CENTRE LINE KEELSONS, AND THEIR ANGLE-IRONS

main body of this Table shows the width of Stringers and Tie Plates for Upper Decks of Ve

DEPTH OF VESSEL IN FEET, MEASURED FROM TO

13½		14		14½		15		15½		16		16½		17		17½		18		18½		19	
....		....		....		....		....		....		....		....		....		....		....		....	
....		....		....		....		....		....		....		....		....		....		....		....	
25	6½	....		....		....		....		....		....		....		....		....		....		....	
25	6½	25	6½	....		....		....		....		....		....		....		....		....		....	
26	7½	26	7½	26	7½	....		....		....		....		....		....		....		....		....	
26	7½	26	7½	26	7½	26	7½	....		....		....		....		....		....		....		....	
27	9	27	9	27	9	27	9	27	9	....		....		....		....		....		....		....	
27	9	27	9	27	9	27	9	27	9	27	9	....		....		....		....		....		....	
31	10½	28	9½	28	9½	28	9½	28	9½	28	9½	28	9½	....		....		....		....		....	
31	10½	31	10½	28	9½	28	9½	28	9½	28	9½	28	9½	28	9½	....		....		....		....	
33½	11½	31	10½	31	10½	31	10½	28	9½	28	9½	28	9½	28	9½	28	9½	....		....		....	
35	12	35	12	32	11	32	11	29	10	29	10	29	10	29	10	29	10	29	10	....		....	
36	12½	36	12½	36	12½	33	11½	33	11½	30	10½	30	10½	30	10½	30	10½	30	10½	30	10½	....	
*36	12½	36	12½	36	12½	33	11½	33	11½	33	11½	30	10½	30	10½	30	10½	30	10½	30	10½	30	10½
*37	13¼	*37	13¼	37	13¼	37	13¼	34	12	34	12	34	12	31	11	31	11	31	11	31	11	31	11
*37	13¼	*37	13¼	*37	13¼	37	13¼	37	13¼	34	12	34	12	34	12	31	11	31	11	31	11	31	11
*37	13¼	*37	13¼	*37	13¼	37	13¼	37	13¼	34	12	34	12	34	12	34	12	31	11	31	11	31	11
*45	15½	*45	15½	*38½	13¾	*38½	13¾	38½	13¾	38½	13¾	38½	13¾	35	12½	35	12½	32	12½	32	12½	32	11½
....		*46	16¾	*46	16¾	*39½	14½	*39½	14½	39½	14½	39½	14½	39½	14½	36	13	36	13	36	13	33	12
....		*46	16¾	*46	16¾	*39½	14½	*39½	14½	*39½	14½	39½	14½	39½	14½	39½	14½	35	13	35	13	35	13
....		....		*47½	17½	*47½	17½	*41	14¾	*41	14¾	*41	14¾	41	14¾	41	14¾	41	14¾	37½	13¾	37½	13¾
....		....		....		*47½	17½	*47½	17½	*41	14¾	*41	14¾	*41	14¾	41	14¾	41	14¾	41	14¾	37½	13¾
....		....		....		*49	18	*49	18	*49	18	*42	15½	*42	15½	*42	15½	42	15½	42	15½	42	15½
....		....		....		....		*49	18	*49	18	*49	18	*42	15½	*42	15½	42	15½	42	15½	42	15½
....		....		....		....		....		*49	18	*49	18	*42	15½	*42	15½	*42	15½	42	15½	42	15½
....		....		....		....		....		....		*50½	18½	*50½	18½	*43	16	*43	16	*43	16	43	16
....		....		....		....		....		....		*52	19½	*52	19½	*52	19½	*44	16½	*44	16½	*44	16½
....		....		....		....		....		....		....		*52	19½	*52	19½	*44	16½	*44	16½	*44	16½
....		....		....		....		....		....		....		....		*53	20¼	*53	20¼	*45½	17½	*45½	17½
....		....		....		....		....		....		....		....		*53	20¼	*53	20¼	*53	20¼	*45½	17½
....		....		....		....		....		....		....		....		....		*54½	21	*54½	21	*54½	21
....		....		....		....		....		....		....		....		....		....		*54½	21	*54½	21
....		....		....		....		....		....		....		....		....		....		....		*56	21½
....		....		....		....		....		....		....		....		....		....		....		*56	21½
....		....		....		....		....		....		....		....		....		....		....		....	
....		....		....		....		....		....		....		....		....		....		....		....	
....		....		....		....		....		....		....		....		....		....		....		....	
....		....		....		....		....		....		....		....		....		....		....		....	
....		....		....		....		....		....		....		....		....		....		....		....	
....		....		....		....		....		....		....		....		....		....		....		....	
....		....		....		....		....		....		....		....		....		....		....		....	
....		....		....		....		....		....		....		....		....		....		....		....	
....		....		....		....		....		....		....		....		....		....		....		....	
....		....		....		....		....		....		....		....		....		....		....		....	
....		....		....		....		....		....		....		....		....		....		....		....	
....		....		....		....		....		....		....		....		....		....		....		....	
....		....		....		....		....		....		....		....		....		....		....		....	
....		....		....		....		....		....		....		....		....		....		....		....	
....		....		....		....		....		....		....		....		....		....		....		....	
....		....		....		....		....		....		....		....		....		....		....		....	
13½		14		14½		15		15½		16		16½		17		17½		18		18½		19	

bles of Scantling. See also provision for excessive proportions, sections 16 and 17. For the 18 years' grade a red of one-third in width and one-sixteenth of an inch in thickness is also permitted. For the reduction of Keelsons at

S FOR VESSELS (STEAMERS AND SAILING VESSELS) TAKING T

essels within the proportion of 14 depths to length, and of any of the undermentioned dimensi

P OF FLOOR TO TOP OF UPPER-DECK BEAM.

First Figures in each column indicate Stringers; Second Figures indicate Tie Plates. As for example:—For a Vessel 100 feet long and 10 feet deep the Stringer will be 25 inches, and the Tie Plates 6½ inches.

19½	20	21	22	23	24	25	26	27	28	29	30
....	....	....	....	....	....	....	....	....	....	....	....
....	....	....	....	....	....	....	....	....	....	....	....
....	....	....	....	....	....	....	....	....	....	....	....
....	....	....	....	....	....	....	....	....	....	....	....
....	....	....	....	....	....	....	....	....	....	....	....
....	....	....	....	....	....	....	....	....	....	....	....
....	....	....	....	....	....	....	....	....	....	....	....
....	....	....	....	....	....	....	....	....	....	....	....
....	....	....	....	....	....	....	....	....	....	....	....
....	....										
....	....										
....	....										
....	....	....	....	....	....	....	....	....	....	....	....
....	....	....	....	....	....	....	....	....	....	....	....
31 11	....	....	....	....	....	....	....	....	....	....	....
31 11	31 11	....	....	....	....	....	....	....	....	....	....
31 11	31 11	31 11	....	....	....	....	....	....	....	....	....
32 11½	32 11½	32 11½	....	....	....	....	....	....	....	....	....
33 12	33 12	33 12	33 12	....	....	....	....	....	....	....	....
33 12	33 12	33 12	33 12	33 12	....	....	....	....	....	....	....
37½ 13¾	34 12½	34 12½	34 12½	34 12½	....	....	....	....	....	....	....
37½ 13¾	37½ 13¾	34 12½	34 12½	34 12½	34 12½	....	....	....	....	....	....
38½ 14¼	38½ 14¼	35 13	35 13	35 13	35 13	35 13	....	....	....	....	....
42 15½	38½ 14¼	38½ 14¼	35 13	35 13	35 13	35 13	....	....	....	....	....
42 15½	38½ 14¼	38½ 14¼	38½ 14¼	35 13	35 13	35 13	35 13	....	....	....	....
43 16	43 16	39½ 14¾	39½ 14¾	36 13½	36 13½	36 13½	36 13½	....	....	....	....
44 16½	44 16½	41 15¼	41 15¼	37 14	37 14	37 14	37 14	37 14	....	....	....
*44 16½	44 16½	44 16½	41 15¼	41 15¼	37 14	37 14	37 14	37 14	....	....	....
*45½ 17½	*45½ 17½	45½ 17½	42 15¾	42 15¾	38 14½	38 14½	38 14½	38 14½	38 14½	....	....
*45½ 17½	*45½ 17½	45½ 17½	45½ 17½	42 15¾	38 14½	38 14½	38 14½	38 14½	38 14½	....	....
*47 18	*47 18	47 18	47 18	43 16½	43 16½	39 15	39 15	39 15	39 15	39 15	....
*54½ 21	*47 18	*47 18	47 18	47 18	43 16½	43 16½	39 15	39 15	39 15	39 15	....
*56 21½	*48 18½	*48 18½	48 18½	48 18½	44 17	44 17	40 15½	40 15½	40 15½	40 15½	40 15½
*56 21½	*56 21½	*48 18½	*48 18½	48 18½	48 18½	44 17	44 17	40 15½	40 15½	40 15½	40 15½
*56 21½	*56 21½	*48 18½	*48 18½	48 18½	48 18½	44 17	44 17	40 15½	40 15½	40 15½	40 15½
....	*57½ 22	*57½ 22	*49 19	49 19	49 19	45 17½	45 17½	45 17½	41 16	41 16	41 16
....	*59 22	*59 22	*50½ 19	*50½ 19	50½ 19	50½ 19	46 17½	46 17½	42 16	42 16	42 16
....	....	*59 22	*50½ 19	*50½ 19	50½ 19	50½ 19	46 17½	46 17½	46 17½	42 16	42 16
....	....	*60 22½	*60 22½	*51½ 19½	*51½ 19½	51½ 19½	51½ 19½	47½ 18	47½ 18	43 16½	43 16½
....	....	....	*60 23½	*51½ 19½	*51½ 19½	51½ 19½	51½ 19½	47½ 18	47½ 18	47½ 18	43 16½
....	....	....	*61½ 23	*61½ 23	*53 20	*53 20	53 20	53 20	48½ 18½	48½ 18½	44 17
....	....	....	*61½ 23	*61½ 23	*53 20	*53 20	53 20	53 20	48½ 18½	48½ 18½	48½ 18½
....	....	....	....	*63 24½	*63 24½	*54 21	54 21	54 21	54 21	49½ 19¼	49½ 19¼
....	....	....	....	*63 24½	*63 24½	*54 21	*54 21	54 21	54 21	49½ 19¼	49½ 19¼
....	....	....	....	*63 24½	*63 24½	*54 21	*54 21	54 21	54 21	49½ 19¼	49½ 19¼
....	....	....	....	....	*64½ 25	*64½ 25	*55 21½	*55 21½	55 21½	55 21½	50½ 19¾
....	....	....	....	....	*66 25	*66 25	*56½ 21½	*56½ 21½	56½ 21½	56½ 21½	52 19¾
....	....	....	....	....	....	*67 25	*67 25	*57½ 21½	*57½ 21½	57½ 21½	57½ 21½
....	....	....	....	....	....	....	*68½ 25	*59 21½	*59 21½	*59 21½	59 21½
....	....	....	....	....	....	....	*70 25	*70 25	*60 21½	*60 21½	*60 21½
19½	20	21	22	23	24	25	26	27	28	29	30

luction in lower Stringers is permitted of one-sixteenth of an inch in thickness and one-twelfth in width. For the 1 ends, see section 10. For the 18 years' and 16 years' grades a reduction not exceeding two-sixteenths in the total t

THE TWENTY YEARS' GRADE.

ions.

31	32	33	34	35	36	37	Depth of Centre Keelson above Floors. Inches.	Thickness of Plates for Box and Double Centre Plate Keelsons in 16ths of an inch.	Sizes of Flanges of Angle-Irons on Stringers, Tie Plates, and Keelsons.	Length of Vessel in Feet.
....	....	....	....	....	....	....	10	6	3 × 3	**100**
....	....	....	....	....	....	....	10	6	3 × 3	**105**
....	....	....	....	....	....	....	10	6	3 × 3	**110**
....	....	....	....	....	....	....	10	6	3 × 3	**115**
....	....	....	....	....	....	....	10	7	3½ × 3	**120**
....	....	....	....	....	....	....	10	7	3½ × 3	**125**
....	....	....	....	....	....	....	12	7	4 × 3½	**130**
....	....	....	....	....	....	....	12	7	4 × 3½	**135**
....	....	....	....	....	....	....	12	7	4 × 3½	**140**
....	....	....	....	....	....	....	12	7	4 × 3½	**145**
....	....	....	....	....	....	....	14	8	4 × 4	**150**
....	....	....	....	....	....	....	14	8	4 × 4	**155**
....	....	....	....	....	....	....	14	8	4 × 4	**160**
....	....	....	....	....	....	....	14	8	4 × 4	**165**
....	....	....	....	....	....	....	16	8	4 × 4	**170**
....	....	....	....	....	....	....	16	8	4 × 4	**175**
....	....	....	....	....	....	....	16	9	5 × 3½	**180**
....	....	....	....	....	....	....	16	9	5 × 3½	**185**
....	....	....	....	....	....	....	16	9	5 × 3½	**190**
....	....	....	....	....	....	....	16	9	5 × 3½	**195**
....	....	....	....	....	....	....	18	9	5 × 3½	**200**
....	....	....	....	....	....	....	18	9	5 × 3½	**205**
....	....	....	....	....	....	....	18	9	5 × 3½	**210**
....	....	....	....	....	....	....	18	9	5 × 3½	**215**
....	....	....	....	....	....	....	19	10	5 × 4	**220**
....	....	....	....	....	....	....	19	10	5 × 4	**225**
....	....	....	....	....	....	....	19	10	5 × 4	**230**
....	....	....	....	....	....	....	19	10	5 × 4	**235**
....	....	....	....	....	....	....	20	10	5 × 4	**240**
....	....	....	....	....	....	....	20	10	5 × 4	**245**
....	....	....	....	....	....	....	20	10	5 × 4	**250**
....	....	....	....	....	....	....	20	10	5 × 4	**255**
....	....	....	....	....	....	....	21	10	5 × 4	**260**
....	....	....	....	....	....	....	21	10	5 × 4	**265**
40 15½	....	....	....	....	....	....	21	11	5 × 4	**270**
41 16	....	....	....	....	....	....	21	11	5 × 4	**275**
42 16	42 16	....	....	....	....	....	22	11	5 × 4	**280**
42 16	42 16	....	....	....	....	....	22	11	5 × 4	**285**
43 16½	43 16½	43 16½	....	....	....	....	22	11	5 × 4	**290**
43 16½	43 16½	43 16½	....	....	....	....	22	11	5 × 4	**295**
44 17	44 17	44 17	44 17	....	....	....	23	11	5 × 4	**300**
44 17	44 17	44 17	44 17	....	....	....	23	11	5 × 4	**305**
45 17½	45 17½	45 17½	45 17½	45 17½	....	....	23	11	5 × 4	**310**
49½ 19¼	45 17½	45 17½	45 17½	45 17½	....	....	23	11	5 × 4	**315**
49½ 19¼	45 17½	45 17½	45 17½	45 17½	....	....	24	12	5 × 5	**320**
50½ 19¾	50½ 19¾	46 18	46 18	46 18	46 18	....	24	12	5 × 5	**325**
52 19¾	52 19¾	47 18	47 18	47 18	47 18	....	24	12	5 × 5	**330**
53 19¾	53 19¾	53 19¾	48 18	48 18	48 18	48 18	25	12	5 × 5	**340**
59 21½	54 19¾	54 19¾	54 19¾	49 18	49 18	49 18	25	12	5 × 5	**350**
60 21½	60 21½	55 19¾	55 19¾	55 19¾	50 18	50 18	25	12	5 × 5	**360**
31	32	33	34	35	36	37				

16 years' grade a reduction in width is permitted of one twelfth in Lower beam Stringers, and of one-
thickness of Centre Plates or Plate, and a slight reduction in the Angle-Irons are also allowed.

TABLE III.

BEAMS AND FLOORS.

Breadth of Vessel.	*Depth of Beam upon alternate Frames, in inches.	Size of Angles for each side of upper Edge of Beams, in inches.	Centre depth of Floors upon every Frame, in inches.	Thickness, in inches.
12 feet	$3\frac{1}{2}$	$2 \times 2 \times$ 4-16ths	8	3-16ths
15 feet	4	$2 \times 2\frac{1}{4} \times$ 4-16ths	10	4-16ths
$17\frac{1}{2}$ feet	$4\frac{1}{2}$	$2\frac{1}{4} \times 2\frac{1}{4} \times$ 5-16ths	12	5-16ths
20 feet	5	$2\frac{1}{2} \times 2\frac{1}{4} \times$ 5-16ths	$13\frac{1}{2}$	5-16ths
21 feet	$5\frac{1}{4}$	$2\frac{1}{2} \times 2\frac{1}{2} \times$ 5-16ths	14	5-16ths
22 feet	$5\frac{1}{2}$	$2\frac{1}{2} \times 2\frac{3}{4} \times$ 5-16ths	$14\frac{1}{2}$	6-16ths
23 feet	$5\frac{3}{4}$	$2\frac{1}{2} \times 2\frac{3}{4} \times$ 5-16ths	15	6-16ths
24 feet	6	$2\frac{3}{4} \times 2\frac{3}{4} \times$ 5-16ths	16	6-16ths
25 feet	$6\frac{1}{4}$	$2\frac{3}{4} \times 2\frac{3}{4} \times$ 5-16ths	$16\frac{1}{2}$	6-16ths
26 feet	$6\frac{1}{2}$	$2\frac{3}{4} \times 2\frac{3}{4} \times$ 6-16ths	$17\frac{1}{2}$	7-16ths
27 feet	$6\frac{3}{4}$	$2\frac{3}{4} \times 2\frac{3}{4} \times$ 6-16ths	18	7-16ths
28 feet	7	$3 \times 2\frac{3}{4} \times$ 6-16ths	$18\frac{1}{2}$	7-16ths
29 feet	$7\frac{1}{4}$	$3 \times 2\frac{3}{4} \times$ 6-16ths	19	8-16ths
30 feet	$7\frac{1}{2}$	$3 \times 3 \times$ 6-16ths	20	8-16ths
31 feet	$7\frac{3}{4}$	$3 \times 3\frac{1}{4} \times$ 6-16ths	$20\frac{1}{2}$	8-16ths
32 feet	8	$3 \times 3\frac{1}{4} \times$ 6-16ths	21	8-16ths
33 feet	$8\frac{1}{4}$	$3\frac{1}{4} \times 3\frac{1}{4} \times$ 6-16ths	$21\frac{1}{2}$	9-16ths
34 feet	$8\frac{1}{2}$	$3\frac{1}{4} \times 3\frac{1}{4} \times$ 6-16ths	$22\frac{1}{2}$	9-16ths
36 feet	9	$3\frac{1}{4} \times 3\frac{1}{2} \times$ 7-16ths	$23\frac{1}{2}$	9-16ths
37 feet	$9\frac{1}{4}$	$3\frac{1}{4} \times 3\frac{1}{2} \times$ 7-16ths	$24\frac{1}{2}$	9-16ths
39 feet	$9\frac{3}{4}$	$3\frac{1}{2} \times 3\frac{1}{2} \times$ 7-16ths	$25\frac{1}{2}$	10-16ths
40 feet	10	$3\frac{1}{2} \times 3\frac{1}{2} \times$ 7-16ths	$26\frac{1}{2}$	10-16ths
42 feet	$10\frac{1}{2}$	$3\frac{1}{2} \times 3\frac{3}{4} \times$ 7-16ths	$27\frac{1}{2}$	10-16ths
44 feet	11	$3\frac{3}{4} \times 3\frac{3}{4} \times$ 8-16ths	$28\frac{1}{2}$	10-16ths
46 feet	$11\frac{1}{2}$	$3\frac{3}{4} \times 3\frac{3}{4} \times$ 8-16ths	30	11-16ths
48 feet	12	$4 \times 3\frac{3}{4} \times$ 8-16ths	32	11-16ths

* The full depth and thickness of the beams to be carried over 3-5ths of vessel's length amidships, and may be reduced thence to ends 1-16th of an inch in thickness.

TABLE IV.

STANCHIONS FOR BEAMS.

TWIXT DECK STANCHIONS.

6 feet between decks	.. .. ..	$2\frac{1}{2}$ inches diameter.
7 ,, ,,	.. .. ..	$2\frac{3}{4}$,, ,,
8 ,, ,,	.. .. ..	$2\frac{3}{4}$,, ,,

LOWER HOLD STANCHIONS.

9 feet hold	.. ..	3 inches diameter.
10 ,, ,,	.. ..	$3\frac{1}{8}$,, ,,
11 ,, ,,	.. ..	$3\frac{1}{8}$,, ,,
12 ,, ,,	.. ..	$3\frac{1}{4}$,, ,,
13 ,, ,,	.. ..	$3\frac{1}{4}$,, ,,
14 ,, ,,	.. ..	$3\frac{3}{8}$,, ,,
15 ,, ,,	.. ..	$3\frac{3}{8}$,, ,,
16 ,, ,,	.. ..	$3\frac{1}{2}$,, ,,

In vessels with three tiers of beams, stanchions between middle and lower-deck beams, to be of a diameter intermediate between those of the upper and lower stanchions.

TABLE VI.

Diameter of rivets, in sixteenths of an inch.	8	10	12	13	13	14	14	15	16	18	19	20
Thickness of plates, in sixteenths of an inch.	5	6	7	8	9	10	11	12	13	14	15	16

TABLE VII.

Masts.

Length.	Diameter.	Body. Thickness.	Head.		Angle Irons.
60	20	$\frac{3}{8}$	$\frac{5}{16}$	Three angle-irons.	$3 \times 2\frac{1}{2} \times \frac{5}{16}$
72	24	$\frac{3}{8}$	$\frac{5}{16}$		$3 \times 3 \times \frac{3}{8}$
78	26	$\frac{3}{8}$	$\frac{5}{16}$		$3 \times 4 \times \frac{7}{16}$
84	28	$\frac{7}{16}$	$\frac{3}{8}$	Four angle-irons.	$3 \times 4 \times \frac{1}{2}$
90	30	$\frac{7}{16}$	$\frac{3}{8}$		$5 \times 3 \times \frac{1}{2}$
96	32	$\frac{1}{2}$	$\frac{7}{16}$		$5 \times 3 \times \frac{1}{2}$

Yards of Steel.

Length.	Diameter.	Centre.	Thicknesses. To Arms.	Thicknesses. Arms.		Angle Irons.
60	15	$\frac{1}{4}$	$\frac{3}{16}$	$\frac{1}{8}$		——
64	16	$\frac{1}{4}$	$\frac{3}{16}$	$\frac{1}{8}$		——
68	17	$\frac{1}{4}$	$\frac{3}{16}$	$\frac{1}{8}$	Three angle steels.	$2\frac{1}{2} \times 2 \times \frac{1}{4}$
		middle plates.				$3 \times 3 \times \frac{1}{4}$
72	18	$\frac{5}{16}$ and $\frac{1}{4}$	$\frac{3}{16}$	$\frac{1}{8}$		$3 \times 3 \times \frac{1}{4}$
76	19	$\frac{5}{16}$ and $\frac{1}{4}$	$\frac{3}{16}$	$\frac{1}{8}$		$3 \times 3 \times \frac{1}{4}$
80	20	$\frac{5}{16}$ and $\frac{1}{4}$	$\frac{3}{16}$	$\frac{1}{8}$		$3 \times 3 \times \frac{1}{4}$
84	21	$\frac{5}{16}$ and $\frac{1}{4}$	$\frac{3}{16}$	$\frac{1}{8}$		

All angles in masts and yards to extend the whole length.

When angle-irons are omitted in masts and spars the following compensations are required:—

An addition in the plating of one-sixteenth of an inch in thickness beyond that given in the Table.

The thickness of the plates at the partners to be continued to the heel of the mast.

Lower masts to be composed of not less than three plates in circumference.

Butt-strips to be one-sixteenth of an inch thicker than mast plates; to have the fibre of the iron in the direction of the mast or spar; and to be wide enough to admit of treble or quadruple riveting in way of caps, trusses, and partners. No butts at these parts to be less than treble-riveted.

Laps of seams in all lower masts, bowsprits, topmasts and large yards to be wide enough to admit of double-riveting.

All yards to be doubled (in thickness of plating) in the way of truss hoops and over centre.

All topmasts to be doubled in way of lower-mast cap.

All bowsprits to be doubled where resting at knight or stem heads.

All lower masts to be doubled in way of wedging.

INDEX.

LONDON: PRINTED BY W. CLOWES AND SONS, DUKE STREET, STAMFORD STREET, AND CHARING CROSS.

For EU product safety concerns, contact us at Calle de José Abascal, 56–1°,
28003 Madrid, Spain or eugpsr@cambridge.org.

www.ingramcontent.com/pod-product-compliance
Ingram Content Group UK Ltd.
Pitfield, Milton Keynes, MK11 3LW, UK
UKHW040225290926
15471UKWH00001B/7
9781108026451